THE HYPERCIRCLE IN MATHEMATICAL PHYSICS

THE HYPERCIRCLE IN MATHEMATICAL PHYSICS

A METHOD FOR THE APPROXIMATE SOLUTION OF BOUNDARY VALUE PROBLEMS

BY

J. L. SYNGE, Sc.D., F.R.S.

Senior Professor, School of Theoretical Physics
Dublin Institute for Advanced Studies

CAMBRIDGE
AT THE UNIVERSITY PRESS
1957

CAMBRIDGE UNIVERSITY PRESS
Cambridge, New York, Melbourne, Madrid, Cape Town,
Singapore, São Paulo, Delhi, Tokyo, Mexico City

Cambridge University Press
The Edinburgh Building, Cambridge CB2 8RU, UK

Published in the United States of America by Cambridge University Press, New York

www.cambridge.org
Information on this title: www.cambridge.org/9781107666559

First published 1957
First paperback edition 2011

A catalogue record for this publication is available from the British Library

ISBN 978-1-107-66655-9 Paperback

CONTENTS

PREFACE

This book describes a technique for the approximate solution of certain boundary value problems of mathematical physics. This technique involves concepts of function-space. These are developed *ab initio*, so that no special knowledge beyond the calculus is required on the part of the reader. Thus it is an elementary book in a mathematical sense, but the arguments are, I hope, mathematically exact, although I have tried to avoid the rather bleak axiomatics which repel mathematical physicists and engineers for whom the book is intended and for whom function-space will remain a means to an end and not an end in itself.

The book has been ten years in the making, during which time I have given isolated lectures, and in some cases short courses, on various aspects of the subject in a number of places: Princeton University, Massachusetts Institute of Technology, University of Leeds, Harvard University, Brown University, Carnegie Institute of Technology, Institute for Fluid Dynamics and Applied Mathematics (University of Maryland), Severi Jubilee Celebration (Rome, 1950), St Andrews Mathematical Colloquium (1951), University of Trieste, Henri Poincaré Colloquium (Paris, 1954), and the Dublin Institute for Advanced Studies. The experience so gained has been valuable because it brought home to me how reluctant mathematicians and physicists are in this age of analysis to use geometrical intuition as a guide, particularly when the geometry is that of a multidimensional space, or (worse) a space with an infinity of dimensions. Since this intuitional approach seems to me the first essential (for suggestion, not for proof), I have developed it very slowly in the early part of the book in order to establish a common understanding with the reader.

Outside this special field of application my knowledge of function-space is slight, and I am much indebted to Dr F. Smithies, mathematical adviser of the Cambridge University Press, whose kindly advice has kept me from wandering too far from current usages.

As stated in the Introduction, Professor W. Prager was joint originator of the hypercircle method in 1946; my thanks are due to him not only for the stimulating collaboration at that time but also for reading most of the manuscript of this book. Professor A. J. McConnell, Provost of Trinity College, Dublin, has taken an active interest in the work, his approach through variational principles (see Chapter 5) throwing light on the range of applicability of the method. Critical discussions with Professors J. B. Diaz and A. Weinstein have helped me very much, and Dr J. McMahon and Mr V. G. Hart, while Scholars at the Dublin Institute for Advanced Studies, gave me assistance in the preparation of the manuscript; the computations of Chapter 5 were done by Mr Hart, who also wrote more than a third of the text of that chapter.

Mr Hart has given invaluable assistance in proof-reading. In a book of this sort, a complete absence of errors is not to be expected; the most we can hope is that they are few and venial. Our task has been greatly lightened by the efficiency of the Cambridge University Press in handling a rather difficult job.

DUBLIN J.L.S.

August 1956

INTRODUCTION

The physicist and the engineer feel at a loss when a process of mathematical analysis carries them out of contact with physical reality, for they like each symbol to be physically identifiable and each step to be guided by physical intuition. Mathematicians with a geometrical turn of mind feel the same revulsion against pure analysis, and there is a famous example of this in the deliberate revolt of Poinsot, in his study of the motion of rigid bodies, against the uncompromising attitude of Lagrange, who excluded all diagrams from his *Mécanique analytique*.

But there are complex situations which baffle the intuition, as in the theory of elasticity when we try to keep track of six components of stress and three components of displacement, and in such cases there seems to be nothing for it but to throw oneself into the mathematical formulae and hope for the best without intuitive guidance. It was to escape this fate that Professor W. Prager and I evolved what we called the method of the hypercircle in function-space (Prager and Synge (1)†).

In the hypercircle method as applied to elasticity we substitute for the direct intuition of stress and displacement the intuition of Euclidean geometry, extended from three dimensions to an infinity of dimensions, an extension much less troublesome than one would at first suppose. To each state of stress of the body (six functions of three coordinates) there corresponds a single point of function-space, and this space is endowed with a metric in a way which seems very natural physically, the square of the distance of a point from the origin of function-space (the state of zero stress) being twice the strain energy of the state corresponding to that point. This representation of states of stress by points of function-space would be merely a rather trivial game were it not for the fact that the geometrical picture fits together in a remarkable way, the minimum principles which hold in elastic equilibrium taking on a simple geometrical interpretation analogous to the fact that the perpendicular dropped from a point on a plane is the shortest distance to that plane. The method is called the method of the hypercircle because a certain geometrical figure (called a hypercircle by analogy with the circle of ordinary geometry) appears in the theory as the locus of possible positions of the point corresponding to the unknown solution of a problem of elastic equilibrium.

† See Bibliography at end for all references.

It was the complexity of the theory of elasticity which forced us to invent this geometrical approach, but Professor Prager rightly suggested at the time that the method of the hypercircle had a much wider range of applicability. It can, indeed, be used for many boundary value problems (Synge (1)), including the comparatively simple ones associated with Laplace's equation in a plane (torsion, electrostatic capacity, etc.), the essential condition being that the analytic problem can be presented as the geometrical problem of finding the point of intersection of two orthogonal linear subspaces in a suitably chosen function-space (the analogues of two perpendicular straight lines in ordinary space). This is linked with the possibility of deriving the partial differential equations of the problems considered from variational principles (McConnell (1)).

A wide class of boundary value problems of mathematical physics may thus be given a geometrical form, a wedding of analysis to geometry which is pleasant to contemplate. It is by no means a sterile marriage, for though this geometrical picture of two intersecting subspaces with the goal at the point of intersection may not tell us how to attain that goal, it does suggest that we should advance towards it in both of the subspaces instead of in one only, a great improvement on standard procedures because the use of both subspaces enables us to estimate our error accurately at any stage of the advance.

Ordinarily one works in one linear subspace. For example, to solve the torsion problem for a square section one works in the linear subspace of harmonic functions, setting up a series of such functions with coefficients chosen to satisfy the boundary conditions. For a square this is an excellent plan, because we get an exact solution in this way. But our success is due to the simplicity of the square, and we cannot get the solution in this simple way for a more general section, say a square with one corner knocked off.

The hypercircle method is applied in detail to the torsion problem in Chapter 4, and here it is enough to say that, when we advance towards the solution through both the linear subspaces, we get a controlled approximation. Knowing that exact formal solutions can be obtained only in a few special cases, we are prepared from the first to break off with an approximation, but when we do break off we know just where we stand in the sense that we know how far we are away from the solution in terms of the metric of function-space, or equivalently in terms of an integral of the square of the error.

The method of the hypercircle makes contact on one side with the theory of Hilbert spaces and on the other with equations in finite differences and the relaxation method of Sir Richard Southwell.

Mathematicians will recognize at once that (when the metric is positive-definite, as it is throughout Part II) the function-space is a Hilbert space, shorn of those refinements which we do not need because we are not concerned with existence theorems; we know or assume that the solution exists, and are interested only in finding it.† The Hilbert space is further simplified by our concentration on real functions; this makes our function-space much easier to think about than the Hilbert space of quantum mechanics.

But Part III introduces a less familiar type of function-space. It is not a Hilbert space because the metric is now indefinite, and its geometry is analogous, not to Euclidean geometry, but to the geometry of Minkowski in the space-time of special relativity. The method of the hypercircle (now the method of the pseudohypercircle) softens and serves less firmly as a guide to approximate solutions, arithmetical bounds being no longer available and minimum principles changing to stationary principles. This type of function-space occurs in the geometrization of problems of forced vibrations, mechanical and electromagnetic. Free vibrations are touched on only lightly and the determination of eigenvalues is not discussed, for although a geometrical approach is powerful here, it is not the geometry appropriate to the method of the book.

The method of the hypercircle may be called a relaxation method because points in the two linear subspaces in which we advance towards the solution represent solutions of problems (sometimes physically artificial) in which some of the conditions of the original problem are relaxed. Equations in finite differences are by no means an essential part of the method of the hypercircle; they come in when we use a certain technique, the method of pyramid functions, to get points on these subspaces of relaxation. This technique enables us to handle bounding curves of any form in plane problems.

As commonly used, equations in finite differences are substituted for the differential equation of the problem, and one proceeds with a general confidence that the solution of the equations in finite differences will not be far off the solution of the differential equation, provided the grid is fine. The method of the hypercircle is much more precise. We never replace a differential equation by equations in finite differences,

† For an introduction to the mathematical theory of Hilbert space, see Stone (1) or Halmos (1).

but use the latter within the framework of the hypercircle method, the solution of the equations in finite differences giving us accurately, not the solution to our original problem, but the centre and radius of the hypercircle on which the point representing that solution lies. The finer the grid, the closer we are to the solution, but with any grid at all we know how far we are away from the solution. Thus the torsional rigidity of a regular hexagon is given on p. 268 with an accuracy of 0·4 %, and we know exactly what we mean by this statement of error, upper and lower bounds being established. We might say that the method of the hypercircle, when combined with pyramid functions, is a refined finite difference method with rigorously controlled error.

It is impossible to write a book which carries every reader along at the right speed; it is bound to be too slow for some and too fast for others. I have thought it wiser to err on the side of slowness, particularly at the beginning, for if the reader fails to get hold of the geometry of function-space with a feeling of security in its use, then the book has failed in its purpose. I have filled in a good deal of detail in the calculations, knowing that one can grasp the full significance of a general argument only by seeing cases worked out in full detail. My own mind works that way, and I assume that it is true of many others. A few exercises are inserted at the end of each section, mostly of a very simple nature; they serve to keep one's feet on the ground.

This book has been written on the assumption that frequent appeal to geometrical intuition, not for proof but for suggestion, will please the reader as much as it pleases the author. Those who prefer to take their analysis neat will find other treatments of the problem of bounding the solutions of boundary value problems elsewhere, without diagrams or geometrical ideas, notably in papers by Diaz (1, 2, 3) and Cooperman (1).

We must reconcile ourselves to the fact that, in mathematics, there is no single universal mode of thought which reveals in a flash the inner meaning of an argument so that it becomes easy and almost self-evident to everyone. What is natural and easy for one man is often artificial and difficult for another. Anyone who believes in the simplicity and uniqueness of mathematical thought would be well advised to read Hadamard's (1) book on *The Psychology of Invention in the Mathematical Field*; incidentally, his remarks (*op. cit.* p. 88) on Hilbert's use of diagrams in setting up the logical principles of geometry are apposite to the question of diagrams of function-space.

PART I

NO METRIC

CHAPTER 1

GEOMETRY OF FUNCTION-SPACE WITHOUT A METRIC

1·1. INTRODUCTORY IDEAS

Representation of numbers

Logic rules mathematics, but we are human beings and the ways in which we see or understand things in mathematics are not always logical ways. The fascination of mathematics lies in the interplay of intuition and logic—the discipline of intuition by logic; neither intuition nor logic alone suffices.

Logically, the concept of a real number is independent of the idea of a point on a line. But mathematicians habitually think of real numbers as points on a line; in complicated situations this representation is essential, for we are human beings with limited

Fig. 1·11. Good representation.

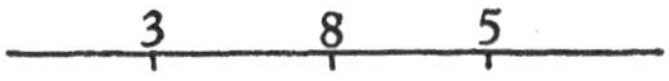

Fig. 1·12. Bad representation.

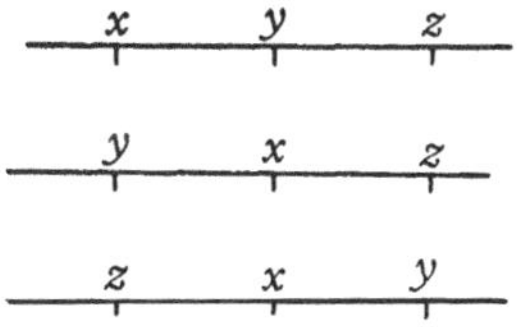

Fig. 1·13. Representations of three unknown real numbers.

facilities of thought (otherwise all mathematics would be obvious to us immediately). The extension of the representation into the complex plane is equally essential. These are things we cannot do without, although we are at all times ready to admit the ultimate authority of logic.

Even in the representation of real and complex numbers there are snares for the unwary. Suppose we are asked to think of the numbers 3, 5, 8. Immediately we think of a representation as in Fig. 1·11. No one would be so foolish as to make a representation as in Fig. 1·12; the order is wrong. But suppose we are asked to think of three *unknown* real numbers, x, y, z. Three different representations suggest themselves (Fig. 1·13); there are others also, in some of which two or all of the three points coincide. Which are good representations and which are bad? If we were given more information (e.g. $x < y < z$) then we might be able to decide, but if no such information is available, all the representations are bad, because each of them commits us further than we ought to be committed at this stage of ignorance.

In mathematics we are constantly being asked to find an unknown number or a set of unknown numbers. To think of the number or numbers, we crave a representation. Any representation we make is dangerous—it may indicate something that is actually false. It is like forming a detailed mental image of John Smith when he appears on page one of a novel, only to find that this mental image is entirely wrong when we read the description of him on page two.

In disgust at the unreliability of representations based on insufficient data, some people try to get on without them. When they have to deal with three numbers x, y, z, they do not attempt to represent them, preferring to carry out the formal manipulation of symbols according to established rules. But those who follow this cautious policy forego a powerful aid to thought, and the wisest course seems to be one of compromise, in which we use representations in a fluid and tentative way, preferring representations which do not say too much.

To sum up: if we are asked to solve a problem involving unknown real numbers, we have two options:

(i) Carry out formal manipulations according to established rules, and make no representation at all.

(ii) Make tentative representations as guides for thought, modifying them as further information becomes available. The standard representation of real numbers is by points on a line, and that of complex numbers by points in a plane (Argand diagram).

Representation of functions

In the problems with which we shall be concerned, we sometimes seek unknown numbers, but usually it is unknown *functions* that we seek. What is a function, and how are we to represent it?

In the eighteenth century a function of x meant a formula involving the letter x (like $1+x^2$ or $\sin x$). Now we say that $f(x)$ is a function of x if there exists a rule by which a value of $f(x)$ corresponds to each value of x in an assigned range.

As for the representations of functions, three are familiar:

(i) a formula,
(ii) a graph,
(iii) a tabulation of values.

Let us now suppose that we have before us a problem in which we are required to find an unknown function $f(x)$ which satisfies a given differential equation for some range of values of x, with sufficient data concerning the end-values of the function and its derivatives to make the solution unique. This is the situation we shall face again and again in this book, complicated by the presence of several unknowns in some cases, by the presence of several independent variables instead of the single x, and by the change from an ordinary differential equation to one or more partial differential equations.

Suppose that the problem is an easy one. By familiar manipulations (not bothering about a representation) we solve the equation and get a formula for the function $f(x)$. From it we can prepare a graph and a tabulation of values. These representations are quite satisfactory, revealing the true properties of the solution.

But suppose that the problem is not an easy one, and that we despair of finding a formula for the solution $f(x)$. Nor can we make a graph or tabulation of values. Nevertheless, we hope to find out *some* facts about $f(x)$. But while we are doing this, how are we to *think* about this unknown function? To think about anything, we must have a representation for it.

We try first to represent $f(x)$ by a formula. What formula? Since we do not know the function, we can merely write $f(x)$ and perform formal manipulations with this symbol. This is like manipulating an unknown number as the letter x, and for some purposes this is the simplest and best thing to do.

What about a graph? What sort of graph should we draw, the function $f(x)$ being unknown? Any graph we draw will have properties—positive slope here, negative slope there, and so on. These properties may be the properties of $f(x)$, but it is probable that they will not. In fact, a graph commits us too much, and a tabulation of values of an unknown function would be absurd, saying far too much about a function of which we are so ignorant.

Thus we have on the one hand a symbol $f(x)$ which says too little and on the other hand a graph or tabulation which says too much. Is there a middle way? For our purposes a function-space representation provides the middle way, and to it we shall now proceed.

The idea of function-space

Let $f(x)$ and $g(x)$ be two functions of x for the range $x_1 \leqslant x \leqslant x_2$. Suppose we know that these functions exist (perhaps as solutions of differential equations), but suppose we do not know what the functions are, i.e. we have no formulae, no graphs, no tabulations of values. We are dealing, in fact, with two *unknown* functions.

We take a sheet of paper and mark two points at random, labelling one point $f(x)$ and the other $g(x)$. This is merely a scheme for mental concentration; when we think of either function, we associate it in our minds with the corresponding point.

Now we bring in a third function, the *zero function* which vanishes for all values of x in the range. For it we mark a third point O on the paper. We have now a *representation* of three functions by three points on a plane (Fig. 1·14).

Fig. 1·14. Representation of functions by points.

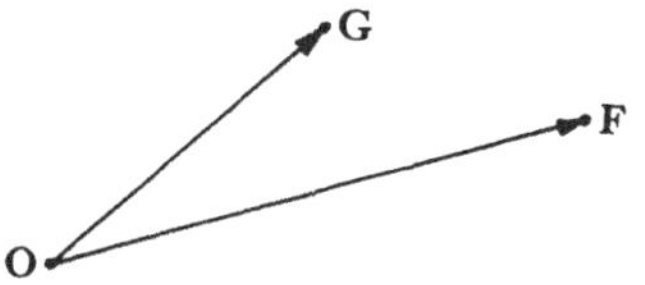

Fig. 1·15. Representation of functions by vectors.

Next we join the zero point (O) to the other points and put arrows on the joins as shown in Fig. 1·15. Regarding these directed joins as vectors, we change the notation, using heavy capital letters, with **O** for the zero vector. We have then the following correspondence between vectors and functions:

$$\mathbf{O} \leftrightarrow 0, \quad \mathbf{F} \leftrightarrow f(x), \quad \mathbf{G} \leftrightarrow g(x).$$

We shall employ the symbol $\leftrightarrow$ to indicate correspondences of this type.

We have used very little of our paper, and the rest of it is available for the representation of other functions. How should we proceed? Remember that we are not now in the domain of logical deductions —we are playing with the problem of representation. In this spirit we consider the vector $\mathbf{F}+\mathbf{G}$ (obtained by the usual parallelogram law) and ask: To what function shall we make $\mathbf{F}+\mathbf{G}$ correspond?

An obvious suggestion is that we should let it correspond to the function $f(x)+g(x)$, and further that we should let the vector $a\mathbf{F}$ (where a is any number, or scalar) correspond to the function $af(x)$. Thus we commit ourselves to the correspondence

$$a\mathbf{F}+b\mathbf{G} \leftrightarrow af(x)+bg(x), \tag{1·101}$$

a and b being any two real numbers, positive or negative.

As a and b take all real values, the extremity of the vector $a\mathbf{F}+b\mathbf{G}$ covers the whole plane, as indicated in Fig. 1·16. This means that we have, in the plane, representations of all functions of the form $af(x)+bg(x)$, where $f(x)$ and $g(x)$ are two definite (if unknown) functions and a and b arbitrary constants, with a one-to-one correspondence between vectors and functions.

If, for example, $f(x)=x^2$ and $g(x)=x^3$, we have representations of all functions of the form ax^2+bx^3, but no others; x^4, for example, will not be represented. Thus, though the vectors in the plane yield what may be called tentatively a function-space representation, it is a feeble representation, since it includes only functions which are linear combinations of two initial functions. This we shall have to remedy.

Linear dependence and independence

The assertion that two functions are *equal* means that they are equal for all values of x in the assigned range. Thus $h(x)=0$ means that $h(x)$ is the zero function.

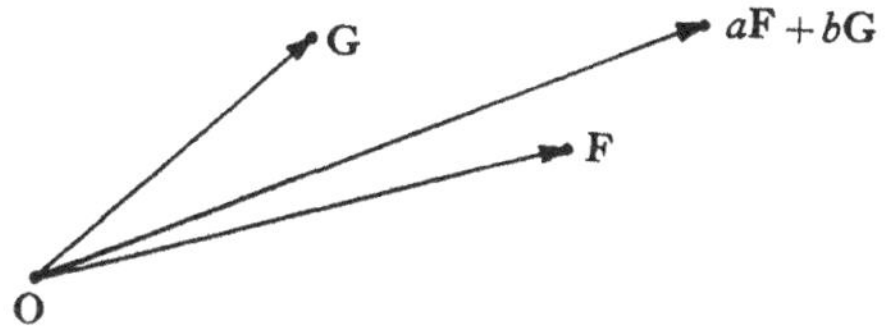

Fig. 1·16. The plane is covered by $a\mathbf{F}+b\mathbf{G}$.

Two functions $f(x)$ and $g(x)$ are said to be *linearly dependent* if there exist two constants (a and b), not both zero, such that

$$af(x)+bg(x)=0. \tag{1·102}$$

If such constants do not exist, then the functions are *linearly independent.*

Generally, n functions $f_1(x), f_2(x), \ldots, f_n(x)$ are said to be *linearly dependent* if there exist constants $a_1, a_2, \ldots, a_n$, not all zero, such that

$$a_1f_1(x)+a_2f_2(x)+\ldots+a_nf_n(x)=0. \tag{1·103}$$

If such constants do not exist, the functions are *linearly independent.*

The property of linear dependence is a very special one. If n functions are picked at random, there is zero probability that they are linearly dependent.

For any range of values of x, any selection of the functions

$$1, x, x^2, x^3, \ldots$$

forms a linearly independent set.

In (1·303) and (1·306) we shall meet more general definitions of linear dependence.

The number of dimensions in function-space is infinite

Let $f(x)$ and $g(x)$ be two linearly independent functions. Consider the plane described above (Fig. 1·16) in which we represent all functions of the form $af(x)+bg(x)$. Let $h(x)$ be a third function, such that f, g and h are linearly independent. Then $h(x)$ is not of the form $af(x)+bg(x)$, and so it cannot be represented in the plane according to our plan. But we can extend our representation to include $h(x)$ (and indeed all functions of the form $af(x)+bg(x)+ch(x)$) by making a *three*-dimensional model, in which we set up *three* non-coplanar vectors **F**, **G**, **H** to represent $f(x)$, $g(x)$, $h(x)$, respectively.

To represent *four* linearly independent functions we need a *four*-dimensional space. To represent n linearly independent functions $f_1(x), f_2(x), \ldots, f_n(x)$ we need a space of n dimensions; in it we can represent all functions of the form

$$a_1 f_1(x) + a_2 f_2(x) + \ldots + a_n f_n(x),$$

but no others.

But no matter how large n may be, it is always possible to find $n+1$ linearly independent functions (e.g. powers of x). Thus, no matter how large n may be, a space of n dimensions will not suffice for the representation of all functions. This we summarize by saying that *the number of dimensions in function-space is infinite.* We are left then with a vague mental picture of a function-space with an infinity of dimensions such that there is a one-to-one correspondence between the vectors or points of this space and all possible functions of x for a given range of x.

Pictorial representation of function-space

We are creating the concept of function-space for our convenience in handling certain problems, not as an end in itself. At the moment the outlook is gloomy—a space with an infinite number of dimensions!

Actually, the infinity of dimensions will trouble us little, because at any one time we shall be dealing with only a finite number of vectors, and for their representation a finite number of dimensions is sufficient, and often that number is quite small. In mathematical treatments of abstract spaces it is customary to borrow the language of geometry, but to refrain from making pictures, at least in public. That is not the policy pursued in this book, for to suppress the pictures is to suppress a powerful source of suggestion. Time and again a simple geometrical figure in function-space suggests an analytical result; it may even go further, suggesting the form of proof.

Pictorial representation is essential for discovery and rapid understanding; formal proofs will make sure that our intuition has not erred, as it may occasionally. The procedures of ordinary elementary geometry are not different from this. We follow lines with our pencils and talk in terms of a figure drawn on a piece of paper. Thus we construct a mode of thought, which we then refer to formal logic based on axioms. But we cannot conveniently do geometry without the piece of paper, our minds being what they are.

The pictorial representation of function-space becomes difficult only when a great number of vectors are involved, or when the infinity of dimensions plays a fundamental part in the argument. If we have to deal only with the linear combinations of two vectors in function-space, the geometrical representation on a piece of paper is precisely as good or as bad as the representation of the Euclidean plane in the same manner. If there are three linearly independent vectors involved, we encounter precisely the same difficulties of representation as we encounter in dealing with Euclidean space of three dimensions, and we meet those difficulties in the same way—namely, by making a sketch showing a two-dimensional projection of the three-dimensional figure. Similarly, if four or more linearly independent vectors are involved, we can still use a projection on a plane, as discussed more fully on pp. 26–7.

Physical examples

To introduce the idea of function-space as above, we took the simplest case, namely, that in which a vector in function-space corresponds to a single function of a single variable. We shall enlarge this concept formally in § 1·2, but now we shall consider some physical examples, the first of which illustrates the simple case considered above while the others introduce the more general concept.

(i) *Vibrating string.* Consider a stretched string, vibrating laterally with fixed end points ($x=0, x=b$). For the gravest mode, the displacement is of the form

$$y = a \cos(nt + c) \sin(\pi x/b). \qquad (1\cdot104)$$

At any instant t, the form of the string is a sine curve, and the form at any instant is obtained from the form at $t=0$ by magnification in the direction of the y-axis.

For any fixed t, y as given by (1·104) is a function of x and corresponds to a function-space vector, say **Y**. As the string moves, the vector changes, but only through change in a scalar factor multiplying the function $\sin(\pi x/b)$. If, then, we write **F** for the function-space vector corresponding to $\sin(\pi x/b)$, the history of the vector **Y** in function-space is very simply described: its extremity oscillates up and down a straight segment joining the ends of the vectors a**F** and $-a$**F** (Fig. 1·17).

A vibration compounded of the three gravest modes will be of the form

$$y = a_1 \cos(nt + c_1) \sin(\pi x/b) + a_2 \cos(2nt + c_2) \sin(2\pi x/b) + a_3 \cos(3nt + c_3) \sin(3\pi x/b). \quad (1\cdot105)$$

The corresponding vector **Y** in function-space may be written

$$\mathbf{Y} = f_1(t)\,\mathbf{F}_1 + f_2(t)\,\mathbf{F}_2 + f_3(t)\,\mathbf{F}_3, \quad (1\cdot106)$$

where the coefficients are regarded as variable scalars,

$$f_1(t) = a_1 \cos(nt + c_1), \quad f_2(t) = a_2 \cos(2nt + c_2),$$
$$f_3(t) = a_3 \cos(3nt + c_3), \quad (1\cdot107)$$

and where

$$\mathbf{F}_1 \leftrightarrow \sin(\pi x/b), \quad \mathbf{F}_2 \leftrightarrow \sin(2\pi x/b), \quad \mathbf{F}_3 \leftrightarrow \sin(3\pi x/b). \quad (1\cdot108)$$

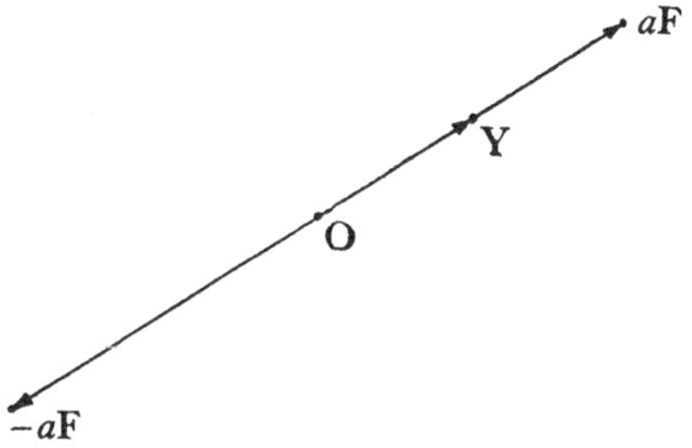

Fig. 1·17. Function-space representation of a string vibrating in its gravest mode [cf. (1·104)].

Let us draw these three linearly independent vectors (Fig. 1·18), our diagram being a projection on the plane of the paper of a three-dimensional figure. **Y** is a linear combination of these three vectors, and so lies in what we call the *linear* 3-*space* determined by them. Thus, although function-space has an infinity of dimensions, we are here concerned only with a three-dimensional subspace of it. Our diagrammatic representation, as in Fig. 1·18, is as good or as bad as the diagrammatic representation of any three-dimensional situation.

The motion of the string, i.e. the sequence of forms for changing t, is represented by a curve (C) in function-space. In the case of Fig. 1·17 the motion is very simple—an oscillation on the segment with period $2\pi/n$. It is much more complicated in the case of Fig. 1·18. However, we can assert at once that the curve C is *closed*, since the motion given by (1·105) is periodic, with period $2\pi/n$.

For a more general development of these elementary ideas see Michal (1), (2).

(ii) *A gas.* Consider a box containing a gas. Let p, ρ, T be pressure, density and temperature, respectively. These quantities we shall suppose to vary from point to point in the box, and also to vary with time. They are, in fact, three functions of x, y, z and t.

Let a vector **S** in function-space correspond to an instantaneous state of the gas, i.e. to the set of three functions

$$p(x, y, z, t), \quad \rho(x, y, z, t), \quad T(x, y, z, t),$$

considered as functions of x, y, z for fixed t. As t changes, these functions change, and so the extremity of **S** describes a curve in function-space. This curve represents the history of the gas. It depends on the differential equations of motion of the gas, on its

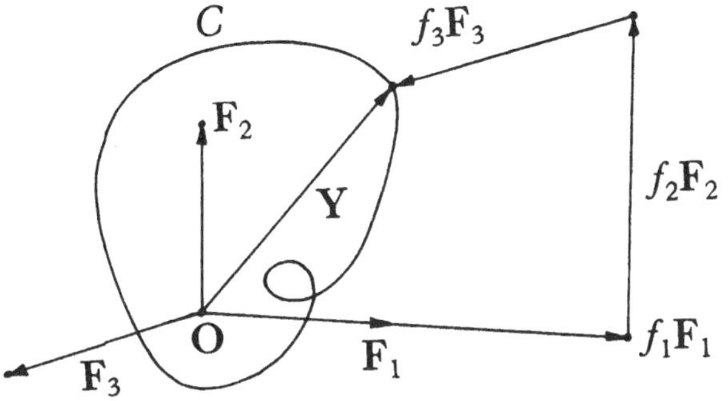

Fig. 1·18. Function-space representation of a string vibrating as in (1·105). (The curve C is not quantitatively drawn.)

initial state, and also on the 'equation of state' of the gas, this last being some thermodynamic relationship connecting p, ρ and T. Once the equation of state is given, we do not have to envisage the possibility of **S** ranging through the whole of function-space; the equation of state confines the extremity of **S** to a *subspace* (analogue of surface) in function-space.

(iii) *A condenser.* Consider a condenser in the form of a hollow sphere, the surface being divided into a number of portions insulated from one another. These portions are changed to various electrostatic potentials, and we are interested in the distribution of potential inside the cavity of the condenser. The determination of this potential is a problem of Dirichlet—to determine a function harmonic in a given region when the values of the function on the boundary of the region are prescribed. It is known that the solution is unique.

To discuss this problem in terms of function-space, we let a function-space vector correspond to any function $v(x, y, z)$ in the cavity of the condenser. To be the actual potential, this function v must

(a) satisfy the boundary condition,

(b) satisfy Laplace's partial differential equation.

From the totality of all functions $v(x, y, z)$, let us pick out those which satisfy the boundary condition. The extremities of the corresponding function-space vectors form a subspace (F' in Fig. 1·19). This subspace contains the extremity of the vector corresponding to the solution (say **S**), since the solution satisfies the boundary condition.

Now from the totality of all function $v(x, y, z)$ let us pick out those that are harmonic (i.e. satisfy Laplace's equation). The extremities

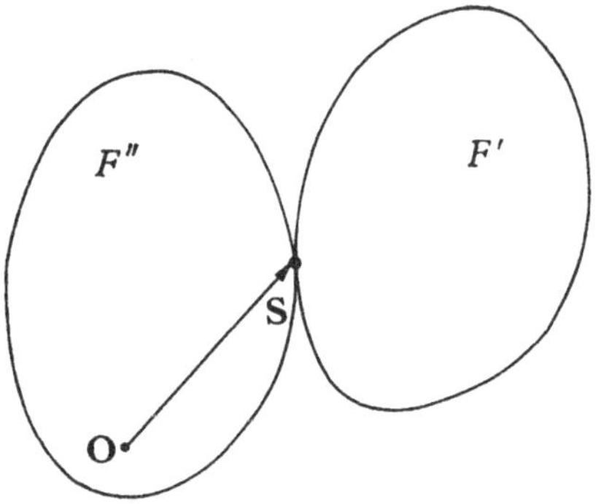

Fig. 1·19. Function-space for the Dirichlet problem. In this diagram the representation of the vector **S** is good, but the representation of the subspaces F' and F'' is faulty; the oval curves suggest some sort of boundedness which does not actually exist. If the reader can think of a more satisfactory way of representing a subspace, let him use it by all means. Meanwhile the representation shown seems worth using, with appropriate caution.

of function-space vectors corresponding to harmonic functions form another subspace of function-space (F'' in Fig. 1·19). Since the solution is a harmonic function, **S** has its extremity in F''.

The fact that the solution of the Dirichlet problem is unique tells us that the subspaces F' and F'' have just one point in common; that point is the extremity of **S**. We may note in passing that F'' contains the origin **O**, corresponding to $v = 0$, but F' does not, except in the trivial case where the condenser is uncharged. In that case we know that $\mathbf{S} = \mathbf{O}$, so that **O** is then the unique common point of F' and F''.

The practical reader may object that such vague ideas really tell us nothing about the solution of the problem. As will be seen later, that is not true; the pictorial representation gives us ideas for planning methods of solution or approximate solution.

(iv) *An elastic body.* Consider an elastic body subject to certain external stresses applied to its surface (e.g. a bar in torsion). Here we take a vector in function-space to correspond to a *state of stress* throughout the body, and we seek that vector which corresponds to a state of equilibrium.

The solution of a problem of elastic equilibrium involves the satisfaction of certain partial differential equations of equilibrium and compatibility, together with boundary conditions. As in the case of the preceding example, we may pick out subspaces of function-space, each composed of the extremities of vectors satisfying *some* of the conditions of the problem. The solution then corresponds to the common point of these subspaces.

The problem of elastic equilibrium will be treated in Chapter 5, and we shall not pursue it further here. We may note that, in considering this problem, we have gone a long way from the initial idea that a vector in function-space corresponds to a single function of a single variable. In elasticity, a vector in function-space corresponds to *six* functions (stress components) of *three* variables (the coordinates).

Exercises

1. Draw a function-space picture of the following functions:

$$1+x, \quad 1-x, \quad 2+3x, \quad 5-7x.$$

2. How many dimensions do you need to make a function-space representation of the following functions?

$$\sin x, \quad \cos x, \quad \sin 2x, \quad \sin(x+a).$$

3. Sketch the four-dimensional set of vectors:

$$1, \quad x, \quad x^2, \quad x^3, \quad 1-x+2x^2-x^3.$$

4. If, in each case, a function-space vector corresponds to a configuration of the system, in what essential respect do the function-spaces of the following systems differ?

(i) A flexible cable suspended from one end and allowed to oscillate in a vertical plane.

(ii) A chain of ten rods suspended from one end and allowed to oscillate in a vertical plane.

5. Prove that the vanishing of the Wronskian is a necessary and sufficient condition for the linear dependence of $f_1(x), \ldots, f_n(x)$.

1·2. F-VECTORS

We shall now make a new start on a more general basis, setting up terminology and notation to suit our needs.

P-space and F-space distinguished

Suppose we are trying to solve some boundary value problem in mathematical physics. We must at all times avoid confusion between the *physical space* in which our problem is set up and the *function-space* which we create for purposes of representation. We shall use the expressions *P-space* to denote the domain of physical space in which our problem is set up and *F-space* to denote function-space.

For vectors in the ordinary sense in P-space (*P-vectors*) we shall generally use indicial notation (V^i or V_i); for vectors in F-space (*F-vectors*) we shall use heavy type ($\mathbf{V}$).

In most of the problems of classical physics, P-space is a domain in Euclidean 3-space (e.g. Newtonian potential), a domain in the Euclidean plane (e.g. logarithmic potential), or a segment of a line (e.g. vibrating string). The domain may be bounded externally by a surface (or a curve or a pair of points according to the dimensionality), or it may be bounded internally and extend to infinity. In the usual eigenvalue problems of quantum mechanics, the domain is the whole of Euclidean 3-space.

Apart from a mathematical desire to generalize for the sake of generalization, there is a certain economy in working with a Euclidean N-space (E_N) without specifying N. Then we get results for the plane by putting $N=2$ and results for ordinary space by putting $N=3$. In this way we may avoid a type of duplication found in books on potential theory, where Newtonian and logarithmic potential are treated separately.

A more general P-space, including a domain in E_N as a special case, is a domain in Riemannian N-space (V_N). If we wished to deal with the general theory of relativity, we would certainly use V_4, but general relativity, being a non-linear theory, lies outside the present scope of the hypercircle method. In classical physics we might find it convenient to use V_2 in a problem involving a curved surface, not as a boundary but as a domain.

In certain cases the theory goes as easily for V_N as for E_N, and there is an advantage in using V_N, for then we are compelled to use tensor notation and so have our results in a form suitable for curvilinear coordinates in E_N.

However, in Part I we are not really concerned with the properties of P-space. It is only in Parts II and III that these properties become significant.

F-vectors defined

Let us leave P-space unspecified except to say that it is of N dimensions and that $x^1, x^2, \ldots, x^N$ are coordinates in it. Consider a set of M functions of these coordinates, say $s_1, s_2, \ldots, s_M$. *We define a function-space vector* (*F-vector*) $\mathbf{S}$ *by saying that it corresponds to this set of functions.* We indicate this correspondence by writing

$$\mathbf{S} \leftrightarrow (s_1, s_2, \ldots, s_M). \qquad (1{\cdot}201)$$

It is understood that the correspondence between F-vectors and sets of functions is one-to-one.

We define the *zero F-vector* **O** by

$$\mathbf{O} \leftrightarrow (0, 0, \ldots, 0), \qquad (1{\cdot}202)$$

so that **O** corresponds to a set of M functions each of which vanishes at every point of P-space.

Two F-vectors,

$$\mathbf{S} \leftrightarrow (s_1, s_2, \ldots, s_M), \quad \mathbf{S}' \leftrightarrow (s_1', s_2', \ldots, s_M'), \qquad (1{\cdot}203)$$

are said to be *equal* ($\mathbf{S} = \mathbf{S}'$) if, and only if,

$$s_1 = s_1', \quad s_2 = s_2', \quad \ldots, \quad s_M = s_M' \qquad (1{\cdot}204)$$

throughout P-space. (Remember that P-space means a domain, and not necessarily a complete space extending in all directions to infinity with no internal boundary.)

To illustrate these ideas, in treating an elastic body we would have $N = 3$ (dimensionality of ordinary space) and $M = 6$ (number of stress components). The F-vector **O** corresponds to the absence of stress, and two vectors are equal if, and only if, they correspond to the same state of stress throughout the body.

Addition, subtraction and multiplication by a scalar

We define *addition* of F-vectors by the following scheme:

$$\left.\begin{aligned} &\mathbf{S} \leftrightarrow (s_1, s_2, \ldots, s_M), \quad \mathbf{S}' \leftrightarrow (s_1', s_2', \ldots, s_M'),\\ &\mathbf{S} + \mathbf{S}' \leftrightarrow (s_1 + s_1', s_2 + s_2', \ldots, s_M + s_M'). \end{aligned}\right\} \qquad (1{\cdot}205)$$

Thus for the example given above (elasticity), addition corresponds to superposition of stresses.

Just as we have P-vectors and F-vectors, so we have *P-scalars* and *F-scalars*. A P-scalar is the ordinary concept (e.g. pressure or density of a gas); *an F-scalar is a constant number.*

We define the *multiplication of an F-vector* (**S**) *by an F-scalar* (a) by the following scheme:

$$\mathbf{S} \leftrightarrow (s_1, s_2, \ldots, s_M), \quad a\mathbf{S} = \mathbf{S}a \leftrightarrow (as_1, as_2, \ldots, as_M). \qquad (1{\cdot}206)$$

Thus, if **S** corresponds to a state of stress in an elastic body, 2**S** (or equivalently **S**2) corresponds to a state in which all the components of stress are doubled.

We define $-\mathbf{S}$ by $-\mathbf{S}=(-1)\mathbf{S}$. The rule for *subtraction* then follows:

$$\mathbf{S}-\mathbf{S}'=\mathbf{S}+(-1)\mathbf{S}'\leftrightarrow(s_1-s_1', s_2-s_2', \ldots, s_M-s_M'). \quad (1{\cdot}207)$$

By virtue of the definitions we have

$$\mathbf{S}+\mathbf{O}=\mathbf{S}, \quad a\mathbf{O}=\mathbf{O}. \quad (1{\cdot}208)$$

We note that the following algebraic laws are satisfied:

$$\left.\begin{array}{l}\text{Commutative: } \mathbf{S}+\mathbf{S}'=\mathbf{S}'+\mathbf{S}, \quad a\mathbf{S}=\mathbf{S}a;\\ \text{Distributive: } a(\mathbf{S}+\mathbf{S}')=a\mathbf{S}+a\mathbf{S}', \quad (a+b)\mathbf{S}=a\mathbf{S}+b\mathbf{S};\\ \text{Associative: } (\mathbf{S}+\mathbf{S}')+\mathbf{S}''=\mathbf{S}+(\mathbf{S}'+\mathbf{S}''), \quad (ab)\mathbf{S}=a(b\mathbf{S})=b(a\mathbf{S}).\end{array}\right\} \quad (1{\cdot}209)$$

The general conclusion is that, for these simple operations, *we may follow the ordinary rules of algebra*, noting however that we do not at present (until Part II) speak of the multiplication of F-vectors by one another.

We shall use only real F-scalars and F-vectors corresponding to real functions in P-space. These are sufficient for our purposes, but we might include the complex without difficulty, and so come into line with the procedure of quantum mechanics.

Logic and intuition

When we study elementary geometry, we start with a body of crude intuitions about space, and our mathematical education consists largely in the ordering of these intuitions into a logical structure. We attain competence in geometry by disciplining our intuitions, not by destroying them. A geometer deprived of his intuitions would be a poor geometer.

In studying function-space, the order of procedure is reversed. We do not start with natural intuitions about function-space; we start with formal definitions and proceed logically. If we are to have intuitions, we must build them up as we go along. Our success in building them up depends on the fact that function-space is very like the ordinary space in which we live, and if we exercise reasonable caution we shall be able to think quickly in function-space by means of ordinary intuitions derived from our experience.

We should throughout try to exploit our intuitive faculties to the full. In all matters of *proof*, however, we must fall back on formal logical procedure.

F-points and bound vectors

In ordinary three-dimensional vector theory we deal with the following things:

points of space, free vectors, bound vectors.

Often the adjective 'free' is omitted, and we use the single word 'vector' for a free vector. A bound vector is a combination of a free vector and a point, viz. a vector applied at a point. If we liked we could talk about points only and get on without the word 'vector', because a bound vector is an ordered pair of points and a free vector is a class of bound vectors.

In (1·201) we defined an F-vector. As we shall have occasion to distinguish between free F-vectors and bound F-vectors, let us revise our definition by saying that (1·201) defines a *free F-vector*. In fact, a free F-vector corresponds to a set of M functions.

We now define an *F-point* (point of function-space) by saying that an F-point A corresponds to a set of M functions,

$$A \leftrightarrow (t_1, t_2, \ldots, t_M), \tag{1·210}$$

the correspondence being one-to-one. The *origin* O of function-space is that point which corresponds to the set of zero functions,

$$O \leftrightarrow (0, 0, \ldots, 0). \tag{1·211}$$

If we compare (1·201) and (1·210), we see that the right-hand sides are the same—a set of M functions in each case. Are we then to consider that a free F-vector and an F-point are the same thing? No more and no less than we are to consider an ordinary free vector and an ordinary point as the same thing—each corresponds to a set of three numbers, the components of the vector and the coordinates of the point. There is a subtle difference which we know well is worth preserving.

We now have free F-vectors and F-points. Next we define a *bound F-vector* as a free F-vector associated with *a point of application*. If the free vector is $\mathbf{S}$ and the point A, we may refer to the bound vector as '$\mathbf{S}$ at A'.

A free F-vector defines an infinity of bound F-vectors, since the point of application may be chosen arbitrarily. A bound F-vector determines a unique free F-vector.

An F-point determines a set of functions; that set determines a free F-vector. An F-point determines also a certain bound vector, viz. the free F-vector just referred to, associated with the origin. The bound vector so determined is called the *position-vector* of the point.

Consider now an ordered pair of F-points, A and B in that order, where

$$\left.\begin{aligned} A &\leftrightarrow (s_1, s_2, \ldots, s_M), \\ B &\leftrightarrow (t_1, t_2, \ldots, t_M). \end{aligned}\right\} \tag{1·212}$$

This ordered point-pair determines a unique free F-vector $\mathbf{V}$, where

$$\mathbf{V} \leftrightarrow (t_1 - s_1, t_2 - s_2, \ldots, t_M - s_M); \tag{1·213}$$

note that the functions corresponding to the *second* point B are put *first*. If we now associate A with $\mathbf{V}$ as point of application, we have a unique bound vector ($\mathbf{V}$ at A) determined by the ordered point-pair (A, B). The point B is called the *extremity* of the bound vector, which might also be written $\overrightarrow{AB}$.

All this may be tedious to the reader who recognizes that we are merely engaged in validating in function-space words and ideas already familiar in ordinary space. The validation lacks excitement because it is monotonously easy. But it is necessary, because a large measure of primitive space intuition is used in understanding the elements of ordinary vector theory, and unless we build up the ideas of function-space on a logical basis, we shall suffer later from a feeling of insecurity when we use an elementary word with a function-space meaning.

The parallelogram law of addition

Consider two free F-vectors, $\mathbf{S}$ and $\mathbf{T}$. We want to find their sum, $\mathbf{S}+\mathbf{T}$. The functions corresponding to this sum are of course given by the rule (1·205), but we have something else in mind. Can we get $\mathbf{S}+\mathbf{T}$ by following the elementary parallelogram or triangle construction?

Start from the origin O, and consider the bound F-vector $\mathbf{S}$ at O (Fig. 1·21). If

$$\mathbf{S} \leftrightarrow (s_1, s_2, \ldots, s_M),$$

the extremity of $\mathbf{S}$ at O is, by (1·213), the point

$$C \leftrightarrow (s_1, s_2, \ldots, s_M).$$

Now take the bound vector $\mathbf{T}$ at C. If

$$\mathbf{T} \leftrightarrow (t_1, t_2, \ldots, t_M),$$

then, again by (1·213), the extremity of $\mathbf{T}$ at C is the point

$$D \leftrightarrow (s_1 + t_1, s_2 + t_2, \ldots, s_M + t_M).$$

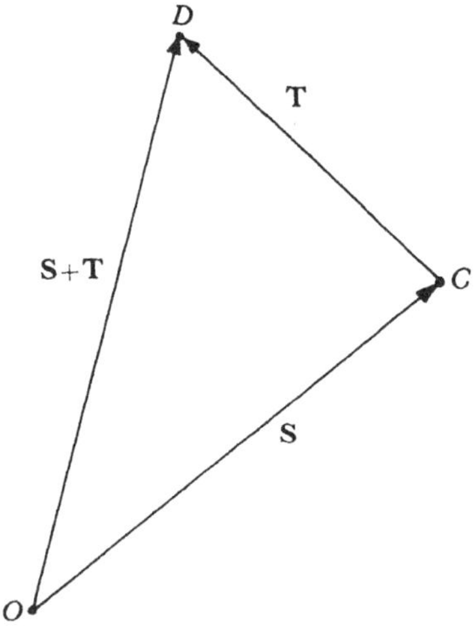

Fig. 1·21. Addition of F-vectors by the parallelogram law.

Thus the position-vector of D is the bound F-vector $\mathbf{S}+\mathbf{T}$ at O, and so we may obtain the sum $\mathbf{S}+\mathbf{T}$ in the elementary way, by first drawing **S** at O and then drawing **T** at the extremity of **S** at O. (If the undefined use of the word 'drawing' has passed unnoticed, it is a sign that the process of intuition in function-space is beginning to work.)

As one gains confidence, pedantic use of exact terminology becomes tedious, and so we shall use (unless there is possible confusion) the expression 'the point **S**' to mean 'the point with position-vector **S**' or 'the extremity of the vector **S** at O', which is the same thing. Heavy type will henceforth be used for F-points as well as F-vectors.

Translational transformations

For simplicity, we have defined position-vectors relative to the origin. We could, of course, have used any F-point as base for position-vectors, and in some cases it is convenient to use a point other than the origin.

The effect of changing the base point of position-vectors (i.e. their common point of application) is to change all position-vectors by the subtraction of the same vector. Thus, if **X** is a position-vector relative to **O**, and **X**′ the position-vector of the *same* F-point relative to **A** as base, then

$$\mathbf{X}' = \mathbf{X} - \mathbf{A}. \tag{1·214}$$

This transformation may be called a *translational transformation*; it corresponds precisely to a translation in Euclidean 3-space.

Exercises

1. Suppose we are studying the motion of a rigid body with a fixed point. The motion can be discussed in a representative space of three dimensions in which the Eulerian angles are taken as coordinates. Is this representative space to be regarded as a P-space or as an F-space? If it is a P-space, suggest a suitable F-space for the discussion of the motion of the body.

2. If a, b, c, d are four constants such that $ad - bc \neq 0$, and we are told that two F-vectors **S** and **T** satisfy

$$a\mathbf{S} + b\mathbf{T} = 0, \quad c\mathbf{S} + d\mathbf{T} = 0,$$

what can we deduce about **S** and **T**?

3. For an elastic solid **S** and **T** correspond to two states of stress, each satisfying the equations of equilibrium. Show that **S** + **T** corresponds to a state of stress which also satisfies the equations of equilibrium if there are no body forces. What if body forces are present?

4. A perfect incompressible fluid is in motion under no body forces. An F-vector corresponds to a distribution of velocity and pressure throughout the fluid. Let **S** correspond to a state satisfying the equations of steady motion. Show that 2**S** does not in general correspond to a state satisfying those equations.

5. In discussing electromagnetic radiation in vacuo, we might take for P-space a portion of space-time, and let an F-vector $\mathbf{S}$ correspond to six functions of the four space-time coordinates x, y, z, t:

$$\mathbf{S} \leftrightarrow (E_x, E_y, E_z, H_x, H_y, H_z).$$

Show that if $\mathbf{S}$ and $\mathbf{T}$ both satisfy Maxwell's equations, then so does $a\mathbf{S}+b\mathbf{T}$, where a and b are any constants.

1·3. STRAIGHT LINES AND LINEAR SUBSPACES

Straight lines

Let $\mathbf{A}$ and $\mathbf{B}$ be two F-points and a and b two real numbers, arbitrary except for the relation $a+b=1$. Consider the totality of points given by

$$\mathbf{X}=a\mathbf{A}+b\mathbf{B} \quad (a+b=1). \qquad (1\cdot301)$$

These points are said to form *the straight line AB* (Fig. 1·31).

If we put $a=1$, $b=0$, we get $\mathbf{X}=\mathbf{A}$; if we put $a=0$, $b=1$, we get $\mathbf{X}=\mathbf{B}$. Thus the straight line does in fact include the points $\mathbf{A}$ and $\mathbf{B}$. The set of points for which a and b are positive may be called the *segment AB*.

If we define $\mathbf{C}=\mathbf{B}-\mathbf{A}$, then (since $a+b=1$) (1·301) may be written in the alternative form

$$\mathbf{X}=\mathbf{A}+b\mathbf{C} \quad (-\infty<b<\infty). \qquad (1\cdot302)$$

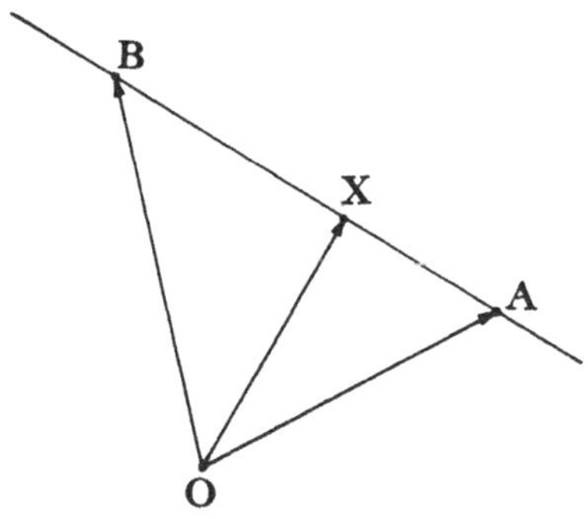

Fig. 1·31. The straight line through the points **A** and **B**.

Just as an ordinary straight line consists of a single infinity of points, so a function-space straight line consists of a single infinity of F-points.

Linear dependence of two F-vectors

In (1·103) we defined linear dependence for a set of functions of a single variable. Now we shall be more general and deal with a set of M functions of N variables, but the actual complexity of the definition will be hidden by the vector notation, and the new definition will look as simple as the old one.

Two F-vectors, $\mathbf{S}_1$ and $\mathbf{S}_2$, are said to be *linearly dependent* if there exist two numbers, a_1 and a_2, not both zero, such that

$$a_1\mathbf{S}_1+a_2\mathbf{S}_2=\mathbf{O}. \qquad (1\cdot303)$$

Otherwise they are *linearly independent.*

Let us see what this means. If

$$\left.\begin{aligned} \mathbf{S}_1 &\leftrightarrow (s_{11}, s_{12}, \ldots, s_{1M}), \\ \mathbf{S}_2 &\leftrightarrow (s_{21}, s_{22}, \ldots, s_{2M}), \end{aligned}\right\} \tag{1·304}$$

then the single vector equation (1·303) is equivalent to the M equations

$$\left.\begin{aligned} a_1 s_{11} + a_2 s_{21} &= 0, \\ a_1 s_{12} + a_2 s_{22} &= 0, \\ \cdots\cdots\cdots&\cdots\cdots \\ a_1 s_{1M} + a_2 s_{2M} &= 0, \end{aligned}\right\} \tag{1·305}$$

these M equations to be satisfied at each point of N-dimensional P-space. It is clear that linear dependence is a very special property

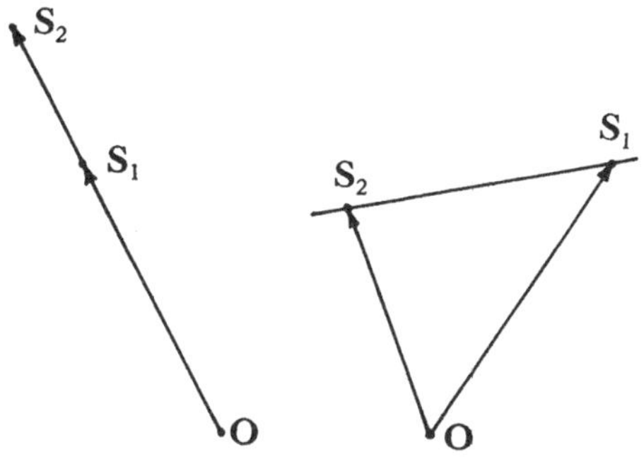

$\mathbf{S}_1$ and $\mathbf{S}_2$ linearly dependent $\mathbf{S}_1$ and $\mathbf{S}_2$ linearly independent

Fig. 1·32

of two F-vectors; it means that at each point of P-space the functions corresponding to one F-vector are proportional to those corresponding to the other, and this proportionality holds with the same ratios as we range through P-space.

Although the concept of linear dependence is complicated when viewed analytically, it is very simple when viewed geometrically. It is clear from (1·301) and (1·303) that *two F-vectors are linearly dependent if, and only if, the origin of F-space lies on the straight line joining the points of which they are position-vectors* (Fig. 1·32).

Linear dependence of n F-vectors

A set of n F-vectors, $\mathbf{S}_1, \mathbf{S}_2, \ldots, \mathbf{S}_n$, are said to be *linearly dependent* if there exist constants $a_1, a_2, \ldots, a_n$, not all zero, such that

$$a_1 \mathbf{S}_1 + a_2 \mathbf{S}_2 + \ldots + a_n \mathbf{S}_n = \mathbf{O}. \tag{1·306}$$

Otherwise they are *linearly independent.*

Analytically, linear dependence implies a set of M equations like (1·305) but more complicated, each equation containing n terms instead of two. We shall not trouble to write them out.

What is the geometrical meaning of the linear dependence of n F-vectors?

Let us start with $n=3$, so that we have

$$a_1\mathbf{S}_1+a_2\mathbf{S}_2+a_3\mathbf{S}_3=\mathbf{O}. \qquad (1\cdot307)$$

One of the a's is not zero; suppose $a_1 \neq 0$. Divide across by it and get

$$\mathbf{S}_1=b_2\mathbf{S}_2+b_3\mathbf{S}_3, \qquad (1\cdot308)$$

where $b_2=-a_2/a_1$, $b_3=-a_3/a_1$. Comparing this with (1·301), we see that this equation would tell us that the point $\mathbf{S}_1$ lies on the straight line through the points $\mathbf{S}_2$ and $\mathbf{S}_3$ provided $b_2+b_3=1$. But we have no reason to suppose that this last relation is satisfied. However, we certainly can write (1·308) in the form

$$\mathbf{S}_1=k(c_2\mathbf{S}_2+c_3\mathbf{S}_3), \quad c_2+c_3=1. \qquad (1\cdot309)$$

This tells us that the extremity of a vector proportional to $\mathbf{S}_1$ (in fact $k^{-1}\mathbf{S}_1$) lies on the straight line through the points $\mathbf{S}_2$ and $\mathbf{S}_3$. Thus we are led to the diagram Fig. 1·33 depicting the relationship between three linearly dependent F-vectors. The relationship is that of coplanarity in ordinary space, and we shall presently interpret the linear dependence of n F-vectors in terms of a linear $(n-1)$-space, which is a generalization of the Euclidean plane. But before proceeding to this, some remarks on pictorial representation of F-space will be useful, supplementing those of pp. 12–13.

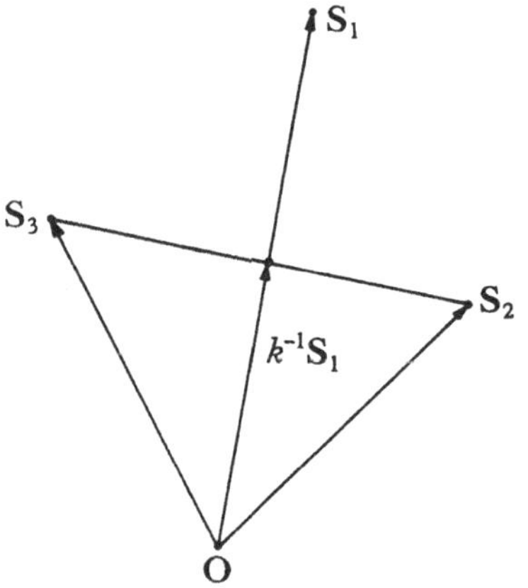

Fig. 1·33. The F-vectors $\mathbf{S}_1$, $\mathbf{S}_2$, $\mathbf{S}_3$ are linearly dependent [cf. (1·309)].

Pictorial representation of function-space

It is very difficult (in fact we might say that it is impossible) to make a thoroughly satisfactory representation of a solid body on a sheet of paper. The trouble lies in the reduction of three dimensions to two, so that a line of points in the solid body collapses into a single point in the representation.

It is much easier to get a satisfactory representation when the space figure is not a solid but a set of separate points. Consider, for example, the eight corners of a cube. If we project orthogonally on a plane parallel to a face of the cube, we get four points, each of which corresponds to two corners of the cube. That is unsatisfactory. But by altering the direction of projection a little, we can separate the points and obtain eight points in the plane, each representing a corner of the cube. This is satisfactory.

No matter how many points we are given in ordinary space, we can choose an orthogonal projection which throws them into distinct positions on a plane. The representation preserves some features of the space figure formed by the original points, for collinear points project into collinear points and parallel equal segments into parallel equal segments. But the representation ceases to be satisfactory if the number of points is infinite, although curves and straight lines present no great difficulty provided we do not make the mistake of thinking that an apparent intersection means a real intersection.

These ideas carry over to the representation on a plane of a configuration of points in a Euclidean space of any number of dimensions or a configuration of points in F-space. We can in fact draw pictures of F-space satisfying the following rules:

(i) Each of a finite number of F-points is represented by one and only one point in the picture.

(ii) Each of a finite number of bound F-vectors is represented by one and only one directed segment, and equal F-vectors are represented by parallel equal segments drawn in one sense.

In dealing with an infinite number of F-points or F-vectors we may have to resort to a vaguer and more symbolic representation, as we did in Fig. 1·19. If judiciously drawn, such symbolic representations can be real aids to thought.

It is advisable to have some notation to indicate linear dependence in a diagram. Thus linearly dependent vectors may be linked together as shown in Fig. 1·34, the number written next the link showing the number of linearly independent vectors involved. We might, for example, have ten vectors, all expressible in terms of two of them; in that case we would put the number 2 next the link.

Linear n-spaces

Let us take n linearly independent F-vectors, $\mathbf{T}_1, \mathbf{T}_2, \ldots, \mathbf{T}_n$, and write down the equation

$$\mathbf{X} = a_1\mathbf{T}_1 + a_2\mathbf{T}_2 + \ldots + a_n\mathbf{T}_n, \qquad (1·310)$$

the a's being any real numbers (F-scalars). If we hold the $\mathbf{T}$'s fixed and let the a's take all real values, the point of which $\mathbf{X}$ is the position-vector traces out a subspace of F-space. This subspace we call *the linear n-space through* $\mathbf{O}$ *of the* $\mathbf{F}$*-vectors* $\mathbf{T}_1, \mathbf{T}_2, \ldots, \mathbf{T}_n$. It is obvious (on putting all the a's equal to zero) that the subspace does in fact contain $\mathbf{O}$ (the origin).

More generally, if we take an additional fixed F-vector $\mathbf{S}_0$, and write

$$\mathbf{X} = \mathbf{S}_0 + a_1 \mathbf{T}_1 + a_2 \mathbf{T}_2 + \ldots + a_n \mathbf{T}_n, \tag{1·311}$$

we get a subspace which we call *the linear n-space through the point* $\mathbf{S}_0$ *of the vectors* $\mathbf{T}_1, \mathbf{T}_2, \ldots, \mathbf{T}_n$. It is obvious that it does contain the point $\mathbf{S}_0$.

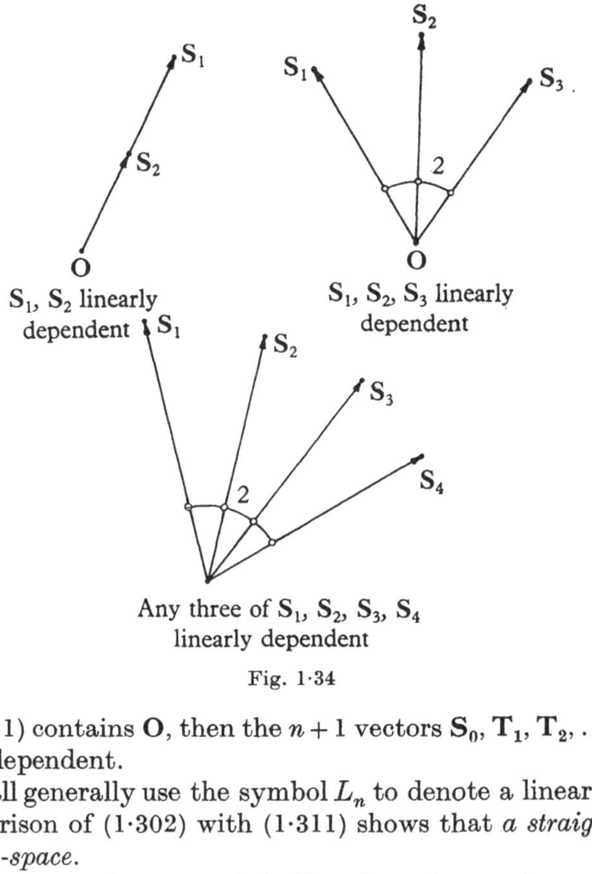

Fig. 1·34

If (1·311) contains $\mathbf{O}$, then the $n+1$ vectors $\mathbf{S}_0, \mathbf{T}_1, \mathbf{T}_2, \ldots, \mathbf{T}_n$ are linearly dependent.

We shall generally use the symbol L_n to denote a linear n-space.

Comparison of (1·302) with (1·311) shows that *a straight line is a linear* 1*-space.*

Just as we can draw a straight line through two given points, as in (1·301), so we can draw a linear n-space through $n+1$ given points. To see this, take any set of $n+1$ F-points, $\mathbf{S}_1, \mathbf{S}_2, \ldots, \mathbf{S}_{n+1}$, and consider the equation

$$\mathbf{X} = a_1 \mathbf{S}_1 + a_2 \mathbf{S}_2 + \ldots + a_{n+1} \mathbf{S}_{n+1},$$
$$a_1 + a_2 + \ldots + a_{n+1} = 1, \tag{1·312}$$

the a's being allowed to take all real values subject to the condition that their sum shall be unity. Putting $a_{n+1} = 1 - a_1 - a_2 - \ldots - a_n$, we may write (1·312) in the equivalent form

$$\mathbf{X} = \mathbf{S}_{n+1} + a_1(\mathbf{S}_1 - \mathbf{S}_{n+1}) + a_2(\mathbf{S}_2 - \mathbf{S}_{n+1}) + \ldots + a_n(\mathbf{S}_n - \mathbf{S}_{n+1}), \tag{1·313}$$

where $a_1, a_2, \ldots, a_n$ are completely unrestricted. Comparison with (1·311) shows that this is a linear n-space L_n through the point $\mathbf{S}_{n+1}$. If we choose all the a's zero except $a_1 = 1$, we get $\mathbf{X} = \mathbf{S}_1$; therefore L_n contains the point $\mathbf{S}_1$. Similarly, it contains the points $\mathbf{S}_2, \ldots, \mathbf{S}_n$, and *so in fact* (1·312) *is a linear n-space through the $n+1$ points* $\mathbf{S}_1, \mathbf{S}_2, \ldots, \mathbf{S}_{n+1}$.

In Euclidean 3-space three points define a unique plane in general, but not when the three points are collinear. A similar degeneracy occurs in F-space when all the $n+1$ points lie in a linear $(n-1)$-space.

Intersections of linear m-spaces and linear n-spaces

In Euclidean 3-space we have two types of linear n-space, the straight line ($n = 1$) and the plane ($n = 2$), in addition of course to the 3-space itself, which is a linear 3-space. The general rules of intersection are very simple:

(i) Two straight lines
  (*a*) do not meet at all, or
  (*b*) meet in a single point, or
  (*c*) coincide.

(ii) A straight line and a plane
  (*a*) do not meet at all (parallel), or
  (*b*) meet in a single point, or
  (*c*) the straight line is contained in the plane.

(iii) Two planes
  (*a*) do not meet at all (parallel), or
  (*b*) meet in a straight line, or
  (*c*) coincide.

Let us make a quick examination of the corresponding situations in F-space.

First, as regards two straight lines in F-space, the statement (i) above holds without change. To prove this, we write the equations of the two straight lines as in (1·302),

$$\mathbf{X} = \mathbf{A} + b\mathbf{C}, \quad \mathbf{X} = \mathbf{A}' + b'\mathbf{C}', \tag{1·314}$$

the vectors on the right-hand sides being fixed and b, b' being variable parameters. For an intersection, we require values of b and b' such that

$$(\mathbf{A}-\mathbf{A}')+b\mathbf{C}-b'\mathbf{C}'=\mathbf{O}. \qquad (1\cdot315)$$

If the three vectors $(\mathbf{A}-\mathbf{A}')$, $\mathbf{C}$, $\mathbf{C}'$ are linearly independent, then no values of b and b' can make this equation true; so in that case there is no intersection. This corresponds to a pair of skew lines in Euclidean 3-space. Suppose now that there is a linear dependence:

$$a(\mathbf{A}-\mathbf{A}')+c\mathbf{C}+c'\mathbf{C}'=\mathbf{O}. \qquad (1\cdot316)$$

But it becomes tedious to go through the argument in detail, and the reader will easily satisfy himself (using a diagram as a guide) of the truth of the following statements:

(i) *Intersections of two straight lines* (1·314).

(*a*) If $\mathbf{A}-\mathbf{A}'$, $\mathbf{C}$, $\mathbf{C}'$ are linearly independent, then there is no intersection (skew lines); if $\mathbf{C}$, $\mathbf{C}'$ are linearly dependent, and $\mathbf{A}-\mathbf{A}'$ linearly independent of each of them, then there is no intersection (parallel lines).

(*b*) If $\mathbf{A}-\mathbf{A}'$, $\mathbf{C}$, $\mathbf{C}'$ are linearly dependent, and $\mathbf{C}$, $\mathbf{C}'$ linearly independent, then there is one intersection.

(*c*) If $\mathbf{A}-\mathbf{A}'$, $\mathbf{C}$, $\mathbf{C}'$ are linearly dependent in pairs, then the two straight lines coincide.

As regards (ii) and (iii) above, we modify them slightly for F-space, leaving the proof to the reader:

(ii) *Intersections of a straight line and a linear n-space* ($n>1$).

(*a*) Do not meet at all (skew or parallel), or
(*b*) meet in a single point, or
(*c*) the straight line is contained in the linear n-space.

(iii) *Intersections of a linear m-space and a linear n-space* ($m \geqslant n > 1$).

(*a*) Do not meet at all (skew or parallel), or
(*b*) meet in a single point, or
(*c*) meet in a linear p-space ($p<n$), or
(*d*) the n-space is contained in the m-space ($m>n$), or they coincide ($m=n$).

All this is complicated and not very much to the purpose of our later work. However, two important statements emerge:

A. The intersection of a linear m-space with a linear n-space (and here we include straight lines by putting $m \geqslant n \geqslant 1$) consists of 0, 1 or ∞ points, but never of 2, 3, ...

B. Two linear 2-spaces may intersect in one point without coinciding.

This last result is so surprising to the denizen of Euclidean 3-space that we shall prove it. Consider four linearly independent F-vectors, $\mathbf{S}_1, \mathbf{S}_2, \mathbf{S}_3, \mathbf{S}_4$, and two linear 2-spaces

$$\mathbf{X}=a_1\mathbf{S}_1+a_2\mathbf{S}_2, \quad \mathbf{X}=a_3\mathbf{S}_3+a_4\mathbf{S}_4, \tag{1·317}$$

the a's being variable parameters as usual. Obviously $\mathbf{X}=\mathbf{O}$ is a point of intersection. It is the only one, for if there were another we would have values of the a's (not all zero) satisfying

$$a_1\mathbf{S}_1+a_2\mathbf{S}_2-a_3\mathbf{S}_3-a_4\mathbf{S}_4=\mathbf{O}, \tag{1·318}$$

and this cannot be on account of the linear independence of the $\mathbf{S}$'s.

Linear subspaces in general

The word *linear* has a general and important meaning in mathematics, and we have anticipated a little in speaking of linear n-spaces. Let us now define *linear subspaces* in general; we shall see that a linear n-space is one.

A subspace L of F-space is said to be a linear subspace if, given any two points in L, the straight line through these points lies entirely in L, and this holds for every pair of points in L.

To show that a linear n-space is in fact a linear subspace as just defined, we turn to the equation (1·311) and take two points, $\mathbf{B}$ and $\mathbf{C}$, satisfying it:

$$\left.\begin{aligned}\mathbf{B}&=\mathbf{S}_0+b_1\mathbf{T}_1+b_2\mathbf{T}_2+\ldots+b_n\mathbf{T}_n,\\ \mathbf{C}&=\mathbf{S}_0+c_1\mathbf{T}_1+c_2\mathbf{T}_2+\ldots+c_n\mathbf{T}_n.\end{aligned}\right\} \tag{1·319}$$

A general point on the straight line joining $\mathbf{B}$ and $\mathbf{C}$ is, by (1·301),

$$\mathbf{X}=p\mathbf{B}+q\mathbf{C}, \quad p+q=1. \tag{1·320}$$

Substituting for $\mathbf{B}$ and $\mathbf{C}$ from (1·319), we get

$$\mathbf{X}=\mathbf{S}_0+(pb_1+qc_1)\mathbf{T}_1+(pb_2+qc_2)\mathbf{T}_2+\ldots+(pb_n+qc_n)\mathbf{T}_n. \tag{1·321}$$

Thus $\mathbf{X}$ satisfies (1·311) with $a_1=pb_1+qc_1$, etc., and this proves that *a linear n-space, as defined by* (1·311), *is a linear subspace in the sense of the above definition.*

The linear n-space is a generalization to F-space of the concept of the plane in Euclidean 3-space. In Chapter 2 we shall meet the hyperplane, another generalization of the same basic idea. The linear n-space has n dimensions and the hyperplane has an infinity of dimensions, but both are linear subspaces, as we have just seen in the case of the linear n-space and as we shall see later for the hyperplane.

There are linear subspaces which are not linear n-spaces, as the following two examples show.

Let P-space be the interior of a circle in the Euclidean plane, and let a point or vector in F-space correspond to a function u of the coordinates in P-space. Consider those functions u which take on assigned values on the circumference of the circle. The F-points corresponding to such functions define a subspace F' of F-space, and it is easy to see that it is a linear subspace. For let u_1 and u_2 be any two functions each taking on the circumference the assigned value f (some function of position on the circumference), and write

$$\mathbf{S}_1 \leftrightarrow u_1, \quad \mathbf{S}_2 \leftrightarrow u_2.$$

Then, by our basic rules (1·205) and (1·206),

$$a\mathbf{S}_1 + b\mathbf{S}_2 \leftrightarrow au_1 + bu_2.$$

But if $a+b=1$, the function on the right takes the value f on the circumference, and so $a\mathbf{S}_1 + b\mathbf{S}_2$ is in the subspace F'. Thus the straight line through the points $\mathbf{S}_1$ and $\mathbf{S}_2$ lies in F', which is therefore a linear subspace. But F' is not a linear n-space, since all functions u such that $u=f$ on the circumference cannot be expressed linearly in terms of any finite number of functions.

Similarly it is easy to see that the solutions of any linear partial differential equation, such as

$$\partial^2 u/\partial x^2 + \partial^2 u/\partial y^2 = g(x, y),$$

without regard to boundary conditions, define a linear subspace which is not a linear n-space. (Here g is any given function of position in P-space.)

The following important fact is obvious from the definition of a linear subspace: *The intersection of two linear subspaces is itself a linear subspace.*

It is convenient to speak of *a vector lying in a linear subspace L*. This means a bound vector formed by joining two points of L, and it must of course be carefully distinguished from *the position-vector of a point in L*. The position-vector lies in L if, and only if, L contains the origin of F-space.

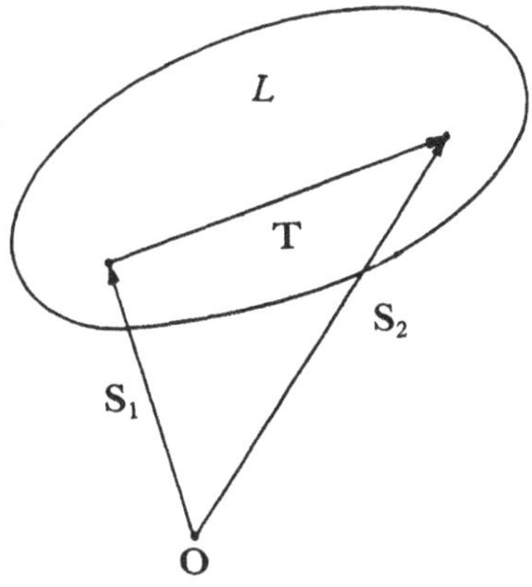

Fig. 1·35. $\mathbf{S}_1$ and $\mathbf{S}_2$ are position-vectors of two points in a linear subspace L; $\mathbf{T}=\mathbf{S}_2-\mathbf{S}_1$ at $\mathbf{S}_1$ is a vector lying in L.

If $\mathbf{S}_1$ and $\mathbf{S}_2$ are points in L, then the vector $\mathbf{T}=\mathbf{S}_2-\mathbf{S}_1$ at $\mathbf{S}_1$ lies in L (Fig. 1·35). Where there is no risk of confusion we shall not trouble to refer explicitly to the bound character of $\mathbf{T}$. Thus in the case of the linear n-space (1·311), we say that the *point* $\mathbf{S}_0$ is in the linear n-space and that the vectors $\mathbf{T}_1$, $\mathbf{T}_2$, ... *lie in* the linear n-space.

Is an F-vector nothing but a certain set of functions?

The great charm and power of the classical vector notation lies in its independence of the choice of coordinates. We think of a vector as a *thing*, just as in geometry we think of a point as a *thing*. This way of thinking is at once profound and elemental. It is profound because, if challenged, we cannot define these *things*; they obey certain laws, that is all. It is elemental because it is in tune with physical space intuitions which are sometimes indispensable for discovery and rapid understanding.

To avoid dealing with undefined things, the mathematician may prefer to introduce a coordinate system from the very beginning; then he defines a vector as a triad of numbers (its components) and a point as a triad of numbers also (its coordinates). In doing so, however, he violates the spirit of vector theory by introducing the very coordinates which it is the purpose of the theory to avoid.

To practical people this matter may appear of slight importance in Euclidean 3-space, where logic and space-intuition go hand in hand. But it becomes more serious in F-space where our intuition needs all the help it can get. In (1·201) we took what seemed the clearest course, defining an F-vector as a set of M functions of N variables. The time has come to offer a partial apology for that definition and to point out that in the long run a more abstract view is perhaps more powerful. This abstract view does not destroy or alter the definition already given; it merely supplements it.

To be definite, let us think of an elastic body. If we choose rectangular Cartesian coordinates x_i (suffixes here take the range 1, 2, 3), we may represent any state of stress in the body by six functions of the coordinates—the stress components E_{ij}, where $E_{ij} = E_{ji}$. These six functions provide us with an F-vector, according to the definition already given.

But we might have chosen a different system of coordinates x'_1; then the *same* state of stress would have been described by six functions E'_{ij}, these *functions* being different from the E_{ij} which corresponded to the coordinates x_i. Are these two *different* representations of the *same* state of stress to be regarded as different F-vectors?

This is indeed a deep question. It is the same as the question whether, in a geometrical transformation of coordinates, we are to regard such a transformation as, on the one hand, a mere change of labels attached to an invariable thing (a point), or, on the other hand, a motion of the point in a fixed coordinate system. Some geometers like one point of view, others the other.

We shall not pursue the question here, but merely point out that, in the applications of the method of function-space, it is often a great mental assistance to regard an F-vector as a *state* of the system under consideration, this state being capable of many *representations* by different sets of functions according to the coordinates used in P-space. Indeed, even without bringing in transformations of the P-coordinates, we might find it convenient to represent a given state of stress, not by the six components of stress, but by suitable linear combinations of them.

In this flux of different possible representations it is a psychological aid to have something firm to hold on to, namely, that there is a *thing* called a *state of the system*, capable of many representations, and that this thing corresponds to an F-vector.

Exercises

1. Show that if the two F-vectors, $\mathbf{S}_1$ and $\mathbf{S}_2$, are linearly dependent, then so are the three F-vectors $\mathbf{S}_1, \mathbf{S}_2, \mathbf{S}_3$. Is the converse true?

2. If a vector in function-space corresponds to a single function of x, then the three functions 1, $\sin x$, $\cos x$ form a triangle in function-space. Write down expressions to include all those functions which correspond to F-points on the sides of the triangle, not produced. At the present stage in the development of the theory of function-space, would there be any sense in asking whether this triangle is equilateral?

3. If $\mathbf{S}_0, \mathbf{S}_1, \ldots, \mathbf{S}_n$ are $n+1$ linearly independent F-vectors, show that the two linear n-spaces

$$\mathbf{X} = a_1\mathbf{S}_1 + a_2\mathbf{S}_2 + \ldots + a_n\mathbf{S}_n,$$
$$\mathbf{X} = \mathbf{S}_0 + a_1\mathbf{S}_1 + a_2\mathbf{S}_2 + \ldots + a_n\mathbf{S}_n,$$

do not intersect (parallel linear n-spaces).

4. In a linear n-space L_n we take m F-points, where $m < n$. Show that the linear $(m-1)$-space containing these points is contained in L_n.

5. If the $m+n$ F-vectors

$$\mathbf{S}_1, \mathbf{S}_2, \ldots, \mathbf{S}_m, \mathbf{T}_1, \mathbf{T}_2, \ldots, \mathbf{T}_n$$

are linearly dependent, show that the linear m-space

$$\mathbf{X} = a_1\mathbf{S}_1 + a_2\mathbf{S}_2 + \ldots + a_m\mathbf{S}_m$$

and the linear n-space

$$\mathbf{X} = a_1\mathbf{T}_1 + a_2\mathbf{T}_2 + \ldots + a_n\mathbf{T}_n$$

have at least a straight line in common.

6. Show that the F-vector joining any two points of the linear n-space (1·312) is of the form

$$b_1\mathbf{S}_1 + b_2\mathbf{S}_2 + \ldots + b_{n+1}\mathbf{S}_{n+1},$$

where the sum of the b's is zero.

PART II

POSITIVE-DEFINITE METRIC

CHAPTER 2

GEOMETRY OF FUNCTION-SPACE WITH POSITIVE-DEFINITE METRIC

2·1. THE SCALAR PRODUCT AND METRIC IN *F*-SPACE

As long as we restrict the algebraic operations in F-space to addition and substraction of F-vectors and multiplication of them by F-scalars (as we have done in Part I), the geometry of F-space remains somewhat meagre. It is only with the *multiplication of F-vectors by one another* that the geometry becomes rich enough to act as a guide in dealing with boundary value problems. This multiplication of two F-vectors gives a scalar (i.e. a number), not another F-vector, and so it is analogous to the scalar product of ordinary vector theory, not the vector product. We shall call it the *scalar product* and use the dot notation for it.

Conditions to be satisfied by the scalar product

In (1·205) and (1·206) we defined the sum of two F-vectors and the product of an F-vector by a scalar. These definitions are so simple and natural that we can hardly conceive of anyone wishing to replace them by others, although, since they are only *definitions*, he would be at liberty to do so if he chose.

It is not so simple in the case of the scalar product. We cannot write down one definition which will serve all our purposes; the appropriate definition depends on the physical problem involved. Our immediate task is to find out what can be said about the geometry of function-space for which a scalar product exists, without committing ourselves to any particular definition of it.

This appears to be an impossible task—we have not enough to go on—and that is true; we must say *something* definitive about the scalar product if we are to talk about it. But what we say must be true of all the particular scalar products we shall use afterwards.

Let us then consider only those definitions of the scalar product $\mathbf{S}.\mathbf{S}'$ of two F-vectors $\mathbf{S}$ and $\mathbf{S}'$ which satisfy the following algebraic laws:

$$\left.\begin{aligned}&\text{Commutative law: } \mathbf{S}.\mathbf{S}'=\mathbf{S}'.\mathbf{S},\\&\text{Distributive law: } \mathbf{S}.(\mathbf{S}'+\mathbf{S}'')=\mathbf{S}.\mathbf{S}'+\mathbf{S}.\mathbf{S}'',\\&\text{Multiplication by zero: } \mathbf{S}.\mathbf{O}=0,\\&\text{Associative law for scalars: } (a\mathbf{S}).\mathbf{S}'=a(\mathbf{S}.\mathbf{S}').\end{aligned}\right\}\qquad(2\cdot101)$$

If at any future time we define a scalar product, and wish to use the results we are about to develop now, we must verify that our definition is such that (2·101) are all satisfied.

Examples of scalar products

To fix the ideas, we should have before us some actual definitions of scalar products. Here are some simple examples, which it is convenient to refer to as the scalar products of Hilbert, Dirichlet and Minkowski:

(i) *Hilbert scalar product.* An F-vector corresponds to a single function of a single variable x in the range $x_1 \leqslant x \leqslant x_2$. The scalar product $\mathbf{S}.\mathbf{T}$ is defined as follows:

If $\mathbf{S} \leftrightarrow s(x)$ and $\mathbf{T} \leftrightarrow t(x)$, then

$$\mathbf{S}.\mathbf{T} = \int_{x_1}^{x_2} s(x)\, t(x)\, dx. \tag{2·102}$$

(ii) *Dirichlet scalar product.* Let P-space be a domain in the Euclidean plane, and let an F-vector $\mathbf{S}$ correspond to a vector field (p_1, p_2) in this domain. (Thus we have $N = 2$, $M = 2$ in (1·201).) The scalar product is defined as follows:

$$\left.\begin{aligned} &\mathbf{S} \leftrightarrow (p_1, p_2), \quad \mathbf{T} \leftrightarrow (q_1, q_2), \\ &\mathbf{S}.\mathbf{T} = \iint (p_1 q_1 + p_2 q_2)\, dA, \end{aligned}\right\} \tag{2·103}$$

where the integral is taken over P-space (i.e. over the given domain), dA being an element of area.

(iii) *Minkowski scalar product.* We take P-space as in (ii) above and let an F-vector correspond to a vector field together with a scalar field. (Thus we have $N = 2$, $M = 3$ in (1·201).) With an obvious notation for the vector and scalar fields, we define the scalar product as follows:

$$\left.\begin{aligned} &\mathbf{S} \leftrightarrow (p_1, p_2, u), \quad \mathbf{T} \leftrightarrow (q_1, q_2, v), \\ &\mathbf{S}.\mathbf{T} = \iint (p_1 q_1 + p_2 q_2 - uv)\, dA. \end{aligned}\right\} \tag{2·104}$$

It is easy to verify that each of these three definitions of the scalar product satisfies the four laws in (2·101). Other more complicated scalar products will be defined as required, all satisfying (2·101).

As an example of a scalar product which does *not* satisfy (2·101), let an F-vector correspond to a single function of a single variable as in the definition of the Hilbert scalar product (i) above, and let the scalar product be defined by

$$\mathbf{S}.\mathbf{T} = \int_{x_1}^{x_2} [s(x) + t(x)]\, dx, \tag{2·105}$$

instead of as in (2·102). The first of (2·101)—the commutative law—is satisfied, but the second—the distributive law—is not. For we have

$$\left.\begin{aligned} \mathbf{S}.\mathbf{T} &= \int_{x_1}^{x_2} [s(x)+t(x)]\,dx, \\ \mathbf{S}.\mathbf{W} &= \int_{x_1}^{x_2} [s(x)+w(x)]\,dx, \\ \mathbf{S}.(\mathbf{T}+\mathbf{W}) &= \int_{x_1}^{x_2} [s(x)+t(x)+w(x)]\,dx, \end{aligned}\right\} \tag{2·106}$$

and so

$$\mathbf{S}.(\mathbf{T}+\mathbf{W}) - \mathbf{S}.\mathbf{T} - \mathbf{S}.\mathbf{W} = -\int_{x_1}^{x_2} s(x)\,dx, \tag{2·107}$$

which does not vanish for an arbitrary function $s(x)$. Thus the definition (2·105) does not satisfy our requirements.

Another unsatisfactory definition is

$$\mathbf{S}.\mathbf{T} = \int_{x_1}^{x_2} (s^2+t^2)\,dx. \tag{2·108}$$

It is by no means implied that (2·102), (2·103) and (2·104) are the only suitable scalar products for the P-spaces in question. As a matter of fact, we are not in a position to decide on a definition of a scalar product useful for our purposes until we are assigned a boundary value problem to deal with.

Differentiation of a scalar product

A scalar product may be differentiated according to the ordinary rule of differential calculus. By changing the functions which correspond to them, we change F-vectors $\mathbf{S}$ and $\mathbf{S}'$ into $\mathbf{S}+\Delta\mathbf{S}$ and $\mathbf{S}'+\Delta\mathbf{S}'$ respectively. Then, by virtue of the distributive law in (2·101), the increment in the scalar product $\mathbf{S}.\mathbf{S}'$ may be developed as follows:

$$\left.\begin{aligned} \Delta(\mathbf{S}.\mathbf{S}') &= (\mathbf{S}+\Delta\mathbf{S}).(\mathbf{S}'+\Delta\mathbf{S}') - \mathbf{S}.\mathbf{S}' \\ &= \mathbf{S}.\Delta\mathbf{S}' + \mathbf{S}'.\Delta\mathbf{S} + \Delta\mathbf{S}.\Delta\mathbf{S}'. \end{aligned}\right\} \tag{2·109}$$

If the variations are infinitesimal, we get, as relation between principal parts,

$$d(\mathbf{S}.\mathbf{S}') = \mathbf{S}.d\mathbf{S}' + \mathbf{S}'.d\mathbf{S}, \tag{2·110}$$

the usual formula for the differential of a product. Just as in the differential calculus, we may avoid all vagueness in respect of 'infinitesimals' by taking $d\mathbf{S}$ and $d\mathbf{S}'$ to be the F-vectors corresponding to the actual changes in the functions which correspond to $\mathbf{S}$ and $\mathbf{S}'$ and by accepting (2·110) as the *definition* of the 'differential' $d(\mathbf{S}.\mathbf{S}')$.

Metric in F-space; definite, semi-definite and indefinite

The scalar product of a vector by itself may be written

$$\mathbf{S}\,.\,\mathbf{S} = \mathbf{S}^2. \tag{2·111}$$

This square or self-product is the basis on which we construct a metrical geometry of F-space, and for that reason we shall call it the *metric* of F-space (F-metric).

We must not read into the notation $\mathbf{S}^2$ more than is actually implied by this definition. Although $\mathbf{S}$ corresponds to a real function or functions, there is no reason to assume that $\mathbf{S}^2$ is necessarily positive.

We shall classify metrics by the following scheme:

The metric in F-space is said to be	
positive-definite	if $\mathbf{S}^2 > 0$ for all F-vectors except the zero vector $\mathbf{O}$. (By (2·101) we always have $\mathbf{O}^2 = 0$)
positive-semi-definite	if $\mathbf{S}^2 \geqslant 0$ for all F-vectors. (Now $\mathbf{S}^2 = 0$ does not imply $\mathbf{S} = \mathbf{O}$, as it did in the positive-definite case)
negative-definite	if $\mathbf{S}^2 < 0$ for all F-vectors except the zero vector $\mathbf{O}$
negative-semi-definite	if $\mathbf{S}^2 \leqslant 0$ for all F-vectors
indefinite	if $\mathbf{S}^2$ is positive, zero, or negative according to the choice of $\mathbf{S}$

We shall not discuss negative-definite or negative-semi-definite metrics, because, by merely reversing the sign in the definition of the scalar product, we can change them into positive-definite and positive-semi-definite respectively.

That leaves us then with three types to consider: positive-definite, positive-semi-definite and indefinite. The positive-semi-definite is of rare occurrence, and we shall not bother with it further. We are left then with just two important types of metric:

positive-definite,
indefinite.

In some respects the distinction between these two is trivial, in other respects it is vital, for on this turns the distinction between *minimum* principles (for the positive-definite metric) and *merely stationary* principles (for the indefinite metric).

From now on, in this Part II of the book, we shall assume the metric *to be positive-definite, reserving to Part III the discussion of indefinite metrics.**

Let us examine the examples given earlier to see which correspond to positive-definite metrics. We shall consider here only functions continuous in P-space, regarded as a closed domain, i.e. the boundary included.

The Hilbert scalar product (2·102) gives

$$\mathbf{S}^2 = \int_{x_1}^{x_2} [s(x)]^2 \, dx. \tag{2·112}$$

This is positive unless $s(x) = 0$, i.e. $\mathbf{S} = \mathbf{O}$; the metric is positive-definite. So is the Dirichlet metric

$$\mathbf{S}^2 = \iint (p_1^2 + p_2^2) \, dA, \tag{2·113}$$

given by (2·103). On the other hand, the Minkowski metric

$$\mathbf{S}^2 = \iint (p_1^2 + p_2^2 - u^2) \, dA, \tag{2·114}$$

given by (2·104), is indefinite, for we can make $\mathbf{S}^2$ positive by choosing $u = 0$, negative by choosing $p_1 = p_2 = 0$, and zero by choosing $u = p_1$, $p_2 = 0$.

Accordingly, for the rest of the present chapter, and indeed throughout Part II, we may take the Hilbert metric (2·112) or the Dirichlet metric (2·113) as illustrative examples, but not the Minkowski metric (2·114); this last is an illustrative example for Part III.

As regards the distinction between physical problems which lead to positive-definite F-metrics and those which lead to indefinite F-metrics, we are to remember that the scalar product which we impose in any particular problem is a matter of our choice. However, it turns out that the suitable scalar product in problems of electrostatics or elastic equilibrium leads to a positive-definite F-metric, and the suitable scalar product in problems of vibration (mechanical or electrical) leads to an indefinite F-metric.

Scalar product derived from metric

Since we obtained the metric in (2·111) by a specialization in the scalar product ($\mathbf{S}' = \mathbf{S}$ in $\mathbf{S} . \mathbf{S}'$), it might appear that the scalar product is a more general concept than metric. This is not

* In the definition of a *Hilbert space* (cf. Stone(1), p. 3) the positive-definite character of the metric is postulated. This means that in Part II of the book we are dealing essentially with a Hilbert space, and in Part III we are not. However, since we shall almost always restrict our attention to a finite number of points, our subject in Part II is for the most part *real unitary space*, a much simpler concept than Hilbert space (cf. Stone(1), p. 16; Halmos (2), p. 90).

the case, for we can define the scalar product in terms of the metric by the laws (2·101) and the identity

$$\mathbf{S}.\mathbf{S}' = \tfrac{1}{2}[\mathbf{S}^2 + \mathbf{S}'^2 - (\mathbf{S} - \mathbf{S}')^2]. \tag{2·115}$$

Admissible functions

In (1·201) we defined an F-vector as a set of M functions in P-space. Nothing was said about the class of the functions, and so (in the absence of a statement to the contrary) we admitted such curious functions as the following:

(i) a function which takes the value 1 at one point of P-space and is zero elsewhere;

(ii) a function which takes the value 1 at all points of P-space which have rational coordinates, and is zero elsewhere.

There is no reason to reject such functions as far as Chapter 1 is concerned, but we may find ourselves in grave difficulties if we admit them into scalar products. Thus, for the function (i) above, (2·112) gives $\mathbf{S}^2 = 0$, but $\mathbf{S} \neq \mathbf{O}$; the metric ceases to be positive-definite if such a function is admitted.

In the physical problems of interest to us we do not meet such strange functions as those described above. But we do encounter functions with poles on the boundary (e.g. in the torsion of a hollow square) and we shall deliberately introduce functions with discontinuities (pyramid functions). We cannot therefore get rid of our difficulties by admitting only continuous functions; the class is not wide enough.

The situation is a common one in mathematics. We have set up certain formal machinery and we are a little uncertain what to feed into it; if we feed in the wrong elements, the machine may jam or explode in statements obviously false. We encounter this in elementary algebra: $a/a = 1$ is a useful algebraic formula, generally true, but we must not feed a zero value of a into it. Though our field consists of all the real numbers, special precautions must be taken against division by zero.

In our work we have two alternatives. The first is heuristic. We may go ahead and use the machine to get results; if it jams or explodes, we examine the particular case and make any special adjustments required to overcome the difficulty. This was the method of eighteenth-century mathematics and is widely used in the mathematical physics of the present day. The second plan is to lay down restrictions from the beginning which will ensure the correctness of all future statements. This is the way of modern mathematics, the only way acceptable to pure mathematicians, but it is distasteful to mathematical physicists, who would rather live dangerously than have the broad sweep of a theory obscured by details which are of importance only in exceptional cases.

The aim of this book is to present to mathematical physicists and engineers a practical method based on a theory usually available only in the precise terms of pure mathematics. We must compromise between logical nonsense and excessive precision, and the best plan seems to be to tell the reader at this point about the actual range of application which he will find in this book. Be it understood then that

(i) P-space is a portion of Euclidean space (of any number of dimensions), bounded by surfaces composed of a finite number of portions, each with continuous normals.

(ii) Admissible functions are piecewise continuous, except possibly for a finite number of singular points, and the squares of the functions are integrable. At discontinuities a function has an unassigned value, and two functions are regarded as identical if their values agree at all points of continuity.

The exploration of such wider scope as the theory may have is left to the discretion of the reader.

Exercises

1. Take for P-space a domain in Euclidean 3-space and let an F-vector correspond to a function of the three coordinates, $\mathbf{S} \leftrightarrow f(x, y, z)$. Take for F-metric

$$\mathbf{S}^2 = \iiint (f_x^2 + f_y^2 + f_z^2)\, dx\, dy\, dz,$$

and use (2·115) to obtain the scalar product $\mathbf{S}\,.\,\mathbf{S}'$. Verify that it satisfies the laws (2·101).

2. Show that

$$(\mathbf{A}+\mathbf{B})\,[(\mathbf{A}+\mathbf{B}).(\mathbf{A}-\mathbf{B})] + (\mathbf{B}+\mathbf{C})\,[(\mathbf{B}+\mathbf{C}).(\mathbf{B}-\mathbf{C})] + (\mathbf{C}+\mathbf{A})\,[(\mathbf{C}+\mathbf{A}).(\mathbf{C}-\mathbf{A})]$$
$$= \mathbf{A}(\mathbf{C}^2 - \mathbf{B}^2) + \mathbf{B}(\mathbf{A}^2 - \mathbf{C}^2) + \mathbf{C}(\mathbf{B}^2 - \mathbf{A}^2).$$

3. Using the Hilbert scalar product (2·102), take the range of x to be $(-\pi, \pi)$ and $\mathbf{S} \leftrightarrow \sin x$. Write down an infinite sequence of functions $f_n(x) \leftrightarrow \mathbf{T}_n$ such that $\mathbf{S}\,.\,\mathbf{T}_n = 0$.

4. If P-space is a domain in Euclidean 3-space and an F-vector corresponds to an electromagnetic field given by the P-vectors $\overrightarrow{E}, \overrightarrow{H}$ (functions of position for a definite time t), does the definition

$$\mathbf{S}\,.\,\mathbf{S}' = \iiint (\overrightarrow{E}.\overrightarrow{H'} + \overrightarrow{E'}.\overrightarrow{H})\, dx\, dy\, dz$$

satisfy the laws (2·101)? (Under the integral sign the dot indicates the ordinary scalar product.)

Does this give a positive-definite metric? If not, can you alter it so that it does?

5. If for the Hilbert scalar product (2·102) the range of x is $(0, \pi)$ and there is a set of F-vectors $\mathbf{S}(a) \leftrightarrow \sin ax$, a being a parameter, to what function does $d\mathbf{S}/da$ correspond, if we define $d\mathbf{S}/da$ as the limit of $\Delta\mathbf{S}/\Delta a$ as Δa tends to zero? Verify that

$$d(\mathbf{S}^2)/da = 2\mathbf{S}\,.\,(d\mathbf{S}/da).$$

2·2. LENGTH AND ANGLE IN F-SPACE

The word *metric* has been used in anticipation of the definitions of *length* and *distance* in F-space. These definitions will now be given.

Length and distance in F-space

The *length* or *magnitude* or *norm* of an F-vector $\mathbf{S}$ is defined to be *the positive square root of* $\mathbf{S}^2$, and is denoted by $|\,\mathbf{S}\,|$ or S:

$$|\,\mathbf{S}\,| = S = (\mathbf{S}^2)^{\frac{1}{2}}. \tag{2·201}$$

We are of course to remember that here, and for the rest of Part II, we are committed to a positive-definite metric, so that $\mathbf{S}^2$ cannot be negative; (2·201) cannot yield an imaginary value. We note that the length or magnitude of an F-vector vanishes if, and only if, the vector is the zero vector ($\mathbf{S} = \mathbf{O}$).

If $\mathbf{S}$ and $\mathbf{S}'$ are the position-vectors of two F-points, we define the *distance* between the two points to be $|\,\mathbf{S} - \mathbf{S}'\,|$, i.e. the length of the vector $(\mathbf{S} - \mathbf{S}')$. This definition justifies the use we have made of the word *metric*.

From the positive-definite character of the metric it follows that two F-points coincide if, and only if, the distance between them is zero; in symbols, $|\,\mathbf{S} - \mathbf{S}'\,| = 0$ implies $\mathbf{S} = \mathbf{S}'$, and conversely.

Example

Using the Hilbert scalar product (2·102) and taking for x the range $(-\pi, \pi)$, let us consider the F-points A, B, C with position-vectors

$$\mathbf{A} \leftrightarrow 1, \quad \mathbf{B} \leftrightarrow \sin x, \quad \mathbf{C} \leftrightarrow \cos x. \tag{2·202}$$

By computing the definite integrals for

$$\mathbf{A}^2, \quad \mathbf{B}^2, \quad \mathbf{C}^2, \quad (\mathbf{B} - \mathbf{C})^2, \quad (\mathbf{C} - \mathbf{A})^2, \quad (\mathbf{A} - \mathbf{B})^2,$$

we easily find the distances from the origin to the vertices of the triangle ABC and the lengths of the sides of that triangle. They come out as follows (Fig. 2·21), and in fact $OABC$ is a tetrahedron with three right angles at O, as we can verify when we define 'right angle' a little later:

$$\left.\begin{aligned} OA &= (2\pi)^{\frac{1}{2}}, \quad OB = \pi^{\frac{1}{2}}, \quad OC = \pi^{\frac{1}{2}}, \\ BC &= (2\pi)^{\frac{1}{2}}, \quad CA = (3\pi)^{\frac{1}{2}}, \quad AB = (3\pi)^{\frac{1}{2}}. \end{aligned}\right\} \tag{2·203}$$

Unit F-vectors and normalization

A *unit* F-vector is defined as one of unit length ($|\mathbf{S}|=1$). From any vector $\mathbf{S}$ (except the zero vector) we can form a unit vector $\mathbf{I}$ by the formula

$$\mathbf{I}=\mathbf{S}/|\mathbf{S}|. \tag{2·204}$$

This vector has the same direction as $\mathbf{S}$. The unit vector with direction opposite to that of $\mathbf{S}$ is

$$\mathbf{I}=-\mathbf{S}/|\mathbf{S}|. \tag{2·205}$$

In both of these cases $\mathbf{I}^2=1$, and either of these processes leading from $\mathbf{S}$ to $\mathbf{I}$ may be called *normalization*, although we shall usually understand (2·204) rather than (2·205).

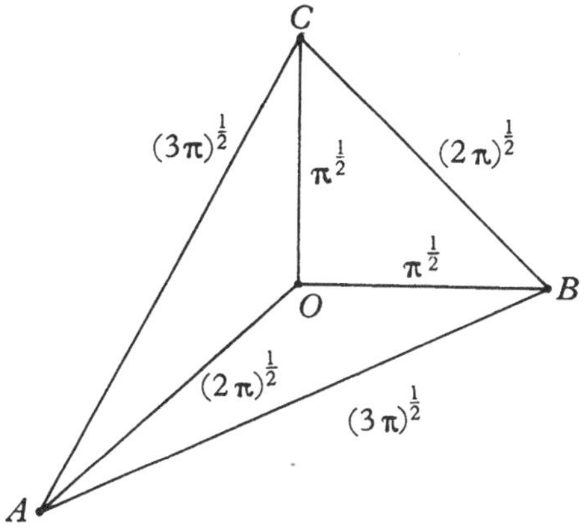

Fig. 2·21. The tetrahedron *OABC* [cf. (2·202)].

For example, normalization of the three F-vectors (2·202) for the Hilbert metric (2·112) for the range $(-\pi,\pi)$ yields the three unit vectors

$$\mathbf{I}\leftrightarrow(2\pi)^{-\frac{1}{2}},\quad \mathbf{J}\leftrightarrow\pi^{-\frac{1}{2}}\sin x,\quad \mathbf{K}\leftrightarrow\pi^{-\frac{1}{2}}\cos x. \tag{2·206}$$

The Schwarz inequality

The Schwarz inequality, established below, is basic in our work. It will be seen that the positive-definite character of the metric is involved in the proof, and, indeed, the inequality does not hold if the metric is indefinite. The consequence is that the minimum principles which occur in the case of a positive-definite metric do not occur when the metric is indefinite.

Let $\mathbf{S}$ and $\mathbf{S}'$ be two non-zero F-vectors. (If either is the zero vector, then the inequality is obvious.) By virtue of the assumed positive-definite character of the metric, we have

$$(a\mathbf{S}+\mathbf{S}')^2\geqslant 0, \tag{2·207}$$

no matter what value is given to the real constant a. Thus

$$(a\mathbf{S}+\mathbf{S}').(a\mathbf{S}+\mathbf{S}')\geqslant 0, \tag{2·208}$$

or

$$a^2\mathbf{S}^2+2a\mathbf{S}.\mathbf{S}'+\mathbf{S}'^2\geqslant 0. \tag{2·209}$$

Since this quadratic expression in a is never negative, it follows from elementary algebra that

$$(\mathbf{S}.\mathbf{S}')^2\leqslant \mathbf{S}^2\mathbf{S}'^2, \tag{2·210}$$

or equivalently

$$|\mathbf{S}.\mathbf{S}'|\leqslant|\mathbf{S}||\mathbf{S}'|, \tag{2·211}$$

where the modulus sign on the left is used in the ordinary sense (absolute value of a quantity which may be positive or negative), while those on the right are used in the technical sense of (2·201).

The inequality (2·210) or (2·211) is called the *Schwarz inequality*.

By reference to (2·207) it is easily seen that the equality sign in (2·210) or (2·211) holds if, and only if, the vectors $\mathbf{S}$ and $\mathbf{S}'$ have the same direction or opposite directions (i.e. they are linearly dependent). If they have the same direction, then $\mathbf{S}'=b\mathbf{S}$ with b positive, and consequently $\mathbf{S}.\mathbf{S}'=|\mathbf{S}||\mathbf{S}'|$; if they have opposite directions, then $\mathbf{S}.\mathbf{S}'=-|\mathbf{S}||\mathbf{S}'|$.

In the case of two unit vectors, $\mathbf{I}$ and $\mathbf{J}$, the Schwarz inequality gives

$$-1\leqslant\mathbf{I}.\mathbf{J}\leqslant 1. \tag{2·212}$$

If $\mathbf{I}.\mathbf{J}=1$, then $\mathbf{I}=\mathbf{J}$; if $\mathbf{I}.\mathbf{J}=-1$, then $\mathbf{I}=-\mathbf{J}$.

Angles in F-space

The *angle* θ between two F-vectors, $\mathbf{S}$ and $\mathbf{S}'$, is defined by

$$\cos\theta=(\mathbf{S}.\mathbf{S}')/(|\mathbf{S}||\mathbf{S}'|) \quad (0\leqslant\theta\leqslant\pi). \tag{2·213}$$

The Schwarz inequality (2·211) assures us that this fraction cannot exceed unity in absolute value, and so the formula always defines a real angle. (We would get into difficulties here in the case of an indefinite metric.) If the vectors have the same direction, then $\theta=0$; if opposite directions, then $\theta=\pi$.

It follows from (2·213) that the angle θ between two unit F-vectors, $\mathbf{I}$ and $\mathbf{J}$, is given by

$$\cos\theta=\mathbf{I}.\mathbf{J} \quad (0\leqslant\theta\leqslant\pi). \tag{2·214}$$

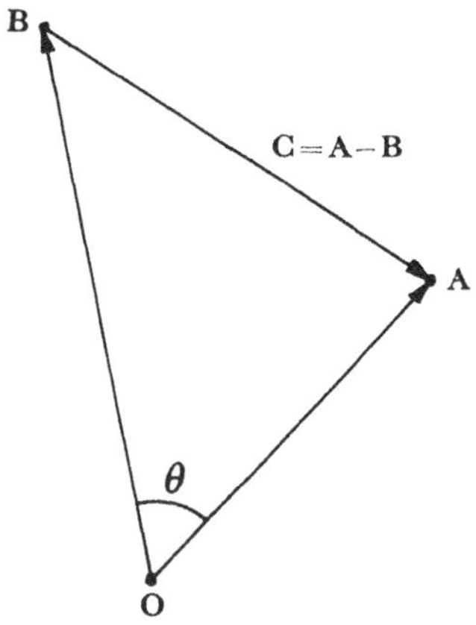

Fig. 2·22. Triangle in F-space.

It is easy to prove the familiar cosine formula for a triangle. Let

A and **B** be drawn from **O**, and let the triangle be completed by drawing $\mathbf{C}=\mathbf{A}-\mathbf{B}$ from the extremity of **B** (Fig. 2·22). Then

$$\begin{aligned}\tfrac{1}{2}[\mathbf{A}^2+\mathbf{B}^2-\mathbf{C}^2]/(|\mathbf{A}|\,|\mathbf{B}|) &= \tfrac{1}{2}[\mathbf{A}^2+\mathbf{B}^2-(\mathbf{A}-\mathbf{B})^2]/(|\mathbf{A}|\,|\mathbf{B}|)\\ &= (\mathbf{A}\,.\,\mathbf{B})/(|\mathbf{A}|\,|\mathbf{B}|)\\ &= \cos\theta, \qquad (2{\cdot}215)\end{aligned}$$

where θ is the angle between **A** and **B**.

Elementary geometry of the triangle

All the usual geometry of the triangle holds for a triangle formed by three F-points. The reader may amuse himself by proving that the bisectors of the angles are concurrent, or the perpendiculars dropped from the vertices on the opposite sides.

By way of example, we shall here find the circumscribed circle of the triangle formed by the F-points

$$\mathbf{A}\leftrightarrow 1, \quad \mathbf{B}\leftrightarrow x, \quad \mathbf{C}\leftrightarrow x^2, \qquad (2{\cdot}216)$$

an F-point corresponding to a function of a single variable x in the range $(0, 1)$ and the scalar product being that of Hilbert,

$$\mathbf{S}\,.\,\mathbf{T}=\int_0^1 s(x)\,t(x)\,dx.$$

The linear 2-space through **A**, **B** and **C** is, by (1·312),

$$\mathbf{X}=a\mathbf{A}+b\mathbf{B}+c\mathbf{C}, \quad a+b+c=1. \qquad (2{\cdot}217)$$

In discussing the circumcircle, we are interested only in points which lie in this linear 2-space, so we put for the position-vector of the circumcentre

$$\mathbf{D}=a\mathbf{A}+b\mathbf{B}+c\mathbf{C}, \quad a+b+c=1. \qquad (2{\cdot}218)$$

In (2·217) a, b, c were variable parameters, now they have definite values which we are to determine.

The conditions that **D** be equidistant from **A**, **B** and **C** are

$$(\mathbf{D}-\mathbf{A})^2=(\mathbf{D}-\mathbf{B})^2=(\mathbf{D}-\mathbf{C})^2. \qquad (2{\cdot}219)$$

From these we obtain the three following equations, of which only two are independent:

$$\left.\begin{aligned}2\mathbf{D}\,.\,(\mathbf{A}-\mathbf{B})&=\mathbf{A}^2-\mathbf{B}^2,\\ 2\mathbf{D}\,.\,(\mathbf{B}-\mathbf{C})&=\mathbf{B}^2-\mathbf{C}^2,\\ 2\mathbf{D}\,.\,(\mathbf{C}-\mathbf{A})&=\mathbf{C}^2-\mathbf{A}^2.\end{aligned}\right\} \qquad (2{\cdot}220)$$

Carrying out the required integrations, we have

$$\left.\begin{aligned}&\mathbf{A}^2=1, \quad \mathbf{B}^2=\tfrac{1}{3}, \quad \mathbf{C}^2=\tfrac{1}{5},\\ &\mathbf{D}.\mathbf{A}=a+\tfrac{1}{2}b+\tfrac{1}{3}c, \quad \mathbf{D}.\mathbf{B}=\tfrac{1}{2}a+\tfrac{1}{3}b+\tfrac{1}{4}c, \quad \mathbf{D}.\mathbf{C}=\tfrac{1}{3}a+\tfrac{1}{4}b+\tfrac{1}{5}c.\end{aligned}\right\} \tag{2·221}$$

Substituting these values in the first two of (2·220) we have, with (2·218), the three linear equations for a, b, c:

$$\left.\begin{aligned}6a+2b+\ c&=4,\\ 10a+5b+3c&=4,\\ a+\ b+\ c&=1.\end{aligned}\right\} \tag{2·222}$$

Hence
$$a=\tfrac{5}{3}, \quad b=-\tfrac{16}{3}, \quad c=\tfrac{14}{3}, \tag{2·223}$$

and so by (2·218) the circumcentre is

$$\mathbf{D}\leftrightarrow\tfrac{1}{3}(5-16x+14x^2). \tag{2·224}$$

The quantities in (2·219) now have a common value, which is the square of the circumradius R. We find

$$R^2=(\mathbf{D}-\mathbf{A})^2=\int_0^1[\tfrac{1}{3}(2-16x+14x^2)]^2\,dx=\tfrac{16}{45}. \tag{2·225}$$

We have now located the centre of the circumcircle at $\mathbf{D}$ as in (2·224) and found its radius R in (2·225). It remains to give an equation for the circumcircle. This may be written in the parametric form

$$\mathbf{X}=\mathbf{D}+R\mathbf{J}, \tag{2·226}$$

where $\mathbf{J}$ is an arbitrary unit vector lying in the linear 2-space determined by the points $\mathbf{A}$, $\mathbf{B}$, $\mathbf{C}$. (Note that it is the vector $\mathbf{J}$, and not merely its extremity, that is to lie in the linear 2-space.) A general vector lying in this 2-space may be written

$$\mathbf{J}=m(\mathbf{A}-\mathbf{B})+n(\mathbf{A}-\mathbf{C}). \tag{2·227}$$

The condition that this be a unit vector is

$$m^2(\mathbf{A}-\mathbf{B})^2+2mn(\mathbf{A}-\mathbf{B}).(\mathbf{A}-\mathbf{C})+n^2(\mathbf{A}-\mathbf{C})^2=1, \tag{2·228}$$

and from (2·216) this turns out to be

$$10m^2+25mn+16n^2=30. \tag{2·229}$$

Thus the circumcircle of the triangle formed by the points $\mathbf{A}$, $\mathbf{B}$ and $\mathbf{C}$ has the parametric equation (2·226), where $\mathbf{J}$ is as in (2·227), the parameters m and n taking all values consistent with (2·229).

The triangle inequality

In a Euclidean triangle the length of any side is less than the sum of the lengths of the other two sides. This is an important property and we shall prove it for a triangle in F-space.

Let the triangle be **A**, **B**, **C**. Then

$$\begin{aligned}(\mathbf{A}-\mathbf{B})^2 &= (\mathbf{A}-\mathbf{C}+\mathbf{C}-\mathbf{B})^2 \\ &= (\mathbf{A}-\mathbf{C})^2 + 2(\mathbf{A}-\mathbf{C}).(\mathbf{C}-\mathbf{B}) + (\mathbf{C}-\mathbf{B})^2. \qquad (2\cdot230)\end{aligned}$$

For any vector **S** we have $\mathbf{S}^2 = |\,\mathbf{S}\,|^2$, and so the above identity may be written

$$|\,\mathbf{A}-\mathbf{B}\,|^2 = |\,\mathbf{A}-\mathbf{C}\,|^2 + 2(\mathbf{A}-\mathbf{C}).(\mathbf{C}-\mathbf{B}) + |\,\mathbf{C}-\mathbf{B}\,|^2. \qquad (2\cdot231)$$

By the Schwarz inequality (2·211) we have for any pair of linearly independent vectors $\mathbf{S}.\mathbf{S}' < |\,\mathbf{S}\,|\,|\,\mathbf{S}'\,|$; if **A**, **B**, **C** are not collinear (i.e. if they form a non-degenerate triangle), then $\mathbf{A}-\mathbf{C}$ and $\mathbf{C}-\mathbf{B}$ are linearly independent, and so (2·231) gives

$$|\,\mathbf{A}-\mathbf{B}\,|^2 < |\,\mathbf{A}-\mathbf{C}\,|^2 + 2\,|\,\mathbf{A}-\mathbf{C}\,|\,\,|\,\mathbf{C}-\mathbf{B}\,| + |\,\mathbf{C}-\mathbf{B}\,|^2, \qquad (2\cdot232)$$

and hence

$$|\,\mathbf{A}-\mathbf{B}\,| < |\,\mathbf{A}-\mathbf{C}\,| + |\,\mathbf{C}-\mathbf{B}\,|. \qquad (2\cdot233)$$

This is the required result, usually known as the *triangle inequality*. The inequality becomes an equality if, and only if, the three points are collinear with **C** between **A** and **B**.

Orthogonality

Two F-vectors, **S** and **S′**, are said to be *orthogonal* if their scalar product vanishes:

$$\mathbf{S}.\mathbf{S}' = 0. \qquad (2\cdot234)$$

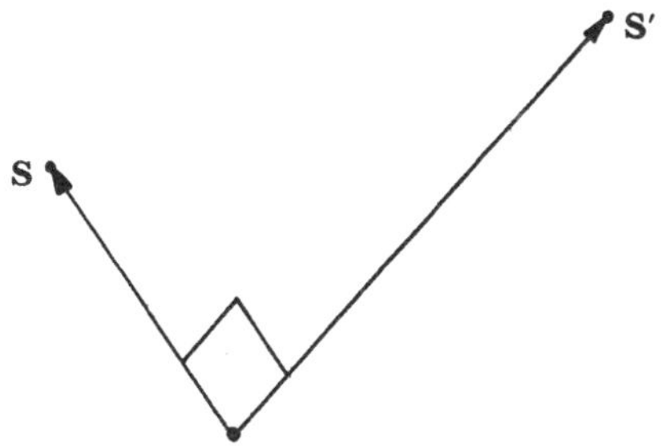

Fig. 2·23. Orthogonal F-vectors ($\mathbf{S}.\mathbf{S}'=0$).

It follows from the definition of angle given in (2·213) that the angle between orthogonal vectors is $\frac{1}{2}\pi$. Orthogonal vectors are, in fact, *at right angles*, and we may legitimately use this expression if we like. In diagrams a right angle may be indicated as in Fig. 2·23.

In F-space orthogonality is a much less restrictive condition than we might at first suppose. In Euclidean 3-space the line of a vector is fixed in direction by two conditions of orthogonality to assigned vectors. But in F-space the line of a vector is not fixed in direction by any finite number of conditions of orthogonality.

Consider, for example, the Hilbert scalar product (2·102). Suppose that an F-vector $\mathbf{S} \leftrightarrow s(x)$ is orthogonal to the n F-vectors $\mathbf{T}_r \leftrightarrow x^r$ $(r = 1, 2, \ldots, n)$, i.e. integer powers of x. The orthogonality conditions read

$$\int xs(x)\,dx = 0, \quad \int x^2 s(x)\,dx = 0, \quad \ldots, \quad \int x^n s(x)\,dx = 0, \qquad (2\cdot235)$$

the range of integration being $x_1 \leqslant x \leqslant x_2$. Now pick *arbitrarily* $n+1$ functions $f_1(x), f_2(x), \ldots, f_n(x), f_{n+1}(x)$, and write

$$s(x) = c_1 f_1(x) + c_2 f_2(x) + \ldots + c_{n+1} f_{n+1}(x). \qquad (2\cdot236)$$

Then (2·235) can be satisfied by proper choice of the $n+1$ constants $c_1, c_2, \ldots, c_{n+1}$, which leaves us with a very general function $s(x)$ satisfying the n conditions of orthogonality.

Exercises

1. Taking for P-space the interior of the circle $x_1^2 + x_2^2 = a^2$ in the Euclidean plane (x_1, x_2 rectangular Cartesian coordinates), and letting an F-vector correspond to a P-vector field (p_1, p_2), normalize $\mathbf{S} \leftrightarrow (p_1 = x_1, p_2 = 0)$ for the Dirichlet metric (2·103) and (2·113).

[*Ans.* $\mathbf{I} \leftrightarrow (2x_1 a^{-2}\pi^{-\frac{1}{2}}, 0)$].

2. If $\mathbf{I}$, $\mathbf{J}$, $\mathbf{K}$ are unit F-vectors, prove that

$$\mathbf{I}.(\mathbf{J}+\mathbf{K}) \leqslant 2.$$

If, further, $\mathbf{J}$ and $\mathbf{K}$ are orthogonal, then

$$\mathbf{I}.(\mathbf{J}+\mathbf{K}) \leqslant 2^{\frac{1}{2}}.$$

Under what circumstances will the equality sign hold in each of these statements?

3. Show that the F-vector $\mathbf{X}$ which minimizes the sum of the squares of the distances from the point $\mathbf{X}$ to given points $\mathbf{A}$, $\mathbf{B}$, $\mathbf{C}$ is $\mathbf{X} = \frac{1}{3}(\mathbf{A}+\mathbf{B}+\mathbf{C})$.

4. Taking P-space and $\mathbf{S}$ as in Exercise 1, write down *any* non-zero F-vector orthogonal to $\mathbf{S}$.

5. Taking P-space to be the segment $-1 \leqslant x \leqslant 1$ with the Hilbert scalar product (2·102), find the constant b so that the three F-vectors

$$\mathbf{X} \leftrightarrow 1, \quad \mathbf{Y} \leftrightarrow x, \quad \mathbf{Z} \leftrightarrow x^2 + b$$

are mutually orthogonal.

2·3. ORTHONORMAL F-VECTORS

Orthonormality and linear independence

Consider a set of n F-vectors $\mathbf{I}_\rho$ $(\rho = 1, 2, \ldots, n)$; they are said to be *orthonormal* (contraction for orthogonal and normalized) if each vector is orthogonal to all the others and each vector is a unit vector. The *conditions of orthonormality* read

$$\mathbf{I}_\rho . \mathbf{I}_\sigma = \delta_{\rho\sigma} \quad (\rho, \sigma = 1, 2, \dots, n), \tag{2·301}$$

$\delta_{\rho\sigma}$ being the Kronecker delta ($=1$ if $\rho=\sigma$; $=0$ if $\rho \neq \sigma$).

It is easy to show that *if n F-vectors are orthonormal, they are linearly independent.* For suppose they are not. Then there exists a set of n constant coefficients, not all zero, such that

$$a_1\mathbf{I}_1 + a_2\mathbf{I}_2 + \dots + a_n\mathbf{I}_n = \mathbf{O}. \tag{2·302}$$

At least one of these coefficients is different from zero. Suppose $a_\rho \neq 0$. Then take the scalar product of (2·302) by $\mathbf{I}_\rho$. By (2·301) we obtain $a_\rho = 0$, a contradiction. So the result is proved.

Orthogonal transformations

Consider an orthonormal set $\mathbf{I}_\rho$ $(\rho = 1, 2, \dots, n)$. These vectors define a linear n-space as in (1·310), consisting of all vectors given by

$$\mathbf{X} = b_1\mathbf{I}_1 + b_2\mathbf{I}_2 + \dots + b_n\mathbf{I}_n, \tag{2·303}$$

the coefficients being variable parameters taking all real values.

If we now pick any set of n^2 real numbers $a_{\rho\mu}$ $(\rho, \mu = 1, 2, \dots, n)$, the n vectors given by

$$\mathbf{J}_\rho = \sum_{\mu=1}^{n} a_{\rho\mu}\mathbf{I}_\mu \quad (\rho = 1, 2, \dots, n) \tag{2·304}$$

lie in the linear n-space (2·303). These n vectors form a new orthonormal set if $\mathbf{J}_\rho . \mathbf{J}_\sigma = \delta_{\rho\sigma}$ $(\rho, \sigma = 1, 2, \dots, n)$. Substituting in these conditions the values of the **J**'s from (2·304) and using the orthonormality conditions (2·301) on the **I**'s, we obtain

$$\sum_{\mu=1}^{n} a_{\rho\mu} a_{\sigma\mu} = \delta_{\rho\sigma} \quad (\rho, \sigma = 1, 2, \dots, n). \tag{2·305}$$

These are the conditions (*necessary and sufficient*) *that the* **J***'s, as given by* (2·304), *form an orthonormal set.*

Any formula which defines a set of n vectors (the **J**'s) in terms of another set (the **I**'s) may be called a *transformation*. A transformation which carries any orthonormal set into another orthonormal set is called an *orthogonal transformation* (as in ordinary Euclidean geometry, where $n = 3$). The conditions for an orthogonal transformation are the conditions (2·305).

It is easily seen from (2·305) that the square of the determinant of the a's is 1. Hence the determinant itself is either $+1$ or -1. Precisely the same situation arises in Euclidean 3-space, where the case $\det(a_{\rho\mu}) = +1$ corresponds to a pure rotation and the case $\det(a_{\rho\mu}) = -1$ to rotation combined with reflexion. We may use the same expressions in F-space.

Orthogonal transformations in a linear 2-space

If $n=2$, the transformation (2·304) reads explicitly

$$\mathbf{J}_1 = a_{11}\mathbf{I}_1 + a_{12}\mathbf{I}_2, \quad \mathbf{J}_2 = a_{21}\mathbf{I}_1 + a_{22}\mathbf{I}_2, \tag{2·306}$$

and the conditions (2·305) that this be an orthogonal transformation are three in number:

$$a_{11}^2 + a_{12}^2 = 1, \quad a_{21}^2 + a_{22}^2 = 1, \quad a_{11}a_{21} + a_{12}a_{22} = 0. \tag{2·307}$$

Since we have only three equations to be satisfied by four quantities, there is one degree of freedom in the general solution; it may be written

$$\left.\begin{aligned} a_{11} &= \cos\theta, & a_{12} &= \sin\theta, \\ a_{21} &= -\zeta\sin\theta, & a_{22} &= \zeta\cos\theta, \end{aligned}\right\} \tag{2·308}$$

where θ may have any value and ζ is chosen equal to $+1$ or -1.

Thus the *general orthogonal transformation for two F-vectors* reads

$$\left.\begin{aligned} \mathbf{J}_1 &= \cos\theta\,\mathbf{I}_1 + \sin\theta\,\mathbf{I}_2, \\ \mathbf{J}_2 &= \zeta(-\sin\theta\,\mathbf{I}_1 + \cos\theta\,\mathbf{I}_2). \end{aligned}\right\} \tag{2·309}$$

All distinct transformations are given by letting θ cover the range $0 \leqslant \theta < 2\pi$. These formulae will be recognized as the usual transformation corresponding to rotation of axes, without reflexion if $\zeta = +1$ and with reflexion if $\zeta = -1$.

We note that the formula

$$\mathbf{J} = \cos\theta\,\mathbf{I}_1 + \sin\theta\,\mathbf{I}_2 \quad (0 \leqslant \theta < 2\pi) \tag{2·310}$$

gives all unit vectors in the linear 2-space of an orthonormal pair, $\mathbf{I}_1, \mathbf{I}_2$.

The process of orthonormalization

A linear n-space through the origin is described, as in (1·310), by a set of n linearly independent F-vectors in it. For many purposes it is most convenient to have these n linearly independent vectors orthonormal. So we are faced with the problem of *orthonormalization*, i.e. of finding an orthonormal set of n vectors in the linear n-space of n vectors which are not orthonormal.

Obviously, the problem does not admit of a unique solution; for once we have found an orthonormal set, we can apply an orthogonal transformation to it and get another orthonormal set. Thus what we require is *some* process which will lead us to an orthonormal set.

We shall now describe the Gram–Schmidt process of orthonormalization (Gram (1), Schmidt (1)).

We start with n linearly independent F-vectors $\mathbf{S}_\rho$ $(\rho = 1, 2, \ldots, n)$. These vectors can be ordered in $n!$ ways. It is not essential that they should be ordered in any particular way, but it is generally found convenient to order them in some way consistent with an obvious law. Thus, if we had four F-vectors corresponding to the functions $x^3, x, 1, x^2$, we would rearrange them in the order $1, x, x^2, x^3$ before

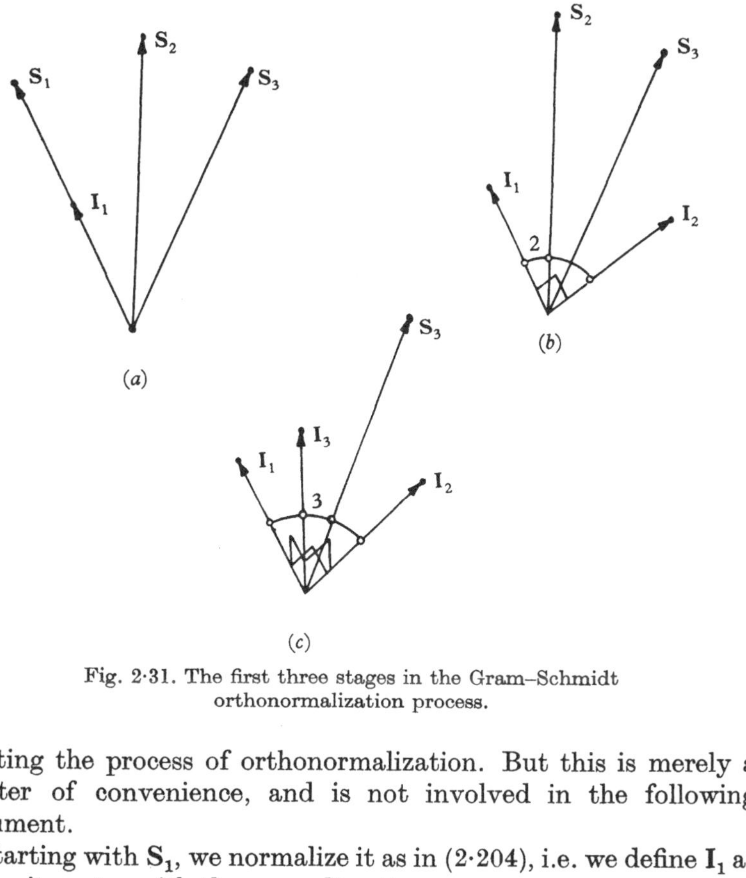

Fig. 2·31. The first three stages in the Gram–Schmidt orthonormalization process.

starting the process of orthonormalization. But this is merely a matter of convenience, and is not involved in the following argument.

Starting with $\mathbf{S}_1$, we normalize it as in (2·204), i.e. we define $\mathbf{I}_1$ as the unit vector with the same direction:

$$\mathbf{I}_1 = \mathbf{S}_1 / |\mathbf{S}_1|. \tag{2·311}$$

This is the first vector of our required orthonormal set (Fig. 2·31 a).

We now forget about $\mathbf{S}_1$; it has been replaced by $\mathbf{I}_1$. For $\mathbf{I}_2$ we take a linear combination of $\mathbf{I}_1$ and $\mathbf{S}_2$,

$$\mathbf{I}_2 = a_{21}\mathbf{I}_1 + a_{22}\mathbf{S}_2, \tag{2·312}$$

and impose the orthonormality conditions

$$\mathbf{I}_1 . \mathbf{I}_2 = 0, \quad \mathbf{I}_2^2 = 1. \tag{2·313}$$

The first of these conditions gives

$$a_{21} + a_{22}(\mathbf{I}_1 . \mathbf{S}_2) = 0, \tag{2·314}$$

and hence

$$a_{21} = -a_{22}(\mathbf{I}_1 . \mathbf{S}_2). \tag{2·315}$$

Substitution in (2·312) gives

$$\mathbf{I}_2 = a_{22}[\mathbf{S}_2 - \mathbf{I}_1(\mathbf{I}_1 . \mathbf{S}_2)]. \tag{2·316}$$

The second orthonormality condition (2·313) now gives for a_{22} the equation

$$1 = a_{22}^2[\mathbf{S}_2 - \mathbf{I}_1(\mathbf{I}_1 . \mathbf{S}_2)]^2 = a_{22}^2[\mathbf{S}_2^2 - (\mathbf{I}_1 . \mathbf{S}_2)^2]. \tag{2·317}$$

This determines a_{22} to within a $\pm$ sign. Just as we chose the + sign in (2·311) (we might have chosen $-$), so we shall choose the + sign here, writing

$$a_{22} = [\mathbf{S}_2^2 - (\mathbf{I}_1 . \mathbf{S}_2)^2]^{-\frac{1}{2}}. \tag{2·318}$$

Substitution of this in (2·315) gives a_{21}, and so we have by (2·312) the second member of the orthonormal pair $\mathbf{I}_1$, $\mathbf{I}_2$ (Fig. 2·31*b*).

The next step is to take

$$\mathbf{I}_3 = a_{31}\mathbf{I}_1 + a_{32}\mathbf{I}_2 + a_{33}\mathbf{S}_3, \tag{2·319}$$

and impose the orthonormality conditions

$$\mathbf{I}_3 . \mathbf{I}_1 = 0, \quad \mathbf{I}_3 . \mathbf{I}_2 = 0, \quad \mathbf{I}_3^2 = 1. \tag{2·320}$$

The first two of these give

$$a_{31} + a_{33}(\mathbf{I}_1 . \mathbf{S}_3) = 0, \quad a_{32} + a_{33}(\mathbf{I}_2 . \mathbf{S}_3) = 0, \tag{2·321}$$

and so (2·319) may be written

$$\mathbf{I}_3 = a_{33}[\mathbf{S}_3 - \mathbf{I}_1(\mathbf{I}_1 . \mathbf{S}_3) - \mathbf{I}_2(\mathbf{I}_2 . \mathbf{S}_3)]. \tag{2·322}$$

The third condition (2·320) then gives (if we choose the + sign)

$$a_{33} = [\mathbf{S}_3^2 - (\mathbf{I}_1 . \mathbf{S}_3)^2 - (\mathbf{I}_2 . \mathbf{S}_3)^2]^{-\frac{1}{2}}. \tag{2·323}$$

We have now got an orthonormal triad $\mathbf{I}_1$, $\mathbf{I}_2$, $\mathbf{I}_3$ (Fig. 2·31*c*).

So the process continues step by step, demanding nothing more complicated than the calculation of scalar products. It is clear that, at the risk of complicating our formulae, we could replace the **I**'s occurring on the right-hand sides of equations such as (2·316) and (2·322) by linear expressions involving the **S**'s only. Indeed, if, before we started the process, we calculated the numerical values of all the $\frac{1}{2}n(n+1)$ scalar products $\mathbf{S}_\rho . \mathbf{S}_\sigma$ $(\rho, \sigma = 1, 2, \ldots, n)$, then the process of orthonormalization might be treated as a purely numerical one. A technique for this procedure has been given by Peach (1).

For reference, let us note here the general recurrence relations of the Gram-Schmidt process, of which (2·322) and (2·323) form a sample and which are easily established by mathematical induction:

$$\mathbf{I}_\rho = a_{\rho\rho}[\mathbf{S}_\rho - \mathbf{I}_1(\mathbf{I}_1 . \mathbf{S}_\rho) - \dots - \mathbf{I}_{\rho-1}(\mathbf{I}_{\rho-1} . \mathbf{S}_\rho)], \tag{2·324}$$

$$a_{\rho\rho} = [\mathbf{S}_\rho^2 - (\mathbf{I}_1 . \mathbf{S}_\rho)^2 - \dots - (\mathbf{I}_{\rho-1} . \mathbf{S}_\rho)^2]^{-\frac{1}{2}}, \tag{2·325}$$

$$(\rho = 1, 2, \dots, n).$$

These formulae serve to determine $\mathbf{I}_\rho$ when we have already found $\mathbf{I}_1, \mathbf{I}_2, \dots, \mathbf{I}_{\rho-1}$.

Examples of orthonormalization

The Gram–Schmidt process is direct and its application mechanical. It is a question of keeping one's work in order and different people will prefer different ways of doing this. But sometimes a problem presents special features; if we take advantage of these, effort may be saved. Of the four examples which follow, the first is routine and the other three have some special features.

Example 1. Let the P-space be $0 \leqslant x \leqslant 1$ and the metric that of the Hilbert scalar product (2·102). We wish to orthonormalize the three F-vectors

$$\mathbf{S}_1 \leftrightarrow 1, \quad \mathbf{S}_2 \leftrightarrow x, \quad \mathbf{S}_3 \leftrightarrow x^2.$$

The calculations for orthonormalization might run as follows:

$$\mathbf{S}_1^2 = \int_0^1 1^2 dx = 1, \quad \mathbf{I}_1 \leftrightarrow 1$$

$$\mathbf{I}_1 . \mathbf{S}_2 = \int_0^1 x\,dx = \tfrac{1}{2}, \quad \mathbf{S}_2^2 = \int_0^1 x^2 dx = \tfrac{1}{3},$$

$$a_{22} = [\mathbf{S}_2^2 - (\mathbf{I}_1 . \mathbf{S}_2)^2]^{-\frac{1}{2}} = 2 . 3^{\frac{1}{2}},$$

$$\mathbf{S}_2 - \mathbf{I}_1(\mathbf{I}_1 . \mathbf{S}_2) \leftrightarrow x - \tfrac{1}{2}, \quad \mathbf{I}_2 \leftrightarrow 3^{\frac{1}{2}}(2x-1).$$

$$\mathbf{I}_1 . \mathbf{S}_3 = \int_0^1 x^2 dx = \tfrac{1}{3}, \quad \mathbf{I}_2 . \mathbf{S}_3 = \int_0^1 3^{\frac{1}{2}}(2x-1)\,x^2 dx = \tfrac{1}{2} . 3^{-\frac{1}{2}}, \quad \mathbf{S}_3^2 = \int_0^1 x^4 dx = \tfrac{1}{5},$$

$$a_{33} = [\mathbf{S}_3^2 - (\mathbf{I}_1 . \mathbf{S}_3)^2 - (\mathbf{I}_2 . \mathbf{S}_3)^2]^{-\frac{1}{2}} = 6 . 5^{\frac{1}{2}},$$

$$\mathbf{S}_3 - \mathbf{I}_1(\mathbf{I}_1 . \mathbf{S}_3) - \mathbf{I}_2(\mathbf{I}_2 . \mathbf{S}_3) \leftrightarrow x^2 - \tfrac{1}{3} - 3^{\frac{1}{2}}(2x-1)\,\tfrac{1}{2}3^{-\frac{1}{2}},$$

$$\mathbf{I}_3 \leftrightarrow 5^{\frac{1}{2}}(6x^2 - 6x + 1).$$

Collecting the results, we have

$$\mathbf{I}_1 \leftrightarrow 1, \quad \mathbf{I}_2 \leftrightarrow 3^{\frac{1}{2}}(2x-1), \quad \mathbf{I}_3 \leftrightarrow 5^{\frac{1}{2}}(6x^2 - 6x + 1). \tag{2·326}$$

Here, as in all cases of normalization, the accuracy of the calculations may be checked by applying the test $\mathbf{I}_\rho . \mathbf{I}_\sigma = \delta_{\rho\sigma}$ to the final result.

Example 2. Let the P-space be $-1 \leqslant x \leqslant 1$, and let the metric be as in the preceding example. We wish to orthonormalize

$$\mathbf{S}_1 \leftrightarrow 1, \quad \mathbf{S}_2 \leftrightarrow x, \quad \mathbf{S}_3 \leftrightarrow x^2.$$

This looks very like Example 1—only the range of x has been changed. But we can now take a short cut. In view of the evenness of the range and the evenness and oddness of the functions, it is evident that

$$\mathbf{S}_1 . \mathbf{S}_2 = 0, \quad \mathbf{S}_2 . \mathbf{S}_3 = 0.$$

Thus $\mathbf{S}_2$ is orthogonal to $\mathbf{S}_1$ and $\mathbf{S}_3$, and the situation is as shown in Fig. 2·32.

We shall get an orthogonal triad if we replace $\mathbf{S}_3$ by $\mathbf{S}'_3 = \mathbf{S}_3 + a\mathbf{S}_1$ and choose a so that $\mathbf{S}_1 . \mathbf{S}'_3 = 0$; then we can normalize $\mathbf{S}_1$, $\mathbf{S}_2$, $\mathbf{S}'_3$ separately. The calculations run as follows:

$$\mathbf{S}_1^2 = 2, \quad \mathbf{S}_2^2 = \tfrac{2}{3}, \quad \mathbf{S}_3^2 = \tfrac{2}{5}, \quad \mathbf{S}_1 . \mathbf{S}_3 = \tfrac{2}{3},$$

$$a = -\mathbf{S}_1 . \mathbf{S}_3 / \mathbf{S}_1^2 = -\tfrac{1}{3}, \quad \mathbf{S}'_3 \leftrightarrow x^2 - \tfrac{1}{3}, \quad \mathbf{S}'^2_3 = \tfrac{8}{45}.$$

Our orthonormal triad is

$$\left.\begin{aligned} &\mathbf{I}_1 = \mathbf{S}_1/|\mathbf{S}_1| \leftrightarrow (\tfrac{1}{2})^{\frac{1}{2}}, \quad \mathbf{I}_2 = \mathbf{S}_2/|\mathbf{S}_2| \leftrightarrow (\tfrac{3}{2})^{\frac{1}{2}}\, x, \\ &\mathbf{I}_3 = \mathbf{S}'_3/|\mathbf{S}'_3| \leftrightarrow (\tfrac{5}{2})^{\frac{1}{2}}\, \tfrac{1}{2}(3x^2 - 1). \end{aligned}\right\} \qquad (2{\cdot}327)$$

We recognize the Legendre polynomials.

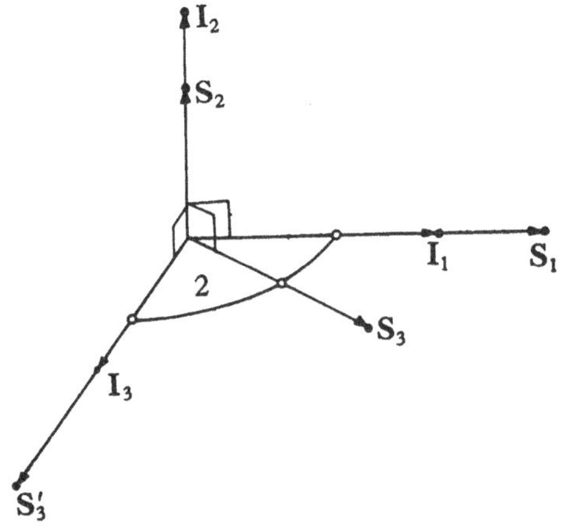

Fig. 2·32. Orthonormalization as in Example (2).

Do the three F-vectors (2·327) form a right-handed or left-handed triad? That is a meaningless question; but it is meaningful to ask whether the orientations of the two triads $\mathbf{S}_1$, $\mathbf{S}_2$, $\mathbf{S}_3$ and $\mathbf{I}_1$, $\mathbf{I}_2$, $\mathbf{I}_3$ are the same. They are.

Example 3. Let P-space be the interior of the square bounded by the lines $x = \pm 1$, $y = \pm 1$; let an F-vector correspond to a P-vector field (p_1, p_2); and let the scalar product be the Dirichlet product as in (2·103). We wish to orthonormalize eight F-vectors corresponding to the gradients of the eight harmonic functions

$$x, \quad y, \quad \tfrac{1}{2}(x^2 - y^2), \quad xy, \quad \tfrac{1}{3}(x^3 - 3xy^2), \quad \tfrac{1}{3}(3x^2y - y^3),$$

$$\tfrac{1}{4}(x^4 - 6x^2y^2 + y^4), \quad x^3y - xy^3.$$

Explicitly, these eight F-vectors are

$$\mathbf{S}_1 \leftrightarrow (1, 0), \qquad \mathbf{S}_2 \leftrightarrow (0, 1),$$
$$\mathbf{S}_3 \leftrightarrow (x, -y), \qquad \mathbf{S}_4 \leftrightarrow (y, x),$$
$$\mathbf{S}_5 \leftrightarrow (x^2 - y^2, -2xy), \qquad \mathbf{S}_6 \leftrightarrow (2xy, x^2 - y^2),$$
$$\mathbf{S}_7 \leftrightarrow (x^3 - 3xy^2, -3x^2y + y^3), \qquad \mathbf{S}_8 \leftrightarrow (3x^2y - y^3, x^3 - 3xy^2).$$

In view of the symmetries possessed by a square, certain integrals obviously vanish, e.g. if the integrand is odd in x or odd in y. Thus, without any calculation we see that all the scalar products $\mathbf{S}_\rho . \mathbf{S}_\sigma (\rho \neq \sigma)$ vanish except $\mathbf{S}_3 . \mathbf{S}_7$ and $\mathbf{S}_4 . \mathbf{S}_8$. But actually symmetry makes these vanish also:

$$\mathbf{S}_3 . \mathbf{S}_7 = \mathbf{S}_4 . \mathbf{S}_8 = \iint [y(3x^2y - y^3) + x(x^3 - 3xy^2)\, dx\, dy$$
$$= \iint (x^4 - y^4)\, dx\, dy = 0.$$

Thus all the scalar products formed by distinct vectors of the set vanish; the eight vectors are orthogonal, and so orthonormalization requires only normalization, which is easily carried out.

Obviously the eight vectors $\mathbf{S}_1, \ldots, \mathbf{S}_8$ are orthogonal even if P-space is not a square but any region having the same symmetries, i.e. it must be unchanged by reflexion in each axis or by a rotation of a right angle about the origin.

Example 4. P-space will now be any finite part of the Euclidean plane. An F-vector will correspond to a P-vector field, but now we shall use *polar coordinates* r, θ and write the components of a P-vector (p_r, p_θ). The scalar product will be

$$\mathbf{S} . \mathbf{S}' = \iint (p_r p_r' + p_\theta p_\theta')\, r\, d\theta\, dr. \tag{2·328}$$

Let us take the four harmonic functions

$$\tfrac{1}{4}r^4 \cos 4\theta, \quad \tfrac{1}{4}r^4 \sin 4\theta, \quad \tfrac{1}{8}r^8 \cos 8\theta, \quad \tfrac{1}{8}r^8 \sin 8\theta,$$

and consider the four F-vectors corresponding to their gradients, expressed in polar coordinates:

$$\left.\begin{aligned} &\mathbf{S}_1 \leftrightarrow (r^3 \cos 4\theta, -r^3 \sin 4\theta), \quad \mathbf{S}_2 \leftrightarrow (r^3 \sin 4\theta, r^3 \cos 4\theta), \\ &\mathbf{S}_3 \leftrightarrow (r^7 \cos 8\theta, -r^7 \sin 8\theta), \quad \mathbf{S}_4 \leftrightarrow (r^7 \sin 8\theta, r^7 \cos 8\theta). \end{aligned}\right\} \tag{2·329}$$

We wish to orthonormalize these four F-vectors.

It is evident at once, from (2·328) and the P-orthogonality of the P-vector fields involved, that

$$\mathbf{S}_1 . \mathbf{S}_2 = 0, \quad \mathbf{S}_3 . \mathbf{S}_4 = 0.$$

Thus we have to deal with two pairs of orthogonal F-vectors, as shown in Fig. 2·33. As for the other scalar products, we have

$$\mathbf{S}_1 . \mathbf{S}_3 = \mathbf{S}_2 . \mathbf{S}_4 = \iint r^{11} \cos 4\theta\, dr\, d\theta = A, \quad \text{say},$$
$$\mathbf{S}_1 . \mathbf{S}_4 = -\mathbf{S}_2 . \mathbf{S}_3 = \iint r^{11} \sin 4\theta\, dr\, d\theta = B, \quad \text{say}.$$

Further,

$$\mathbf{S}_1^2 = \mathbf{S}_2^2 = \iint r^7\, dr\, d\theta = C, \quad \text{say},$$
$$\mathbf{S}_3^2 = \mathbf{S}_4^2 = \iint r^{15}\, dr\, d\theta = D, \quad \text{say}.$$

Instead of using the Gram–Schmidt process, we may appeal to the symmetry of the tetrad of F-vectors; this suggests that we leave $\mathbf{S}_1$ and $\mathbf{S}_2$ alone, and change $\mathbf{S}_3$ and $\mathbf{S}_4$ into $\mathbf{S}_3'$ and $\mathbf{S}_4'$, where

$$\mathbf{S}_3' = a\mathbf{S}_1 + b\mathbf{S}_2 + c\mathbf{S}_3 + d\mathbf{S}_4,$$
$$\mathbf{S}_4' = -b\mathbf{S}_1 + a\mathbf{S}_2 - d\mathbf{S}_3 + c\mathbf{S}_4,$$

the four coefficients being determined (if possible) to satisfy the conditions of orthogonality

$$\mathbf{S}_1 . \mathbf{S}_3' = 0, \quad \mathbf{S}_1 . \mathbf{S}_4' = 0, \quad \mathbf{S}_2 . \mathbf{S}_3' = 0, \quad \mathbf{S}_2 . \mathbf{S}_4' = 0, \quad \mathbf{S}_3' . \mathbf{S}_4' = 0.$$

The last is identically satisfied. The other four reduce to two:

$$aC + cA + dB = 0,$$

$$bC - cB + dA = 0.$$

Choosing $d = 0$, $c = 1$, we satisfy these equations by

$$a = -A/C, \quad b = B/C,$$

and so $\quad \mathbf{S}_3' = C^{-1}(-A\mathbf{S}_1 + B\mathbf{S}_2 + C\mathbf{S}_3), \quad \mathbf{S}_4 = C^{-1}(-B\mathbf{S}_1 - A\mathbf{S}_2 + C\mathbf{S}_4).$

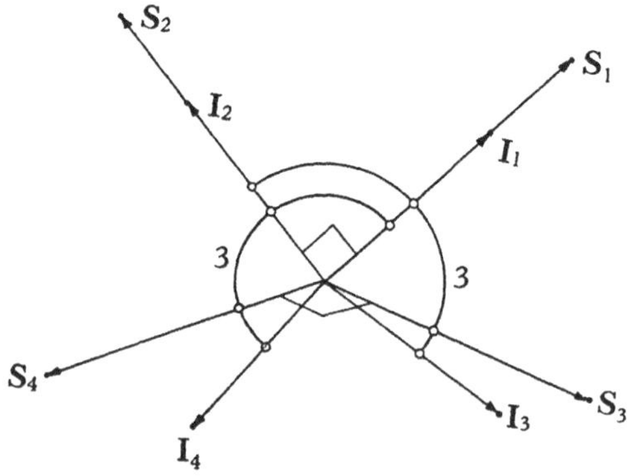

Fig. 2·33. The orthonormal tetrad (2·330).

Now we have an orthogonal tetrad $\mathbf{S}_1, \mathbf{S}_2, \mathbf{S}_3', \mathbf{S}_4'$. It remains only to normalize. We get

$$\left.\begin{aligned} &\mathbf{I}_1 = C^{-\frac{1}{2}}\mathbf{S}_1, \quad \mathbf{I}_2 = C^{-\frac{1}{2}}\mathbf{S}_2, \quad \mathbf{I}_3 = C^{\frac{1}{2}}(CD - A^2 - B^2)^{-\frac{1}{2}}\mathbf{S}_3', \\ &\qquad\qquad\qquad\qquad\qquad\quad\; \mathbf{I}_4 = C^{\frac{1}{2}}(CD - A^2 - B^2)^{-\frac{1}{2}}\mathbf{S}_4', \\ &\mathbf{S}_3' = \mathbf{S}_3 - \mathbf{S}_1 A/C + \mathbf{S}_2 B/C, \quad \mathbf{S}_4' = \mathbf{S}_4 - \mathbf{S}_1 B/C - \mathbf{S}_2 A/C. \end{aligned}\right\} \qquad (2\cdot330)$$

This is an orthonormal tetrad formed from the tetrad (2·329).

To employ the result practically we have of course to compute, for the particular P-domain, the four definite integrals A, B, C, D.

If P-space is a square, or a region with the symmetry of a square as described above in Example 3, the result simplifies, provided we take the line $\theta = 0$ to be an axis of symmetry. Then the constant B vanishes and we get instead of (2·330)

$$\left.\begin{aligned} &\mathbf{I}_1 = C^{-\frac{1}{2}}\mathbf{S}_1, \quad \mathbf{I}_2 = C^{-\frac{1}{2}}\mathbf{S}_2, \\ &\mathbf{I}_3 = C^{\frac{1}{2}}(CD - A^2)^{-\frac{1}{2}}(\mathbf{S}_3 - \mathbf{S}_1 A/C), \quad \mathbf{I}_4 = C^{\frac{1}{2}}(CD - A^2)^{-\frac{1}{2}}(\mathbf{S}_4 - \mathbf{S}_2 A/C). \end{aligned}\right\} \qquad (2\cdot330a)$$

The extended Schwarz inequality

Let $\mathbf{S}$ be any F-vector and $\mathbf{I}_\rho\ (\rho = 1, 2, \ldots, n)$ an orthonormal set of F-vectors. Then from the positive-definite character of the F-metric assumed throughout this part of the book, we have

$$\left(\mathbf{S} - \sum_{\rho=1}^{n} a_\rho \mathbf{I}_\rho\right)^2 \geqslant 0, \tag{2·331}$$

for any choice of the real constant a's. Expanding, we have

$$\mathbf{S}^2 - 2\sum_{\rho=1}^{n} a_\rho(\mathbf{S}\,.\,\mathbf{I}_\rho) + \sum_{n=1}^{\rho} a_\rho^2 \geqslant 0. \tag{2·332}$$

If we choose $a_\rho = \mathbf{S}\,.\,\mathbf{I}_\rho$, the above inequality reads

$$\mathbf{S}^2 - \sum_{\rho=1}^{n} (\mathbf{S}\,.\,\mathbf{I}_\rho)^2 \geqslant 0. \tag{2·333}$$

This is *Bessel's inequality*, but we may also refer to it as the *extended Schwarz inequality*. If we put $n = 1$, it is essentially the same as the Schwarz inequality (2·210).

On following back the reasoning from (2·333) to (2·331), we see that if the sign of equality holds in (2·333), then $\mathbf{S}$ lies in the linear n-space of the $\mathbf{I}$'s, i.e. it is a linear combination of them.

Conversely, if $\mathbf{S}$ lies in the linear n-space of the $\mathbf{I}$'s, there exist scalars b_ρ such that

$$\mathbf{S} = \sum_{\rho=1}^{n} b_\rho \mathbf{I}_\rho. \tag{2·334}$$

From this we deduce $\mathbf{S}\,.\,\mathbf{I}_\rho = b_\rho$, and we have

$$\mathbf{S}^2 = \sum_{\rho=1}^{n} b_\rho^2 = \sum_{\rho=1}^{n} (\mathbf{S}\,.\,\mathbf{I}_\rho)^2, \tag{2·335}$$

and so the sign of equality holds in (2·333).

Thus *the sign of equality holds in* (2·333) *if, and only if,* $\mathbf{S}$ *lies in the linear n-space of the orthonormal vectors* $\mathbf{I}_1, \mathbf{I}_2, \ldots, \mathbf{I}_n$.

The equation (2·335) is a generalization of the familiar

$$r^2 = x^2 + y^2 + z^2,$$

connecting the radius vector r of a point with its three coordinates x, y, z. Fig. 2·34 shows an orthonormal tetrad $\mathbf{I}_1, \mathbf{I}_2, \mathbf{I}_3, \mathbf{I}_4$ and the resolution along their lines of any vector $\mathbf{S}$ contained in their linear 4-space. $\mathbf{S}^2$ is the sum of the squares of these four components, just as the square on the diagonal of a cuboid is the sum of the squares on the three edges.

The process of orthonormalization cannot fail

The process of orthonormalization, which we explained from (2·311) on, could fail in one way only—it might demand division by zero. We shall now show that this cannot happen.

The first danger point occurs at (2·318). The process would break down if the content of the square brackets vanished, i.e. if $\mathbf{S}_2^2 - (\mathbf{I}_1 . \mathbf{S}_2)^2 = 0$. As we have seen above, this could happen only if $\mathbf{S}_2$ lay in the 1-space of $\mathbf{I}_1$, which is in fact the 1-space of $\mathbf{S}_1$. But this would mean the linear dependence of $\mathbf{S}_1$ and $\mathbf{S}_2$, and we have guarded against this by starting off with n linearly independent vectors. So no breakdown is possible at (2·318).

The same line of argument tells us that the process cannot break down at (2·323), nor generally at (2·325). In fact, the process cannot break down at all, and the Gram–Schmidt method is bound to

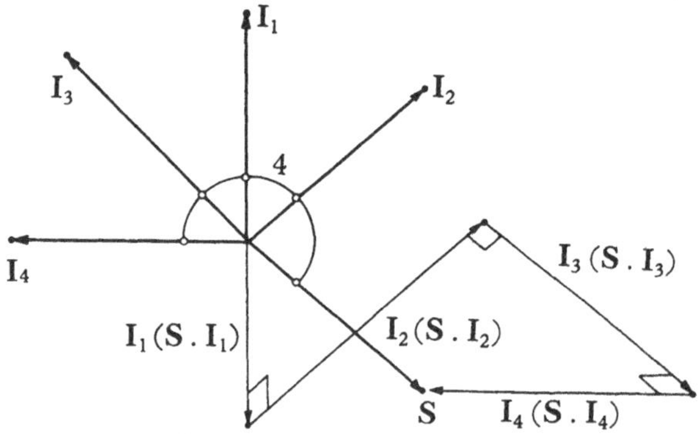

Fig. 2·34. $\mathbf{S}^2 = \sum_{\rho=1}^{4} (\mathbf{S} . \mathbf{I}_\rho)^2$ if $\mathbf{S}$ lies in the linear 4-space of the orthonormal tetrad $\mathbf{I}_1$, $\mathbf{I}_2$, $\mathbf{I}_3$, $\mathbf{I}_4$.

deliver an orthonormal set of n vectors. The question is not so simple if the F-metric is indefinite, but with that we shall not be concerned until Part III.

Resolution of an F-vector in, and orthogonal to, a linear n-space L_n

In Euclidean 3-space a vector can be resolved in a unique way into a component in, and a component perpendicular to, a given plane. The same type of resolution may be carried out in F-space, as we shall now see.

Let $\mathbf{S}$ be an F-vector and let there be a linear n-space L_n

$$\mathbf{X} = \mathbf{A} + a_1 \mathbf{I}_1 + a_2 \mathbf{I}_2 + \ldots + a_n \mathbf{I}_n, \qquad (2·336)$$

containing the orthonormal set $\mathbf{I}_1, \mathbf{I}_2, \ldots, \mathbf{I}_n$. Now we can certainly resolve $\mathbf{S}$ into the sum of a vector lying in L_n and another vector, by writing

$$\mathbf{S} = \mathbf{S}_0 + \sum_{\rho=1}^{n} b_\rho \mathbf{I}_\rho, \tag{2·337}$$

where the b's are given *any* values and $\mathbf{S}_0$ is *defined* by this equation. The question is: Can we make $\mathbf{S}_0$ orthogonal to L_n, i.e. orthogonal to the n vectors $\mathbf{I}_\rho$?

The conditions of orthogonality are $\mathbf{S}_0 . \mathbf{I}_\rho = 0$ $(\rho = 1, 2, \ldots, n)$ and these conditions are satisfied by $\mathbf{S}_0$ as defined in (2·337) provided $b_\rho = \mathbf{S} . \mathbf{I}_\rho$ $(\rho = 1, 2, \ldots, n)$. Thus, *if* $\mathbf{S}_0$ *is defined by*

$$\mathbf{S}_0 = \mathbf{S} - \sum_{\rho=1}^{n} \mathbf{I}_\rho (\mathbf{S} . \mathbf{I}_\rho), \tag{2·338}$$

then $\mathbf{S}_0$ *is orthogonal to* L_n *and* $\mathbf{S}$ *is resolved by the formula*

$$\mathbf{S} = \mathbf{S}_0 + \sum_{\rho=1}^{n} \mathbf{I}_\rho (\mathbf{S} . \mathbf{I}_\rho) \tag{2·339}$$

into a vector $\mathbf{S}_0$ *orthogonal to* L_n *and a vector* $\sum_{\rho=1}^{n} \mathbf{I}_\rho (\mathbf{S} . \mathbf{I}_\rho)$ *in* L_n. This resolution is shown in Fig. 2·35.

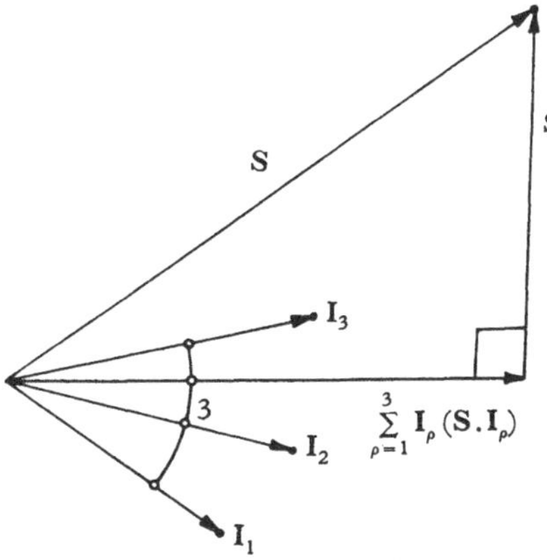

Fig. 2·35. Resolution of S in, and orthogonal to, a n-linear space $(n = 3)$.

It follows from (2·338) that

$$\mathbf{S}_0^2 = \mathbf{S}^2 - \sum_{\rho=1}^{n} (\mathbf{S} . \mathbf{I}_\rho)^2. \tag{2·340}$$

Hence
$$|\mathbf{S}_0| \leqslant |\mathbf{S}|; \tag{2·341}$$

the component of an F-vector orthogonal to a linear n-space is less than (or at most equal to) the magnitude of the vector itself.

The quantity $\mathbf{S} . \mathbf{I}_\rho$ may be called the *scalar component of* $\mathbf{S}$ *on* $\mathbf{I}_\rho$, and $\sum_{\rho=1}^{n} \mathbf{I}_\rho (\mathbf{S} . \mathbf{I}_\rho)$ may be called the *vector component of* $\mathbf{S}$ *on* L_n, or the *orthogonal projection of* $\mathbf{S}$ *on* L_n.

Example

As an example of the above resolution, let us take the linear 3-space of the orthonormal triad (2·327), viz.

$$\mathbf{I}_1 \leftrightarrow (\tfrac{1}{2})^{\frac{1}{2}}, \quad \mathbf{I}_2 \leftrightarrow (\tfrac{3}{2})^{\frac{1}{2}}\, x, \quad \mathbf{I}_3 \leftrightarrow (\tfrac{5}{2})^{\frac{1}{2}}\, \tfrac{1}{2}(3x^2-1), \tag{2·342}$$

and resolve in, and orthogonal to, this L_3 the F-vector

$$\mathbf{S} \leftrightarrow x^3. \tag{2·343}$$

We find $$\mathbf{S}\,.\,\mathbf{I}_1 = 0, \quad \mathbf{S}\,.\,\mathbf{I}_2 = \tfrac{2}{5}(\tfrac{3}{2})^{\frac{1}{2}}, \quad \mathbf{S}\,.\,\mathbf{I}_3 = 0. \tag{2·344}$$

Thus the vector component of $\mathbf{S}$ on L_3 is

$$\sum_{\rho=1}^{3} \mathbf{I}_\rho(\mathbf{S}.\mathbf{I}_\rho) \leftrightarrow \tfrac{3}{5}x, \tag{2·345}$$

and the vector component of $\mathbf{S}$ orthogonal to L_3 is

$$\mathbf{S}_0 \leftrightarrow x^3 - \tfrac{3}{5}x. \tag{2·346}$$

We may check by calculating the two sides of (2·340):

$$\mathbf{S}_0^2 = \tfrac{8}{175}, \quad \mathbf{S}^2 - \sum_{\rho=1}^{3} (\mathbf{S}\,.\,\mathbf{I}_\rho)^2 = \tfrac{2}{7} - \tfrac{6}{25} = \tfrac{8}{175}.$$

The normal drawn to a linear n-space L_n and its minimum property

The results which follow can be deduced from the resolution of a vector as in (2·339), but the following alternative treatment is short and interesting.

Consider an L_n with equation

$$\mathbf{X} = \mathbf{A} + \sum_{\rho=1}^{n} a_\rho \mathbf{I}_\rho, \tag{2·347}$$

where, as usual, the a's are parameters and the $\mathbf{I}$'s a set of fixed orthonormal F-vectors. Let us find the point on L_n which is closest to the origin.

It is a question of minimizing $\mathbf{X}^2$ by choosing the a's appropriately. Now

$$\begin{aligned} \mathbf{X}^2 &= \mathbf{A}^2 + 2\sum_{\rho=1}^{n} a_\rho \mathbf{A}\,.\,\mathbf{I}_\rho + \sum_{\rho=1}^{n} a_\rho^2 \\ &= \mathbf{A}^2 - \sum_{\rho=1}^{n} (\mathbf{A}\,.\,\mathbf{I}_\rho)^2 + \sum_{\rho=1}^{n} (a_\rho + \mathbf{A}\,.\,\mathbf{I}_\rho)^2, \end{aligned} \tag{2·348}$$

and this is minimized by taking

$$a_\rho = -\mathbf{A}\,.\,\mathbf{I}_\rho \quad (\rho = 1, 2, \ldots, n). \tag{2·349}$$

Thus, if we denote by $\mathbf{N}$ that vector $\mathbf{X}$ which minimizes $\mathbf{X}^2$, we have

$$\left.\begin{aligned} \mathbf{N} &= \mathbf{A} - \sum_{\rho=1}^{n} \mathbf{I}_\rho(\mathbf{A}\,.\,\mathbf{I}_\rho), \\ \mathbf{N}^2 &= \mathbf{A}^2 - \sum_{\rho=1}^{n} (\mathbf{A}\,.\,\mathbf{I}_\rho)^2. \end{aligned}\right\} \tag{2·350}$$

Since
$$\mathbf{N}.\mathbf{I}_\rho = 0 \quad (\rho = 1, 2, \ldots, n), \tag{2·351}$$
it is clear that **N** *is orthogonal to* L_n. Thus we naturally call **N** *the normal drawn to* L_n *from the origin* (Fig. 2·36).

We note that the normal has the important property of *minimum length*. This is the prototype of the minimum principles we encounter in an F-space with positive-definite metric.

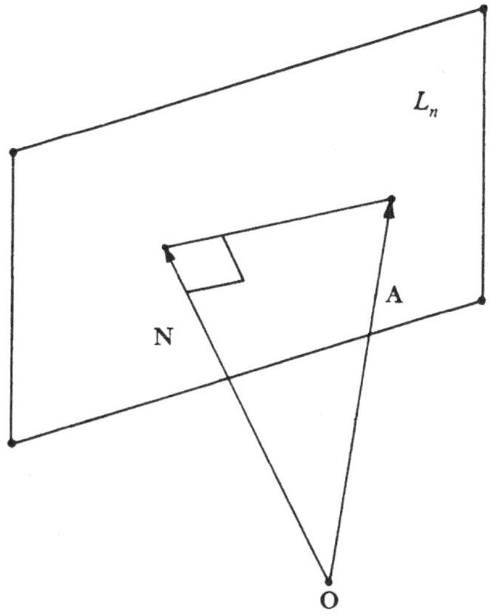

Fig. 2·36. The normal **N** from **O** to a linear n-space L_n.

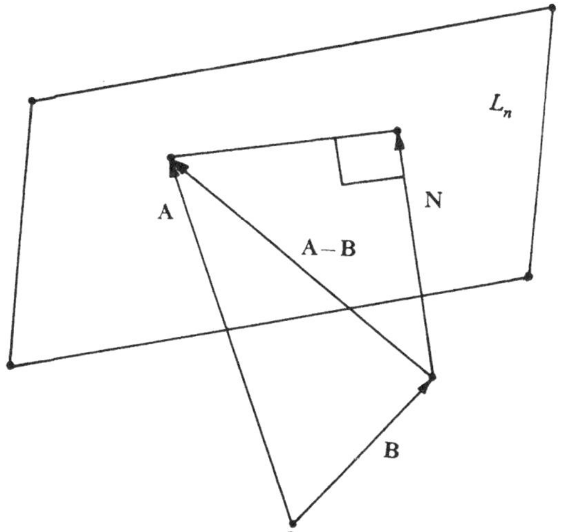

Fig. 2·37. The normal **N** from the point **B** to a linear n-space L_n.

It is evident that we can write the equation of L_n in the form

$$\mathbf{X}=\mathbf{N}+\sum_{\rho=1}^{n} a_\rho \mathbf{I}_\rho; \tag{2·352}$$

for some purposes this is more convenient than (2·347), on account of the orthogonality relations (2·351).

The normal $\mathbf{N}$ drawn to L_n from any point $\mathbf{B}$ may be found similarly by minimizing $(\mathbf{X}-\mathbf{B})^2$, or more simply by translating the origin to the point $\mathbf{B}$ and using the above result. We get

$$\left.\begin{aligned}\mathbf{N}&=\mathbf{A}-\mathbf{B}-\sum_{\rho=1}^{n}\mathbf{I}_\rho[(\mathbf{A}-\mathbf{B}).\mathbf{I}_\rho],\\ \mathbf{N}^2&=(\mathbf{A}-\mathbf{B})^2-\sum_{\rho=1}^{n}[(\mathbf{A}-\mathbf{B}).\mathbf{I}_\rho]^2.\end{aligned}\right\} \tag{2·353}$$

But, indeed, this is obvious (Fig. 2·37), since $\mathbf{N}$ is one side of a right-angled triangle having $(\mathbf{A}-\mathbf{B})$ for hypotenuse and, for the remaining side, the vector component of $(\mathbf{A}-\mathbf{B})$ on L_n.

Example: approximation by a step-function

As an illustration of the analytical meaning of what we have been doing, consider the following problem.

Let $F(x)$ be a function given on the range $(0, 1)$. What is the 'best' approximation to $F(x)$ in the form of a step-function consisting of n steps on equal bases? (A step-function is shown in Fig. 2·38; its graph consists of segments parallel to the x-axis.)

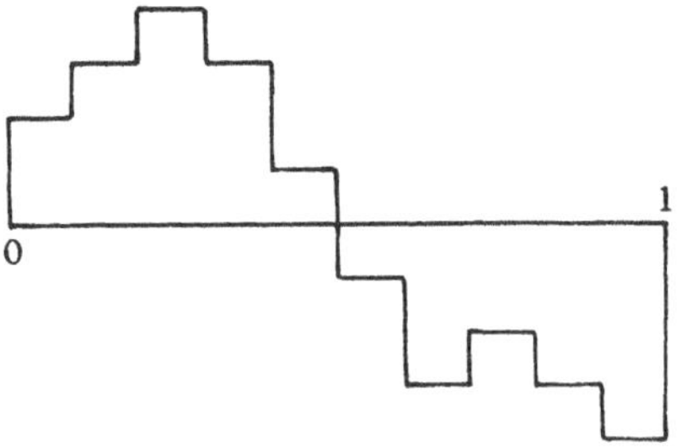

Fig. 2·38. A step-function ($n=10$). $\sum_{\rho=1}^{n} a_\rho \mathbf{I}_\rho \leftrightarrow \sum_{\rho=1}^{n} a_\rho f_\rho(x)$.

To geometrize the problem in F-space, let an F-vector correspond to any function in the range $(0, 1)$, and let the scalar product be that of Hilbert (2·102). Let us divide the range equally at the points

$$0=x_0, \quad x_1, \quad x_2, \quad \ldots, \quad x_n=1,$$

each segment being of length $1/n$. Let $f_\rho(x)$ be that function which is zero outside the range $(x_{\rho-1}, x_\rho)$ and equal to $n^{\frac{1}{2}}$ in that range (Fig. 2·39). Let

$$\mathbf{I}_\rho \leftrightarrow f_\rho(x) \quad (\rho=1, 2, \ldots, n).$$

Then $\mathbf{I}_\rho^2=\int_0^1 [f_\rho(x)]^2\,dx=1,\quad \mathbf{I}_\rho.\mathbf{I}_\sigma=\int_0^1 f_\rho(x)f_\sigma(x)\,dx=0\quad \text{if}\quad \rho\neq\sigma.$

Thus $\mathbf{I}_\rho$ form a set of n orthonormal F-vectors.

Let $\mathbf{B}\leftrightarrow F(x)$, the given function.

Any one of the step-functions which we may consider is of the form $\sum_{\rho=1}^{n} a_\rho f_\rho(x)$ where the a's are constants. This corresponds to $\sum_{\rho=1}^{n} a_\rho \mathbf{I}_\rho$, and the problem of getting the 'best' approximation may be regarded as the problem of choosing the a's so as to minimize the distance between the adjustable F-point $\sum_{\rho=1}^{n} a_\rho \mathbf{I}_\rho$ and the fixed F-point $\mathbf{B}$. In other words, *we seek the normal from the point* $\mathbf{B}$ *to the linear n-space* $\mathbf{X}=\sum_{\rho=1}^{n} a_\rho \mathbf{I}_\rho$.

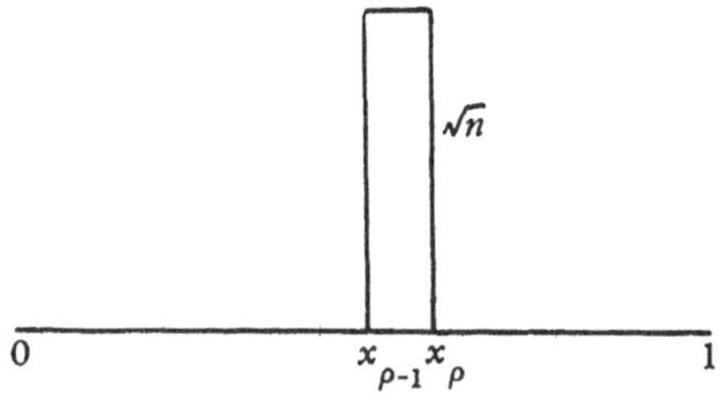

Fig. 2·39. The function $f_\rho(x)\leftrightarrow\mathbf{I}_\rho$.

To find this normal, we put $\mathbf{A}=\mathbf{O}$ in (2·353). The normal drawn from $\mathbf{B}$ is

$$\mathbf{N}=-\mathbf{B}+\sum_{\rho=1}^{n}\mathbf{I}_\rho(\mathbf{B}.\mathbf{I}_\rho),\quad \mathbf{N}^2=\mathbf{B}^2-\sum_{\rho=1}^{n}(\mathbf{B}.\mathbf{I}_\rho)^2,$$

and the best approximation is given (cf. Fig. 2·37) by the foot of the normal, viz. the point

$$\mathbf{B}+\mathbf{N}=\sum_{\rho=1}^{n}\mathbf{I}_\rho(\mathbf{B}.\mathbf{I}_\rho).$$

Consequently the best approximation is the step-function

$$\phi(x)=\sum_{\rho=1}^{n} a_\rho f_\rho(x),$$

where
$$a_\rho=\mathbf{B}.\mathbf{I}_\rho=\int_0^1 F(x)f_\rho(x)\,dx=n^{\frac{1}{2}}\int_{x_{\rho-1}}^{x_\rho} F(x)\,dx.$$

The 'error' is

$$\int_0^1 [F(x)-\phi(x)]^2\,dx=\mathbf{N}^2=\int_0^1 [F(x)]^2\,dx-n\sum_{\rho=1}^{n}\left[\int_{x_{\rho-1}}^{x_\rho} F(x)\,dx\right]^2.$$

This is the least possible value of

$$\int_0^1\left[F(x)-\sum_{\rho=1}^{n} a_\rho f_\rho(x)\right]^2 dx$$

for any choice of the a's.

As a particular case, if we wish to approximate $F(x)=x^2$ by a step-function in this way, we are to choose

$$a_\rho = n^{\frac{1}{2}}\tfrac{1}{3}(x_\rho^3 - x_{\rho-1}^3) = n^{-\frac{1}{2}}\tfrac{1}{3}(x_\rho^2 + x_\rho x_{\rho-1} + x_{\rho-1}^2);$$

the height of the step is then $\frac{1}{3}(x_\rho^2 + x_\rho x_{\rho-1} + x_{\rho-1}^2)$.

Orthogonal linear subspaces

On p. 31 we defined a linear subspace L, the linear n-space being a particular case. We now define the orthogonality of two linear subspaces L, L' as follows:

L and L' are orthogonal if, and only if, every F-vector lying in L is orthogonal to every F-vector lying in L'.

We must remember the distinction between vectors lying in a subspace and the position-vectors of points in the subspace (Fig. 1·35, p. 32). The orthogonality of linear subspaces concerns vectors *lying in* the subspaces.

Since we are inclined to think, often profitably, of analogies in Euclidean 3-space, a warning should be sounded here. Two planes in Euclidean 3-space are linear subspaces, but they cannot be orthogonal in the above sense, for it is impossible that every vector lying in one of them should be orthogonal to every vector lying in the other. Two straight lines may be orthogonal, or a straight line and a plane, but not two planes, if the word 'orthogonal' is used in the new sense. This limitation of 3-space is unfortunate, but there is nothing we can do about it. To get an adequate simple model for certain situations in F-space, we are compelled to use a space of at least four dimensions. This happens in the discussion of the hypercircle method in § 2·7.

Restricting ourselves now to linear subspaces of finite dimensionality, let us consider the orthogonality of a linear m-space L_m and a linear n-space L_n, with equations

$$\left.\begin{aligned} L_m:\quad \mathbf{X} &= \mathbf{A} + a_1\mathbf{I}_1 + a_2\mathbf{I}_2 + \dots + a_m\mathbf{I}_m,\\ L_n:\quad \mathbf{X} &= \mathbf{B} + b_1\mathbf{J}_1 + b_2\mathbf{J}_2 + \dots + b_n\mathbf{J}_n,\end{aligned}\right\} \tag{2·354}$$

where the a's and b's are variable parameters and the $\mathbf{I}$'s are orthonormal and the $\mathbf{J}$'s are orthonormal.

Any vector lying in L_m is of the form

$$\mathbf{S} = a_1\mathbf{I}_1 + a_2\mathbf{I}_2 + \dots + a_m\mathbf{I}_m, \tag{2·355}$$

and any vector lying in L_n is of the form

$$\mathbf{T} = b_1\mathbf{J}_1 + b_2\mathbf{J}_2 + \dots + b_n\mathbf{J}_n, \tag{2·356}$$

and the condition of orthogonality of L_m and L_n is

$$\mathbf{S}.\mathbf{T} = 0 \tag{2·357}$$

for every choice of the $m+n$ parameters (the a's and the b's). Hence a *sufficient* condition for orthogonality is

$$\mathbf{I}_\mu . \mathbf{J}_\nu = 0 \quad (\mu = 1, 2, \ldots, m;\ \nu = 1, 2, \ldots, n). \tag{2·358}$$

This means that every **I** shall be orthogonal to every **J**. It is easy to see that this condition is also *necessary*.

We used orthonormal vectors in (2·354) because expression in terms of orthonormal vectors may be regarded as the standard form for a linear n-space. Actually, we did not use the orthonormality, and (2·358) is a necessary and sufficient condition for the orthogonality of L_m and L_n if the **I**'s form any linearly independent set and the **J**'s any linearly independent set.

It is obvious that the orthogonality condition (2·358) is satisfied if the **I**'s and **J**'s together form an orthonormal set of $m+n$ vectors.

Exercises

1. If $\mathbf{I}_1, \mathbf{I}_2, \mathbf{I}_3$ form an orthonormal triad, show that all unit F-vectors lying in their linear 3-space are included in the formula

$$\mathbf{J} = \sin\theta\cos\phi\,\mathbf{I}_1 + \sin\theta\sin\phi\,\mathbf{I}_2 + \cos\theta\,\mathbf{I}_3,$$
$$0 \leqslant \theta \leqslant \pi, \quad 0 \leqslant \phi < 2\pi.$$

2. Taking $n = 3$, write down the general orthogonal transformation of the form (2·304) in terms of the Eulerian angles θ, ϕ, ψ commonly employed in the kinematics of a rigid body. Distinguish the cases where a reflexion does or does not occur.

3. With the Hilbert scalar product (2·102) and P-space $-\pi \leqslant x \leqslant \pi$, show that the infinite sequence of F-vectors corresponding to 1, $\sin nx$, $\cos nx$ ($n = 1, 2, \ldots$) are orthogonal. Complete the orthonormalization.

4. Let P-space be the interior of the square $x = \pm 1, y = \pm 1$; let an F-vector correspond to a function of x and y, and let the scalar product be defined by

$$\mathbf{S} \leftrightarrow s(x,y), \quad \mathbf{T} \leftrightarrow t(x,y), \quad \mathbf{S}.\mathbf{T} = \iint s(x,y)\,t(x,y)\,dx\,dy.$$

Orthonormalize the six F-vectors corresponding to the following six functions:

$$1, \quad x, \quad y, \quad x^2, \quad xy, \quad y^2.$$

[We know that orthonormalization is not unique. One answer is

$$\tfrac{1}{2}, \quad \tfrac{1}{2}.3^{\frac{1}{2}}x, \quad \tfrac{1}{2}.3^{\frac{1}{2}}y, \quad (\tfrac{1}{4}.5^{\frac{1}{2}})(3x^2-1), \quad (\tfrac{1}{4}.5^{\frac{1}{2}})(3y^2-1), \quad \tfrac{1}{2}.3xy.]$$

5. Given a set of $n+1$ orthonormal F-vectors,

$$\mathbf{I}_1, \mathbf{I}_2, \ldots, \mathbf{I}_n, \mathbf{J},$$

orthonormalize the n F-vectors

$$\mathbf{I}_1 - \mathbf{J}, \quad \mathbf{I}_2 - \mathbf{J}, \ldots, \mathbf{I}_n - \mathbf{J}.$$

Symmetry suggests we try

$$\mathbf{K}_\rho = a(\mathbf{I}_\rho - \mathbf{J}) + b(\mathbf{I}_1 + \mathbf{I}_2 + \ldots + \mathbf{I}_n - n\mathbf{J}) \quad (\rho = 1, 2, \ldots, n);$$

the conditions of orthonormality then give just two equations for a and b, and the answer is

$$\mathbf{K}_\rho = \mathbf{I}_\rho - n^{-1}[1 \pm (n+1)^{-\frac{1}{2}}](\mathbf{I}_1 + \mathbf{I}_2 + \ldots + \mathbf{I}_n) \pm (n+1)^{-\frac{1}{2}}\,\mathbf{J}.$$

6. Find the 'best' approximation in the form of a step-function (n steps with equal bases) for the function $\sin x$ in the range $(-\pi, \pi)$.

7. **I**, **J**, **K** form an orthonormal triad of F-vectors. Using intuitive constructions of ordinary space, but verifying their validity in F-space, show that the normal drawn from the origin to the linear 2-space through the points with position-vectors **I**, **J**, **K** is $\mathbf{N} = \frac{1}{3}(\mathbf{I}+\mathbf{J}+\mathbf{K})$, so that $\mathbf{N}^2 = \frac{1}{3}$.

8. $\mathbf{I}_1, \mathbf{I}_2, \ldots, \mathbf{I}_{n+1}$ are orthonormal. A linear n-space L_n passes through the $n+1$ points of which the **I**'s are position-vectors, as in (1·312). Verify that the normal drawn from the origin to L_n is

$$\mathbf{N} = (n+1)^{-1}(\mathbf{I}_1 + \mathbf{I}_2 + \ldots + \mathbf{I}_{n+1}),$$

so that $\mathbf{N}^2 = (n+1)^{-1}$.

2·4. HYPERPLANES

Hyperplanes of class 1 (H_1)

As a simple analogy to what we are about to discuss, consider Euclidean 3-space. Let **A** be the position-vector of a point and let P be a plane through this point (Fig. 2·41).

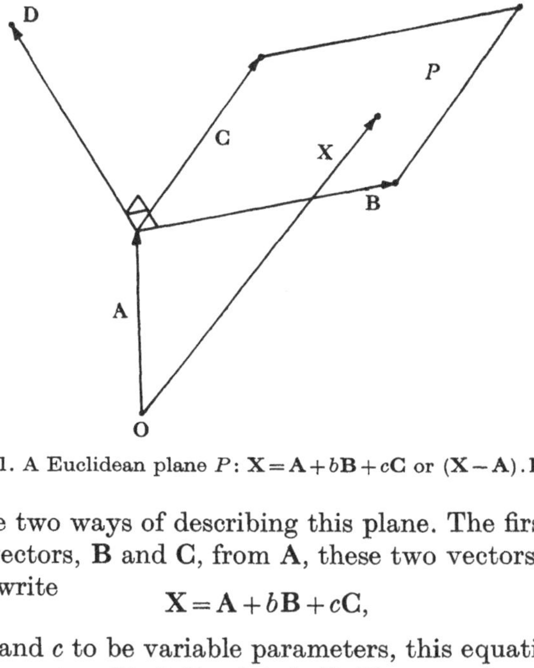

Fig. 2·41. A Euclidean plane P: $\mathbf{X} = \mathbf{A} + b\mathbf{B} + c\mathbf{C}$ or $(\mathbf{X} - \mathbf{A}).\mathbf{D} = 0$.

There are two ways of describing this plane. The first way is to draw two vectors, **B** and **C**, from **A**, these two vectors lying in P. If we then write

$$\mathbf{X} = \mathbf{A} + b\mathbf{B} + c\mathbf{C}, \tag{2·401}$$

and take b and c to be variable parameters, this equation gives us the position-vectors **X** of all points in P. The second way is to draw a vector **D** normal to P, and write

$$(\mathbf{X} - \mathbf{A}).\mathbf{D} = 0. \tag{2·402}$$

The equations (2·401) and (2·402) are equivalent descriptions of the plane P in Euclidean 3-space.

The generalizations of these two equations to F-space lead to two different generalizations of the plane.

We have already generalized (2·401) in (1·311). There we had, as the equation of a linear n-space L_n,

$$\mathbf{X} = \mathbf{S}_0 + \sum_{\rho=1}^{n} a_\rho \mathbf{T}_\rho, \tag{2·403}$$

where $\mathbf{S}_0$ is a fixed F-vector, $\mathbf{T}_\rho$ a set of linearly independent F-vectors and a_ρ a set of variable parameters. This concept was introduced before we put a metric into F-space.

If $\mathbf{A}$ and $\mathbf{D}$ are fixed F-vectors, the equation (2·402) has a meaning in F-space. We find it convenient to write it in an equivalent form

$$\mathbf{X} . \mathbf{S}_1 = b_1, \tag{2·404}$$

where $\mathbf{S}_1$ is a fixed F-vector and b_1 a fixed number. The subspace of F-space defined by (2·404), i.e. the totality of F-points with position-vectors $\mathbf{X}$ satisfying this equation, we shall call a *hyperplane of class* 1 and use the general symbol H_1 to denote it.

The equivalence of (2·401) and (2·402) in Euclidean 3-space might lead us to expect a similar equivalence of (2·403) and (2·404). Actually, they are far from being equivalent. The L_n of (2·403) is a subspace of n dimensions, the parameters a_ρ being the coordinates of the points in it. The H_1 of (2·404) is of infinite dimensionality, because the equation imposes only one condition on the infinite variety of functions which may correspond to the variable F-vector $\mathbf{X}$. This is made clear by considering, for example, the Hilbert scalar product (2·102); the equation (2·404) imposes on the function $f(x)$ corresponding to $\mathbf{X}$ a single condition

$$\int_{x_1}^{x_2} f(x)\, s_1(x)\, dx = b_1, \tag{2·405}$$

where $s_1(x)$ is a given function.

We might proceed at this point to discuss the properties of the hyperplane of class 1 (H_1). However, these may at once be obtained (by putting $n=1$) from the properties of the hyperplane of class n (H_n), which we shall now define.

Hyperplanes of class n (H_n).*

Consider the n equations

$$\mathbf{X} . \mathbf{S}_\rho = b_\rho \quad (\rho = 1, 2, \ldots, n), \tag{2·406}$$

where $\mathbf{S}_\rho$ are n linearly independent fixed F-vectors and b_ρ are n fixed numbers. The F-points with position-vectors $\mathbf{X}$ satisfying

* Sometimes called 'codimension n'.

these equations form a subspace which we shall call a *hyperplane of class n*, and denote generally by H_n.

To repeat what has been said in the case of H_1, we must avoid confusion between L_n and H_n; L_n is of n dimensions, while H_n is of infinite dimensionality.

One of the most important properties of H_n is its linear character; *it is a linear subspace*, as defined on p. 31. This is easy to see. Let $\mathbf{X}_1$ and $\mathbf{X}_2$ be two points in H_n, so that, by (2·406),

$$\mathbf{X}_1 . \mathbf{S}_\rho = b_\rho, \quad \mathbf{X}_2 . \mathbf{S}_\rho = b_\rho \quad (\rho = 1, 2, ,\ldots n). \qquad (2{\cdot}407)$$

Then for any constants p, q, satisfying $p+q=1$, we have

$$(p\mathbf{X}_1 + q\mathbf{X}_2) . \mathbf{S}_\rho = (p+q)\, b_\rho = b_\rho \quad (\rho = 1, 2, \ldots, n), \qquad (2{\cdot}408)$$

and so the straight line joining any two points of H_n lies entirely in H_n; this shows that it is a linear subspace.

For some purposes it is convenient to replace the vectors $\mathbf{S}_\rho$ in (2·406) by an orthonormal set $\mathbf{I}_\rho$, and this can always be done. For let $\mathbf{I}_\rho$ be any orthonormal set in the linear n-space of the $\mathbf{S}$'s. Then each $\mathbf{I}$ is a linear function of the $\mathbf{S}$'s, and so the equations (2·406) lead to the equations

$$\mathbf{X} . \mathbf{I}_\rho = c_\rho \quad (\rho = 1, 2, \ldots, n), \qquad (2{\cdot}409)$$

where c_ρ are n constants. This set of equations is completely equivalent to (2·406), and we may refer to (2·409) as the *standard equations of a hyperplane of class n* (H_n).

The equations of H_n may also be written in parametric form. Let $\mathbf{S}_0$ be any point in H_n; then, by (2·409),

$$\mathbf{S}_0 . \mathbf{I}_\rho = c_\rho \quad (\rho = 1, 2, \ldots, n). \qquad (2{\cdot}410)$$

Corresponding to any point $\mathbf{X}$ in H_n, let us define $\mathbf{Y}$ by

$$\mathbf{Y} = \mathbf{X} - \mathbf{S}_0. \qquad (2{\cdot}411)$$

Then, by virtue of (2·409) and (2·410), we have

$$\mathbf{Y} . \mathbf{I}_\rho = 0 \quad (\rho = 1, 2, \ldots, n),$$

and so we may write the equations

$$\mathbf{X} = \mathbf{S}_0 + \mathbf{Y}, \quad \mathbf{Y} . \mathbf{I}_\rho = 0 \quad (\rho = 1, 2, \ldots, n). \qquad (2{\cdot}412)$$

These are the *parametric equations of a hyperplane of class n* (H_n). They are to be understood as follows. We obtain the points of H_n by adding to a fixed vector $\mathbf{S}_0$ a vector $\mathbf{Y}$ which is arbitrary except for the n orthogonality conditions contained in (2·412).

The normal drawn to a hyperplane H_n and its minimum property

Consider the hyperplane H_n with equations (2·409). It contains the point

$$\mathbf{N}=\sum_{\rho=1}^{n} c_\rho \mathbf{I}_\rho; \tag{2·413}$$

this is verified at once by substitution in (2·409), using the orthonormality of the **I**'s. We may therefore substitute **N** for $\mathbf{S}_0$ in (2·412), and obtain the following parametric equation for H_n:

$$\mathbf{X}=\mathbf{N}+\mathbf{Y}, \quad \mathbf{Y}.\mathbf{I}_\rho=0 \quad (\rho=1,2,\ldots,n). \tag{2·414}$$

We note that
$$\mathbf{Y}.\mathbf{N}=\sum_{\rho=1}^{n} c_\rho \mathbf{Y}.\mathbf{I}_\rho=0, \tag{2·415}$$

and so, if we square the first equation in (2·414), we get

$$\mathbf{X}^2=\mathbf{N}^2+\mathbf{Y}^2. \tag{2·416}$$

This gives the square of the distance from the origin to any point of H_n. It is clear that this distance is a minimum when $\mathbf{Y}=\mathbf{O}$, and

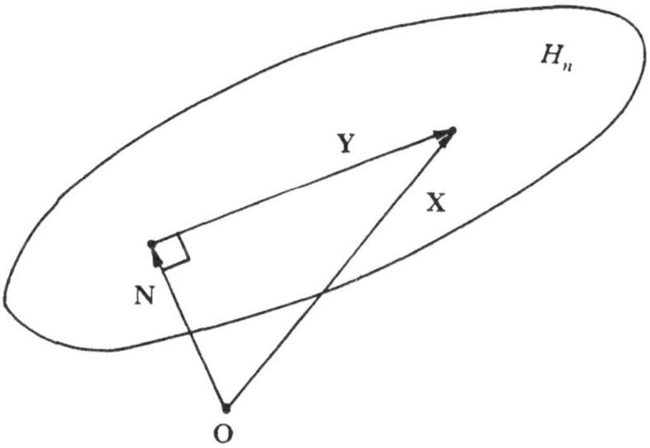

Fig. 2·42. The normal **N** drawn from **O** to the hyperplane H_n with equations $\mathbf{X}.\mathbf{I}_\rho=c_\rho$ $(\rho=1, 2, \ldots, n)$:

$$\mathbf{N}=\sum_{\rho=1}^{n} c_\rho \mathbf{I}_\rho, \quad \mathbf{N}^2=\sum_{\rho=1}^{n} c_\rho^2.$$

so *the point* **N**, *as given by* (2·413), *is the closest to the origin of all points of H_n, the hyperplane of class n with equations* (2·409).

The vector **N** is in fact the normal drawn from the origin to H_n, in the sense that **N** is orthogonal to all vectors lying in H_n. This is easy to see, because any vector **T** lying in H_n is the difference $\mathbf{X}_1-\mathbf{X}_2$ of the position-vectors of two points in H_n, and so, by (2·414), $\mathbf{T}=\mathbf{Y}_1-\mathbf{Y}_2$. Thus, by (2·415), $\mathbf{T}.\mathbf{N}=0$, and the result is proved. It is illustrated in Fig. 2·42.

To sum up: *the normal*

$$\mathbf{N}=\sum_{\rho=1}^{n} c_\rho \mathbf{I}_\rho, \quad \mathbf{N}^2=\sum_{\rho=1}^{n} c_\rho^2, \tag{2·417}$$

is the shortest vector that can be drawn from the origin to the hyperplane H_n with equations

$$\mathbf{X}.\mathbf{I}_\rho=c_\rho \quad (\rho=1,2,\ldots,n). \tag{2·418}$$

By shifting the origin in F-space, we may similarly discuss the normal drawn to H_n from a general point $\mathbf{B}$. This normal $\mathbf{N}$ is the vector $\mathbf{X}-\mathbf{B}$, when $\mathbf{X}$ has been chosen to satisfy (2·418) and to minimize $(\mathbf{X}-\mathbf{B})^2$. We find

$$\mathbf{N}=\sum_{\rho=1}^{n}(c_\rho-\mathbf{B}.\mathbf{I}_\rho)\,\mathbf{I}_\rho, \quad \mathbf{N}^2=\sum_{\rho=1}^{n}(c_\rho-\mathbf{B}.\mathbf{I}_\rho)^2; \tag{2·419}$$

This vector $\mathbf{N}$ is orthogonal to all vectors lying in H_n (Fig. 2·43).

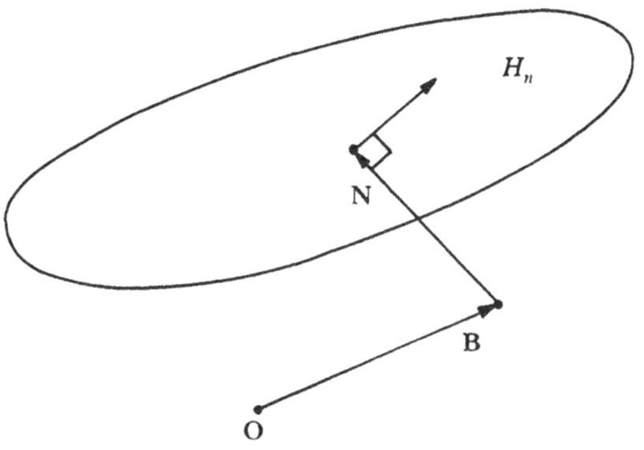

Fig. 2·43. The normal $\mathbf{N}$ drawn from the point $\mathbf{B}$ to the hyperplane H_n with equations $\mathbf{X}.\mathbf{I}_\rho=c_\rho$ $(\rho=1, 2, \ldots, n)$:

$$\mathbf{N}=\sum_{\rho=1}^{n}(c_\rho-\mathbf{B}.\mathbf{I}_\rho)\,\mathbf{I}_\rho, \quad \mathbf{N}^2=\sum_{\rho=1}^{n}(c_\rho-\mathbf{B}.\mathbf{I}_\rho)^2.$$

Orthogonality of a linear n-space L_n and a hyperplane H_n

For the sake of analogy, consider a plane in Euclidean 3-space and a straight line perpendicular to it. The plane determines the line in direction, or, in other words, it determines an infinite family of parallel lines, all perpendicular to the plane; conversely, the line determines an infinite family of parallel planes. This type of relationship exists between a linear n-space L_n and a hyperplane H_n of class n in F-space.

Consider L_n as in (2·336):

$$\mathbf{X}=\mathbf{A}+\sum_{\rho=1}^{n} a_\rho \mathbf{I}_\rho. \tag{2·420}$$

Here $\mathbf{A}$ is a fixed vector, $\mathbf{I}_\rho$ a set of n fixed orthonormal vectors and a_ρ variable parameters. If we change $\mathbf{A}$ but keep the $\mathbf{I}$'s unchanged, we get a family of parallel linear n-spaces. The simplest member of this family is the one through the origin:

$$\mathbf{X} = \sum_{\rho=1}^{n} a_\rho \mathbf{I}_\rho. \tag{2·421}$$

Consider now all vectors orthogonal to L_n. Any such vector, say $\mathbf{Z}$, must satisfy

$$\mathbf{Z}.\left(\sum_{\rho=1}^{n} a_\rho \mathbf{I}_\rho\right) = 0 \tag{2·422}$$

for all values of the a's, and so

$$\mathbf{Z}.\mathbf{I}_\rho = 0 \quad (\rho = 1, 2, \ldots, n). \tag{2·423}$$

Conversely, if these equations are satisfied, then $\mathbf{Z}$ is orthogonal to L_n—not only to (2·421), but of course to the whole parallel family (2·420).

Now (2·423) expresses the fact that the point $\mathbf{Z}$ is in a certain hyperplane of class n, or equivalently that the vector $\mathbf{Z}$ lies in any one of the hyperplanes H_n with equations

$$\mathbf{X}.\mathbf{I}_\rho = c_\rho \quad (\rho = 1, 2, \ldots, n), \tag{2·424}$$

the c's being any constants; if they are changed in value, we pass from one hyperplane of class n to a parallel one.

The result may be stated as follows: *The linear n-spaces*

$$L_n: \quad \mathbf{X} = \mathbf{A} + \sum_{\rho=1}^{n} a_\rho \mathbf{I}_\rho, \tag{2·425}$$

and the hyperplanes of class n

$$H_n: \quad \mathbf{X}.\mathbf{I}_\rho = c_\rho \quad (\rho = 1, 2, \ldots, n) \tag{2·426}$$

are orthogonal to one another.

We note that

(*a*) the same orthonormal set $\mathbf{I}_\rho$ occurs in (2·425) and (2·426);
(*b*) in (2·425) $\mathbf{A}$ is a fixed vector and a_ρ variable parameters;
(*c*) in (2·426) c_ρ are constants.

If L_n is given as in (2·425) with any particular $\mathbf{A}$, it does not determine a unique H_n, but only a family of parallel hyperplanes obtained by giving all values to c_ρ. Conversely, if H_n is given as in (2·426) with particular c_ρ, it does not determine a unique L_n, but only a family of parallel linear n-spaces obtained by giving all values to the vector $\mathbf{A}$ in (2·425). We can say that a family of parallel linear n-spaces determines a family of parallel hyperplanes of class n, orthogonal to it, and conversely. This is the F-space version of the Euclidean situation described on p. 72.

As we saw in (2·339), any vector $\mathbf{S}$ can be resolved into a component in L_n and a component orthogonal to it. We can make a similar resolution relative to H_n. If L_n and H_n are orthogonal as in (2·425) and (2·426), then the component of $\mathbf{S}$ orthogonal to L_n is in fact the component of $\mathbf{S}$ in H_n, and by (2·339) we have

$$\left.\begin{aligned}\text{Component of } \mathbf{S} \text{ in } L_n &= \sum_{\rho=1}^{n} \mathbf{I}_\rho(\mathbf{S}\,.\,\mathbf{I}_\rho),\\ \text{Component of } \mathbf{S} \text{ in } H_n &= \mathbf{S} - \sum_{\rho=1}^{n} \mathbf{I}_\rho(\mathbf{S}\,.\,\mathbf{I}_\rho).\end{aligned}\right\} \qquad (2{\cdot}427)$$

The non-orthogonality of hyperplanes

Can two hyperplanes, H_m and H'_n, be orthogonal to one another?

The orthogonality of linear subspaces was defined on p. 66; the condition for the orthogonality of H_m and H'_n is that every vector lying in the one shall be orthogonal to every vector lying in the other.

Now the totality of vectors orthogonal to all vectors lying in H_m form a family of linear m-spaces L_m. Thus the orthogonality of H_m and H'_n would imply that a linear m-space L_m was the same as a hyperplane H'_n. This cannot be, since one of these is of m dimensions and the other of infinite dimensionality. Thus *two hyperplanes cannot be orthogonal to one another.*

Examples

The fact that the normal to a hyperplane has a minimum length is the basis of some of the minimum principles used in solving physical problems. However, we shall restrict ourselves here to two simple artificial examples to illustrate the theory.

Example 1. Let P-space be $-\pi \leqslant x \leqslant \pi$ and let the scalar product be Hilbert's (2·102). Take the orthonormal pair of F-vectors

$$\mathbf{I}_1 \leftrightarrow (2\pi)^{-\frac{1}{2}}, \quad \mathbf{I}_2 \leftrightarrow \pi^{-\frac{1}{2}} \sin x.$$

Then the two equations $\qquad \mathbf{X}\,.\,\mathbf{I}_1 = 1, \quad \mathbf{X}\,.\,\mathbf{I}_2 = 1, \qquad (2{\cdot}428)$

define a hyperplane of class 2. It consists of all F-points corresponding to functions $f(x)$ satisfying the two conditions

$$\int_{-\pi}^{\pi} (2\pi)^{-\frac{1}{2}} f(x)\,dx = 1, \quad \int_{-\pi}^{\pi} \pi^{-\frac{1}{2}} \sin x\, f(x)\,dx = 1. \qquad (2{\cdot}429)$$

Comparing (2·428) with (2·409) and (2·417), we see that the normal from the origin to this hyperplane is the F-vector $\mathbf{I}_1 + \mathbf{I}_2$; hence the function

$$f(x) = (2\pi)^{-\frac{1}{2}} + \pi^{-\frac{1}{2}} \sin x \qquad (2{\cdot}430)$$

produces the minimum

$$\mathbf{N}^2 = [\mathbf{X}^2]_{\text{min.}} = \left[\int_{-\pi}^{\pi} (f(x))^2\,dx\right]_{\text{min.}} = 2, \qquad (2\cdot431)$$

when we consider as integrands the squares of all functions $f(x)$ satisfying (2·429). In other words, if $f(x)$ is arbitrary except for (2·429), then

$$\int_{-\pi}^{\pi} (f(x))^2\,dx \geqslant 2, \qquad (2\cdot432)$$

and the minimum is attained for the function (2·430).

The result may also be easily established, without the machinery of F-space, by using the fact that

$$\int_{-\pi}^{\pi} [f(x) - a - b\sin x]^2\,dx \geqslant 0$$

for all real values of a and b.

Example 2. A function $f(x)$ is such that

$$\int_{-1}^{0} f(x)\,dx = -1, \quad \int_{0}^{1} f(x)\,dx = 1. \qquad (2\cdot433)$$

We wish to find the least possible value for $\int_{-1}^{1} [f(x)]^2\,dx$.

To discuss this problem in terms of F-space, we let P-space be $-1 \leqslant x \leqslant 1$ and use the Hilbert scalar product (2·102). We introduce an orthonormal pair

$$\left.\begin{aligned} &\mathbf{I}_1 \leftrightarrow 1 \text{ if } x \text{ is negative and } 0 \text{ if } x \text{ is positive,}\\ &\mathbf{I}_2 \leftrightarrow 0 \text{ if } x \text{ is negative and } 1 \text{ if } x \text{ is positive.}\end{aligned}\right\} \qquad (2\cdot434)$$

(The orthonormality can be verified immediately.) Then the conditions (2·433) may be written

$$\mathbf{X}.\mathbf{I}_1 = -1, \quad \mathbf{X}.\mathbf{I}_2 = 1, \qquad (2\cdot435)$$

where $\mathbf{X} \leftrightarrow f(x)$. Thus from (2·416) and (2·417)

$$\mathbf{X}^2 = \int_{-1}^{1} [f(x)]^2\,dx \geqslant 1^2 + 1^2 = 2, \qquad (2\cdot436)$$

and the minimum is attained for that function which is -1 for negative x and $+1$ for positive x.

Exercises

1. If point $\mathbf{P}$ is on the hyperplane $\mathbf{X}.\mathbf{I} = a$, where $\mathbf{I}^2 = 1$, and the point $\mathbf{Q}$ is on the hyperplane $\mathbf{X}.\mathbf{I} = b$, prove that $(\mathbf{P} - \mathbf{Q})^2 \geqslant (a - b)^2$.

2. $\mathbf{I}_1, \mathbf{I}_2, \ldots, \mathbf{I}_{n+1}$ are given orthonormal F-vectors and $a_1, a_2, \ldots, a_{n+1}$ are given constants. Consider the two hyperplanes:

$$\begin{aligned} &H_n \text{ (class } n\text{):} && \mathbf{X}.\mathbf{I}_\rho = a_\rho \quad (\rho = 1, 2, \ldots, n);\\ &H_{n+1} \text{ (class } n+1\text{):} && \mathbf{X}.\mathbf{I}_\rho = a_\rho \quad (\rho = 1, 2, \ldots, n+1).\end{aligned}$$

Show that, in general, the normal from the origin on H_{n+1} is greater than the normal from the origin on H_n.

Under what circumstances will the normals be equal in length? Can they be equal in length without coinciding?

3. Show that, in general, a straight line $\mathbf{X} = \mathbf{A} + c\mathbf{J}$ ($\mathbf{A}$, $\mathbf{J}$ fixed, c variable, $\mathbf{J}^2 = 1$) meets the hyperplane of class 1 with equation $\mathbf{X}.\mathbf{I} = a$ ($\mathbf{I}$ and a fixed,

$\mathbf{I}^2 = 1$) in just one point. Under what conditions on the fixed quantities does the straight line

(*a*) fail to meet the hyperplane?

(*b*) lie in the hyperplane?

4. Show that in general a straight line does not meet a hyperplane of class 2.

5. If $f(x)$ is an unknown function and $g(x)$ a known function, then the condition

$$\int_0^1 f(x)\, g(x)\, dx = 1$$

defines a hyperplane of class 1 if we use the Hilbert scalar product (2·102) for the range (0, 1). What function corresponds to the normal drawn from the origin to this hyperplane?

6. For the range (0, 1) find the function $f(x)$ which is orthogonal in the Hilbert sense to the functions

$$1, \quad x, \quad x^2, \quad x^3,$$

and approximates as closely as possible to the function x^4. (It is a question of finding the normal drawn from a point to a hyperplane of class 4.)

7. Find that function $f(x, y)$ inside the unit circle $x^2 + y^2 = 1$ which makes

$$\iint [f(x,y)]^2\, dx\, dy$$

as small as possible, given

$$\iint f(x,y)\, dx\, dy = 1.$$

2·5. HYPERSPHERES

Neighbourhoods

In Euclidean 3-space we can construct a box by drawing six planes, parallel in pairs. The box has an inside and an outside. The inside defines a *neighbourhood* in the sense that the distance between any two points lying inside the box has an upper bound—a diagonal of the box. Similarly, we can construct a neighbourhood by means of a sphere or an ellipsoid, or indeed by means of any closed surface.

If we pass from Euclidean 3-space to Euclidean n-space no difficulty arises; neighbourhoods can be similarly defined. In particular, we can make a box out of $2n$ planes, parallel in pairs. But in function-space the situation is different, as we shall now see.

No closed boxes in function-space

Our first thought might be that we could make boxes in F-space by using linear n-spaces. The attempt is as futile as the attempt to build a box in Euclidean 3-space out of straight lines. The straight line fails here because it has not got two sides like a plane—it does not divide space into two parts. In F-space a linear n-space has not got two sides; we can pass continuously from any point of F-space to any other without passing through an assigned linear n-space.

Our next thought is that we might make boxes in F-space out of hyperplanes. A hyperplane of class 1 ($\mathbf{X}.\mathbf{I}=a$) has two sides. It divides F-space into two parts, namely, a part consisting of those points $\mathbf{X}$ making $\mathbf{X}.\mathbf{I}>a$ and a part consisting of those points $\mathbf{X}$ making $\mathbf{X}.\mathbf{I}<a$. We cannot pass continuously from the one part to the other without passing through the hyperplane, for we cannot change the quantity $(\mathbf{X}.\mathbf{I}-a)$ continuously from positive to negative values without passing through the number zero.

Nevertheless, we cannot make a satisfactory box in F-space by means of any finite number of hyperplanes. It was pointed out in connexion with equation (2·236) how little a condition of orthogonality really restricts a function; similarly the information that a point lies on a hyperplane is really only a slight restriction on the function or functions corresponding to the point. It is true that we can form a sort of box by the inequalities

$$a_\rho<\mathbf{X}.\mathbf{I}_\rho<b_\rho \quad (\rho=1,2,\ldots,n); \tag{2·501}$$

this box has an inside for which the inequalities are satisfied; the 'surface' of the box consists of those points for which at least one of the equations

$$\mathbf{X}.\mathbf{I}_\rho=a_\rho, \quad \mathbf{X}.\mathbf{I}_\rho=b_\rho \quad (\rho=1,2,\ldots,n) \tag{2·502}$$

is satisfied, and the outside consists of all other F-points. We cannot pass from the inside to the outside of the box continuously without crossing the surface. *But the box extends to infinity*, like any box we attempt to make in Euclidean 3-space out of only *two* planes. The box does not define a neighbourhood, and in that sense we say that *there are no closed boxes in F-space.*

If we want to define a neighbourhood in F-space (and we do), we must use something other than hyperplanes. We use a *hypersphere.*

n-Spheres and hyperspheres defined

Since there is danger of confusion between n-spheres and hyperspheres, we shall discuss them both here for contrast, although the hypersphere is much more important for later purposes.

An *n-sphere* is defined as a subspace of F-space consisting of all F-points which satisfy the following two conditions:

(i) They lie in a linear n-space.

(ii) They are equidistant (radius R) from a point $\mathbf{C}$ (the centre) lying in the linear n-space.

Let us for simplicity consider the case where the linear n-space passes through the origin and the centre of the n-sphere is at the origin ($\mathbf{C}=\mathbf{O}$). Then the equation of the linear n-space may be written

$$\mathbf{X}=c_1\mathbf{I}_1+c_2\mathbf{I}_2+\ldots+c_n\mathbf{I}_n, \tag{2·503}$$

where the c's are variable parameters and the $\mathbf{I}$'s fixed orthonormal vectors. The conditions that $\mathbf{X}$ should be a point on the n-sphere are that the vector $\mathbf{X}$ should be of the form (2·503) and that

$$\mathbf{X}^2=R^2, \tag{2·504}$$

where R is the radius of the n-sphere. This is equivalent to

$$c_1^2+c_2^2+\ldots+c_n^2=R^2. \tag{2·505}$$

Thus the n-sphere has the *parametric equation* (2·503), the parameters taking all values consistent with (2·505).

We note that a 1-sphere is a pair of points, a 2-sphere is an ordinary circle, and a 3-sphere is an ordinary sphere.

An n-sphere divides the linear n-space containing it into two parts, an inside and an outside. But, by wandering out of the linear n-space, we can get from the inside to the outside without crossing the n-sphere. Thus an n-sphere does not define a neighbourhood in F-space.

A *hypersphere* is defined as a subspace of F-space consisting of all F-points equidistant (radius R) from some fixed F-point $\mathbf{C}$ (the centre). Its equation is simply

$$(\mathbf{X}-\mathbf{C})^2=R^2, \tag{2·506}$$

where $\mathbf{C}$ and R are given. We shall always understand R to be positive. We must of course remember that throughout this part of the book we are dealing exclusively with the case of a positive-definite F-metric. If the metric were indefinite, the equation (2·506) would still be meaningful, but the properties of the subspace defined by it would be very different, and so we would use a different name, *pseudohypersphere* (cf. §6·2).

We shall call (2·506) the *standard equation* of a hypersphere. The equivalent *parametric equation* is

$$\mathbf{X}=\mathbf{C}+R\mathbf{J}, \quad \mathbf{J}^2=1; \tag{2·507}$$

it is understood that $\mathbf{J}$ is completely arbitrary except for the condition that its magnitude is unity.

A hypersphere divides F-space into an *inside*

$$(\mathbf{X}-\mathbf{C})^2<R^2, \quad \text{or equivalently} \quad \mathbf{X}=\mathbf{C}+M\mathbf{J} \quad (0<M<R), \tag{2·508}$$

and an *outside*

$$(\mathbf{X}-\mathbf{C})^2>R^2, \quad \text{or equivalently} \quad \mathbf{X}=\mathbf{C}+M\mathbf{J} \quad (M>R). \tag{2·509}$$

We cannot pass from the inside to the outside continuously without passing through the hypersphere.

The inside of a hypersphere is a neighbourhood. If $\mathbf{X}$ and $\mathbf{X}'$ both lie inside, then

$$|\mathbf{X}-\mathbf{X}'|<2R. \qquad (2\cdot510)$$

To show this, we write, as in (2·508),

$$\mathbf{X}=\mathbf{C}+M\mathbf{J}, \quad \mathbf{X}'=\mathbf{C}+M'\mathbf{J}'; \qquad (2\cdot511)$$

then

$$\begin{aligned}(\mathbf{X}-\mathbf{X}')^2 &= (M\mathbf{J}-M'\mathbf{J}')^2 \\ &= M^2+M'^2-2MM'\mathbf{J}\,.\,\mathbf{J}' \\ &< R^2+R^2+2R^2=4R^2, \end{aligned} \qquad (2\cdot512)$$

since M and M' are both less than R and $|\mathbf{J}\,.\,\mathbf{J}'|\leqslant 1$ by (2·212).

The inequality (2·510) is just what one would guess from the analogy of the Euclidean sphere. This is an example of the way in which ordinary intuition can help us in discussing the hypersphere, in spite of its infinite dimensionality. Needless to say, such intuitions must be regarded as suggestions, not as proofs.

Bounds on $\mathbf{X}^2$ *as* $\mathbf{X}$ *ranges on a hypersphere*

Consider a hypersphere with centre $\mathbf{C}$ and radius R. We want to know the upper and lower bounds on $\mathbf{X}^2$ as the point $\mathbf{X}$ ranges over the hypersphere.

If this problem were presented to us in ordinary space, with the hypersphere replaced by an ordinary sphere, our space intuition would give us the answer immediately. To find the bounds in F-space we can either rationalize our space intuition (a very useful method, frequently used), or we can proceed by pure algebra. We shall here follow the second course, which is certainly the simpler, once one knows what to do!

From the parametric equation (2·507) we have

$$\mathbf{X}^2=\mathbf{C}^2+2R\mathbf{C}\,.\,\mathbf{J}+R^2. \qquad (2\cdot513)$$

Now $\mathbf{C}^2=|\mathbf{C}|^2$, and by the Schwarz inequality (2·211)

$$-|\mathbf{C}|\leqslant\mathbf{C}\,.\,\mathbf{J}\leqslant|\mathbf{C}|, \qquad (2\cdot514)$$

since $\mathbf{J}$ is a unit vector. Hence (2·513) gives

$$(|\mathbf{C}|-R)^2\leqslant\mathbf{X}^2\leqslant(|\mathbf{C}|+R)^2. \qquad (2\cdot515)$$

These are the required bounds. The lower bound is attained when $\mathbf{J}=-\mathbf{C}/|\mathbf{C}|$, and the upper bound when $\mathbf{J}=\mathbf{C}/|\mathbf{C}|$.

The problem is pictured in Fig. 2·51 for the case $|\mathbf{C}| > R$ (origin outside the hypersphere) and in Fig. 2·52 for the case $|\mathbf{C}| < R$ (origin inside the hypersphere).

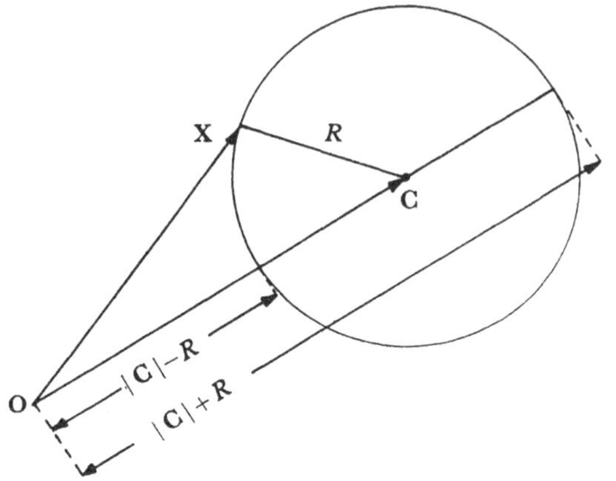

Fig. 2·51. Hypersphere with **O** outside it; bounds on | **X** |.

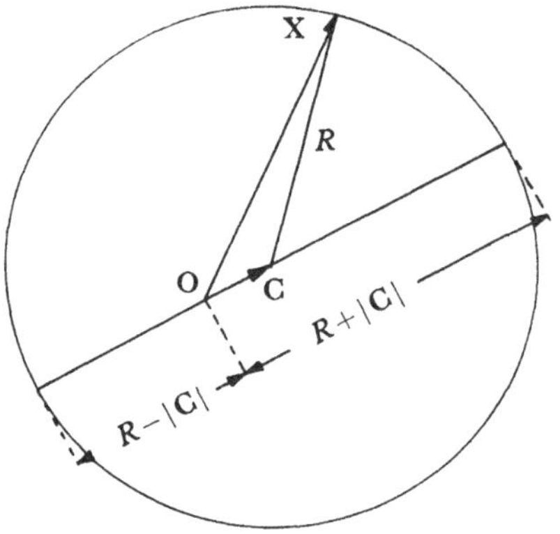

Fig. 2·52. Hypersphere with **O** inside it; bounds on | **X** |.

Bounds on **X . G** *as* **X** *ranges on a hypersphere*

Let **G** be any fixed vector. We seek bounds on the scalar product **X . G** as the point **X** ranges on a hypersphere with centre **C** and radius R.

Here again we might appeal to space intuition for the answer. The algebraic procedure is as follows.

By (2·507) we have

$$\mathbf{X}.\mathbf{G} = \mathbf{C}.\mathbf{G} + R\mathbf{J}.\mathbf{G}. \tag{2·516}$$

Now **C . G** is fixed and $|\mathbf{J}.\mathbf{G}| \leqslant |\mathbf{G}|$. Thus we have the required bounds

$$\mathbf{C}.\mathbf{G} - R|\mathbf{G}| \leqslant \mathbf{X}.\mathbf{G} \leqslant \mathbf{C}.\mathbf{G} + R|\mathbf{G}|. \tag{2·517}$$

The lower bound is attained when $\mathbf{J} = -\mathbf{G}/|\mathbf{G}|$ and the upper bound when $\mathbf{J} = \mathbf{G}/|\mathbf{G}|$.

If $\mathbf{G}$ is a unit vector (write $\mathbf{G} = \mathbf{I}$), then (2·517) reads

$$\mathbf{C}.\mathbf{I} - R \leqslant \mathbf{X}.\mathbf{I} \leqslant \mathbf{C}.\mathbf{I} + R. \qquad (2\cdot518)$$

Since $\mathbf{C}.\mathbf{I}$ is the projection of $\mathbf{C}$ on $\mathbf{I}$, it is very easy (see Fig. 2·53) to reconcile these inequalities with space intuition.

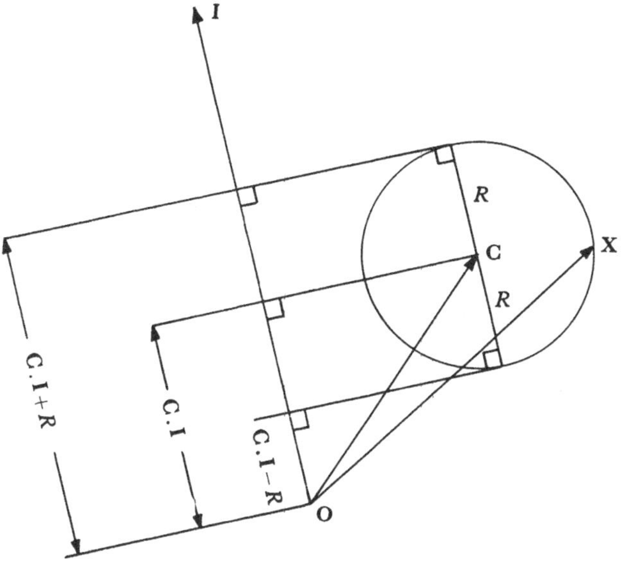

Fig. 2·53. Bounds on **X**.**I** when **X** ranges on a hypersphere.

Examples

In the results just established emphasis has been laid on the geometry rather than on the analytical nature of the subspaces and bounds. Examples (3), (4) and (5) below may help to remedy this. It is in the interplay of geometry and analysis that the method appears at its best.

Example 1. Consider the P-space $-1 \leqslant x \leqslant 1$ and the Hilbert scalar product (2·102). Then $\mathbf{X} \leftrightarrow c_1 + c_2 x$ is a linear 2-space. The totality of functions

$$2^{-\frac{1}{2}}(\sin\theta + 3^{\frac{1}{2}} x \cos\theta) \quad (0 \leqslant \theta < 2\pi),$$

considered of course as functions of x for constant θ, correspond to a 2-sphere in the linear 2-space, with centre at the origin and radius 1.

Example 2. Just as we use co-latitude and longitude to describe the position of a point on an ordinary sphere, so we can express any point on an n-sphere in terms of $n-1$ angles. For the case $n=4$ in (2·503) and (2·505), we may write

$$\mathbf{X} = R(\sin\theta_1 \sin\theta_2 \sin\theta_3 \mathbf{I}_1 + \sin\theta_1 \sin\theta_2 \cos\theta_3 \mathbf{I}_2 + \sin\theta_1 \cos\theta_2 \mathbf{I}_3 + \cos\theta_1 \mathbf{I}_4), \qquad (2\cdot519)$$

where θ_1 and θ_2 have the range $(0, \pi)$, θ_3 the range $(0, 2\pi)$, and the $\mathbf{I}$'s form the orthonormal tetrad of (2·503). That this is in fact a 4-sphere of radius R is verified by calculating $\mathbf{X}^2$ and getting $\mathbf{X}^2 = R^2$. It is easy to see that the ranges for the three angles suffice to cover the whole 4-sphere.

Example 3. Let $f(x)$ be an unknown function about which we have no information except that contained in the inequality

$$\int_{-1}^{1} [f(x) - x^3]^2 \, dx < 0{\cdot}0001. \tag{2·520}$$

We might then say that x^3 is a fairly good approximation to $f(x)$ for the range $(-1, 1)$, in the mean square sense. We should not assert dogmatically that it is even a moderately good approximation at *every* point of the range, for the graph of $f(x)$ may have a hill on it, high but yet so narrow that it contributes very little to the integral (2·520).

To interpret (2·520) geometrically, we take $-1 \leqslant x \leqslant 1$ as P-space and use the Hilbert scalar product (2·102); then (2·520) may be written

$$(\mathbf{X} - \mathbf{C})^2 < (0{\cdot}01)^2, \tag{2·521}$$

where $\mathbf{X} \leftrightarrow f(x)$, $\mathbf{C} \leftrightarrow x^3$. The inequality tells us that the point $\mathbf{X}$ lies *inside* a hypersphere of radius 0·01, and it is on account of the smallness of this radius that we may regard the point $\mathbf{C}$ as a fairly good approximation to the unknown point $\mathbf{X}$. But, let us repeat, it is a fairly good approximation only in the mean square sense.

Example 4. Are the following inequalities consistent?

$$\left.\begin{aligned} \int_{-1}^{1} [f(x) - x^3]^2 \, dx &< \tfrac{1}{9}, \\ \int_{-1}^{1} [f(x) - x^4]^2 \, dx &< \tfrac{1}{9}. \end{aligned}\right\} \tag{2·522}$$

Function-space suggests the attack at once. The question is this: Does a hypersphere with centre $\mathbf{C} \leftrightarrow x^3$ and radius $\frac{1}{3}$ overlap a hypersphere with centre $\mathbf{C}' \leftrightarrow x^4$ and radius $\frac{1}{3}$?

They will overlap if, and only if, the sum of the radii, namely, $\frac{2}{3}$, exceeds the distance between the centres, namely, $|\mathbf{C} - \mathbf{C}'|$. This we know intuitively in ordinary space, and it is easy to verify it in F-space. Now

$$(\mathbf{C} - \mathbf{C}')^2 = \int_{-1}^{1} (x^3 - x^4)^2 \, dx = \tfrac{32}{63},$$

and $\frac{2}{3} < (\frac{32}{63})^{\frac{1}{2}}$. Thus the inequalities (2·522) are inconsistent; no function $f(x)$ can satisfy them both at the same time.

Example 5. If

$$\int_{-1}^{1} [f(x) - x]^2 \, dx = 4, \tag{2·523}$$

what are the bounds of

$$\int_{-1}^{1} f(x) \, x^2 \, dx\,? \tag{2·524}$$

This is a straight question on the bounds of $\mathbf{X}.\mathbf{G}$ when the point $\mathbf{X}$ is on a hypersphere. We use (2·517) with

$$\left.\begin{gathered} \mathbf{C} \leftrightarrow x, \quad \mathbf{G} \leftrightarrow x^2, \quad R = 2, \\ \mathbf{C}.\mathbf{G} = \int_{-1}^{1} x \,.\, x^2 \, dx = 0, \quad \mathbf{G}^2 = \int_{-1}^{1} x^4 \, dx = \tfrac{2}{5}, \end{gathered}\right\} \tag{2·525}$$

and so the integral (2·524) has the bounds

$$-2(\tfrac{2}{5})^{\frac{1}{2}} \quad \text{and} \quad 2(\tfrac{2}{5})^{\frac{1}{2}}.$$

These bounds are attained respectively by

$$f(x) = x - x^2(10)^{\frac{1}{2}}, \quad f(x) = x + x^2(10)^{\frac{1}{2}}.$$

Intersection of a hypersphere and a linear n-space

Some of the things we can say about an ordinary sphere can be said about a hypersphere also; others cannot. Hence the necessity for caution in using space intuition in dealing with a hypersphere. But there is of course no harm in using intuition for the purpose of guessing things which may be tested rationally afterwards; if we do not guess, we shall have nothing to test.

A straight line meets an ordinary sphere in 0, 1 or 2 points. If there is just one point of intersection, the line is tangent to the sphere, and orthogonal to the radius drawn to the point of contact. All this is true of a hypersphere also; the proof is easy and is left to the reader.

A linear n-space may not meet a hypersphere at all; if it does meet the hypersphere, the intersection is either one point or an n-sphere. This we shall now prove.

Let

$$\mathbf{X} = \mathbf{A} + \sum_{\rho=1}^{n} c_\rho \mathbf{I}_\rho \tag{2·526}$$

be the linear n-space. (The $\mathbf{I}$'s are orthonormal as usual.) Let

$$(\mathbf{X} - \mathbf{C})^2 = R^2 \tag{2·527}$$

be the hypersphere. To investigate intersections, we substitute from (2·526) in (2·527). Writing $\mathbf{D} = \mathbf{A} - \mathbf{C}$, we get

$$\left(\sum_{\rho=1}^{n} c_\rho \mathbf{I}_\rho + \mathbf{D}\right)^2 = R^2. \tag{2·528}$$

Now resolve $\mathbf{D}$ as in (2·339):

$$\mathbf{D} = \mathbf{D}_0 + \sum_{\rho=1}^{n} (\mathbf{D}\,.\,\mathbf{I}_\rho)\,\mathbf{I}_\rho, \quad \mathbf{D}_0\,.\,\mathbf{I}_\rho = 0 \quad (\rho = 1, 2, \ldots, n). \tag{2·529}$$

Substitution in (2·528) gives

$$\sum_{\rho=1}^{n} (c_\rho + \mathbf{D}\,.\,\mathbf{I}_\rho)^2 = R^2 - \mathbf{D}_0^2. \tag{2·530}$$

Here everything is fixed except the c's; to get an intersection, they must satisfy this equation. Since the left-hand side is a sum of squares, there is no intersection if

$$R^2 < \mathbf{D}_0^2 = \mathbf{D}^2 - \sum_{\rho=1}^{n} (\mathbf{D}\,.\,\mathbf{I}_\rho)^2. \tag{2·531}$$

This condition is very simple when viewed geometrically. It means that the radius of the hypersphere is less than the length of the normal from its centre to the linear n-space (cf. (2·353)). (Intuition is correct in this case!)

If $\mathbf{R} = |\mathbf{D}_0|$, there is one point common to the hypersphere and the linear n-space, corresponding to

$$c_\rho = -\mathbf{D} . \mathbf{I}_\rho \quad (\rho = 1, 2, \ldots, n), \tag{2·532}$$

and by (2·526) this point is

$$\mathbf{B} = \mathbf{A} - \sum_{\rho=1}^{n} [(\mathbf{A} - \mathbf{C}) . \mathbf{I}_\rho] \, \mathbf{I}_\rho. \tag{2·533}$$

We say then that the linear n-space *touches* the hypersphere and that $\mathbf{B}$ is the *point of contact*.

There are of course infinitely many linear n-spaces tangent to a hypersphere at any point on it. But we shall not pursue this question further.

If $R > |\mathbf{D}_0|$, then (2·530) can be satisfied by infinitely many values of the c's. Each set of c's satisfying this equation gives a point in the linear n-space when this set of values is substituted in (2·526). Using $\mathbf{B}$ as in (2·533), we have for any such point

$$\mathbf{X} - \mathbf{B} = \sum_{\rho=1}^{n} (c_\rho + \mathbf{D} . \mathbf{I}_\rho) \, \mathbf{I}_\rho, \tag{2·534}$$

and so by (2·530)

$$(\mathbf{X} - \mathbf{B})^2 = R^2 - \mathbf{D}_0^2. \tag{2·535}$$

Comparing with (2·504), we see that the intersection of the hypersphere and the linear n-space is an n-sphere with centre $\mathbf{B}$ and radius $(R^2 - \mathbf{D}_0^2)^{\frac{1}{2}}$.

Intersection of a hypersphere and a hyperplane

We shall now discuss the intersection of a hypersphere and a hyperplane. We might use a line of argument very like that used above in discussing the intersection of a hypersphere and a linear n-space, but for variety a slightly different approach will be made.

Consider a hyperplane of class n with equations

$$\mathbf{X} . \mathbf{I}_\rho = a_\rho \quad (\rho = 1, 2, \ldots, n), \tag{2·536}$$

the $\mathbf{I}$'s being orthonormal, and a hypersphere

$$(\mathbf{X} - \mathbf{C})^2 = R^2, \tag{2·537}$$

or, in parametric form,

$$\mathbf{X} = \mathbf{C} + R\mathbf{J}, \quad \mathbf{J}^2 = 1. \tag{2·538}$$

To investigate the intersection of the hypersphere and the hyperplane, we substitute from (2·538) in (2·536) and get

$$R\mathbf{J}\,.\,\mathbf{I}_\rho = a_\rho - \mathbf{C}\,.\,\mathbf{I}_\rho \quad (\rho = 1, 2, \ldots, n). \tag{2·539}$$

Here **J** is the only variable; for an intersection we need a unit vector **J** satisfying these n equations.

Let us assume an intersection, so that (2·539) are satisfied. Then, squaring and adding, we get

$$\sum_{\rho=1}^{n} (a_\rho - \mathbf{C}\,.\,\mathbf{I}_\rho)^2 = R^2 \sum_{\rho=1}^{n} (\mathbf{J}\,.\,\mathbf{I}_\rho)^2 \leqslant R^2, \tag{2·540}$$

by the extended Schwarz inequality (2·333). Thus a necessary condition for intersection is

$$\sum_{\rho=1}^{n} (a_\rho - \mathbf{C}\,.\,\mathbf{I}_\rho)^2 \leqslant R^2, \tag{2·541}$$

a very natural condition, since, by (2·419), it expresses that the normal drawn from the centre of the hypersphere on the hyperplane shall be less than or equal to the radius.

If (2·541) is satisfied, there is a subspace common to the hypersphere and the hyperplane of class n; this subspace we call a *hypercircle of class n*. As we shall devote the next section to hypercircles, we shall not discuss them further here.

A few words, however, about the case where the sign of equality holds in (2·541). Then equality holds also in (2·540), and by the reasoning following the extended Schwarz inequality (2·333) we know that **J** must be a linear combination of the **I**'s:

$$\mathbf{J} = \sum_{\rho=1}^{n} b_\rho \mathbf{I}_\rho. \tag{2·542}$$

Then the b's are uniquely determined by (2·539):

$$Rb_\rho = a_\rho - \mathbf{C}\,.\,\mathbf{I}_\rho \quad (\rho = 1, 2, \ldots, n). \tag{2·543}$$

When we substitute these values in (2·542) and the result in (2·538), we get a unique point common to the hyperplane and the hypersphere. To summarize: *If between the defining elements of a hyperplane of class n* (2·536) *and a hypersphere* (2·537) *the relation*

$$\sum_{\rho=1}^{n} (a_\rho - \mathbf{C}\,.\,\mathbf{I}_\rho)^2 = R^2 \tag{2·544}$$

holds, then the hyperplane and the hypersphere have just one point in common, viz.

$$\mathbf{B} = \mathbf{C} + \sum_{\rho=1}^{n} (a_\rho - \mathbf{C}\,.\,\mathbf{I}_\rho)\,\mathbf{I}_\rho. \tag{2·545}$$

We say that then the hyperplane *touches* or is *tangent* to the hypersphere, with $\mathbf{B}$ as point of contact. The condition (2·544) means simply that the normal drawn from the centre $\mathbf{C}$ of the hypersphere to the hyperplane is equal in length to the radius of the hypersphere. If

$$\sum_{\rho=1}^{n} (a_\rho - \mathbf{C}.\mathbf{I}_\rho)^2 > R^2, \tag{2·546}$$

then the hyperplane does not intersect the hypersphere at all.

In the case of tangency we suspect that the vector $\mathbf{B}-\mathbf{C}$, drawn from the centre of the hypersphere to the point of contact $\mathbf{B}$, is orthogonal to the hyperplane. This is easily confirmed, for any vector $\mathbf{Y}$ lying in the hyperplane (2·536) satisfies

$$\mathbf{Y}.\mathbf{I}_\rho = 0 \quad (\rho = 1, 2, \ldots, n), \tag{2·547}$$

and so by (2·545)

$$\mathbf{Y}.(\mathbf{B}-\mathbf{C}) = \mathbf{Y}.\left[\sum_{\rho=1}^{n} (a_\rho - \mathbf{C}.\mathbf{I}_\rho)\,\mathbf{I}_\rho\right] = 0, \tag{2·548}$$

as required.

Hyperplanes of class 1 *tangent to a hypersphere*

Let us take the hypersphere

$$(\mathbf{X}-\mathbf{C})^2 = R^2, \tag{2·549}$$

and any point $\mathbf{T}$ on it. As remarked earlier, there are infinitely many linear n-spaces tangent to the hypersphere at $\mathbf{T}$. There is only one tangent hyperplane of class 1; its equation is

$$(\mathbf{X}-\mathbf{T}).(\mathbf{T}-\mathbf{C}) = 0. \tag{2·550}$$

To justify this statement, we remark first that the subspace defined by (2·550) is certainly a hyperplane of class 1. To find its intersection with the hypersphere, we substitute in (2·550) the value of $\mathbf{X}$ given by the parametric form (2·538) of the hypersphere; this gives

$$(\mathbf{C}+R\mathbf{J}-\mathbf{T}).(\mathbf{T}-\mathbf{C}) = 0, \tag{2·551}$$

or, since $(\mathbf{C}-\mathbf{T})^2 = R^2$,

$$\mathbf{J}.(\mathbf{T}-\mathbf{C}) = |\,\mathbf{T}-\mathbf{C}\,|. \tag{2·552}$$

By the Schwarz inequality (2·211) and the remarks following it, this admits a unique solution:

$$\mathbf{J} = (\mathbf{T}-\mathbf{C})/R. \tag{2·553}$$

Thus, by (2·538) again, there is just one point common to the hypersphere and the hyperplane (2·550) and that point is

$$\mathbf{X} = \mathbf{C}+R\mathbf{J} = \mathbf{C}+\mathbf{T}-\mathbf{C} = \mathbf{T}. \tag{2·554}$$

Thus (2·550) is a hyperplane of class 1 tangent to the hypersphere at the point **T**. That it is unique is obvious, because its equation merely expresses a property which we have already seen to belong to all tangent hyperplanes, namely, orthogonality to the radius drawn from the centre to the point of contact (cf. (2·548)).

To summarize: *At any point* **T** *on a hypersphere* (2·549) *there is a unique tangent hyperplane of class* 1 *and its equation is* (2·550).

While the tangent hyperplane of class 1 is unique, the tangent hyperplane of class n $(n>1)$ is not. Consider the equations

$$\left.\begin{aligned} &(\mathbf{X}-\mathbf{T}).(\mathbf{T}-\mathbf{C})=0, \\ &(\mathbf{X}-\mathbf{T}).\mathbf{I}_\rho=0 \quad (\rho=1,2,\ldots,n-1), \end{aligned}\right\} \tag{2·555}$$

the **I**'s being orthonormal but otherwise arbitrary, except that $\mathbf{T}-\mathbf{C}$ is to be linearly independent of them. These equations define

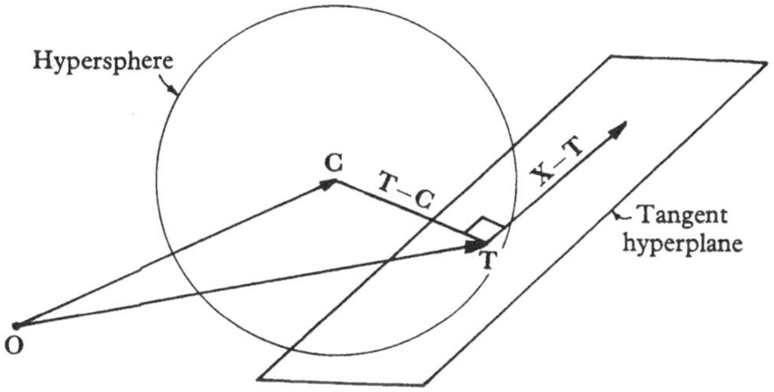

Fig. 2·54. The tangent hyperplane of class 1 at the point **T**.

a hyperplane of class n which obviously passes through the point **T**. If **T** is a point on the hypersphere (2·549), the hyperplane H_n with equations (2·555) meets the hypersphere at **T** and at **T** only, since the equations (2·549) and (2·550) have a unique solution, and (2·550) is the same as the first of (2·555). Thus H_n is tangent to the hypersphere, but it is by no means unique on account of the arbitrariness of the **I**'s.

Fig. 2·54 illustrates the unique tangent hyperplane of class 1.

Tangent hypercones

Consider a hypersphere and a point **A**, not on it. From **A** draw all straight lines tangent to the hypersphere. The points on these straight lines form a subspace of F-space which we call the *tangent hypercone with vertex* **A**. (This common use of the word *vertex* should not be confused with a different use in § 2·7.)

The simplest way to get the equation of a tangent hypercone is to use ordinary space intuition. For simplicity we shall take the vertex at the origin; the hypersphere is $(\mathbf{X}-\mathbf{C})^2 = R^2$. Fig. 2·55 suggests, by projection on the tangent line, the equation

$$\mathbf{C}.\mathbf{X}/|\,\mathbf{X}\,| = (\mathbf{C}^2 - R^2)^{\frac{1}{2}},$$

or equivalently $$(\mathbf{X}.\mathbf{C})^2 = \mathbf{X}^2(\mathbf{C}^2 - R^2). \tag{2·556}$$

This is in fact the correct equation for the tangent hypercone with vertex at the origin; a formal proof is easy to supply.

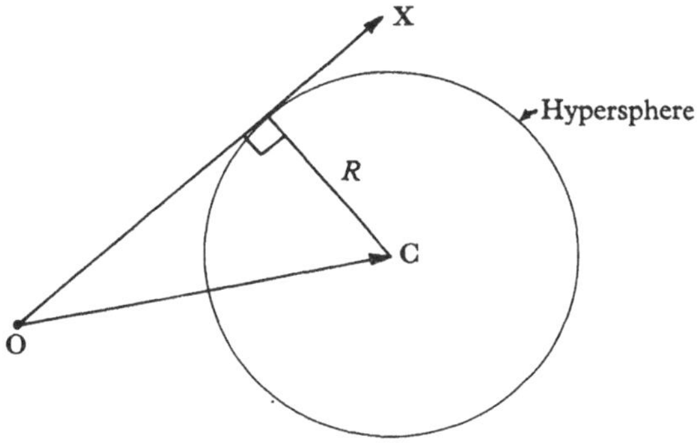

Fig. 2·55. The tangent hypercone from the origin.

Exercises

1. Let **A** and **B** be two given points of F-space. A variable point **X** moves so that the sum of the squares of its distances from **A** and **B** is unity. Show that the locus of **X** does not exist if $(\mathbf{A}-\mathbf{B})^2 > 2$, and that if $(\mathbf{A}-\mathbf{B})^2 < 2$ it is a hypersphere with centre $\frac{1}{2}(\mathbf{A}+\mathbf{B})$ and radius $\frac{1}{2}[2-(\mathbf{A}-\mathbf{B})^2]^{\frac{1}{2}}$.

2. If we change the condition in Exercise 1 to read that the *sum of the distances* (not their squares) from **A** and **B** is to be unity, i.e.

$$|\,\mathbf{X}-\mathbf{A}\,| + |\,\mathbf{X}-\mathbf{B}\,| = 1,$$

and take $|\,\mathbf{A}-\mathbf{B}\,| < 1$, we get a new locus for **X**, which we might call a *hyperellipsoid*. Show that it has an inside and an outside, and that it defines a neighbourhood in the sense that if **X** and **Y** both lie inside it, then $|\,\mathbf{X}-\mathbf{Y}\,|$ has an upper bound, namely, 1.

3. An ordinary plane may be regarded as the limit of a sphere as its radius tends to infinity. Can a hyperplane of class 1 be similarly regarded as the limit of a hypersphere?

4. An infinite number of ordinary spheres can be drawn through three given points, provided they are not collinear. Show that an infinity of hyperspheres can be drawn through n given F-points, provided they do not lie in a linear $(n-2)$-space.

5. Consider the hypersphere $\mathbf{X}^2 = 1$, a point **V** on it, and the hyperplane of class 1 with equation $\mathbf{X}.\mathbf{V} = 0$. Let us project the points of the hypersphere on to the hyperplane by drawing the straight line from **V** through each

point **P** of the hypersphere, this straight line meeting the hyperplane at **Q**. Deduce that

$$\mathbf{Q} = [\mathbf{P} - \mathbf{V}(\mathbf{P}.\mathbf{V})]/(1 - \mathbf{P}.\mathbf{V}).$$

Show that if the point **P** is given two independent infinitesimal displacements $d\mathbf{P}$, $\delta\mathbf{P}$ in the hypersphere, and $d\mathbf{Q}$, $\delta\mathbf{Q}$ are the corresponding infinitesimal displacements of **Q**, then

$$d\mathbf{Q}.\delta\mathbf{Q} = (1 - \mathbf{P}.\mathbf{V})^{-2}\, d\mathbf{P}.\delta\mathbf{P}.$$

Show then that the *angle* between the infinitesimal displacements $d\mathbf{P}$, $\delta\mathbf{P}$ is equal to that between $d\mathbf{Q}$, $\delta\mathbf{Q}$, i.e. the projection is a *conformal* transformation of the hypersphere, a well-known result in ordinary space.

2·6. HYPERCIRCLES

In connexion with the inequality (2·541) we defined a hypercircle of class n as the intersection of a hypersphere and a hyperplane of class n. As the hypercircle plays a fundamental part in the application of function-space to boundary value problems, it will be well to start afresh, dealing first with the simplest case—the hypercircle of class 1.

Hypercircles of class 1

Consider the hypersphere

$$(\mathbf{X} - \mathbf{C}_0)^2 = R_0^2. \tag{2·601}$$

(We use a subscript zero here as we shall want the symbols without a subscript for the centre and radius of the hypercircle.) A *hypercircle of class* 1 is given by the intersection of this hypersphere with a hyperplane of class 1, say

$$\mathbf{X}.\mathbf{I} = a, \tag{2·602}$$

I being a unit vector. By (2·541) the condition for an intersection, and not mere tangency, is

$$|a - \mathbf{C}_0.\mathbf{I}| < R_0; \tag{2·603}$$

we shall suppose this satisfied.

The two equations (2·601) and (2·602), taken together, may be regarded as the equations of the hypercircle, but a *parametric equation* is more useful. This we shall now obtain, using a method suggested by the consideration in ordinary space of the circular intersection of a sphere and a plane.

Starting from the centre $\mathbf{C}_0$ of the hypersphere, we proceed in the direction of **I** (that is, orthogonal to the hyperplane) until we reach the hyperplane; let the point so obtained be **C**. This means that we set up a vector

$$\mathbf{C} = \mathbf{C}_0 + b\mathbf{I}, \tag{2·604}$$

the constant b being chosen so that $\mathbf{C}.\mathbf{I}=a$ or

$$\mathbf{C}_0.\mathbf{I}+b=a. \tag{2·605}$$

Thus $b=a-\mathbf{C}_0.\mathbf{I}$ and so

$$\mathbf{C}=\mathbf{C}_0+(a-\mathbf{C}_0.\mathbf{I})\,\mathbf{I}. \tag{2·606}$$

We check and note for reference that

$$\mathbf{C}.\mathbf{I}=a. \tag{2·607}$$

Now let $\mathbf{X}$ be any point on the hypercircle. By (2·602) and (2·607) we have

$$(\mathbf{X}-\mathbf{C}).\mathbf{I}=a-a=0, \tag{2·608}$$

and hence by (2·606) $\quad(\mathbf{X}-\mathbf{C}).(\mathbf{C}-\mathbf{C}_0)=0. \qquad$ (2·609)

Then by (2·601)

$$R_0^2=(\mathbf{X}-\mathbf{C}_0)^2=(\mathbf{X}-\mathbf{C}+\mathbf{C}-\mathbf{C}_0)^2, \tag{2·610}$$

so that, by (2·609), $\quad R_0^2=(\mathbf{X}-\mathbf{C})^2+(\mathbf{C}-\mathbf{C}_0)^2. \qquad$ (2·611)

Therefore $\quad(\mathbf{X}-\mathbf{C})^2=R_0^2-(a-\mathbf{C}_0.\mathbf{I})^2, \qquad$ (2·612)

in which the right-hand side involves only known quantities.

This last equation means that every point on the hypercircle is at the same distance R from $\mathbf{C}$, where

$$R^2=R_0^2-(a-\mathbf{C}_0.\mathbf{I})^2, \tag{2·613}$$

and (2·608) tells us that the vector drawn from $\mathbf{C}$ to any point on the hypercircle is orthogonal to $\mathbf{I}$. It is therefore clear that we can express in the following *parametric form* the equation of the hypercircle of class 1 which is the intersection of the hypersphere (2·601) and the hyperplane (2·602):

$$\mathbf{X}=\mathbf{C}+R\mathbf{J}, \quad \mathbf{J}.\mathbf{I}=0, \quad \mathbf{J}^2=1. \tag{2·614}$$

Here $\mathbf{C}$ is the *centre of the hypercircle* and is given by (2·606), and R is the *radius of the hypercircle*, given by (2·613); $\mathbf{J}$ is a parametric vector, arbitrary except for the stated conditions, i.e. it is a unit vector orthogonal to $\mathbf{I}$.

The hypercircle is shown in Fig. 2·61.

Hypercircles of class n

The equations

$$(\mathbf{X}-\mathbf{C}_0)^2=R_0^2, \quad \mathbf{X}.\mathbf{I}_\rho=a_\rho \quad (\rho=1,2,\ldots,n), \tag{2·615}$$

define a *hypercircle of class n* as the intersection of a hypersphere and a hyperplane of class n. In these equations $\mathbf{C}_0$ and the $\mathbf{I}$'s are fixed vectors, the **I**'s forming an orthonormal set; R_0 and the a's are constants. Thus there are involved $n+1$ vectors and $n+1$ scalars. We shall suppose that an inequality as in (2·541) is satisfied, so that the hypercircle exists.

We may call (2·615) the *standard equations* of a hypercircle of class n; as above for the hypercircle of class 1, we shall now obtain more useful parametric equations.

We form a vector

$$\mathbf{C}=\mathbf{C}_0+\sum_{\rho=1}^{n} b_\rho \mathbf{I}_\rho, \tag{2·616}$$

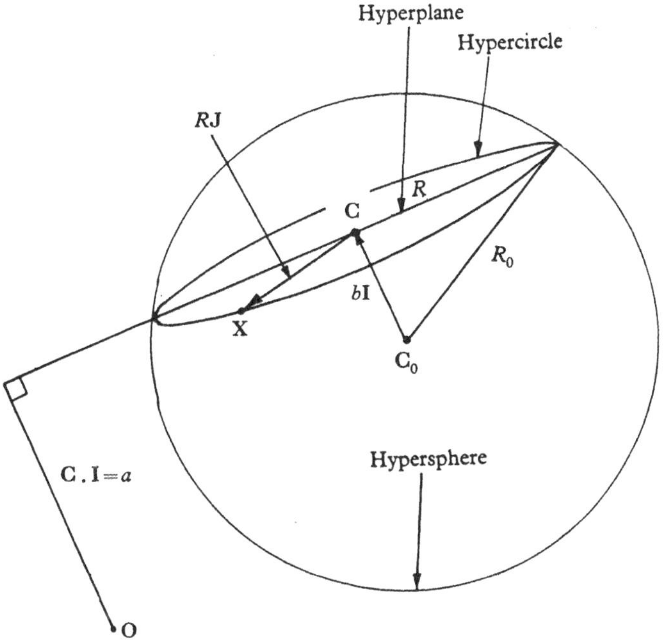

Fig. 2·61. Hypersphere, hyperplane and hypercircle. $\mathbf{C}_0$ is the centre of the hypersphere and R_0 its radius. **C** is the centre of the hypercircle and R its radius.

choosing the b's so that

$$\mathbf{C}.\mathbf{I}_\rho=a_\rho \quad (\rho=1,2,\ldots,n). \tag{2·617}$$

This is done by making

$$b_\rho=a_\rho-\mathbf{C}_0.\mathbf{I}_\rho \quad (\rho=1,2,\ldots,n), \tag{2·618}$$

and we have

$$\mathbf{C}=\mathbf{C}_0+\sum_{\rho=1}^{n}(a_\rho-\mathbf{C}_0.\mathbf{I}_\rho)\,\mathbf{I}_\rho. \tag{2·619}$$

It is easy to see then that for any **X** on the hypercircle we have

$$(\mathbf{X}-\mathbf{C}).\mathbf{I}_\rho=0 \quad (\rho=1,2,\ldots,n), \tag{2·620}$$

and hence $$(\mathbf{X}-\mathbf{C}).(\mathbf{C}-\mathbf{C}_0)=0. \tag{2·621}$$

Therefore $$\begin{aligned}(\mathbf{X}-\mathbf{C}_0)^2 &= (\mathbf{X}-\mathbf{C}+\mathbf{C}-\mathbf{C}_0)^2 \\ &= (\mathbf{X}-\mathbf{C})^2+(\mathbf{C}-\mathbf{C}_0)^2.\end{aligned} \tag{2·622}$$

Now the left-hand side is R_0^2 by (2·615) and the last term may be computed by (2·619); so $$(\mathbf{X}-\mathbf{C})^2=R^2, \tag{2·623}$$

where $$R^2=R_0^2-\sum_{\rho=1}^{n}(a_\rho-\mathbf{C}_0.\mathbf{I}_\rho)^2. \tag{2·624}$$

It is clear from (2·620) and (2·623) that we may write the equation of a hypercircle of class n in *parametric form*:

$$\left.\begin{aligned}&\mathbf{X}=\mathbf{C}+R\mathbf{J}, \quad \mathbf{J}^2=1, \\ &\mathbf{J}.\mathbf{I}_\rho=0 \quad (\rho=1,2,\ldots,n).\end{aligned}\right\} \tag{2·625}$$

Here $\mathbf{J}$ is a parametric vector, arbitrary except for the condition that it shall be a unit vector orthogonal to the orthonormal set of $\mathbf{I}$'s.

We note that (2·625) contains $n+1$ fixed vectors but only one scalar, R, instead of the $n+1$ scalars of (2·615). The reason is that there are many hyperspheres passing through a given hypercircle (as, in ordinary space, many spheres pass through a given circle), and this arbitrariness is reflected in the excess of constants in (2·615). On the other hand, (2·625) are unique equations for a given hypercircle of class n, except for orthogonal transformations of the $\mathbf{I}$'s. Just as in ordinary geometry we often think of a circle as a curve in its own right, and not as the intersection of a sphere and a plane, so we may think of a hypercircle of class n as a subspace of F-space defined by (2·625), without regard to the fact that it is the intersection of a hypersphere and a hyperplane.

In (2·625) we call the point with position-vector $\mathbf{C}$ the *centre* of the hypercircle and R its *radius*. We recall that $\mathbf{C}$ is given in terms of the hypersphere and the hyperplane by (2·619) and R by (2·624).

The hypercircle is contained in a hyperplane of class n which may be described in terms of the elements occurring in (2·625); it has the standard equations

$$\mathbf{X}.\mathbf{I}_\rho=\mathbf{C}.\mathbf{I}_\rho \quad (\rho=1,2,\ldots,n), \tag{2·626}$$

and the parametric equation

$$\mathbf{X}=\mathbf{C}+\mathbf{Y}, \quad \mathbf{Y}.\mathbf{I}_\rho=0 \quad (\rho=1,2,\ldots,n), \tag{2·627}$$

where $\mathbf{Y}$ is a parametric vector.

Here for reference are the main formulae connected with the hypercircle:

Hypercircle of class n

$$\left.\begin{array}{ll}
\text{Standard equations:} & \\
\quad (\mathbf{X}-\mathbf{C}_0)^2 = R_0^2, \quad \mathbf{X}.\mathbf{I}_\rho = a_\rho \quad (\rho = 1, 2, \ldots, n). & \\
\text{Parametric equation:} & \\
\mathbf{X} = \mathbf{C} + R\mathbf{J}, \quad \mathbf{J}^2 = 1, \quad \mathbf{J}.\mathbf{I}_\rho = 0 \quad (\rho = 1, 2, \ldots, n). & \\
\text{Centre:} \qquad \mathbf{C} = \mathbf{C}_0 + \sum_{\rho=1}^{n} (a_\rho - \mathbf{C}_0.\mathbf{I}_\rho)\,\mathbf{I}_\rho. & \\
\text{Radius:} \qquad R^2 = R_0^2 - \sum_{\rho=1}^{n} (a_\rho - \mathbf{C}_0.\mathbf{I}_\rho)^2. &
\end{array}\right\} \quad (2{\cdot}628)$$

Bounds on $\mathbf{X}^2$ as $\mathbf{X}$ ranges on a hypercircle

In (2·515) we found bounds on $\mathbf{X}^2$ as $\mathbf{X}$ ranges on a hypersphere; we now take up the same problem with a hypercircle instead of a hypersphere, using for the hypercircle the parametric equation of (2·628).

Let us resolve $\mathbf{C}$ in, and orthogonal to, the linear n-space of the $\mathbf{I}$'s as in (2·339):

$$\mathbf{C} = \mathbf{C}_0' + \sum_{\rho=1}^{n} (\mathbf{C}.\mathbf{I}_\rho)\,\mathbf{I}_\rho, \quad \mathbf{C}_0'.\mathbf{I}_\rho = 0 \quad (\rho = 1, 2, \ldots, n). \quad (2{\cdot}629)$$

(We have written $\mathbf{C}_0'$ here instead of the more natural $\mathbf{C}_0$ because $\mathbf{C}_0$ has been used in (2·628) for the centre of a hypersphere from which the hypercircle is generated; the vector $\mathbf{C}_0'$ is in fact defined in terms of known quantities by the first of (2·629).) Then, as $\mathbf{X}$ ranges on the hypercircle, we have

$$\mathbf{X}^2 = (\mathbf{C} + R\mathbf{J})^2 = \mathbf{C}^2 + 2R\mathbf{C}_0'.\mathbf{J} + R^2, \quad (2{\cdot}630)$$

and so, by the Schwarz inequality (2·211),

$$R^2 + \mathbf{C}^2 - 2R\,|\,\mathbf{C}_0'\,| \leqslant \mathbf{X}^2 \leqslant R^2 + \mathbf{C}^2 + 2R\,|\,\mathbf{C}_0'\,|, \quad (2{\cdot}631)$$

where

$$|\,\mathbf{C}_0'\,|^2 = \mathbf{C}_0'^2 = \mathbf{C}^2 - \sum_{\rho=1}^{n} (\mathbf{C}.\mathbf{I}_\rho)^2. \quad (2{\cdot}632)$$

The lower and upper bounds for $\mathbf{X}^2$, as given in (2·631), are attained respectively at (see Fig. 2·62)

$$\mathbf{X}_1 = \mathbf{C} - R\mathbf{C}_0'/|\,\mathbf{C}_0'\,|, \quad \mathbf{X}_2 = \mathbf{C} + R\mathbf{C}_0'/|\,\mathbf{C}_0'\,|, \quad (2{\cdot}633)$$

where

$$\mathbf{C}_0' = \mathbf{C} - \sum_{\rho=1}^{n} (\mathbf{C}.\mathbf{I}_\rho)\,\mathbf{I}_\rho. \quad (2{\cdot}634)$$

We note that the bounds (2·631) agree with those obtained in (2·515) for a hypersphere when we put $\mathbf{C}_0' = \mathbf{C}$.

Bounds on **X.G** *as* **X** *ranges on a hypercircle*

In (2·517) we obtained bounds on **X.G** (**G** being any vector) as **X** ranges on a hypersphere. Now we work with a hypercircle instead, using a method which is a combination of those used in getting (2·517) and (2·631). The result is easy to obtain, and will be quoted here without details of proof.

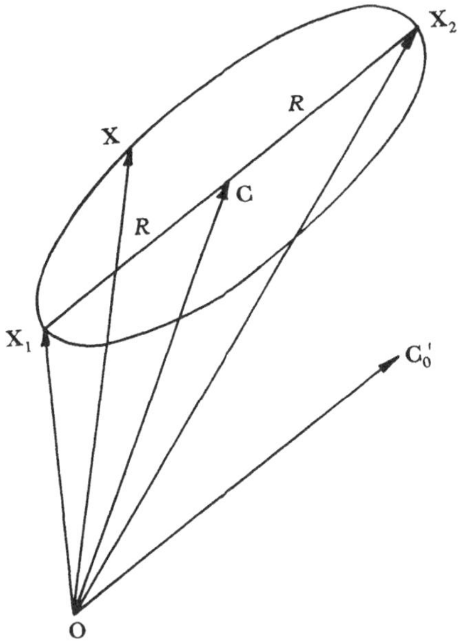

Fig. 2·62. Bounds on $\mathbf{X}^2$ as **X** ranges on hypercircle. $\mathbf{X}_1$ gives lower bound and $\mathbf{X}_2$ upper bound.

As **X** ranges on the hypercircle of class n with the parametric equation as in (2·628), we have, for any fixed vector **G**,

$$\mathbf{C}.\mathbf{G}-R\,|\,\mathbf{G}_0\,|\leqslant\mathbf{X}.\mathbf{G}\leqslant\mathbf{C}.\mathbf{G}+R\,|\,\mathbf{G}_0\,|, \tag{2·635}$$

where
$$|\,\mathbf{G}_0\,|^2=\mathbf{G}_0^2=\mathbf{G}^2-\sum_{\rho=1}^{n}(\mathbf{G}.\mathbf{I}_\rho)^2. \tag{2·636}$$

The lower and upper bounds are attained respectively (see Fig. 2·63) at

$$\mathbf{X}_1=\mathbf{C}-R\mathbf{G}_0/|\,\mathbf{G}_0\,|,\quad \mathbf{X}_2=\mathbf{C}+R\mathbf{G}_0/|\,\mathbf{G}_0\,|, \tag{2·637}$$

where
$$\mathbf{G}_0=\mathbf{G}-\sum_{\rho=1}^{n}(\mathbf{G}.\mathbf{I}_\rho)\,\mathbf{I}_\rho. \tag{2·638}$$

Example

Suppose that we are interested in a function $f(x)$ for the range $-1 \leqslant x \leqslant 1$; we seek to evaluate the function from the values of the two integrals

$$\int_{-1}^{1} [f(x)]^2\,dx = 1, \quad \int_{-1}^{1} xf(x)\,dx = \tfrac{102}{125}. \qquad (2\cdot639)$$

At first sight we seem to have much too little information. However, we recognize in (2·639) the equations of a hypercircle, and that puts us on the track of an approximate evaluation of $f(x)$.

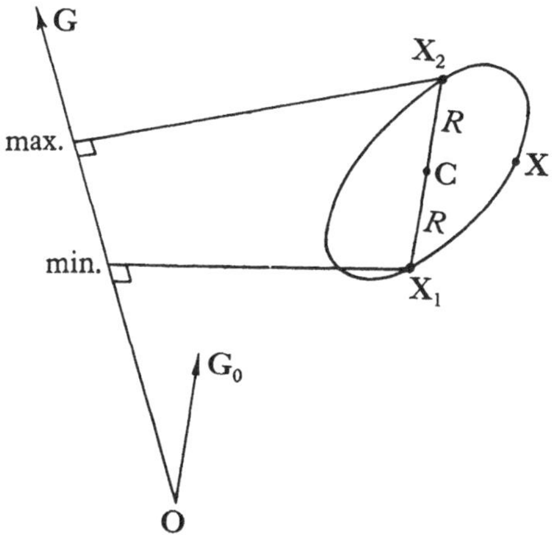

Fig. 2·63. Bounds on **X**.**G** as **X** ranges on hypercircle. $\mathbf{X}_1$ gives lower bound and $\mathbf{X}_2$ upper bound.

For P-space we take $-1 \leqslant x \leqslant 1$, for F-vector a function of x, and for scalar product that of Hilbert (2·102). We write

$$\mathbf{X} \leftrightarrow f(x), \quad \mathbf{S}_1 \leftrightarrow x;$$

then

$$\mathbf{S}_1^2 = \int_{-1}^{1} x^2\,dx = \tfrac{2}{3},$$

and so we normalize: $\mathbf{I}_1 = \mathbf{S}_1/|\mathbf{S}_1| = (\tfrac{3}{2})^{\frac{1}{2}}\,\mathbf{S}_1 \leftrightarrow (\tfrac{3}{2})^{\frac{1}{2}} x.$

Then (2·639) may be written

$$(\mathbf{X} - \mathbf{C}_0)^2 = R_0^2, \quad \mathbf{X}.\mathbf{I}_1 = a_1,$$

$$\mathbf{C}_0 = \mathbf{O}, \quad R_0 = 1, \quad a_1 = \tfrac{102}{125}(\tfrac{2}{3})^{\frac{1}{2}}.$$

Thus **X** is on a hypercircle of class 1, and by (2·628) its parametric equation is

$$\mathbf{X} = \mathbf{C} + R\mathbf{J}, \quad \mathbf{J}^2 = 1, \quad \mathbf{J}.\mathbf{I}_1 = 0,$$

$$\mathbf{C} = a_1 \mathbf{I}_1 \leftrightarrow \tfrac{153}{125}x,$$

$$R^2 = R_0^2 - a_1^2 = \tfrac{38}{31250}.$$

Emerging from function-space, we then assert the following implication of (2·639):

$$f(x) = \tfrac{153}{125}x + (\tfrac{38}{31250})^{\frac{1}{2}}\,g(x),$$

where $g(x)$ is some function such that

$$\int_{-1}^{1} [g(x)]^2 \, dx = 1, \quad \int_{-1}^{1} x g(x) \, dx = 0.$$

We may then say that $f(x)$ is approximately $(\frac{153}{125})\,x$, the meaning of the word *approximately* being clear from the context. Obviously, nothing can be asserted about the pointwise behaviour of $g(x)$ or $f(x)$.

Note that the approximation is linear in x, not through any predilection for a linear function, but because x occurs in the second integrand in (2·639); had it been x^2, then our approximation would have been proportional to x^2.

In practical calculations, integrals such as those in (2·639) may occur as decimals computed to, say, four decimal places: the values might read

$$1{\cdot}0000, \quad 0{\cdot}8160.$$

There are then two procedures. We may carry out our calculations to, say, six decimal places to safeguard the fourth place, or we may regard the data as *inequalities*, the first integral lying between

$$0{\cdot}99995 \quad \text{and} \quad 1{\cdot}00005$$

and the second between

$$0{\cdot}81595 \quad \text{and} \quad 0{\cdot}81605.$$

This information confines the point **X** to what may be called a *hyperring*, a region bounded by two hyperspheres and two hyperplanes, and we can carry out our further reasoning on this basis with absolute precision. The former method is the easier and more practical, but the latter clears up any confusion there may be about significant figures.

Exercises

1. A function $f(x)$ satisfies the following four conditions:

$$\int_{-1}^{1} [f(x)]^2 \, dx = 1, \quad \int_{-1}^{1} f(x) \, dx = (\tfrac{2}{3})^{\frac{1}{2}},$$

$$\int_{-1}^{1} x f(x) \, dx = (\tfrac{2}{9})^{\frac{1}{2}}, \quad \int_{-1}^{1} (3x^2 - 1) f(x) \, dx = (\tfrac{8}{15})^{\frac{1}{2}}.$$

Referring to (2·327) and (2·628), show that an F-point corresponding to $f(x)$ is confined to a hypercircle of zero radius, and hence that

$$f(x) = 6^{-\frac{1}{2}}[1 + 3^{\frac{1}{2}} x + 5^{\frac{1}{2}} \tfrac{1}{2}(3x^2 - 1)].$$

2. Let **A** and **B** be two fixed F-points. Then the equation

$$(\mathbf{X} - \mathbf{A})^2 = (\mathbf{X} - \mathbf{B})^2$$

defines a hyperplane of class 1 equidistant from the two points. Let P be a hypercircle of class 1 passing through **A** and **B**. Show that the centre of P lies on the above hyperplane, but that the centre of a hypercircle cannot be precisely located by drawing any finite number of hyperplanes equidistant from pairs of points on it (cf. p. 69).

3. Show that the distance between any two points on a hypercircle cannot exceed twice its radius.

4. Given three F-points, **A**, **B**, **D**, satisfying

$$(\mathbf{A} - \mathbf{D}) . (\mathbf{B} - \mathbf{D}) = 0,$$

show that an infinite number of hypercircles of class 1 may be constructed to pass through them, and that the centre **C** and radius R of any such hypercircle satisfy $\mathbf{P}\,.\,(\mathbf{A}-\mathbf{D}) = \mathbf{P}\,.\,(\mathbf{B}-\mathbf{D}) = 0$ and $R^2 = \frac{1}{4}(\mathbf{A}-\mathbf{B})^2 + \mathbf{P}^2$, where $\mathbf{P} = \mathbf{C} - \frac{1}{2}(\mathbf{A}+\mathbf{B})$.

5. Given that $f(x)$ satisfies

$$\int_{-1}^{1} [f(x) - 2x^2]^2\,dx = 1, \qquad \int_{-1}^{1} f(x)\,dx = 1,$$

find upper and lower bounds for

$$\int_{-1}^{1} [f(x)]^2\,dx \quad \text{and} \quad \int_{-1}^{1} xf(x)\,dx.$$

[*Answer*: $\frac{1}{45}[97 \pm 8(85)^{\frac{1}{2}}]$ and $\pm(\frac{17}{27})^{\frac{1}{2}}$.]

2·7. THE KEY TO THE HYPERCIRCLE METHOD

As we shall see later, many boundary value problems of mathematical physics may be stated in the form: *It is required to find the intersection of two orthogonal linear subspaces of a function-space.* That statement turns the analytical problem into a geometrical one.

The usual analytical method is to set up an infinite process (such as solution by an infinite series). This amounts to getting an infinite sequence of points in function-space, this sequence having the required intersection as a limit. In general, such a process is carried out in only one of the linear subspaces. For example, if we have to find a harmonic function to satisfy certain boundary conditions, we set up a sequence of harmonic functions (points in one of the two linear subspaces), but the boundary conditions are satisfied only by the limit of the sequence.

In the method of the hypercircle we use only a finite number of points, some in one of the two linear subspaces and some in the other. In general, a limiting process is not used, and we do not actually find the solution, i.e. the point of intersection of the two linear subspaces. But although we do not find it, we learn something about its position, namely, that it is located on a certain hypercircle in function-space. This provides us with inequalities bounding the solution in a mean square sense (and in some cases pointwise), and if the radius of the hypercircle is small, we might accept its centre, or any point on it, as a good approximation to the unknown solution.

There are many cases where it is hard to see how to set up an infinite process to yield an exact solution in a useful form, but where it is easy to apply the method of the hypercircle to obtain information about the solution. There is, of course, no reason why we should not combine the method of the hypercircle with a limiting process. This is done, for, example in §4·2, where the torsion of a rectangle is worked out by the hypercircle method.

Since the method is available for a wide class of problems, it is a real economy to work out the general geometrical theory once for all. In order that everything may be clear, we shall recall here some important concepts about linear subspaces and orthogonality.

Linear subspaces and orthogonality

As always in this Part II, function-space has a positive-definite metric.

The following definitions have already been given; they are summarized here for convenience.

Linear subspace (L): If L contains the points $\mathbf{X}$ and $\mathbf{Y}$, then it also contains the points $a\mathbf{X}+b\mathbf{Y}$ for all a and b satisfying $a+b=1$ (i.e. it contains the straight line joining $\mathbf{X}$ and $\mathbf{Y}$).

Two important particular types of linear subspace are the linear n-space L_n (a linear subspace of finite dimensionality) and the hyperplane.

Linear n-space (L_n):
$$\mathbf{X}=\mathbf{A}+\sum_{\rho=1}^{n} a_\rho \mathbf{T}_\rho, \tag{2·701}$$

where $\mathbf{A}$ is a fixed vector, $\mathbf{T}_\rho$ are n linearly independent fixed vectors, and a_ρ are n variable parameters.

If $\mathbf{T}_\rho$ are not linearly independent, (2·701) still defines a linear subspace, but with dimensionality less than n.

Hyperplane of class n (H_n):

$$\mathbf{X}.\mathbf{S}_\rho=b_\rho \quad (\rho=1, 2, \ldots, n), \tag{2·702}$$

where $\mathbf{S}_\rho$ are n linearly independent fixed vectors and b_ρ are n fixed numbers.

If $\mathbf{S}_\rho$ are not linearly independent, then these equations are either inconsistent (and so define no subspace) or they define a hyperplane of class less than n.

Orthogonality of linear subspaces (L', L''): L' and L'' are orthogonal if, and only if, every vector $\mathbf{T}'$ lying in L' is orthogonal to every vector $\mathbf{T}''$ lying in L''.

A vector $\mathbf{T}'$ lies in L' if it is the difference of the position-vectors of two points in L'.

We are about to discuss several situations which are rather like one another, and there is a danger of getting them mixed up. To keep matters clear, let us summarize the steps crudely as follows, using L'_r and L''_s to denote orthogonal linear subspaces of finite dimensionality and L' and L'' to cover the case where the dimensionalities are in general infinite, but may be finite:

(i) The approach (minimum distance) of L' and L'' (equations (2·703) to (2·711)).

(ii) Specialization of (i) to finite dimensionalities, so that algebraic methods may be used (equations (2·712) to (2·725)).

(iii) Intersection of L' and L'' (equations (2·726) to (2·732)).

(iv) Intersection of L' and L'' discussed in terms of the approach of L'_r and L''_s contained in them respectively (equations (2·733) to (2·756)). This is the essence of the hypercircle method.

(v) Specialization of (iv) to the case where L'' contains the origin (equations (2·757) to (2·770)). This introduces simplifications which are very useful in the applications.

The vertices **V**′, **V**″ *of two non-intersecting orthogonal linear subspaces* L', L''

Fig. 2·71 shows schematically two linear subspaces (L', L'') which are supposed to be orthogonal and to have no point in common. **S**′ is some point in L' and **S**″ some point in L''.

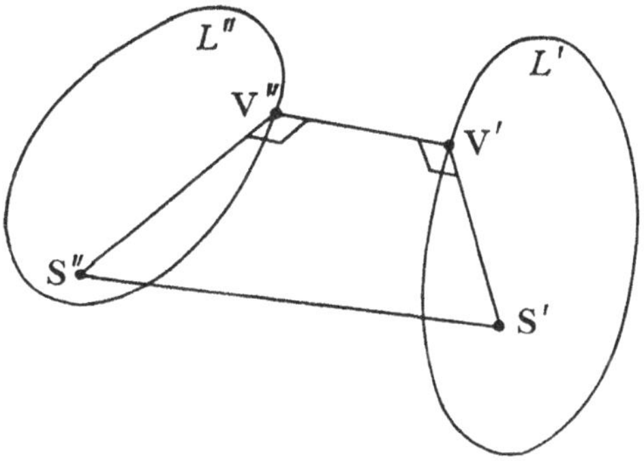

Fig. 2·71. The vertices **V**′, **V**″ of two non-intersecting orthogonal linear subspaces L', L''.

We consider the closest approach of L' and L'' by studying the squared distance $(\mathbf{S}'-\mathbf{S}'')^2$ as **S**′ and **S**″ range through L' and L'' respectively. Since L' and L'' do not intersect, this quantity cannot vanish. It will therefore have a lower bound, greater than zero, and we shall suppose that this lower bound is attained for, say, $\mathbf{S}'=\mathbf{V}'$ and $\mathbf{S}''=\mathbf{V}''$. We call the points **V**′ and **V**″ the *vertices* of L' and L'' respectively; they are the points of closest approach of the two linear subspaces.

If the linear subspaces are straight lines (as they may be), the appropriate diagram is as in Fig. 2·72; we could make a satisfactory model in Euclidean 3-space.

We shall now prove an important result: *If L', L'' are two non-intersecting orthogonal linear subspaces, the line joining their vertices* ($\mathbf{V}'$, $\mathbf{V}''$) *is orthogonal to both L' and L''.*

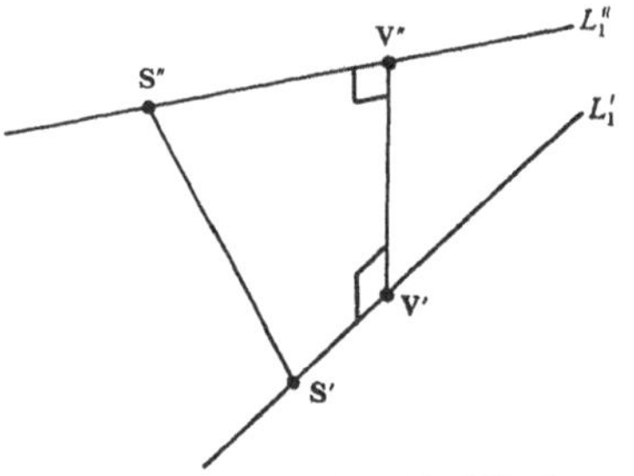

Fig. 2·72. The vertices $\mathbf{V}'$, $\mathbf{V}''$ of two non-intersecting orthogonal lines L', L''.

Two proofs will be given—the first rather intuitive, the second formal.

Let $\mathbf{P}$ at $\mathbf{A}$ (Fig. 2·73) be any vector lying in L'. Then from the linear character of L' we know that the vector $\mathbf{P}$ at $\mathbf{V}'$ also lies in L', and so does the whole line defined by $\mathbf{P}$ at $\mathbf{V}'$. We consider the angle between the vector $\mathbf{V}'-\mathbf{V}''$ and this line. If it is not a right angle, we can find some point $\mathbf{W}$ (other than $\mathbf{V}'$) on the line such that

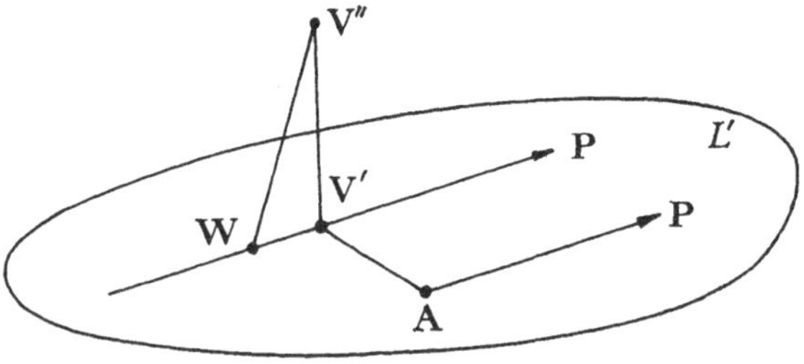

Fig. 2·73. $\mathbf{V}'-\mathbf{V}''$ is orthogonal to every vector $\mathbf{P}$ lying in L'.

$\mathbf{W}-\mathbf{V}''$ makes a right angle with the line, i.e. $(\mathbf{W}-\mathbf{V}'').\mathbf{P}=0$. We have then, as in Fig. 2·73, a triangle with a right angle at $\mathbf{W}$, and consequently

$$(\mathbf{W}-\mathbf{V}'')^2<(\mathbf{V}'-\mathbf{V}'')^2. \qquad (2\cdot703)$$

Thus $\mathbf{W}$ is a point on L' closer to L'' than $\mathbf{V}'$ is; this violates the assumption that $\mathbf{V}'$ and $\mathbf{V}''$ are vertices, and we conclude that $\mathbf{P}$ must be orthogonal to $\mathbf{V}'-\mathbf{V}''$. This proves the theorem, since $\mathbf{P}$ is *any* vector lying in L', and the same result holds for any vector lying in L''.

The formal proof is as follows. Let $\mathbf{S}'$ be any point on L'. Since L' is a linear subspace, the point $a\mathbf{S}'+b\mathbf{V}'$ $(a+b=1)$ is on it, and so, by the minimum property of $\mathbf{V}'$, $\mathbf{V}''$, we have

$$\begin{aligned}(\mathbf{V}'-\mathbf{V}'')^2 &\leqslant (\mathbf{V}''-a\mathbf{S}'-b\mathbf{V}')^2\\ &=[(\mathbf{V}''-\mathbf{V}')-a(\mathbf{S}'-\mathbf{V}')]^2\\ &=(\mathbf{V}''-\mathbf{V}')^2-2a(\mathbf{V}''-\mathbf{V}').(\mathbf{S}'-\mathbf{V}')+a^2(\mathbf{S}'-\mathbf{V}')^2. \qquad (2\cdot704)\end{aligned}$$

Consequently $$a^2(\mathbf{S}'-\mathbf{V}')^2-2a(\mathbf{V}''-\mathbf{V}').(\mathbf{S}'-\mathbf{V}')\geqslant 0 \qquad (2\cdot705)$$

for all real values of a, and so the coefficient of a must vanish; thus

$$(\mathbf{S}'-\mathbf{V}').(\mathbf{V}'-\mathbf{V}'')=0. \qquad (2\cdot706)$$

This expresses the orthogonality of $\mathbf{V}'-\mathbf{V}''$ to all vectors drawn in L' from the point $\mathbf{V}'$, and the orthogonality to all vectors lying in L' follows immediately. Similarly, $\mathbf{V}'-\mathbf{V}''$ is orthogonal to all vectors lying in L'', and so the theorem is proved.

So far we have not used the assumed orthogonality of L' and L''; what we have proved holds even if they are not orthogonal. Let us now use that orthogonality; in view of it and the result established above, we may write the three equations

$$\left.\begin{aligned}(\mathbf{S}'-\mathbf{V}').(\mathbf{V}'-\mathbf{V}'')&=0,\\(\mathbf{S}''-\mathbf{V}'').(\mathbf{V}'-\mathbf{V}'')&=0,\\(\mathbf{S}'-\mathbf{V}').(\mathbf{S}''-\mathbf{V}'')&=0,\end{aligned}\right\} \qquad (2\cdot707)$$

where $\mathbf{S}'$ is any point on L' and $\mathbf{S}''$ any point on L''. (Check these formulae in Fig. 2·72.)

Uniqueness theorem for vertices

We shall now prove a uniqueness theorem: *The vertices* $\mathbf{V}'$, $\mathbf{V}''$ *of two non-intersecting orthogonal linear subspaces are unique.*

Let $\mathbf{V}'$, $\mathbf{V}''$ be vertices and $\mathbf{S}'$, $\mathbf{S}''$ any points on L', L'' respectively. We may write the identity

$$\mathbf{S}'-\mathbf{S}''=(\mathbf{S}'-\mathbf{V}')+(\mathbf{V}'-\mathbf{V}'')+(\mathbf{V}''-\mathbf{S}''). \qquad (2\cdot708)$$

Squaring and using (2·707), we get

$$(\mathbf{S}'-\mathbf{S}'')^2=(\mathbf{V}'-\mathbf{V}'')^2+(\mathbf{S}'-\mathbf{V}')^2+(\mathbf{S}''-\mathbf{V}'')^2. \qquad (2\cdot709)$$

If $\mathbf{S}'$ and $\mathbf{S}''$ are vertices as well as $\mathbf{V}'$ and $\mathbf{V}''$, then

$$(\mathbf{S}'-\mathbf{S}'')^2=(\mathbf{V}'-\mathbf{V}'')^2,$$

the minimum value. Hence the two other squares in (2·709) must vanish, giving $\mathbf{S}'=\mathbf{V}'$, $\mathbf{S}''=\mathbf{V}''$; the uniqueness is proved.

Inequalities and minimum principles for two non-intersecting orthogonal linear subspaces L', L''

The following inequalities are immediate consequences of (2·709), $\mathbf{S}'$ and $\mathbf{S}''$ being any points on L' and L'' respectively:

$$\left.\begin{aligned}(\mathbf{S}'-\mathbf{S}'')^2&\geqslant(\mathbf{V}'-\mathbf{V}'')^2,\\(\mathbf{S}'-\mathbf{S}'')^2&\geqslant(\mathbf{S}'-\mathbf{V}')^2,\\(\mathbf{S}'-\mathbf{S}'')^2&\geqslant(\mathbf{S}''-\mathbf{V}'')^2.\end{aligned}\right\} \qquad (2\cdot710)$$

The following minimum principles are also obvious from (2·709):

$$\left.\begin{array}{l}\text{I. For } \mathbf{S}' \text{ free in } L' \text{ and } \mathbf{S}'' \text{ free in } L'',\ (\mathbf{S}'-\mathbf{S}'')^2 \text{ is minimized by } \mathbf{S}'=\mathbf{V}',\ \mathbf{S}''=\mathbf{V}''.\\ \text{II. For } \mathbf{S}' \text{ free in } L' \text{ and } \mathbf{S}'' \text{ fixed in } L'',\ (\mathbf{S}'-\mathbf{S}'')^2 \text{ (or equivalently } (\mathbf{S}'^2-2\mathbf{S}'.\mathbf{S}'')) \text{ is minimized by } \mathbf{S}'=\mathbf{V}'.\\ \text{III. For } \mathbf{S}' \text{ fixed in } L' \text{ and } \mathbf{S}'' \text{ free in } L'',\ (\mathbf{S}'-\mathbf{S}'')^2 \text{ (or equivalently } (\mathbf{S}''^2-2\mathbf{S}'.\mathbf{S}'')) \text{ is minimized by } \mathbf{S}''=\mathbf{V}''.\end{array}\right\} \quad (2\cdot711)$$

The problem of finding the vertices of two general non-intersecting orthogonal linear subspaces belongs to the calculus of variations. But if the subspaces are of finite dimensionality, their vertices may be found algebraically, as we shall now see.

The vertices $\mathbf{V}'$, $\mathbf{V}''$ *of two non-intersecting orthogonal linear subspaces of finite dimensionality* L'_r, L''_s

For the approximate solution of boundary value problems we need to find the vertices of a pair of linear subspaces of finite dimensionality, orthogonal and non-intersecting. This problem will now be reduced to the solution of a set of linear algebraic equations, or, equivalently, to a question of orthonormalization.

As equations of the two linear subspaces, let us take

$$\left.\begin{array}{ll} L'_r\colon & \mathbf{X}=\mathbf{S}'_0+\mathbf{T}', \quad \mathbf{T}'=\sum_{\rho=1}^{r} a'_\rho \mathbf{T}'_\rho;\\ L''_s\colon & \mathbf{X}=\mathbf{S}''_0+\mathbf{T}'', \quad \mathbf{T}''=\sum_{\sigma=1}^{s} a''_\sigma \mathbf{T}''_\sigma.\end{array}\right\} \quad (2\cdot712)$$

Here $\mathbf{S}'_0$, $\mathbf{S}''_0$ are points on the two subspaces; $\mathbf{T}'_\rho$, $\mathbf{T}''_\sigma$ are vectors lying in them and satisfying the orthogonality conditions resulting from the assumed orthogonality of the two subspaces:

$$\mathbf{T}'_\rho.\mathbf{T}''_\sigma=0 \quad (\rho=1,2,\ldots,r;\ \sigma=1,2,\ldots,s). \qquad (2\cdot713)$$

The parametric vectors $\mathbf{T}'$, $\mathbf{T}''$ have been inserted for notational convenience; $\mathbf{T}'$ is a general vector lying in L'_r and $\mathbf{T}''$ a general vector lying in L''_s. By virtue of (2·713) they satisfy

$$\mathbf{T}'.\mathbf{T}''=0. \qquad (2\cdot714)$$

The algebra of the problem is greatly simplified if $\mathbf{T}'_\rho$ are orthonormalized to $\mathbf{I}'_\rho$ and $\mathbf{T}''_\sigma$ to $\mathbf{I}''_\sigma$. But in practice orthonormalization is a tedious process and on the whole it appears better to proceed without it. In actual problems we are usually presented with sets of non-orthogonal vectors and we have to do the best we can with them.

There is no necessity that the vectors $\mathbf{T}'_\rho$ should be linearly independent, nor need $\mathbf{T}''_\sigma$ be linearly independent. Indeed, in certain problems it is advisable to include, for reasons of symmetry, more vectors of these types than we strictly require, with linear dependences between them. This involves certain verbal changes; for example, if we have r vectors $\mathbf{T}'_\rho$ with one condition of linear dependence connecting them, the first line of (2·712) defines a linear $(r-1)$-space, not a linear r-space. The techniques, however, go through in the same way, with a reduction in the class of the hypercircle finally obtained. On the whole it seems least confusing to regard as standard the case where $\mathbf{T}'_\rho$ and $\mathbf{T}''_\sigma$ are linearly independent, and treat separately as required the slight modifications involved by linear dependences (cf. footnotes on pp. 104, 108, 119).

To return to our problem of finding the vertices $\mathbf{V}'$, $\mathbf{V}''$ of the linear subspaces given in (2·712). We shall find them by using the minimum principle I of (2·711), which is in fact simply the definition of the vertices. This means that we are to minimize $(\mathbf{S}'-\mathbf{S}'')^2$, where

$$\mathbf{S}' = \mathbf{S}'_0 + \mathbf{T}', \quad \mathbf{S}'' = \mathbf{S}''_0 + \mathbf{T}''. \tag{2·715}$$

Using (2·714), we have

$$\begin{aligned}(\mathbf{S}'-\mathbf{S}'')^2 &= (\mathbf{S}'_0-\mathbf{S}''_0)^2 + 2(\mathbf{S}'_0-\mathbf{S}''_0).(\mathbf{T}'-\mathbf{T}'') + \mathbf{T}'^2 + \mathbf{T}''^2 \\ &= (\mathbf{S}'_0-\mathbf{S}''_0+\mathbf{T}')^2 + (\mathbf{S}'_0-\mathbf{S}''_0-\mathbf{T}'')^2 - (\mathbf{S}'_0-\mathbf{S}''_0)^2.\end{aligned} \tag{2·716}$$

This is to be minimized by choice of the parameters a', a'' involved in $\mathbf{T}'$, $\mathbf{T}''$ respectively (see (2·712)). We note that these two sets of parameters are separated in (2·716), and so their determination reduces to two distinct problems:

(i) Minimize $(\mathbf{S}'_0-\mathbf{S}''_0+\mathbf{T}')^2$.

(ii) Minimize $(\mathbf{S}'_0-\mathbf{S}''_0-\mathbf{T}'')^2$.

The minimization may be carried out algebraically, but differentiation is equivalent and quicker. Differentiation of these expressions with respect to a'_ρ, a''_σ respectively gives the minimizing equations

$$\left.\begin{aligned}(\mathbf{S}'_0-\mathbf{S}''_0+\mathbf{T}').\mathbf{T}'_\rho = 0 \quad (\rho=1,2,\ldots,r),\\ (\mathbf{S}'_0-\mathbf{S}''_0-\mathbf{T}'').\mathbf{T}''_\sigma = 0 \quad (\sigma=1,2,\ldots,s),\end{aligned}\right\} \tag{2·717}$$

or, more explicitly,

$$\left.\begin{aligned}\sum_{\mu=1}^{r} a'_\mu \mathbf{T}'_\mu.\mathbf{T}'_\rho + (\mathbf{S}'_0-\mathbf{S}''_0).\mathbf{T}'_\rho = 0 \quad (\rho=1,2,\ldots,r),\\ \sum_{\nu=1}^{s} a''_\nu \mathbf{T}''_\nu.\mathbf{T}''_\sigma - (\mathbf{S}'_0-\mathbf{S}''_0).\mathbf{T}''_\sigma = 0 \quad (\sigma=1,2,\ldots,s).\end{aligned}\right\} \tag{2·718}$$

Here we have r linear equations for a'_ρ and s linear equations for a''_σ. If these are solved and the values inserted in (2·715), we have the vertices

$$\left.\begin{aligned} &\text{Vertex of } L'_r\text{:} \quad \mathbf{V}' = \mathbf{S}'_0 + \mathbf{T}', \quad \mathbf{T}' = \sum_{\rho=1}^{r} a'_\rho \mathbf{T}'_\rho. \\ &\text{Vertex of } L''_s\text{:} \quad \mathbf{V}'' = \mathbf{S}''_0 + \mathbf{T}'', \quad \mathbf{T}'' = \sum_{\sigma=1}^{s} a''_\sigma \mathbf{T}''_\sigma. \end{aligned}\right\} \tag{2·719}$$

It must be clearly understood that, whereas the a', a'' of (2·712) were variable parameters, the a', a'' of (2·719) are definite numbers, the solutions of (2·718).*

The square of the shortest distance from L'_r to L''_s is $(\mathbf{V}'-\mathbf{V}'')^2$. By (2·716) this is

$$(\mathbf{V}'-\mathbf{V}'')^2 = (\mathbf{S}'_0 - \mathbf{S}''_0 + \mathbf{T}')^2 + (\mathbf{S}'_0 - \mathbf{S}''_0 - \mathbf{T}'')^2 - (\mathbf{S}'_0 - \mathbf{S}''_0)^2, \tag{2·720}$$

where $\mathbf{T}'$, $\mathbf{T}''$ are calculated as in (2·719), using the a', a'' which satisfy (2·717) or (2·718). Now if we multiply (2·717) by a'_ρ, a''_σ respectively and sum, we get

$$(\mathbf{S}'_0 - \mathbf{S}''_0 + \mathbf{T}')\,.\,\mathbf{T}' = 0, \quad (\mathbf{S}'_0 - \mathbf{S}''_0 - \mathbf{T}'')\,.\,\mathbf{T}'' = 0, \tag{2·721}$$

and hence $\mathbf{T}'^2 = -\mathbf{T}'\,.\,(\mathbf{S}'_0 - \mathbf{S}''_0), \quad \mathbf{T}''^2 = \mathbf{T}''\,.\,(\mathbf{S}'_0 - \mathbf{S}''_0).$ (2·722)

Using these in (2·720), we get the following equivalent expressions:

$$\left.\begin{aligned} (\mathbf{V}'-\mathbf{V}'')^2 &= (\mathbf{S}'_0 - \mathbf{S}''_0)^2 + (\mathbf{S}'_0 - \mathbf{S}''_0)\,.\,(\mathbf{T}' - \mathbf{T}''), \\ (\mathbf{V}'-\mathbf{V}'')^2 &= (\mathbf{S}'_0 - \mathbf{S}''_0)^2 - \mathbf{T}'^2 - \mathbf{T}''^2. \end{aligned}\right\} \tag{2·723}$$

If, instead of working with general vectors lying in the two subspaces, we had orthonormalized $\mathbf{T}'_\rho$ to $\mathbf{I}'_\rho$ and $\mathbf{T}''_\sigma$ to $\mathbf{I}''_\sigma$, we would have obtained (2·718) in solved form:

$$\left.\begin{aligned} a'_\rho + (\mathbf{S}'_0 - \mathbf{S}''_0)\,.\,\mathbf{I}'_\rho = 0 \quad (\rho = 1, 2, \ldots, r), \\ a''_\sigma - (\mathbf{S}'_0 - \mathbf{S}''_0)\,.\,\mathbf{I}''_\sigma = 0 \quad (\sigma = 1, 2, \ldots, s). \end{aligned}\right\} \tag{2·724}$$

Substitution of these values in (2·719) and (2·723) gives

$$\left.\begin{aligned} &\mathbf{V}' = \mathbf{S}'_0 + \mathbf{T}', \quad \mathbf{T}' = -\sum_{\rho=1}^{r} \mathbf{I}'_\rho[\mathbf{I}'_\rho\,.\,(\mathbf{S}'_0 - \mathbf{S}''_0)], \\ &\mathbf{V}'' = \mathbf{S}''_0 + \mathbf{T}'', \quad \mathbf{T}'' = \sum_{\sigma=1}^{s} \mathbf{I}''_\sigma[\mathbf{I}''_\sigma\,.\,(\mathbf{S}'_0 - \mathbf{S}''_0)], \\ &(\mathbf{V}'-\mathbf{V}'')^2 = (\mathbf{S}'_0 - \mathbf{S}''_0)^2 - \sum_{\rho=1}^{r} [\mathbf{I}'_\rho\,.\,(\mathbf{S}'_0 - \mathbf{S}''_0)]^2 - \sum_{\sigma=1}^{s} [\mathbf{I}''_\sigma\,.\,(\mathbf{S}'_0 - \mathbf{S}''_0)]^2. \end{aligned}\right\} \tag{2·725}$$

* If the vectors $\mathbf{T}'_\rho$, $\mathbf{T}''_\sigma$ are linearly dependent, then the equations (2·718) are not independent; some of the a', a'' may be chosen arbitrarily. This arbitrariness disappears when we write down $\mathbf{V}'$ and $\mathbf{V}''$ as in (2·719), because the vertices are definite points defined by the linear subspaces, irrespective of the vectors which we use in those subspaces.

We recall that in all the above formulae $\mathbf{S}'_0$ and $\mathbf{S}''_0$ are any points on the two subspaces L'_r, L''_s, and $\mathbf{T}'_\rho$ and $\mathbf{T}''_\sigma$ are vectors lying in them; $\mathbf{I}'_\rho$ and $\mathbf{I}''_\sigma$ are orthonormalized vectors lying in the two subspaces.

The intersection of two orthogonal linear subspaces L', L''

We have been considering the closest approach of two linear subspaces which do not intersect. Now we shall discuss the intersection of two linear subspaces L', L''; these are assumed to be orthogonal to one another, and in general of infinite dimensionality.

First we shall prove a uniqueness theorem: *Two orthogonal linear subspaces cannot intersect in more than one point.*

This is a theorem of great importance, the key to many uniqueness theorems in boundary value problems of mathematical physics. It may be regarded as a special case of the uniqueness theorem for vertices established on p. 101, but it admits a very simple direct proof, depending vitally of course on the positive-definite character of the F-metric.

Suppose that L' and L'' have in common *two* points, $\mathbf{S}$ and $\mathbf{T}$. Then $\mathbf{S}-\mathbf{T}$ is a vector lying in L' and also in L''. But, by the assumed orthogonality of L' and L'', every vector lying in the one is orthogonal to every vector lying in the other. Therefore $\mathbf{S}-\mathbf{T}$ is orthogonal to itself, so that $(\mathbf{S}-\mathbf{T})^2=0$. But this is impossible if $\mathbf{S}$ and $\mathbf{T}$ are distinct and the metric is positive-definite. Thus the uniqueness of intersection is established.

An intersection $\mathbf{S}$ may be regarded as the coincidence of vertices $\mathbf{V}'$, $\mathbf{V}''$. On putting $\mathbf{V}'=\mathbf{V}''=\mathbf{S}$ in (2·707) and (2·709), or by an obvious direct argument (Fig. 2·74), we have

$$(\mathbf{S}-\mathbf{S}').(\mathbf{S}-\mathbf{S}'')=0 \tag{2·726}$$

and

$$(\mathbf{S}'-\mathbf{S}'')^2=(\mathbf{S}-\mathbf{S}')^2+(\mathbf{S}-\mathbf{S}'')^2, \tag{2·727}$$

where $\mathbf{S}$ is the intersection and $\mathbf{S}'$, $\mathbf{S}''$ any points on L', L'' respectively.

Inequalities and minimum principles for two intersecting orthogonal linear subspaces L', L''

The following inequalities are immediate consequences of (2·727):

$$\left.\begin{aligned}(\mathbf{S}-\mathbf{S}')^2&\leqslant(\mathbf{S}'-\mathbf{S}'')^2,\\ \mathbf{S}^2-2\mathbf{S}.\mathbf{S}'&\leqslant\mathbf{S}''^2-2\mathbf{S}'.\mathbf{S}'',\\ (\mathbf{S}-\mathbf{S}'')^2&\leqslant(\mathbf{S}'-\mathbf{S}'')^2,\\ \mathbf{S}^2-2\mathbf{S}.\mathbf{S}''&\leqslant\mathbf{S}'^2-2\mathbf{S}'.\mathbf{S}''.\end{aligned}\right\} \tag{2·728}$$

The following minimum principles result (they may also be regarded as particular cases of (2·711) with $\mathbf{V}' = \mathbf{V}'' = \mathbf{S}$):

$$\left.\begin{array}{ll} \text{I.} & \text{For } \mathbf{S}' \text{ free in } L' \text{ and } \mathbf{S}'' \text{ free in } L'',\ (\mathbf{S}'-\mathbf{S}'')^2 \text{ is minimized (to the value zero) by } \mathbf{S}'=\mathbf{S},\ \mathbf{S}''=\mathbf{S}. \\ \text{II.} & \text{For } \mathbf{S}' \text{ free in } L' \text{ and } \mathbf{S}'' \text{ fixed in } L'',\ (\mathbf{S}'-\mathbf{S}'')^2 \text{ (or equivalently } (\mathbf{S}'^2 - 2\mathbf{S}'.\mathbf{S}'')) \text{ is minimized by } \mathbf{S}'=\mathbf{S}. \\ \text{III.} & \text{For } \mathbf{S}' \text{ fixed in } L' \text{ and } \mathbf{S}'' \text{ free in } L'',\ (\mathbf{S}'-\mathbf{S}'')^2 \text{ (or equivalently } (\mathbf{S}''^2 - 2\mathbf{S}'.\mathbf{S}'')) \text{ is minimized by } \mathbf{S}''=\mathbf{S}. \end{array}\right\} \quad (2{\cdot}729)$$

It may seem unnecessary to write out both II and III, for one is obtained from the other by interchange of L' and L'', both linear

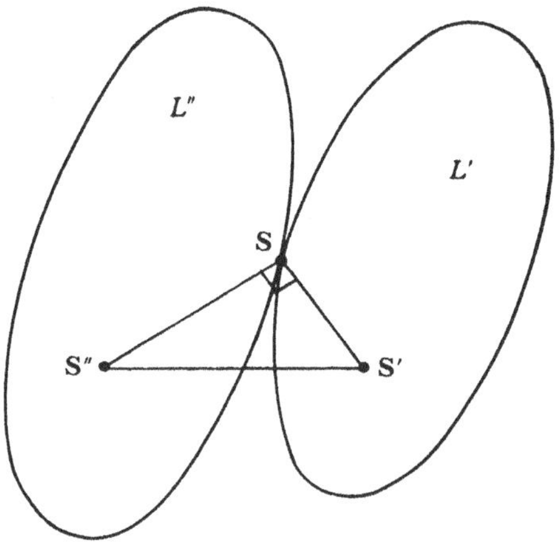

Fig. 2·74. The unique intersection S of two orthogonal linear subspaces L', L''.

subspaces being on the same footing. But these full expressions will be useful for reference later on when L' and L'' will no longer be on the same footing. (We shall take L'' to pass through the origin, so that we may put $\mathbf{S}'' = \mathbf{O}$, but not $\mathbf{S}' = \mathbf{O}$; but this will not be done yet.)

Equations (2·726) and (2·727) are equivalent to one another and to a third equation:

$$[\mathbf{S} - \tfrac{1}{2}(\mathbf{S}' + \mathbf{S}'')]^2 = [\tfrac{1}{2}(\mathbf{S}' - \mathbf{S}'')]^2. \quad (2{\cdot}730)$$

This shows that the point $\mathbf{S}$ lies on a hypersphere with centre $\mathbf{C}_0 = \tfrac{1}{2}(\mathbf{S}' + \mathbf{S}'')$ and radius $R_0 = \tfrac{1}{2}\,|\,\mathbf{S}' - \mathbf{S}''\,|$. (This is of course a hypersphere and not a hypercircle; the hypercircle does not appear until we take at least two points in L' or in L''.)

Application of (2·515) to (2·730) gives bounds for $\mathbf{S}^2$:

$$(\tfrac{1}{2}\,|\,\mathbf{S}'+\mathbf{S}''\,| - \tfrac{1}{2}\,|\,\mathbf{S}'-\mathbf{S}''\,|\,)^2 \leqslant \mathbf{S}^2 \leqslant (\tfrac{1}{2}\,|\,\mathbf{S}'+\mathbf{S}''\,| + \tfrac{1}{2}\,|\,\mathbf{S}'-\mathbf{S}''\,|)^2. \tag{2·731}$$

Also (2·517) gives, $\mathbf{G}$ being any vector,

$$|[\mathbf{S}-\tfrac{1}{2}(\mathbf{S}'+\mathbf{S}'')].\mathbf{G}\,| \leqslant \tfrac{1}{2}\,|\,\mathbf{S}'-\mathbf{S}''\,|\;|\,\mathbf{G}\,|. \tag{2·732}$$

The method of the hypercircle

Having now prepared the way, we proceed to the method of the hypercircle.

Here is the problem. In function-space with positive-definite metric, there are two intersecting orthogonal linear subspaces L', L''. We want to find their point of intersection, $\mathbf{S}$.

Suppose that we can find a finite number of points in L' and in L''—say $r+1$ points in L' and $s+1$ points in L''. The former define a linear r-space L'_r immersed in L' and the latter define a linear s-space L''_s immersed in L''. We may find the vertices $\mathbf{V}'$, $\mathbf{V}''$ of L'_r, L''_s as in (2·719) or (2·725). It is extremely unlikely that either of these vertices will actually be the solution $\mathbf{S}$, but we may hope that by making r and s fairly large we may make both $\mathbf{V}'$ and $\mathbf{V}''$ lie close to $\mathbf{S}$. This rather vague hope initiates the method, which is in fact logical and precise.

Once we have found $\mathbf{V}'$ and $\mathbf{V}''$, then (as will be shown below) we can assert that the unknown point $\mathbf{S}$ must lie on a certain hypercircle having the join of $\mathbf{V}'$ and $\mathbf{V}''$ for diameter. This fact puts bounds on $\mathbf{S}^2$, and in certain cases puts pointwise bounds on the solution of an analytic problem (for of course in the applications of the method the linear subspaces L', L'' are defined analytically in terms of a boundary value problem). We know then precisely the value of $(\mathbf{S}-\mathbf{C})^2$, where $\mathbf{C}$ is the centre of the hypercircle, for its value is R^2, where R is the radius of the hypercircle.

If we seek a good approximation, we must work to make R small. If this is done, then $\mathbf{C}$ (or indeed any point on the hypercircle) is a good approximation to the solution $\mathbf{S}$ in the mean square sense.

These remarks explain the method in general terms; we shall now fill in the details.

The existence of the hypercircle and its equations

We take $r+1$ points in L', say $\mathbf{S}'_0, \mathbf{S}'_1, \ldots, \mathbf{S}'_r$, and define r vectors by

$$\mathbf{T}'_\rho = \mathbf{S}'_\rho - \mathbf{S}'_0 \quad (\rho = 1, 2, \ldots, r); \tag{2·733}$$

we suppose these r vectors to be linearly independent.* The $r+1$ points in L' define a linear r-space L'_r, immersed in L', with equation

$$\mathbf{X} = \mathbf{S}'_0 + \sum_{\rho=1}^{r} a'_\rho \mathbf{T}'_\rho, \tag{2·734}$$

a'_ρ being variable parameters; the vectors $\mathbf{T}'_\rho$ lie in L'_r, and of course in L' also.

We take also $s+1$ points in L'', say $\mathbf{S}''_0, \mathbf{S}''_1, \ldots, \mathbf{S}''_s$, define

$$\mathbf{T}''_\sigma = \mathbf{S}''_\sigma - \mathbf{S}''_0 \quad (\sigma = 1, 2, \ldots, s), \tag{2·735}$$

and obtain the linear s-space L''_s, immersed in L'', with the equation

$$\mathbf{X} = \mathbf{S}''_0 + \sum_{\sigma=1}^{s} a''_\sigma \mathbf{T}''_\sigma. \tag{2·736}$$

If L' happened to contain the origin $\mathbf{O}$, we would probably choose $\mathbf{S}'_0 = \mathbf{O}$ for simplicity, or $\mathbf{S}''_0 = \mathbf{O}$ if L'' contained $\mathbf{O}$. If both L' and L'' contained $\mathbf{O}$, our problem would be solved: $\mathbf{S} = \mathbf{O}$.

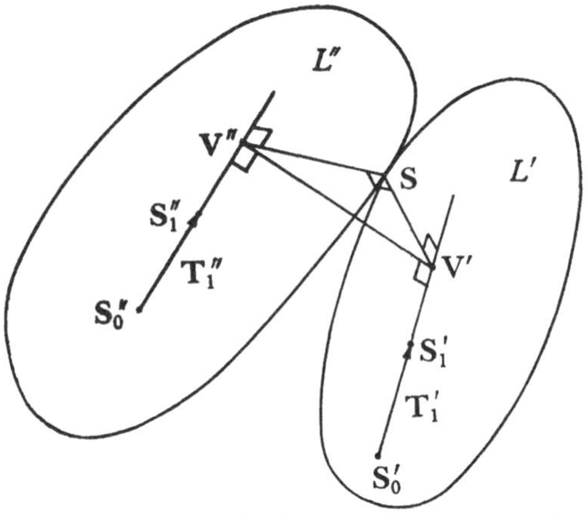

Fig. 2·75. Hypercircle method: diagram shows the vertices $\mathbf{V}'$, $\mathbf{V}''$ of L'_r, L''_s (for $r=s=1$) immersed in L', L'' respectively.

We have now two linear subspaces of finite dimensionality, L'_r, L''_s, immersed respectively in L', L''. Let $\mathbf{V}'$ be the vertex of L'_r and $\mathbf{V}''$ the vertex of L''_s, these being the points of closest approach of these two linear subspaces of finite dimensionality. These vertices are to be calculated as in (2·719) or (2·725).

The situation is shown in Fig. 2·75, where for the sake of simplicity we have taken $r=s=1$, so that L'_r, L''_s are the straight lines L'_1, L''_1.

* If $\mathbf{T}'_\rho$ are not linearly independent, the linear subspace (2·734) has less than r dimensions, and the class of the hypercircle (see later) is reduced (cf. p. 103 and footnotes on pp. 104, 119).

Now although we do not know the vector $\mathbf{S}$, we know that $\mathbf{S}-\mathbf{V}''$ is a vector lying in L'', and as such is orthogonal to every vector lying in L'. Also, $\mathbf{S}-\mathbf{V}'$ is orthogonal to every vector lying in L''. Therefore we can write

$$\left.\begin{aligned}(\mathbf{S}-\mathbf{V}').(\mathbf{S}-\mathbf{V}'')&=0,\\ (\mathbf{S}-\mathbf{V}'').\mathbf{T}'_\rho&=0 \quad (\rho=1,2,\ldots,r),\\ (\mathbf{S}-\mathbf{V}').\mathbf{T}''_\sigma&=0 \quad (\sigma=1,2,\ldots,s),\end{aligned}\right\} \qquad (2\cdot737)$$

using the fact that $\mathbf{T}'_\rho$ lies in L'_r and hence in L', while $\mathbf{T}''_\sigma$ lies in L''_s and hence in L''.

These are key formulae. The first of (2·737) locates the point $\mathbf{S}$ on a hypersphere with centre $\mathbf{C}$ and radius R:

$$\left.\begin{aligned}(\mathbf{X}-\mathbf{C})^2&=R^2,\\ \mathbf{C}=\tfrac{1}{2}(\mathbf{V}'+\mathbf{V}''),\quad R^2&=[\tfrac{1}{2}(\mathbf{V}'-\mathbf{V}'')]^2.\end{aligned}\right\} \qquad (2\cdot738)$$

The other equations in (2·737) confine the point $\mathbf{S}$ to a hyperplane of class $r+s$. Thus *the point of intersection* $\mathbf{S}$ *of the two orthogonal linear subspaces* L', L'' *is situated on the hypercircle of class* $r+s$ *with the equations**

$$\left.\begin{aligned}(\mathbf{X}-\mathbf{V}').(\mathbf{X}-\mathbf{V}'')&=0,\\ (\mathbf{X}-\mathbf{V}'').\mathbf{T}'_\rho&=0 \quad (\rho=1,2,\ldots,r),\\ (\mathbf{X}-\mathbf{V}').\mathbf{T}''_\sigma&=0 \quad (\sigma=1,2,\ldots,s).\end{aligned}\right\} \qquad (2\cdot739)$$

To put these equations of the hypercircle into more convenient parametric form, we go back to (2·706) and note that the vector $\mathbf{V}'-\mathbf{V}''$, joining the vertices of L'_r and L''_s, is orthogonal to both these subspaces, so that

$$\left.\begin{aligned}(\mathbf{V}'-\mathbf{V}'').\mathbf{T}'_\rho&=0 \quad (\rho=1,2,\ldots,r),\\ (\mathbf{V}'-\mathbf{V}'').\mathbf{T}''_\sigma&=0 \quad (\sigma=1,2,\ldots,s).\end{aligned}\right\} \qquad (2\cdot740)$$

Then the second line of (2·739) gives

$$\begin{aligned}0=(\mathbf{X}-\mathbf{V}'').\mathbf{T}'_\rho&=[\mathbf{X}-\tfrac{1}{2}(\mathbf{V}'+\mathbf{V}'')+\tfrac{1}{2}(\mathbf{V}'-\mathbf{V}'')].\mathbf{T}'_\rho\\ &=(\mathbf{X}-\mathbf{C}).\mathbf{T}'_\rho,\end{aligned} \qquad (2\cdot741)$$

and the third line of (2·739) gives similarly

$$(\mathbf{X}-\mathbf{C}).\mathbf{T}''_\sigma=0. \qquad (2\cdot742)$$

* In deriving (2·739) we have not used the fact that $\mathbf{V}'$, $\mathbf{V}''$ are vertices; indeed, (2·739) hold if we replace $\mathbf{V}'$, $\mathbf{V}''$ by any two points $\mathbf{S}'$, $\mathbf{S}''$ which are on L', L'' respectively.

If we combine these last equations ($r+s$ equations in all) with (2·738), we get the following *parametric form* for the hypercircle on which the point **S** is located:

$$\left.\begin{aligned} \mathbf{X} &= \mathbf{C} + R\mathbf{J}, \\ \mathbf{J}^2 &= 1, \\ \mathbf{J}\,.\,\mathbf{T}'_\rho &= 0 \quad (\rho = 1, 2, \ldots, r), \\ \mathbf{J}\,.\,\mathbf{T}''_\sigma &= 0 \quad (\sigma = 1, 2, \ldots, s). \end{aligned}\right\} \tag{2·743}$$

To make the geometrical interpretation clearer, let us note that

$$\left.\begin{aligned} (\mathbf{S} - \mathbf{V}')\,.\,\mathbf{T}'_\rho &= 0 \quad (\rho = 1, 2, \ldots, r), \\ (\mathbf{S} - \mathbf{V}'')\,.\,\mathbf{T}''_\sigma &= 0 \quad (\sigma = 1, 2, \ldots, s). \end{aligned}\right\} \tag{2·744}$$

These equations are actually consequences of the preceding argument, but they may easily be established directly as follows. By the second of (2·737) we have

$$\begin{aligned} 0 = (\mathbf{S} - \mathbf{V}'')\,.\,\mathbf{T}'_\rho &= (\mathbf{S} - \mathbf{V}' + \mathbf{V}' - \mathbf{V}'')\,.\,\mathbf{T}'_\rho \\ &= (\mathbf{S} - \mathbf{V}')\,.\,\mathbf{T}'_\rho + (\mathbf{V}' - \mathbf{V}'')\,.\,\mathbf{T}'_\rho, \end{aligned} \tag{2·745}$$

and the last scalar product vanishes by (2·740). Thus we have the first of (2·744), and the second follows similarly.

If we combine (2·737) and (2·744), we see that each of the vectors $\mathbf{S} - \mathbf{V}'$ and $\mathbf{S} - \mathbf{V}''$ is orthogonal to both L'_r and L''_s. Thus *the linear 2-space containing the three points* **S**, **V′**, **V″** *is orthogonal to both* L'_r *and* L''_s. These orthogonalities are responsible for four of the five right angles shown in Fig. 2·75 and also in Fig. 2·76.

Bounds involving **S**

The simplest and most important formula putting bounds on **S** is that given by (2·738):

$$\boxed{\begin{gathered} (\mathbf{S} - \mathbf{C})^2 = R^2, \\ \mathbf{C} = \tfrac{1}{2}(\mathbf{V}' + \mathbf{V}''), \quad 4R^2 = (\mathbf{V}' - \mathbf{V}'')^2. \end{gathered}} \tag{2·746}$$

Here **V′** and **V″** are calculated as in (2·719) after the linear equations (2·718) have been solved. The general indication of (2·746) is that if R is small, then $\mathbf{S} = \mathbf{C}$ approximately.

Since $\mathbf{S}$ is on the hypercircle (2·743), (2·631) gives the following bounds for $\mathbf{S}^2$:

$$\left.\begin{aligned} &|\,\mathbf{S}^2-\mathbf{C}^2-R^2\,| \leqslant 2R\,|\,\mathbf{C}_0'\,|, \\ &\mathbf{C}=\tfrac{1}{2}(\mathbf{V}'+\mathbf{V}''), \quad R=\tfrac{1}{2}\,|\,\mathbf{V}'-\mathbf{V}''\,|, \quad \mathbf{C}^2+R^2=\tfrac{1}{2}(\mathbf{V}'^2+\mathbf{V}''^2), \\ &\mathbf{C}_0'^2=\mathbf{C}^2-\sum_{\rho=1}^{r}(\mathbf{C}\,.\,\mathbf{I}_\rho')^2-\sum_{\sigma=1}^{s}(\mathbf{C}\,.\,\mathbf{I}_\sigma'')^2. \end{aligned}\right\} \quad (2·747)$$

Here $\mathbf{I}_\rho'$, $\mathbf{I}_\sigma''$ are orthonormalized vectors lying in L_r', L_s'' respectively, as in (2·725).

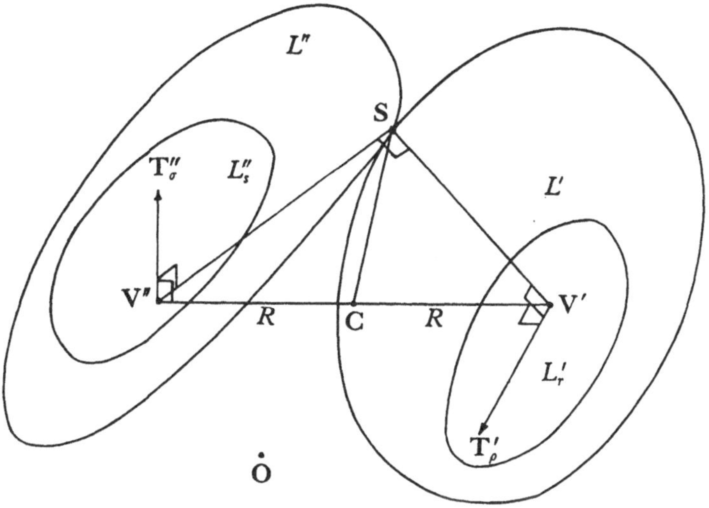

Fig. 2·76. Basic diagram for the hypercircle in the general case where neither L' nor L'' contains the origin $\mathbf{O}$. $\mathbf{S}$ = solution (intersection of orthogonal L', L''). $\mathbf{V}'$, $\mathbf{V}''$ = vertices of L_r', L_s'' immersed in L', L'' respectively, $\mathbf{T}_\rho'$, $\mathbf{T}_\sigma''$ = vectors lying in L_r', L_s'' respectively. $\mathbf{C}=\frac{1}{2}(\mathbf{V}'+\mathbf{V}'')$ = centre of hypercircle, $R=\frac{1}{2}\,|\,\mathbf{V}'-\mathbf{V}''\,|$. $(\mathbf{S}-\mathbf{C})^2=R^2$, $\mathbf{S}-\mathbf{C}$ orthogonal to L_r' and L_s''. The linear 2-space containing the point $\mathbf{S}$, $\mathbf{V}'$, $\mathbf{V}''$ is orthogonal to L_r' and L_s''.

We get other bounds very simply by inspecting right-angled triangles in Fig. 2·76. These results are easily verified by formal algebra.

Looking at the right-angled triangle formed by $\mathbf{S}$, $\mathbf{V}'$, $\mathbf{S}'$ (the last being any point in L_r'), we have

$$(\mathbf{S}-\mathbf{S}')^2=(\mathbf{S}'-\mathbf{V}')^2+(\mathbf{S}-\mathbf{V}')^2.$$

But we see from the right-angled triangle formed by the points $\mathbf{S}$, $\mathbf{V}'$, $\mathbf{V}''$ that $(\mathbf{S}-\mathbf{V}')^2 \leqslant (\mathbf{V}'-\mathbf{V}'')^2$. Thus we get a lower and an upper bound for $(\mathbf{S}-\mathbf{S}')^2$, and similarly there are bounds for

$(\mathbf{S}-\mathbf{S}'')^2$. In fact, if $\mathbf{S}'$, $\mathbf{S}''$ are any points in L'_r, L''_s respectively, we have

$$\left.\begin{aligned}(\mathbf{S}'-\mathbf{V}')^2 \leqslant (\mathbf{S}-\mathbf{S}')^2 \leqslant (\mathbf{S}'-\mathbf{V}')^2+(\mathbf{V}'-\mathbf{V}'')^2,\\ (\mathbf{S}''-\mathbf{V}'')^2 \leqslant (\mathbf{S}-\mathbf{S}'')^2 \leqslant (\mathbf{S}''-\mathbf{V}'')^2+(\mathbf{V}'-\mathbf{V}'')^2.\end{aligned}\right\} \qquad (2\cdot748)$$

These bounds are actually attained if we regard $\mathbf{S}$ as a variable point on the hypercircle; two of them are attained if we write $\mathbf{V}'$ for $\mathbf{S}$, and two when we write $\mathbf{V}''$ for $\mathbf{S}$.

We notice how the bounds draw together as $|\mathbf{V}'-\mathbf{V}''|$ tends to zero, i.e. as the radius R of the hypercircle tends to zero.

These bounds will become simpler and more significant when we discuss later the case where one of the subspaces L', L'' contains the origin of F-space.

Consider now the bounds of $\mathbf{S}.\mathbf{G}$, where $\mathbf{G}$ is any given vector. By (2·635) we have

$$|\mathbf{S}.\mathbf{G}-\mathbf{C}.\mathbf{G}| \leqslant R\,|\mathbf{G}_0|, \qquad (2\cdot749)$$

$$\mathbf{G}_0^2=\mathbf{G}^2-\sum_{\rho=1}^{r}(\mathbf{G}.\mathbf{I}'_\rho)^2-\sum_{\sigma=1}^{s}(\mathbf{G}.\mathbf{I}''_\sigma)^2. \qquad (2\cdot750)$$

This is a convenient formulae if we have orthonormalized vectors $\mathbf{I}'_\rho$, $\mathbf{I}''_\sigma$ in L'_r, L''_s respectively. But if we wish to avoid this orthonormalization, we require a formula involving non-orthonormalized vectors $\mathbf{T}'_\rho$, $\mathbf{T}''_\sigma$. To get such a formula, we note that $\mathbf{G}$ may be resolved into three components: one in L'_r, one in L''_s, and the third normal to both these subspaces. This third component is in fact the vector $\mathbf{G}_0$ occurring in (2·749). Equivalently, $-\mathbf{G}_0$ is the normal drawn from the point $\mathbf{G}$ to the linear $(r+s)$-space

$$\mathbf{X}=\sum_{\rho=1}^{r}c'_\rho\mathbf{T}'_\rho+\sum_{\sigma=1}^{s}c''_\sigma\mathbf{T}''_\sigma, \qquad (2\cdot751)$$

the c's being variable parameters. This normal may be found by minimizing $(\mathbf{G}-\mathbf{X})^2$, and so

$$\mathbf{G}_0^2=(\mathbf{G}-\mathbf{X})^2_{\text{min.}}, \qquad (2\cdot752)$$

where $\mathbf{X}$ is of the form (2·751). This means that the c's are to be chosen to satisfy the equations

$$\left.\begin{aligned}\sum_{\mu=1}^{r}c'_\mu\mathbf{T}'_\mu.\mathbf{T}'_\rho-\mathbf{G}.\mathbf{T}'_\rho=0 \quad (\rho=1,2,\ldots,r),\\ \sum_{\nu=1}^{s}c''_\nu\mathbf{T}''_\nu.\mathbf{T}''_\sigma-\mathbf{G}.\mathbf{T}''_\sigma=0 \quad (\sigma=1,2,\ldots,s).\end{aligned}\right\} \qquad (2\cdot753)$$

When the c's have been so found and $\mathbf{X}$ obtained by inserting their values in (2·751), we have from (2·753)

$$\mathbf{X}^2-\mathbf{G}.\mathbf{X}=0, \qquad (2\cdot754)$$

and so, by (2·752),

$$\begin{aligned}\mathbf{G}_0^2 &= \mathbf{G}^2 - \mathbf{G}.\mathbf{X} \\ &= \mathbf{G}^2 - \sum_{\rho=1}^{r} c'_\rho \mathbf{G}.\mathbf{T}'_\rho - \sum_{\sigma=1}^{s} c''_\sigma \mathbf{G}.\mathbf{T}''_\sigma,\end{aligned} \tag{2·755}$$

or, equivalently,

$$\begin{aligned}\mathbf{G}_0^2 &= \mathbf{G}^2 - \mathbf{X}^2 \\ &= \mathbf{G}^2 - \left(\sum_{\rho=1}^{r} c'_\rho \mathbf{T}'_\rho\right)^2 - \left(\sum_{\sigma=1}^{s} c''_\sigma \mathbf{T}''_\sigma\right)^2.\end{aligned} \tag{2·756}$$

To summarize, *the bounds of* $\mathbf{S}.\mathbf{G}$ *are given by* (2·749), *where* $|\mathbf{G}_0|$ *is the square root of* (2·755) *or* (2·756), *wherein the c's are to be evaluated by solving the linear equations* (2·753); *if orthonormal vectors are available, we use* (2·750).

The case where L'' contains the origin **O**

In many boundary value problems it naturally occurs that one of the linear subspaces (say L'') contains the origin of function-space, **O**. Even if this is not directly the case, we can make L'' pass through the origin by translating the origin to a point on L''. This means subtracting from all vectors the position-vector of some point on L'', and, as we shall verify amply later in special cases, it is easy to find a point on L''. (We cannot translate the origin so as to lie on *both* L' and L'', because their point of intersection is the very thing we do not know, but are trying to find.)

Let us suppose then that L'' contains **O**. We shall now review the general formulae obtained earlier and see how they can be simplified by this fact. Symmetry is lost, since L' and L'' now play different roles, but this loss of symmetry is more than compensated by increased simplicity. These simpler formulae are the ones we shall use later.

The formulae (2·707), (2·710) and (2·711) concern two *non-intersecting* linear subspaces L', L''. If L'' contains **O**, as we now assume, we can choose $\mathbf{S}'' = \mathbf{O}$. Accordingly, we have from (2·707), if $\mathbf{S}'$ is any point of L' and $\mathbf{V}'$, $\mathbf{V}''$ the vertices of L', L'' (their points of closest approach),

$$\left.\begin{aligned}(\mathbf{S}'-\mathbf{V}').(\mathbf{V}'-\mathbf{V}'') &= 0, \\ \mathbf{V}''.(\mathbf{V}'-\mathbf{V}'') &= 0, \\ (\mathbf{S}'-\mathbf{V}').\mathbf{V}'' &= 0.\end{aligned}\right\} \tag{2·757}$$

The second of these equations tells us that $\mathbf{V}''$ *is orthogonal to* $\mathbf{V}'-\mathbf{V}''$. That is indeed obvious, since $\mathbf{V}''$ is now a vector lying in L''. (Cf. Fig. 2·72, p. 100, where $\mathbf{S}''$ is now the origin.)

From (2·710) we get (on putting $\mathbf{S}''=\mathbf{O}$) the inequalities

$$\left.\begin{aligned}\mathbf{S}'^2 &\geqslant (\mathbf{V}'-\mathbf{V}'')^2,\\ \mathbf{S}'^2 &\geqslant (\mathbf{S}'-\mathbf{V}')^2,\\ \mathbf{S}'^2 &\geqslant \mathbf{V}''^2.\end{aligned}\right\} \tag{2·758}$$

As for (2·711), if $\mathbf{S}''=\mathbf{O}$ only II applies, and it reads

$$\text{For } \mathbf{S}' \text{ free in } L',\ \mathbf{S}'^2 \text{ is minimized by } \mathbf{S}'=\mathbf{V}'. \tag{2·759}$$

In other words, *the vertex* $\mathbf{V}'$ *of* L' *is the point of closest approach of* L' *to the origin*. (This is immediately verified intuitively in the simple case shown in Fig. 2·72, if we remember that the two lines L_1', L_1'' are orthogonal.)

From (2·712) to (2·725) we discussed the determination of the vertices of two non-intersecting orthogonal linear subspaces of *finite* dimensionality, L_r' and L_s''. We now assume that L'' contains $\mathbf{O}$ and choose $\mathbf{S}_0''=\mathbf{O}$ in (2·712). At first sight this appears to give only a slight simplification in the formulae and no significant change. However, when we put $\mathbf{S}_0''=\mathbf{O}$, all reference to L_s'' disappears from the first line of (2·718). This means that *the determination of the vertex* $\mathbf{V}'$ *of* L_r' *does not involve* L_s''; in fact, it is easy to see that $\mathbf{V}'$ is simply that point of L_r' which is closest to the origin $\mathbf{O}$.

So far we have been considering two non-intersecting linear subspaces. Let us now see what simplifications result in the case of two *intersecting* linear subspaces L', L'', when we assume that L'' contains $\mathbf{O}$.

Consider (2·726), where $\mathbf{S}$ is the point of intersection and $\mathbf{S}'$, $\mathbf{S}''$ any points in L', L'' respectively. Since L'' contains $\mathbf{O}$, we may put $\mathbf{S}''=\mathbf{O}$ and get

$$(\mathbf{S}-\mathbf{S}').\mathbf{S}=0. \tag{2·760}$$

Using this in (2·726), we have

$$(\mathbf{S}-\mathbf{S}').\mathbf{S}''=0, \tag{2·761}$$

where $\mathbf{S}'$, $\mathbf{S}''$ are now any points in L', L'' respectively.

These are useful formulae. By (2·760) the point $\mathbf{S}$ is on a hypersphere with centre $\frac{1}{2}\mathbf{S}'$ and radius $\frac{1}{2}\,|\,\mathbf{S}'\,|$; by (2·761) the point $\mathbf{S}$ is on a hyperplane of class 1. Therefore the point $\mathbf{S}$ is on a *hypercircle of class* 1, of which the centre $\mathbf{C}$ and the radius R are [cf. (2·628)] given by

$$\left.\begin{aligned}\mathbf{C}&=\tfrac{1}{2}[\mathbf{S}'+\mathbf{I}''(\mathbf{S}'.\mathbf{I}'')],\\ 4R^2&=\mathbf{S}'^2-(\mathbf{S}'.\mathbf{I}'')^2,\end{aligned}\right\} \tag{2·762}$$

where $\mathbf{I}''$ is the unit vector in the direction of $\mathbf{S}''$, so that

$$\mathbf{I}''=\mathbf{S}''/|\,\mathbf{S}''\,|. \tag{2·763}$$

Although the general formulae (2·743) are more restrictive on **S**, confining it to a hypercircle of class $r+s$, the simplicity of the above result makes it worth noting. It is depicted in Fig. 2·77; we easily verify that any chord of the hypercircle is orthogonal to the vector **S**″, or to **I**″, which has the same direction.

The merit of Fig. 2·77 is that we could construct a model of it in ordinary space of three dimensions. Such a model would contain all the vectors and points shown. The only defect would lie in the representation of the hypercircle, which (being of infinite dimensionality) cannot of course be faithfully depicted in three dimensions.

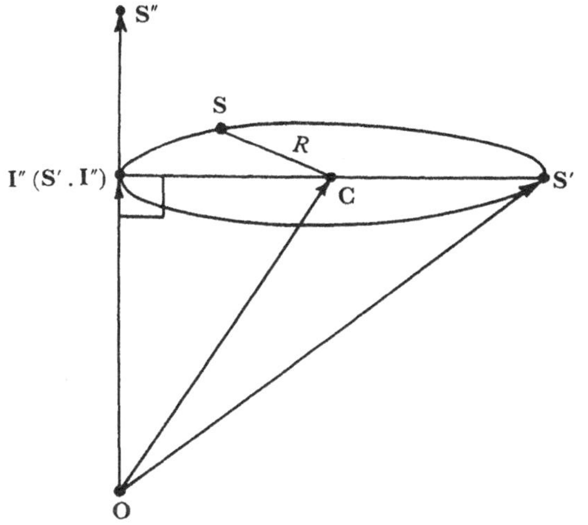

Fig. 2·77. Hypercircle of class 1 obtained from one point **S**′ on L' and one point **S**″ on L'', when L'' contains **O**.

In our model the hypercircle would appear as a circle, as shown in Fig. 2·77, this circle being the intersection of the hypercircle with the linear 3-space containing the vectors **S**, **S**′, **S**″.

We now consider the inequalities (2·728). The fact that L'' contains **O** may be used in either of two ways: (*a*) we may put $\mathbf{S}''=\mathbf{O}$, (*b*) we may use (2·760).

The former plan leads to

$$\left.\begin{aligned}(\mathbf{S}-\mathbf{S}')^2 &\leqslant \mathbf{S}'^2,\\ \mathbf{S}^2-2\mathbf{S}.\mathbf{S}' &\leqslant 0,\\ \mathbf{S}^2 &\leqslant \mathbf{S}'^2.\end{aligned}\right\} \qquad (2{\cdot}764)$$

The last of these tells us that, *when L'' contains* **O**, *the intersection* **S** *of L' and L'' is closer to the origin than any other point of L'.*

To follow plan (*b*), we note the second of (2·728):

$$\mathbf{S}^2 - 2\mathbf{S}\,.\,\mathbf{S}' \leqslant -2\mathbf{S}'\,.\,\mathbf{S}'' + \mathbf{S}''^2. \tag{2·764a}$$

But by (2·760) we have $\mathbf{S}\,.\,\mathbf{S}' = \mathbf{S}^2$, and so this inequality may be written

$$\mathbf{S}^2 \geqslant 2\mathbf{S}'\,.\,\mathbf{S}'' - \mathbf{S}''^2. \tag{2·764b}$$

Now since L'' contains $\mathbf{O}$, the vector $\mathbf{S}''$ lies in L'', and so does the vector $k\mathbf{S}''$, where k is arbitrary. Thus (2·764*b*) is true if we substitute $k\mathbf{S}''$ for $\mathbf{S}''$:

$$\mathbf{S}^2 \geqslant 2k\mathbf{S}'\,.\,\mathbf{S}'' - k^2\mathbf{S}''^2. \tag{2·764c}$$

The right-hand side attains its maximum for

$$k = (\mathbf{S}'\,.\,\mathbf{S}'')/\mathbf{S}''^2, \tag{2·764d}$$

and this maximum is $(\mathbf{S}'\,.\,\mathbf{S}'')^2/\mathbf{S}''^2 = (\mathbf{S}'\,.\,\mathbf{I}'')^2$. Hence by (2·764*c*) and (2·764*b*) we have the inequalities

$$\mathbf{S}^2 \geqslant (\mathbf{S}'\,.\,\mathbf{I}'')^2 \geqslant 2\mathbf{S}'\,.\,\mathbf{S}'' - \mathbf{S}''^2, \quad \mathbf{I}'' = \mathbf{S}''/|\,\mathbf{S}''\,|. \tag{2·765}$$

The first of these inequalities is obvious intuitively from Fig. 2·77.

In (2·764) and (2·765) we have simple upper and lower bounds for $\mathbf{S}^2$.

If we put $\mathbf{S}'' = \mathbf{O}$ in the minimum principle II of (2·729), we get the following:

$$\text{For } \mathbf{S}' \text{ free in } L',\ \mathbf{S}'^2 \text{ is minimized by } \mathbf{S}' = \mathbf{S}. \tag{2·766}$$

This is essentially the same as the last of (2·764). The equation (2·760) has no effect on the minimum principles (2·729). If we use (2·761) these minimum principles give us (2·766) again, and also the following [cf. (2·765)]:

For $\mathbf{S}'$ fixed in L' and $\mathbf{S}''$ free in L'' (which contains $\mathbf{O}$),

$$\mathbf{S}''^2 - 2\mathbf{S}'\,.\,\mathbf{S}'' \text{ is minimized (to } -\mathbf{S}^2) \text{ by } \mathbf{S}'' = \mathbf{S}. \tag{2·767}$$

We consider now the effect of the assumption that L'' contains $\mathbf{O}$ on the discussion, from (2·733) to (2·745), of the hypercircle of class $r+s$. We shall put $\mathbf{S}''_0 = \mathbf{O}$ in (2·736), so that L''_s contains $\mathbf{O}$. At first sight the simplification looks rather trivial, but we must not overlook the important condition of orthogonality given by the second of (2·757), applied to L'_r and L''_s:

$$\mathbf{V}''\,.\,(\mathbf{V}' - \mathbf{V}'') = 0. \tag{2·768}$$

There are now more orthogonalities than are shown in Fig. 2·76, and it is therefore advisable to prepare a new diagram as in Fig. 2·78. *This is the basic diagram for use in the applications of the hypercircle method*. It is true that all the method is contained in the formulae, but for rapid application, with quick understanding of what we are doing, a diagram of this sort is a very real assistance.

Finally, we come to the most important simplification arising from the assumption that L'' contains **O**; this simplification is in the inequalities (2·748). It is understood that we have chosen L''_s to

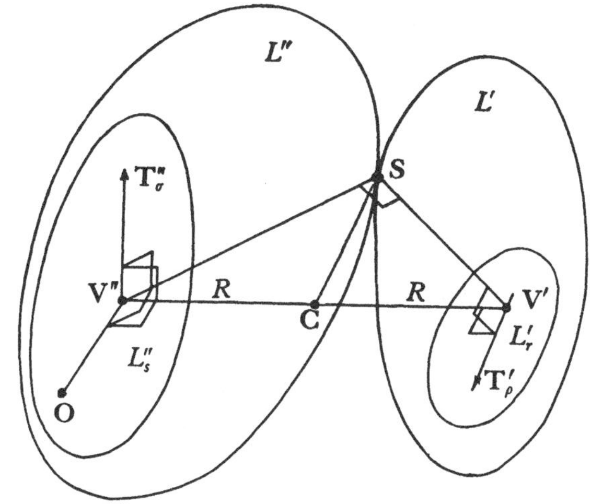

Fig. 2·78. Basic diagram for the hypercircle of class $r+s$ in the case where L'' contains **O**. **S** = solution (intersection of orthogonal L', L''). **V**$'$, **V**$''$ = vertices of L'_r, L''_s immersed in L', L'' (L''_s contains **O**). $\mathbf{T}'_\rho$, $\mathbf{T}''_\sigma$ = vectors lying in L'_r, L''_s respectively. $\mathbf{C} = \frac{1}{2}(\mathbf{V}' + \mathbf{V}'')$ = centre of hypercircle, $R = \frac{1}{2}|\mathbf{V}' - \mathbf{V}''|$. $(\mathbf{S} - \mathbf{C})^2 = R^2$; $\mathbf{S} - \mathbf{C}$ orthogonal to L'_r and L''_s. The linear 2-space containing the points **S**, **V**$'$, **V**$''$ is orthogonal to L'_r and L''_s, and in particular to the position-vector **V**$''$.

contain **O** by putting $\mathbf{S}''_0 = \mathbf{O}$ in (2·736). Then we can put $\mathbf{S}'' = \mathbf{O}$ in (2·748), and the last term in these inequalities then reads

$$\mathbf{V}''^2 + (\mathbf{V}' - \mathbf{V}'')^2 = (\mathbf{V}'' + \mathbf{V}' - \mathbf{V}'')^2 - 2\mathbf{V}''.(\mathbf{V}' - \mathbf{V}'') = \mathbf{V}'^2, \quad (2·769)$$

by (2·768). Thus the last line of inequalities in (2·748) becomes

$$\boxed{\mathbf{V}''^2 \leqslant \mathbf{S}^2 \leqslant \mathbf{V}'^2.} \quad (2·770)$$

This important result may also be obtained more simply by putting $\mathbf{S}' = \mathbf{V}'$, $\mathbf{S}'' = \mathbf{V}''$ in the last of (2·764) and in (2·765) and using (2·768). It is also fairly obvious from Fig. 2·78.

Let us state (2·770) in words: *If L', L'' are two orthogonal linear subspaces intersecting at the point* **S**, *and if L'' contains the origin* **O** *of function-space, then the magnitude of the vector* **S** *lies between the* *magnitudes of the vectors* $\mathbf{V}'$, $\mathbf{V}''$, *where these are the vertices of any linear subspaces of finite dimensionality* (L'_r, L''_s) *immersed in* L', L'', *respectively*, L''_s *containing* $\mathbf{O}$.

It is clear that $\mathbf{S}^2$ is the lower bound of $\mathbf{V}'^2$ and the upper bound of $\mathbf{V}''^2$ for all choices of L'_r, L''_s, the latter always containing $\mathbf{O}$.

If L'' contains $\mathbf{O}$, the problem of finding $\mathbf{S}$, the intersection of L' and L'', is precisely the problem of dropping a perpendicular from $\mathbf{O}$ on L', the F-point $\mathbf{S}$ being the foot of this perpendicular. That this perpendicular is unique follows from the fact that a triangle cannot have two right angles. This way of looking at the problem seems to make L' more important than L'', but by using L' alone we can get only an *upper* bound for $\mathbf{S}^2$ and we cannot locate $\mathbf{S}$ on a hypercircle.

Summary of formulae for application of the method of the hypercircle

This summary deals only with the case where L'' contains $\mathbf{O}$. If a problem presents itself in which L'' does not contain $\mathbf{O}$, we can either (*a*) use the general symmetric formulae given earlier in this section, or (*b*) change the origin of F-space so as to place it on L'' and then use the formulae which follow. Formulae for the case where L' (not L'') contains $\mathbf{O}$ are of course to be obtained by interchange of single and double primes.

Notation:

L', L'' = pair of intersecting orthogonal linear subspaces, L'' containing $\mathbf{O}$.

$\mathbf{S}$ = their point of intersection.

$\mathbf{S}', \mathbf{S}''$ = any points on L', L'' respectively.

Hypercircle of class 1 (Fig. 2·77):

$$\left.\begin{array}{l} \mathbf{S}^2 = \mathbf{S}.\mathbf{S}', \quad \mathbf{S}.\mathbf{S}'' = \mathbf{S}'.\mathbf{S}''; \\ \mathbf{S} = \mathbf{C} + R\mathbf{J}, \quad \mathbf{J}^2 = 1, \quad \mathbf{J}.\mathbf{I}'' = 0, \quad \mathbf{I}'' = \mathbf{S}''/|\,\mathbf{S}''\,|, \\ \mathbf{C} = \tfrac{1}{2}[\mathbf{S}' + \mathbf{I}''(\mathbf{S}'.\mathbf{I}'')], \quad 4R^2 = \mathbf{S}'^2 - (\mathbf{S}'.\mathbf{I}'')^2. \end{array}\right\} \tag{2·771}$$

Inequalities:
$$2\mathbf{S}'.\mathbf{S}'' - \mathbf{S}''^2 \leqslant (\mathbf{S}'.\mathbf{I}'')^2 \leqslant \mathbf{S}^2 \leqslant \mathbf{S}'^2. \tag{2·772}$$

Vertices (Fig. 2·78):

L'_r, L''_s = linear r-space and linear s-space contained in L', L'' respectively, L''_s containing $\mathbf{O}$.

$\mathbf{V}', \mathbf{V}''$ = vertices of L'_r, L''_s respectively (points of closest approach).

$$\mathbf{V}''.(\mathbf{V}' - \mathbf{V}'') = 0. \tag{2·773}$$

To find vertices without orthonormalization:

$$\mathbf{S}'_0 = \text{some point on } L'_r,$$

$$\mathbf{T}'_\rho\ (\rho = 1, 2, \ldots, r) = \text{linearly independent* vectors lying in } L'_r,$$

$$\mathbf{T}''_\sigma\ (\sigma = 1, 2, \ldots, s) = \text{linearly independent* vectors lying in } L''_s.$$

$$\mathbf{T}'_\rho \cdot \mathbf{T}''_\sigma = 0. \tag{2·774}$$

$$\left.\begin{aligned} \mathbf{V}' &= \mathbf{S}'_0 + \mathbf{T}', & \mathbf{T}' &= \sum_{\rho=1}^{r} a'_\rho \mathbf{T}'_\rho, \\ \mathbf{V}'' &= \mathbf{T}'', & \mathbf{T}'' &= \sum_{\sigma=1}^{s} a''_\sigma \mathbf{T}''_\sigma; \end{aligned}\right\} \tag{2·775}$$

$$\left.\begin{aligned} &\sum_{\mu=1}^{r} a'_\mu \mathbf{T}'_\mu \cdot \mathbf{T}'_\rho + \mathbf{S}'_0 \cdot \mathbf{T}'_\rho = 0 \quad (\rho = 1, 2, \ldots, r), \\ &\sum_{\nu=1}^{s} a''_\nu \mathbf{T}''_\nu \cdot \mathbf{T}''_\sigma - \mathbf{S}'_0 \cdot \mathbf{T}''_\sigma = 0 \quad (\sigma = 1, 2, \ldots, s). \end{aligned}\right\} \tag{2·776}$$

Minimum principles for vertices:

For vectors $\mathbf{U}'$ lying in L'_r, $(\mathbf{S}'_0 + \mathbf{U}')^2$ is minimized by

$$\mathbf{U}' = \mathbf{T}' = \mathbf{V}' - \mathbf{S}'_0.$$

For vectors $\mathbf{U}''$ lying in L''_s, $(\mathbf{S}'_0 - \mathbf{U}'')^2$ is minimized by

$$\mathbf{U}'' = \mathbf{T}'' = \mathbf{V}''.$$

Some useful relations:

$$\left.\begin{aligned} \mathbf{V}'^2 &= \mathbf{S}'^2_0 + \mathbf{S}'_0 \cdot \mathbf{T}' = \mathbf{S}'^2_0 - \mathbf{T}'^2, \\ \mathbf{V}''^2 &= \mathbf{S}'_0 \cdot \mathbf{T}'' \qquad = \mathbf{T}''^2, \\ (\mathbf{V}' - \mathbf{V}'')^2 &= \mathbf{V}'^2 - \mathbf{V}''^2 = \mathbf{V}'^2 - \mathbf{V}' \cdot \mathbf{V}'' \\ &= \mathbf{S}'^2_0 + \mathbf{S}'_0 \cdot (\mathbf{T}' - \mathbf{T}'') = \mathbf{S}'^2_0 - \mathbf{T}'^2 - \mathbf{T}''^2. \end{aligned}\right\} \tag{2·777}$$

Orthonormal method:†

$$\mathbf{S}'_0 = \text{some point on } L'_r,$$

$$\mathbf{I}'_\rho\ (\rho = 1, 2, \ldots, r) = \text{orthonormal vectors lying in } L'_r,$$

$$\mathbf{I}''_\sigma\ (\sigma = 1, 2, \ldots, s) = \text{orthonormal vectors lying in } L''_s.$$

* For simplicity, the case of linear independence is given prominence. If there are r_0 relations of linear dependence between $\mathbf{T}'_\rho$ and s_0 between $\mathbf{T}''_\sigma$, then the linear subspaces are to be considered to have dimensionalities indicated by L_{r-r_0} and L_{s-s_0}. The class of the hypercircle is reduced from $r+s$ to $r+s-r_0-s_0$. In solving (2·776), r_0 of the a' and s_0 of the a'' may be arbitrarily assigned, but the method remains essentially unchanged (cf. remarks on pp. 103, 104, 108).

† Useful for theoretical discussions, but often tedious to carry out in applications.

$$\mathbf{I}'_\mu . \mathbf{I}'_\rho = \delta_{\mu\rho}, \quad \mathbf{I}''_\nu . \mathbf{I}''_\sigma = \delta_{\nu\sigma}, \quad \mathbf{I}'_\rho . \mathbf{I}''_\sigma = 0. \tag{2·778}$$

$$\left.\begin{aligned} \mathbf{V}' &= \mathbf{S}'_0 + \mathbf{T}', \quad \mathbf{T}' = -\sum_{\rho=1}^{r} \mathbf{I}'_\rho (\mathbf{S}'_0 . \mathbf{I}'_\rho), \\ \mathbf{V}'' &= \mathbf{T}'', \qquad \mathbf{T}'' = \sum_{\sigma=1}^{s} \mathbf{I}''_\sigma (\mathbf{S}'_0 . \mathbf{I}''_\sigma). \end{aligned}\right\} \tag{2·779}$$

$$\left.\begin{aligned} \mathbf{V}'^2 &= \mathbf{S}_0'^2 - \sum_{\rho=1}^{r} (\mathbf{S}'_0 . \mathbf{I}'_\rho)^2, \quad \mathbf{V}''^2 = \sum_{\sigma=1}^{s} (\mathbf{S}'_0 . \mathbf{I}''_\sigma)^2, \\ (\mathbf{V}' - \mathbf{V}'')^2 &= \mathbf{V}'^2 - \mathbf{V}''^2 = \mathbf{V}'^2 - \mathbf{V}' . \mathbf{V}'' \\ &= \mathbf{S}_0'^2 - \sum_{\rho=1}^{r} (\mathbf{S}'_0 . \mathbf{I}'_\rho)^2 - \sum_{\sigma=1}^{s} (\mathbf{S}'_0 . \mathbf{I}''_\sigma)^2. \end{aligned}\right\} \tag{2·780}$$

The hypercircle of class $r+s$ on which the point $\mathbf{S}$ is located:

$$\left.\begin{aligned} &\mathbf{X} = \mathbf{C} + R\mathbf{J}, \quad \mathbf{J}^2 = 1, \\ &\mathbf{J} . \mathbf{T}'_\rho = 0, \ \mathbf{J} . \mathbf{T}''_\sigma = 0 \quad \text{or} \quad \mathbf{J} . \mathbf{I}'_\rho = 0, \ \mathbf{J} . \mathbf{I}''_\sigma = 0 \\ &(\rho = 1, 2, \ldots, r; \ \sigma = 1, 2, \ldots, s) \\ &\mathbf{C} = \tfrac{1}{2}(\mathbf{V}' + \mathbf{V}'') \quad (\text{see } (2·775) \text{ or } (2·779)), \\ &R = \tfrac{1}{2} \,|\, \mathbf{V}' - \mathbf{V}'' \,| \quad (\text{see } (2·777) \text{ or } (2·780)), \\ &4\mathbf{C}^2 = \mathbf{V}'^2 + 3\mathbf{V}''^2, \quad 4R^2 = \mathbf{V}'^2 - \mathbf{V}''^2. \end{aligned}\right\} \tag{2.781}$$

Inequalities:

$$\left.\begin{aligned} \mathbf{V}''^2 \leqslant \mathbf{S}^2 &\leqslant \mathbf{V}'^2, \\ \mathbf{T}''^2 \leqslant \mathbf{S}^2 &\leqslant \mathbf{S}_0'^2 - \mathbf{T}'^2, \\ \mathbf{S}'_0 . \mathbf{T}'' \leqslant \mathbf{S}^2 &\leqslant \mathbf{S}_0'^2 + \mathbf{S}'_0 . \mathbf{T}', \\ \sum_{\sigma=1}^{s} (\mathbf{S}'_0 . \mathbf{I}''_\sigma)^2 \leqslant \mathbf{S}^2 &\leqslant \mathbf{S}_0'^2 - \sum_{\rho=1}^{r} (\mathbf{S}'_0 . \mathbf{I}'_\rho)^2. \end{aligned}\right\} \tag{2·782}$$

Bounds on $\mathbf{S} . \mathbf{G}$ where $\mathbf{G}$ is any vector:

$$|\, \mathbf{S} . \mathbf{G} - \mathbf{C} . \mathbf{G} \,| \leqslant R \,|\, \mathbf{G}_0 \,| \quad (\mathbf{C}, R \text{ as in } (2·781)), \tag{2·783}$$

$$\left.\begin{aligned} \mathbf{G}_0^2 &= \mathbf{G}^2 - \sum_{\rho=1}^{r} c'_\rho \mathbf{G} . \mathbf{T}'_\rho - \sum_{\sigma=1}^{s} c''_\sigma \mathbf{G} . \mathbf{T}''_\sigma \\ &= \mathbf{G}^2 - \left(\sum_{\rho=1}^{r} c'_\rho \mathbf{T}'_\rho\right)^2 - \left(\sum_{\sigma=1}^{s} c''_\sigma \mathbf{T}''_\sigma\right)^2, \end{aligned}\right\} \tag{2·784}$$

$$\left.\begin{aligned} &\sum_{\mu=1}^{r} c'_\mu \mathbf{T}'_\mu . \mathbf{T}'_\rho - \mathbf{G} . \mathbf{T}'_\rho = 0 \quad (\rho = 1, 2, \ldots, r) \\ &\sum_{\nu=1}^{s} c''_\nu \mathbf{T}''_\nu . \mathbf{T}''_\sigma - \mathbf{G} . \mathbf{T}''_\sigma = 0 \quad (\sigma = 1, 2, \ldots, s); \end{aligned}\right\} \tag{2·785}$$

$$\mathbf{G}_0^2 = \mathbf{G}^2 - \sum_{\rho=1}^{r} (\mathbf{G} . \mathbf{I}'_\rho)^2 - \sum_{\sigma=1}^{s} (\mathbf{G} . \mathbf{I}''_\sigma)^2. \tag{2·786}$$

Use of non-linear subspaces

The systematic method described above involves the use of *linear* subspaces L'_r, L''_s; as a consequence it is a set of *linear* equations we have to solve in (2·776). Although this is usually the most convenient procedure, we can do otherwise.

Suppose we are seeking, as usual, the intersection $\mathbf{S}$ of two *linear* subspaces L', L''. Take two *non-linear* subspaces of finite dimensionality (F'_r, F''_s), immersed respectively in L', L''. We may denote general F-points on them as follows:

$$\left.\begin{aligned} F'_r:\quad & \mathbf{S}'(a'_1, a'_2, \ldots, a'_r), \\ F''_s:\quad & \mathbf{S}''(a''_1, a''_2, \ldots, a''_s), \end{aligned}\right\} \tag{2·787}$$

where the a' and a'' are variable parameters.

We define the vertices $\mathbf{V}'$, $\mathbf{V}''$ of F'_r, F''_s respectively as their points of closest approach, or equivalently by the minimum principle.

For $\mathbf{S}'$ free in F'_r and $\mathbf{S}''$ free in F''_s, $(\mathbf{S}'-\mathbf{S}'')^2$ is minimized by

$$\mathbf{S}'=\mathbf{V}', \quad \mathbf{S}''=\mathbf{V}''. \tag{2·788}$$

Thus to find the values of a', a'' corresponding to $\mathbf{V}'$, $\mathbf{V}''$, we have to solve the $r+s$ equations

$$\left.\begin{aligned} (\mathbf{S}'-\mathbf{S}'').\partial\mathbf{S}'/\partial a'_\rho=0 \quad & (\rho=1,2,\ldots,r), \\ (\mathbf{S}'-\mathbf{S}'').\partial\mathbf{S}''/\partial a''_\sigma=0 \quad & (\sigma=1,2,\ldots,s). \end{aligned}\right\} \tag{2·789}$$

For any choice of $\mathbf{S}'$, $\mathbf{S}''$ on L', L'' respectively, the inequalities (2·728) are available, and $\mathbf{S}$ is confined to a hypersphere as in Fig. 2·74, with consequent upper and lower bounds on $\mathbf{S}^2$ which are easy to write down. If we choose $\mathbf{S}'=\mathbf{V}'$ and $\mathbf{S}''=\mathbf{V}''$, we get the smallest possible hypersphere consistent with choice of $\mathbf{S}'$ on F'_r and $\mathbf{S}''$ on F''_s.

Bounds on $\mathbf{S}^2$ *obtained without solving exactly the equations for the vertices* $\mathbf{V}'$, $\mathbf{V}''$

It will be recalled that the standard procedure requires us to solve *exactly* a set of linear algebraic equations (2·776). The values of a'_ρ, a''_σ found from these equations must be substituted in the $\mathbf{T}'$, $\mathbf{T}''$ of (2·775) so that we may evaluate $\mathbf{V}'^2$ and $\mathbf{V}''^2$ as in (2·777); then we have bounds on $\mathbf{S}^2$ by the first of (2·782), viz. $\mathbf{V}''^2 \leqslant \mathbf{S}^2 \leqslant \mathbf{V}'^2$. And when we say 'bounds', we mean arithmetical values for any particular problem we are considering.

But it is only in very simple cases that we can hope to obtain *exact* arithmetical solutions of a set of simultaneous equations. Usually the work must be done on a machine with a restricted number of digits; for speed of calculation we may work with a smaller number of digits than the machine allows. In any case, we are likely to reject as impracticable the idea of getting an exact solution, and then we are in difficulties, for without an exact solution, the first of (2·782) does not supply bounds for $\mathbf{S}^2$.

Such a situation often arises in arithmetical calculations, and it is usually met by retaining a few more significant figures than are required for the answer, and pretending that the approximate solution is an exact solution. But this is mathematically objectionable, and an alternative procedure is described below. Essentially, the plan is to use (2·772), which supplies bounds for $\mathbf{S}^2$ for any choice of points $\mathbf{S}'$ and $\mathbf{S}''$ on the two linear subspaces, i.e. for any choice of the constants a', a'' in (2·775). We draw the bounds together by making these constants satisfy (2·776) *approximately*, the number of significant figures used depending on the time and machines available. But, whatever values of a', a'' we use, the inequalities (2·796) below are mathematically precise. The details of the procedure are as follows.

We consider only the standard case where L'' contains $\mathbf{O}$. We choose a point $\mathbf{S}'_0$ in L', r vectors $\mathbf{T}'_\rho$ ($\rho = 1, 2, \ldots, r$) lying in L' and s vectors $\mathbf{T}''_\sigma$ ($\sigma = 1, 2, \ldots, s$) lying in L''. We write

$$\mathbf{S}' = \mathbf{S}'_0 + \sum_{\rho=1}^{r} a'_\rho \mathbf{T}'_\rho, \quad \mathbf{S}'' = \sum_{\sigma=1}^{s} a''_\sigma \mathbf{T}''_\sigma. \tag{2·790}$$

Then, no matter what values we give to the coefficients a', a'', the point $\mathbf{S}'$ is in L' and $\mathbf{S}''$ in L''.

The inequalities (2·772) apply:

$$2\mathbf{S}'.\mathbf{S}'' - \mathbf{S}''^2 \leqslant (\mathbf{S}'.\mathbf{I}'')^2 \leqslant \mathbf{S}^2 \leqslant \mathbf{S}'^2, \tag{2·791}$$

where $\mathbf{I}'' = \mathbf{S}''/|\mathbf{S}''|$.

For any choice of a', a'', we define quantities e', e'' by

$$\left.\begin{aligned} e'_\rho &= \sum_{\mu=1}^{r} a'_\mu \mathbf{T}'_\mu.\mathbf{T}'_\rho + \mathbf{S}'_0.\mathbf{T}'_\rho \quad (\rho = 1, 2, \ldots, r),\\ e''_\sigma &= \sum_{\nu=1}^{s} a''_\nu \mathbf{T}''_\nu.\mathbf{T}''_\sigma - \mathbf{S}'_0.\mathbf{T}''_\sigma \quad (\sigma = 1, 2, \ldots, s).\end{aligned}\right\} \tag{2·792}$$

Note that if a', a'' happen to be approximate solutions of (2·776), then e', e'' are small; e', e'' are in fact the *residuals* of these equations.

Using (2·774) (the orthogonality of L' and L''), direct calculation from (2·790) gives

$$\left.\begin{aligned} \mathbf{S}'^2 &= \mathbf{S}_0'^2 + 2\sum_{\rho=1}^{r} a'_\rho \mathbf{S}'_0.\mathbf{T}'_\rho + \sum_{\rho=1}^{r}\sum_{\mu=1}^{r} a'_\rho a'_\mu \mathbf{T}'_\rho.\mathbf{T}'_\mu,\\ \mathbf{S}'.\mathbf{S}'' &= \sum_{\sigma=1}^{s} a''_\sigma \mathbf{S}'_0.\mathbf{T}''_\sigma,\\ \mathbf{S}''^2 &= \sum_{\sigma=1}^{s}\sum_{\nu=1}^{s} a''_\sigma a''_\nu \mathbf{T}''_\sigma.\mathbf{T}''_\nu.\end{aligned}\right\} \tag{2·793}$$

By virtue of (2·792), these may be written

$$\left.\begin{aligned} \mathbf{S}'^2 &= \mathbf{S}_0'^2 + \sum_{\rho=1}^{r} a'_\rho \mathbf{S}'_0 . \mathbf{T}'_\rho + \sum_{\rho=1}^{r} a'_\rho e'_\rho, \\ \mathbf{S}'.\mathbf{S}'' &= \sum_{\sigma=1}^{s} a''_\sigma \mathbf{S}'_0 . \mathbf{T}''_\sigma, \\ \mathbf{S}''^2 &= \sum_{\sigma=1}^{s} a''_\sigma \mathbf{S}'_0 . \mathbf{T}''_\sigma + \sum_{\sigma=1}^{s} a''_\sigma e''_\sigma . \end{aligned}\right\} \qquad (2\cdot794)$$

Hence

$$\left.\begin{aligned} 2\mathbf{S}'.\mathbf{S}'' - \mathbf{S}''^2 &= \sum_{\sigma=1}^{s} a''_\sigma \mathbf{S}'_0 . \mathbf{T}''_\sigma - \sum_{\sigma=1}^{s} a''_\sigma e''_\sigma, \\ (\mathbf{S}'.\mathbf{I}'')^2 &= \left(\sum_{\sigma=1}^{s} a''_\sigma \mathbf{S}'_0 . \mathbf{T}''_\sigma\right)^2 \left(\sum_{\sigma=1}^{s} a''_\sigma \mathbf{S}'_0 . \mathbf{T}''_\sigma + \sum_{\sigma=1}^{s} a''_\sigma e''_\sigma\right)^{-1} . \end{aligned}\right\} \qquad (2\cdot795)$$

We note that if e''_σ are small, these two quantities differ only by a small quantity of the second order. Thus, although it is evident from (2·791) that $(\mathbf{S}'.\mathbf{I}'')^2$ gives us the better lower bound, the improvement is hardly worth the increase in complexity, and so we shall use only the first line of (2·795).

Here is our result: *For any choice of a'_ρ, a''_σ and with e'_ρ, e''_σ defined by* (2·792), *we have for* $\mathbf{S}^2$ *the bounds*

$$\sum_{\sigma=1}^{s} a''_\sigma \mathbf{S}'_0 . \mathbf{T}''_\sigma - \sum_{\sigma=1}^{s} a''_\sigma e''_\sigma \leqslant \mathbf{S}^2 \leqslant \mathbf{S}_0'^2 + \sum_{\rho=1}^{r} a'_\rho \mathbf{S}'_0 . \mathbf{T}'_\rho + \sum_{\rho=1}^{r} a'_\rho e'_\rho . \qquad (2\cdot796)$$

This is a useful formula when the 'errors' e', e'' are small, i.e. when a', a'' are approximate solutions of (2·776).

The following equivalent formula is useful in the torsion problem (Chapter 4):

$$-\sum_{\rho=1}^{r} a'_\rho \mathbf{S}'_0 . \mathbf{T}'_\rho - \sum_{\rho=1}^{r} a'_\rho e'_\rho \leqslant \mathbf{S}_0'^2 - \mathbf{S}^2 \leqslant \mathbf{S}_0'^2 - \sum_{\sigma=1}^{s} a''_\sigma \mathbf{S}'_0 . \mathbf{T}''_\sigma + \sum_{\sigma=1}^{s} a''_\sigma e''_\sigma . \qquad (2\cdot797)$$

Bounds given by (2·796) or (2·797) will be called *reliable*; they are mathematically precise. In contrast, we shall use the word *unreliable* for bounds obtained from (2·782) by using for a', a'' solutions of the linear equations (2·776) obtained by the usual methods of numerical computation to so many significant figures. The disagreement between the reliable and unreliable bounds decreases when the number of significant figures is increased. There is no reason to suppose that the reliable bounds are further apart (or closer together) than the unreliable bounds. In a strict mathematical sense, unreliable bounds have no standing; but then few arithmetical results obtained by practical computation satisfy such strict tests, and to the practical computer (who claims to have *solved* a set of equations when he has not, strictly speaking) the distinction between the two sorts of bounds may seem to be hair-splitting. Nevertheless it is good to have the reliable bounds available if one is worried about rounding-off errors; one can use a small number of significant figures and yet arrive at a meaningful result.

Exercises

1. Let P-space be $-1 \leqslant x \leqslant 1$; let an F-vector $\mathbf{S}$ correspond to a function $s(x)$ in this range, and let the scalar product be $\mathbf{S}.\mathbf{T} = \int_{-1}^{1} s(x)\,t(x)\,dx$. Consider the linear 1-spaces (straight lines):

$$L_1': \quad \mathbf{X} \leftrightarrow ax^2+1,$$
$$L_1'': \quad \mathbf{X} \leftrightarrow bx,$$

where a and b are variable parameters. Show that

(i) L_1'' contains $\mathbf{O}$ but L_1' does not;
(ii) L_1' and L_1'' do not intersect;
(iii) L_1' and L_1'' are orthogonal;
(iv) their vertices are

$$\mathbf{V}' \leftrightarrow -\tfrac{5}{3}x^2+1, \quad \mathbf{V}'' = \mathbf{O}, \quad (\mathbf{V}'-\mathbf{V}'')^2 = \tfrac{8}{9}.$$

$\left(\text{In (iv), minimize } \int_{-1}^{1} (ax^2+1-bx)^2\,dx.\right)$

2. Change the preceding exercise by taking the scalar product

$$\int_{-1}^{1} s'(x)\,t'(x)\,dx$$

instead of the Hilbert product. Show that the statements (i), (ii) and (iii) remain true, but that when we seek the vertices we obtain $\mathbf{V}' \leftrightarrow 1$, $\mathbf{V}'' = \mathbf{O}$, which are at zero distance from one another. (The metric is not positive-definite!)

3. Taking P-space and the scalar product as in Example 1, show that $\mathbf{X} \leftrightarrow \sin ax$ (a being a variable parameter) defines a curve in F-space which is not a straight line, i.e. not a linear subspace, and that this curve is contained inside the hypersphere $\mathbf{X}^2 = 1$.

4. Let P-space be the circular domain $x_1^2 + x_2^2 \leqslant 1$ in the Euclidean plane, in which x_1, x_2 are rectangular Cartesians. Let an F-vector $\mathbf{S}$ correspond to any P-vector field (s_1, s_2), and let the scalar product be

$$\mathbf{S}.\mathbf{T} = \iint (s_1 t_1 + s_2 t_2)\,dx_1\,dx_2.$$

Consider the subspace L' defined by $\mathbf{X} \leftrightarrow \operatorname{grad} u$, where u is any function of the coordinates which equals x_1 on the circle $x_1^2 + x_2^2 = 1$. Show that L' is a linear subspace. Show that the F-point

$$\mathbf{V}' \leftrightarrow (1, 0) = \operatorname{grad} x_1$$

is the point of L' closest to the origin $\mathbf{O}$ of F-space. [For any point $\mathbf{X}$ in L' we can write $\mathbf{X} = \mathbf{V}' + \mathbf{Y}$ where the vector $\mathbf{Y}$ lies in L'. It is easy to prove that $\mathbf{V}'.\mathbf{Y} = 0$ and hence $\mathbf{X}^2 = \mathbf{V}'^2 + \mathbf{Y}^2$, from which the result follows.]

CHAPTER 3

THE DIRICHLET PROBLEM FOR A FINITE DOMAIN IN THE EUCLIDEAN PLANE

The two preceding chapters have dealt with the geometry of function-space, particularly function-space with positive-definite metric. (For indefinite metric, see Part III.) We have now to apply this geometrical theory to the solution of boundary value problems, and in this chapter we shall discuss the simplest and most important of all boundary value problems for partial differential equations—the Dirichlet problem in a plane. In order to show how the hypercircle method may be used in actual arithmetical computation a considerable amount of detail is included in the present chapter and the next one (dealing with the torsion problem), and the reader who would like to get a more general view of the application of the method to boundary value problems may turn at once to Chapter 5.

3·1. THE DIRICHLET PROBLEM IN PHYSICS

The Dirichlet problem will now be defined, and the more important physical instances collected for reference.

Statement of the Dirichlet problem, the Neumann problem, and the mixed problem

Let P-space (physical space) be a finite domain V of the Euclidean plane, bounded by a curve B. The domain may be simply or multiply connected (Fig. 3·11).

The *Dirichlet problem* is the problem of finding a function u to satisfy

$$\Delta u = 0, \quad (u)_B = f, \tag{3·101}$$

where f is a function of position on the boundary B; here Δ is the Laplace operator $\Delta = \partial^2/\partial x_i \partial x_i$, where x_i are rectangular Cartesian coordinates and the suffixes take the values 1, 2 with summation understood for repeated suffixes. (The summation convention will not apply to Greek suffixes occurring later.)

Consider the equations

$$\Delta u = \rho, \quad (u)_B = f, \tag{3·102}$$

where ρ is a given function of position in V. Actually, there is little difference between (3·101) and (3·102), for we can convert the latter into the former by subtracting from u any particular solution of $\Delta u = \rho$. We may regard (3·102) as an alternative, and slightly more general, statement of the Dirichlet problem.

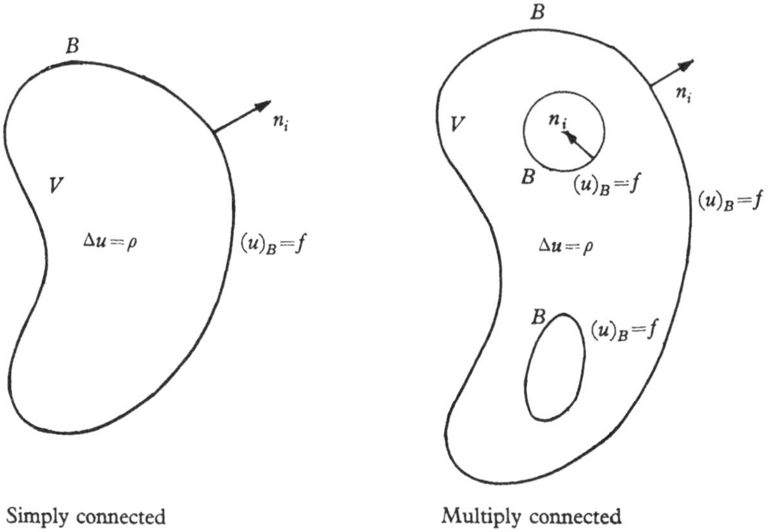

Fig. 3·11. P-space for the Dirichlet problem in a plane.

Another form for the Dirichlet problem is

$$\Delta u = \rho, \quad (u)_B = 0. \tag{3·103}$$

The general form (3·102) may be converted into (3·103) by subtracting from u any particular function which takes the values f on B. And we can of course easily pass from (3·103) to (3·101) just as we passed from (3·102) to (3·101).

The three problems (3·101), (3·102) and (3·103) should be regarded as essentially equivalent. As will appear later, when we have geometrized the Dirichlet problem, the transformations mentioned above (converting one problem into another) are translations in F-space (cf. p. 23). Such translations do not alter F-distances, and they move linear subspaces into parallel positions.*

The *Neumann problem* is that of finding a function v to satisfy

$$\Delta v = 0, \quad (\partial v/\partial n)_B = g, \tag{3·104}$$

* This gives us the idea of an absolute geometry in function-space, rather like *synthetic* geometry in contrast with *analytical* geometry, and this idea is a great help in reconciling the method of the hypercircle with other methods which at first sight seem quite different; cf. Basu(1), (2), Courant(1), Sokolnikoff(1), Chap. 7.

where g is a given function of position on B and ∂n an element of the normal, drawn outwards from V; in other words $\partial v/\partial n = v_{,i} n_i$, the comma denoting partial differentiation with respect to x_i and n_i being the unit normal to B, drawn out from V.

The Neumann problem has no solution unless g satisfies

$$\int g dB = 0, \tag{3·105}$$

where the integral is taken over B, dB being an element of arc. When a solution exists, it is obviously undetermined to within an additive constant.

Other forms of the Neumann problem are

$$\Delta v = \rho, \quad (\partial v/\partial n)_B = g, \tag{3·106}$$

and

$$\Delta v = \rho, \quad (\partial v/\partial n)_B = 0. \tag{3·107}$$

By virtue of the identity

$$\int \Delta\phi dV = \int (\partial\phi/\partial n) dB, \tag{3·108}$$

the conditions for solubility of (3·106) and (3·107) are respectively

$$\int \rho dV = \int g dB \quad \text{and} \quad \int \rho dV = 0. \tag{3·109}$$

The above notation for integrals will be used throughout, except in cases of possible ambiguity. Only a single sign $\int$ is used, although dV is an element of area; an integral with the element dV is to be taken over the domain V.

By the same type of device as those used in the Dirichlet problem, the three Neumann problems (3·104), (3·106) and (3·107) are convertible into one another, and are to be regarded as essentially equivalent.

The forms (3·101) and (3·104) for the Dirichlet and Neumann problems respectively are generally found the most useful, because u and v are then harmonic, and the general theory of harmonic functions becomes available.

The *mixed boundary value problem* is expressed by

$$\Delta w = 0, \quad \alpha(w)_B + \beta(\partial w/\partial n)_B = h, \tag{3·110}$$

where α, β, h are given functions of position on B. The problem may also be stated in the more general, but equivalent, form with $\Delta w = \rho$ instead of $\Delta w = 0$. The Dirichlet and Neumann problems are of course particular cases of (3·110) ($\beta = 0$ and $\alpha = 0$ respectively).

The Dirichlet problem is the most frequently occurring boundary value problem of mathematical physics, and below we shall describe a number of physical instances. These problems will all be stated in two dimensions, although some of them occur in similar form in three dimensions. It should be observed that the equations we have written so far are meaningful in three dimensions if we understand V to be a volume and B the surface that bounds it. In fact, much of the theory which we shall work out for the Dirichlet problem in the plane can be carried over without formal change to the Dirichlet problem in ordinary 3-space, or, indeed, in Euclidean N-space for any N. For simplicity we shall, in the present chapter, speak only of the Euclidean plane, although using for *area* the letter V, more appropriate for *volume*, in order that formulae may pass over easily into three or N dimensions.

Electrostatics

Consider an electrostatic field in a dielectric bounded by a cylindrical conducting surface B. We shall assume independence of the coordinate in space measured parallel to the generators of the cylinder, so that all quantities are functions only of x_i $(i=1,2)$ in the cross-section. There exists then an electrostatic potential ϕ such that, if E_i and D_i are respectively electric intensity and induction, we have

$$\left.\begin{aligned} D_i=\epsilon E_i, \quad E_i=-\phi_{,i},\\ D_{i,i}=0, \quad \Delta\phi=0. \end{aligned}\right\} \tag{3·111}$$

Here ϵ is the dielectric constant, assumed independent of the coordinates.

If the potential is given on the boundary B, then the problem of finding ϕ is the Dirichlet problem

$$\Delta\phi=0, \quad (\phi)_B=f. \tag{3·112}$$

If σ is the surface density of charge on the cylinder, then (for suitable units)

$$2\pi\sigma=-(D_i n_i)_B=\epsilon(\partial\phi/\partial n)_B. \tag{3·113}$$

Suppose now that the domain V of the dielectric is doubly connected, so that the bounding curve B of the cross-section consists of two parts, B_0 outside and B_1 inside. Let them be charged to potentials ϕ_0 and ϕ_1 respectively; then the charges on them (per unit length of cylinder) are

$$e_0=\int\sigma dB_0, \quad e_1=\int\sigma dB_1. \tag{3·114}$$

We have necessarily $e_0+e_1=0$. The capacity C of the condenser (per unit length) is defined as the charge e_0 when the potential difference $\phi_0-\phi_1$ is unity. Now application of Green's theorem gives

$$\begin{aligned}\int\phi_{,i}\phi_{,i}dV&=\int\phi(\partial\phi/\partial n)\,dB\\ &=\phi_0\int(\partial\phi/\partial n)\,dB_0+\phi_1\int(\partial\phi/\partial n)\,dB_1\\ &=2\pi\epsilon^{-1}(\phi_0-\phi_1)\,e_0=2\pi\epsilon^{-1}(\phi_1-\phi_0)\,e_1.\end{aligned} \tag{3·115}$$

Hence $$2\pi C = \epsilon \int \phi_{,i}\phi_{,i}dV. \qquad (3\cdot116)$$

When we come to geometrize the Dirichlet problem in function-space, this Dirichlet integral suggests the positive-definite metric which we shall use. This exemplifies what usually happens—we do not have to hunt round for a suitable metric in function-space; it suggests itself naturally through an integral of fundamental physical importance. In the case of the condenser, the integral in (3·116) represents not only capacity but also energy.

Current flow in a conductor

In plane steady current flow in a conductor of constant conductivity κ, the current vector I_i, the electric intensity E_i and the potential ϕ satisfy

$$I_i = \kappa E_i, \quad E_i = -\phi_{,i}, \quad \Delta\phi = 0. \qquad (3\cdot117)$$

If we are given the potential on the boundary $(\phi)_B = f$, we have a Dirichlet problem. If we are given the normal component of the current on the boundary, we have a Neumann problem. The Dirichlet integral $\int \phi_{,i}\phi_{,i}dV$ is proportional to the rate of generation of heat.

Heat conduction

In plane steady heat flow in a substance of constant conductivity, the temperature T satisfies $\Delta T = 0$, and if the temperature on the boundary is assigned, we have a Dirichlet problem. In this case (alone of those considered) the Dirichlet integral has no elementary and obvious meaning.

Irrotational fluid flow

In plane irrotational flow of an incompressible fluid there is a velocity potential ϕ satisfying $\Delta\phi = 0$. It is not necessary that the motion be steady—the equation holds at every instant.

If the normal velocity of the boundary is assigned, we have $(\partial\phi/\partial n)_B = g$ and hence a Neumann problem. If we use, instead of ϕ, a stream function ψ, this also is harmonic, and an assigned normal velocity on the boundary gives the tangential derivative of ψ. If the region is simply connected, this enables us to state the problem in Dirichlet form

$$\Delta\psi = 0, \quad (\psi)_B = f, \qquad (3\cdot118)$$

but if the region is multiply connected, certain complications arise; these will be dealt with later in this section.

A particular case of some interest is that in which the boundary is rigid and rotates about the origin with angular velocity ω. Then (3·118) reads

$$\Delta\psi = 0, \quad (\psi)_B = \tfrac{1}{2}\omega r^2, \tag{3·119}$$

where r is the distance from the origin.

In irrotational flow the Dirichlet integral has a simple physical meaning. If ρ is the density, the kinetic energy of the fluid (per unit length perpendicular to the plane of the motion) is

$$T = \tfrac{1}{2}\rho \int \phi_{,i}\phi_{,i} dV = \tfrac{1}{2}\rho \int \psi_{,i}\psi_{,i} dV. \tag{3·120}$$

Flow of a viscous fluid through a tube

Let x_i be coordinates in the cross-section of a fixed straight tube through which an incompressible viscous fluid flows under a constant pressure gradient Π. The velocity u (in the direction of the tube) satisfies

$$\Delta u = -\Pi/\mu, \quad (u)_B = 0, \tag{3·121}$$

where μ is the viscosity. This is a Dirichlet problem in the form (3·103). The Dirichlet integral corresponds to the rate of generation of heat per unit length of the tube (cf. Synge (6)).

Elastic membrane

Suppose that an elastic membrane (simply or multiply connected) is attached at its edge to a wire and lies in a plane. A small pressure p is applied on one side of the membrane, the wire being given at the same time a small displacement f perpendicular to the plane. Then the small displacement u of the membrane, perpendicular to the plane, satisfies

$$\Delta u = -p/T, \quad (u)_B = f, \tag{3·122}$$

where T is the tension in the membrane (a constant). Here we have a Dirichlet problem in the form (3·102). The Dirichlet integral $\int u_{,i}u_{,i} dV$ represents, to within a constant factor, the energy in the membrane.

Torsion

The torsion problem will be treated in detail in Chapter 4. Here we shall merely note that it concerns the twisting of an elastic cylinder and that it can be stated, for a simply connected cross-section V, in the form

$$\Delta\psi = 0, \quad (\psi)_B = \tfrac{1}{2}r^2, \tag{3·123}$$

where r is the distance from the origin. This is essentially the same as (3·119) (fluid in rotating container). If we use the fact that $\Delta(\tfrac{1}{2}r^2) = 2$, it is easy to see that (3·121) (viscous flow) is essentially the same mathematical problem, and so also is (3·122) (elastic membrane) if we put either $f=0$ (i.e. do not deform the bounding wire) or $p=0$, $f=\frac{1}{2}r^2$. But for a multiply connected cross-section the torsion problem is not a Dirichlet problem in the ordinary sense.

In the torsion problem the Dirichlet integral is closely related to the torsional rigidity (see (4·120)) and to the energy per unit length in the twisted cylinder.

The conjugate harmonic function and multiple connectivity

If $u+iv$ is an analytic function of the complex variable x_1+ix_2, then u and v are harmonic functions related by

$$u_{,1}=v_{,2}, \quad u_{,2}=-v_{,1}, \tag{3·124}$$

or, more compactly,

$$u_{,i}=\epsilon_{ij}v_{,j}, \quad v_{,i}=-\epsilon_{ij}u_{,j}, \tag{3·125}$$

where
$$\epsilon_{11}=\epsilon_{22}=0, \quad \epsilon_{12}=-\epsilon_{21}=1. \tag{3·126}$$

We say that v is the *conjugate* of u (and $-u$ the conjugate of v). The relations (3·124) or (3·125) imply

$$\frac{\partial u}{\partial s}=\frac{\partial v}{\partial n}, \quad \frac{\partial u}{\partial n}=-\frac{\partial v}{\partial s}, \tag{3·127}$$

where ∂s is the element of any curve C and ∂n the element of its normal, chosen with senses such that the 90° rotation which carries ∂s into ∂n has the same sense as the 90° rotation which carries Ox_1 into Ox_2.

Here are two questions of some importance to us:

(i) Does a given harmonic function u always possess a conjugate v?

(ii) Can every Neumann problem be changed into an equivalent Dirichlet problem?

To answer Question (i), we integrate the second of (3·127) along any curve C. We see that a conjugate harmonic function v exists if the domain V in which u is defined is simply connected. But if V is multiply connected, a (single-valued) conjugate harmonic function v exists if, and only if,

$$\int \frac{\partial u}{\partial n}\,ds=0, \tag{3·128}$$

the integral being taken round each hole in V (Fig. 3·11), so that there are as many conditions in (3·128) as there are holes. If these conditions are not satisfied, it is still true in a sense that a conjugate harmonic function v exists, but it is multiple-valued (though its derivatives are single-valued), v increasing by a certain amount on going round each hole, viz. by the value of the integral in (3·128), proper attention being paid to signs. These facts are important to us in applying the hypercircle method to the Dirichlet problem, because, although we seek one harmonic function, its conjugate appears indirectly in the work.

As for Question (ii), let us consider single-valued conjugate harmonic functions u, v, which are solutions, respectively, of the Dirichlet problem

$$\Delta u = 0, \quad (u)_B = f, \tag{3·129}$$

and the Neumann problem

$$\Delta v = 0, \quad (\partial v/\partial n)_B = g, \quad \int g\,dB = 0. \tag{3·130}$$

If the domain V is simply connected, we can change either problem into the other by putting $g = df/ds$ or $f = \int g\,ds$, the last equation in (3·130) ensuring that this f is single-valued.

Consider a multiply connected V with h holes bounded by curves B_λ ($\lambda = 1, 2, \ldots, h$); let B_0 be the outer boundary. Let v be a single-valued solution of the Neumann problem (3·130), the single-valuedness implying

$$\int_{B_\lambda} \frac{\partial v}{\partial s}\,ds = 0 \quad (\lambda = 1, 2, \ldots, h). \tag{3·131}$$

In general v possesses no single-valued harmonic conjugate $-u$, and so the Neumann problem (3·130) cannot be changed into a Dirichlet problem (3·129). But if the function g satisfies the additional conditions

$$\int_{B_\lambda} g\,ds = 0 \quad (\lambda = 1, 2, \ldots, h), \tag{3·132}$$

then by (3·128) u does exist, its values on B being given by

$$\frac{\partial u}{\partial s} = \frac{\partial v}{\partial n} = g, \tag{3·133}$$

these equations determining u to within additive constants, one constant for each of the curves B_0, B_1, ..., B_h. Suppressing the constant on B_0 without loss of generality, we have then

$$\left.\begin{aligned} \Delta u = 0, \quad (u)_{B_0} = (G)_{B_0}, \\ (u)_{B_\lambda} = (G)_{B_\lambda} + C_\lambda \quad (\lambda = 1, 2, \ldots, h), \end{aligned}\right\} \tag{3·134}$$

where $G = \int g\,ds$ and C_λ are constants. If these constants were given, (3·134) would be a Dirichlet problem. Actually, they are not given, but on putting $\partial v/\partial s = -\partial u/\partial n$ in (3·131), we have

$$\int_{B_\lambda} \frac{\partial u}{\partial n}\,ds = 0 \quad (\lambda = 1, 2, \ldots, h), \tag{3·135}$$

which are h equations to determine C_λ when (3·134) has been solved.

To sum up: *A Neumann problem can be changed at once into a Dirichlet problem when the domain is simply connected, but, when the domain is multiply connected, this can be done only when the boundary value g satisfies the conditions* (3·132), *and then the new problem is not an ordinary Dirichlet problem, because the constants in* (3·134) *are not assigned, but are to be determined by means of* (3·135).

We shall call the problem posed in (3·134) and (3·135) an *extended Dirichlet problem.*

The Dirichlet problem (3·129) has a unique solution u, but in general for a multiply connected domain the conjugate harmonic function v is not single-valued, and so the Dirichlet problem cannot be changed into a Neumann problem.

The present chapter is devoted to ordinary Dirichlet problems as in (3·129). If the domain is simply connected, as it is in some cases, the Dirichlet problem is equivalent to a Neumann problem, and might have been so formulated. In fact Neumann problems may be regarded as covered in this chapter, provided the domain is simply connected.

But when the domain is multiply connected, this is no longer the case. The Dirichlet problems of the present chapter for multiply connected domains are not equivalent to Neumann problems. Conversely, when we come to discuss the torsion problem in multiply connected domains in Chapter 4, we shall have Neumann problems which cannot be transformed into ordinary Dirichlet problems, although they can be transformed into extended Dirichlet problems, and that is in fact the way in which we shall treat them.

In dealing with conjugate harmonic functions it is often helpful to think hydrodynamically. In the irrotational motion of an incompressible fluid the velocity potential ϕ and the stream function ψ are conjugate harmonic functions. A multiple-valued ϕ corresponds to circulation in an irreducible circuit, and a multiple-valued ψ to a source inside such a circuit.

Exercises

1. Transform the problem

$$\Delta u = x_1^3, \quad (u)_B = 0$$

to the form (3·101).

2. Show that the following problems are essentially the same:

$$\text{(i)} \quad \Delta u = 0, \qquad (u)_B = \tfrac{1}{2}(x_1^2 + x_2^2),$$

$$\text{(ii)} \quad \Delta v = -2, \quad (v)_B = 0.$$

3. Convert the Neumann problem

$$\Delta v = 0, \quad (\partial v/\partial n)_B = n_1 x_2 - n_2 x_1,$$

to a Dirichlet problem or to an extended Dirichlet problem.

4. For the region between the concentric circles $r = a$, $r = b\,(a < b)$, solve the Neumann problem

$$\Delta v = 0, \quad (\partial v/\partial n)_{r=a} = a^{-1}, \quad (\partial v/\partial n)_{r=b} = -b^{-1}.$$

Show that the conjugate harmonic function is multiple-valued.

5. For the region of Exercise 4, solve the Dirichlet problem

$$\Delta u = 0, \quad (u)_{r=a} = c + a\cos\theta, \quad (u)_{r=b} = 0,$$

r, θ being polar coordinates. Find c so that the conjugate harmonic function may be single-valued. [$c = 0$.]

3·2. SPLITTING THE DIRICHLET PROBLEM

Having provided a background, we now proceed with our main concern—the application to the Dirichlet problem of the ideas of function-space and, in particular, the method of the hypercircle.

Although we shall presently put $\rho = 0$, let us first consider the Dirichlet problem as in (3·102):

$$\Delta u = \rho, \quad (u)_B = f. \tag{3·201}$$

The domain V (P-space) is a finite portion of the Euclidean plane, simply or multiply connected. We shall suppose ρ and f to be continuous; later we shall see how a piecewise continuous f may be replaced by a continuous one.

Function-space and scalar product

We shall let an F-point (or F-vector) correspond to a P-vector field p_i defined in the given domain and on its boundary ($V + B$). For the present we shall merely assume p_i piecewise continuous; later we shall restrict the discontinuities to what we shall call *permissible discontinuities*.

If $\mathbf{S}$, $\mathbf{S}'$ are two F-vectors, with $\mathbf{S} \leftrightarrow p_i$, $\mathbf{S}'_i \leftrightarrow p'_i$, we define the scalar product by

$$\mathbf{S}.\mathbf{S}' = \int p_i p'_i \, dV. \tag{3·202}$$

We note that the integrand is the ordinary scalar product of the two P-vectors. For the F-metric we have then

$$\mathbf{S}^2 = \int p_i p_i \, dV, \tag{3·203}$$

which is positive-definite, so that $\mathbf{S}^2 = 0$ implies $\mathbf{S} = \mathbf{O}$.

In the above formulae we have used $\mathbf{S}$ and $\mathbf{S}'$ for any two F-vectors, but from now on we shall use these symbols in special senses described below. In particular, **S** will correspond to the *solution* of the Dirichlet problem, or more precisely to its gradient:

$$\mathbf{S} \leftrightarrow p_i, \quad p_i = u_{,i}, \quad \Delta u = \rho, \quad (u)_B = f. \tag{3·204}$$

Here, and throughout, the comma denotes partial differentiation with respect to a coordinate ($u_{,i} = \partial u / \partial x_i$).

The linear subspaces L', L''

We now define a linear subspace L' as follows:*

$$L': \quad \mathbf{S}' \leftrightarrow p'_i, \quad p'_i = u'_{,i}, \quad (u')_B = f. \tag{3·205}$$

Thus L' consists of all F-points corresponding to the gradients of functions which satisfy the boundary condition. The fact that L' is a *linear* subspace can be verified immediately (cf. pp. 31 and 98). We call any F-vector satisfying (3·205) (i.e. any point of L') an *associated* vector.

An F-point in L' (or, equivalently, the position-vector of such a point) must be carefully distinguished from an F-vector *lying in* L'. The latter is obtained by joining two points of L', and so corresponds to the difference of their P-vector fields. Thus, if $\mathbf{T}'$ lies in L', we write

$$\mathbf{T}' \leftrightarrow \bar{p}'_i, \quad \bar{p}'_i = \bar{u}'_{,i}, \quad (\bar{u}')_B = 0. \tag{3·206}$$

Any such F-vector we call a *homogeneous associated* vector.

Next we define a linear subspace L'' as follows:

$$L'': \quad \mathbf{S}'' \leftrightarrow p''_i, \quad p''_{i,i} = \rho. \tag{3·207}$$

Thus L'' consists of all F-points corresponding to P-vector fields with divergence ρ. The linear character of L'' is obvious. We call any F-vector satisfying (3·207) a *complementary* vector.

An F-vector $\mathbf{T}''$ lying in L'' satisfies

$$\mathbf{T}'' \leftrightarrow \bar{p}''_i, \quad \bar{p}''_{i,i} = 0. \tag{3·208}$$

We call such a vector *homogeneous complementary*.

We have now defined two linear subspaces L', L'' as in (3·205) and (3·207) respectively. If these two subspaces intersect, so that $p'_i = p''_i$, this intersection corresponds to a P-vector field p_i which

* Note that the boundary condition must be satisfied by u'. For a method in which only harmonic functions are used, without satisfaction of the boundary condition, see Payne and Weinberger (1, 2); their work deals also with the Neumann problem and the biharmonic equation. The importance of introducing a vector field into what is primarily a scalar problem was emphasized by Weyl (1), but the idea occurs naturally in hydrodynamics, where we think of a field of velocity which is a gradient only if the motion is irrotational (cf. Lamb (1), p. 47).

satisfies the conditions in (3·204), i.e. *the intersection of L' and L'' is the solution F-point* **S**.

Thus we have *split* the Dirichlet problem in the sense that we have reduced it to finding the intersection of two linear subspaces of a suitable function-space with a positive-definite metric. It remains to see whether these subspaces are orthogonal, so that the method of the hypercircle may be applied.

Orthogonality of L' and L''; permissible discontinuities

Let $\mathbf{T}'$ lie in L' and $\mathbf{T}''$ in L''. Then, by (3·206) and (3·208),

$$\mathbf{T}'.\mathbf{T}'' = \int \overline{p}'_i \overline{p}''_i dV = \int \overline{u}'_{,i} \overline{p}''_i dV, \qquad (3\cdot209)$$

or, by Green's theorem,

$$\mathbf{T}'.\mathbf{T}'' = -\int \overline{u}' \overline{p}''_{i,i} dV + \int \overline{u}' \overline{p}''_i n_i dB + \Sigma \int \overline{u}' \overline{p}''_i n_i ds; \qquad (3\cdot210)$$

here the first integral is taken over V and vanishes by (3·208); the second integral is taken over B and vanishes by (3·206); the final term is a sum of integrals taken over both sides of any curves in V across which there are discontinuities, n_i being the unit normal drawn across such a curve from the side on which the integral is taken. Thus the orthogonality of $\mathbf{T}'$, $\mathbf{T}''$ (and hence the orthogonality of L', L'') is assured if we restrict the discontinuities of u' and p'_i (and hence those of $\overline{u}'$ and $\overline{p}''_i$) so as to make the final term in (3·210) disappear, and this is done as follows:

I. In L' (for associated vectors, and hence for homogeneous associated vectors) the function u' (and hence $\overline{u}'$) shall be continuous, but there may exist discontinuities in $\partial u'/\partial n$ (and hence $\partial \overline{u}'/\partial n$) across curves drawn in V.

II. In L'' (for complementary vectors, and hence for homogeneous complementary vectors) the normal component of p''_i (and hence of $\overline{p}''_i$) shall be continuous across all curves drawn in V, the normal component being $p''_i n_i$ where n_i has a single sense for both sides of the curve in question. There may exist discontinuities in the tangential component of p''_i (or $\overline{p}''_i$). (3·211)

The discontinuities allowed in (3·211) we shall call *permissible discontinuities*, and we shall assume in future that only permissible discontinuities occur. *Under this condition the linear subspaces L' and L'' are orthogonal to one another.* We shall make good use of permissible discontinuities in the applications, particularly in connexion with pyramid F-vectors in §3·5.

The fact that **S** is now the intersection of two orthogonal linear subspaces in an F-space with positive-definite metric tells us that the solution of the Dirichlet problem is unique (p. 105). But of course this is well known and easy to prove directly. It is of much greater importance that we have set up a function-space geometry for the Dirichlet problem so that the hypercircle method of §2·7 may be applied. Before proceeding to that, let us clear up the question of possible discontinuities in the boundary value function f of (3·201).

Case where the boundary value function f is only piecewise continuous

In (3·201) we assumed f to be continuous. If we adhered to this restriction, we would rule out an interesting class of problems in which f is only *piecewise* continuous. This occurs, for example, in the case of a hollow circular condenser, split in two by a diameter, the two halves being charged to different electrostatic potentials. Under such circumstances it is impossible to find a continuous function u' satisfying the boundary condition, and so our method breaks down. We save the situation by changing the origin of F-space, as follows.

The polar angle θ is a harmonic function for $x_2 > 0$, piecewise continuous on the x_1-axis with a jump of π at the origin. With this suggestion, it is easy to see that if f is only piecewise continuous on B, we can find a harmonic function v such that $f-(v)_B$ is continuous on B. Such a harmonic function will have infinite derivatives at the points of discontinuity, but that is not a matter of importance at the moment. Defining w by $w = u - v$, we change our original Dirichlet problem (3·201) with f piecewise continuous into the new Dirichlet problem

$$\Delta w = \rho, \quad (w)_B = f - (v)_B = g, \tag{3·212}$$

where g is continuous on B.

In future we shall assume f continuous; if it is only piecewise continuous in any given problem we can make it continuous by the above device, which amounts to displacing the origin of F-space through the vector which corresponds to grad v.

To make L' or L'' pass through **O**

In the theory following p. 113, we found it advantageous to have L'' passing through **O**, the origin of F-space, and the summary of formulae on pp. 118–20 was given for that case. It was merely for the sake of definiteness that we made L'', rather than L', contain **O**; as far as the geometrical theory was concerned, both L' and L'' were on an equal footing. That is no longer the case in the Dirichlet problem, where we have tied L' to the boundary conditions and L'' to the differential equation. It is easy to make either of them pass through **O**.

Our problem is as in (3·201) with L' defined by (3·205) and L'' by (3·207). If $f=0$, L' passes through **O**; if $\rho=0$, L'' passes through **O**.

In general, to make L' pass through **O**, we take any particular function v such that $(v)_B=f$, write $w=u-v$, and so change the original problem (3·201) to

$$\Delta w=\rho-\Delta v, \quad (w)_B=0. \tag{3·213}$$

For this Dirichlet problem L' passes through **O**.

On the other hand, to make L'' pass through **O**, we take any particular solution v of $\Delta v=\rho$, write $w=u-v$, and so change (3·201) to

$$\Delta w=0, \quad (w)_B=f-(v)_B. \tag{3·214}$$

For this Dirichlet problem, L'' passes through **O**.

There is little to choose between the two methods, for in applying the method of the hypercircle we have always to get at least one point on L' and one on L''; having got such a point, we can transfer the origin to it. The problem of finding F-vectors *lying in* the two subspaces is unaffected by the transformations described above, which amount only to translations of F-space.

Summary of splitting

We shall assume that L'' contains **O**, so that the summary of formulae of pp. 118–20 is available without modification. Our Dirichlet problem reads

$$\Delta u=0, \quad (u)_B=f, \tag{3·215}$$

and we assume f to be continuous on B (cf. p. 137). The splitting of the problem is summarized as follows, it being understood that the P-vector fields for L' and L'' may have the permissible discontinuities of (3·211):

$$\left.\begin{array}{l}
\text{Solution:}\quad \mathbf{S}\leftrightarrow p_i=u_{,i},\quad \Delta u=0,\quad (u)_B=f.\\
L':\quad \text{associated:}\quad \mathbf{S}'\leftrightarrow p_i'=u_{,i}',\quad (u')_B=f;\\
\qquad \text{homogeneous associated:}\quad \mathbf{T}'\leftrightarrow \bar{p}_i'=\bar{u}_{,i}',\quad (\bar{u}')_B=0.\\
L'':\quad \text{complementary:}\quad \mathbf{S}''\leftrightarrow p_i'',\quad p_{i,i}''=0;\\
\qquad \text{homogeneous complementary:}\quad \mathbf{T}''=\mathbf{S}''.\\
L', L''\ \text{orthogonal to one another, } L''\ \text{containing } \mathbf{O}.
\end{array}\right\} \tag{3·216}$$

Examples of F-points in L′ and L″

By (3·216) we see that to get an F-point $\mathbf{S}'$ in L' means getting a function u' which satisfies the boundary condition; this function is to be continuous, but may have the permissible discontinuity in $\partial u'/\partial n$ (cf. (3·211)). Thus to get an $\mathbf{S}'$ we have to consider both the shape of the boundary B and also the values of f on it.

On the other hand, to get an F-vector $\mathbf{T}'$ lying in L' (a homogeneous associated vector) only the shape of B is involved. Thus it is possible once for all to set up F-vectors $\mathbf{T}'$ valid for a square, a rectangle, a circle, a semicircle, and so on. But of course some of these may not be as useful as others, because any symmetry in f which coincides with a symmetry of B indicates a symmetry which we should seek in $\mathbf{S}'$ and in $\mathbf{T}'$ if we wish to get a good approximation quickly.

F-vectors $\mathbf{S}''$ or $\mathbf{T}''$ lying in L'' (complementary vectors, necessarily homogeneous since L'' contains $\mathbf{O}$) are defined without reference to the boundary at all. Nevertheless, considerations of symmetry in f and B should guide us in our choice of $\mathbf{S}''$ or $\mathbf{T}''$.

To enter into these questions in more detail, let us return to the problem of getting an $\mathbf{S}'$. The simplest case is that in which the boundary-value function f is actually given as a function of the two coordinates, $f(x_1, x_2)$, continuous and differentiable throughout V. This occurs in the torsion problem as in (3·123). We can then get an $\mathbf{S}'$ at once by writing

$$\mathbf{S}' \leftrightarrow f_{,i}. \tag{3·217}$$

If this plan fails us, we have to resort to some device suited to the particular problem, and the consideration of these is best postponed until they are required.

Having in some way obtained one point $\mathbf{S}'$ on L', all other points on L' may be obtained by adding to that $\mathbf{S}'$ vectors $\mathbf{T}'$ lying in L'. This involves getting functions $\overline{u}'$ which vanish on B. If B is given by an equation $b(x_1, x_2) = 0$ with b a function regular in V, an obvious plan is to put

$$\overline{u}' = b(x_1, x_2)\, w(x_1, x_2), \tag{3·218}$$

and select for w simple functions of the coordinates such as

$$1, \quad x_1, \quad x_2, \quad x_1^2, \quad x_1 x_2, \quad x_2^2, \quad \ldots,$$

or combinations indicated by symmetry.

When we come to pyramid F-vectors in § 3·5, a more systematic method will be given. But we shall here note another device for getting $\mathbf{T}'$ which makes use of the permissible discontinuity in the normal derivative of $\overline{u}'$.

Let C be any closed curve which does not pass outside $V+B$. Then take for $\overline{u}'$ any function which is regular inside C, vanishes on C, and also vanishes in the region between C and B. For example, we might take C to be a circle

$$(x_1-a_1)^2+(x_2-a_2)^2=c^2, \tag{3·219}$$

supposing this contained in $V+B$. Then we might take

$$\left.\begin{aligned}\overline{u}' &= (x_1-a_1)^2+(x_2-a_2)^2-c^2 \text{ inside } C,\\ \overline{u}' &= 0 \text{ outside } C.\end{aligned}\right\} \tag{3·220}$$

The expression might of course be multiplied by any factor regular inside C.

It may be remarked here, as an explanation of ideas which may appear to the reader vague and unsystematic, that the hypercircle method may be used in two different ways. One of these is to get rough bounds on the solution of a problem quickly and with little labour, and for that ingenuity rather than system is needed. The other is to grind out, usually with a good deal of toil, bounds which are close to one another, as 'close' is commonly understood. It is quite true that unsystematic methods are usually unavailing for this goal.

As for complementary vectors $\mathbf{S}''$, a good factory for these is

$$\mathbf{S}''\leftrightarrow p''_i, \quad p''_1=\psi_{,2}, \quad p''_2=-\psi_{,1}, \tag{3·221}$$

obviously satisfying the required relation, $p''_{i,i}=0$. Here ψ is any function continuous in V. By (3·211) ψ may have discontinuous normal derivatives across curves in V, since the continuity of ψ ensures the required continuity of $p''_i n_i$. Thus we might make use of a simple curve C as in (3·219) with a function ψ which vanishes on C and outside it; this sometimes simplifies integrations.

Another plan is to take

$$\mathbf{S}''\leftrightarrow p''_i=\phi_{,i}, \tag{3·222}$$

where ϕ is any harmonic function. On account of the absence of boundary conditions for complementary vectors, it is obviously much easier to get them than it is to get associated vectors.

The pyramid F-vectors of § 3·5 will give us a systematic approach to both types.

Exercises

1. For the problem $\Delta u=1$, $(u)_B=1$, where B is the unit circle $r=1$, find
   (*a*) an associated F-vector,
   (*b*) a homogeneous associated F-vector,
   (*c*) a complementary F-vector,
   (*d*) a homogeneous complementary F-vector.

2. Convert the above problem to the form (3·215) and verify that the homogeneous F-vectors you have obtained still are homogeneous F-vectors for the new problem.

3. For what domains V does the gradient of the following function $\bar{u}'$ give a homogeneous associated F-vector?

$\bar{u}' = (x_1^2 - 1)(x_2^2 - 1)$ inside the square bounded by the four lines $x_1^2 = 1$, $x_2^2 = 1$;

$\bar{u}' = 0$ outside that square.

4. For what value of a does the gradient of $ax_1^2 + x_2^2$ give a complementary F-vector as in (3·216)?

3·3. THE HYPERCIRCLE

The hypercircle of class 1 and associated inequalities

With the Dirichlet problem $\Delta u = 0$, $(u)_B = f$ split as in (3·216), we may apply the general results given on p. 118 which are available

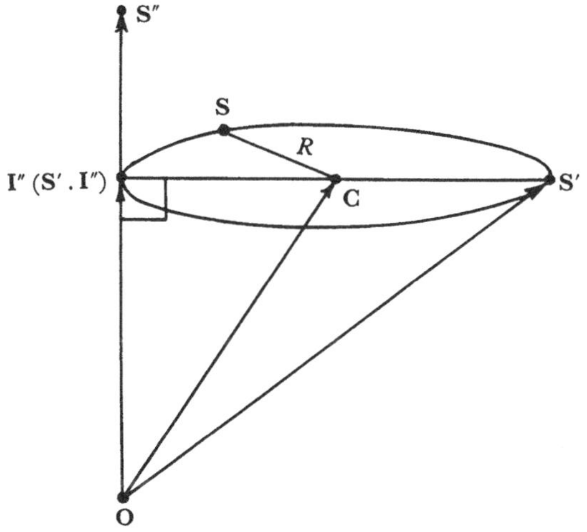

Fig. 3·31. Hypercircle of class 1 obtained from one point $\mathbf{S}'$ on L' and one point $\mathbf{S}''$ on L'', when L'' contains $\mathbf{O}$.

when we have to do with the intersection of two orthogonal linear subspaces L', L'', with L'' containing $\mathbf{O}$. The equations (2·771) locate the solution $\mathbf{S}$ on a hypercircle of class 1, and (2·772) give lower and upper bounds for $\mathbf{S}^2$. It is convenient for reference to quote them here and to reproduce Fig. 2·77 in Fig. 3·31:

Hypercircle of class 1:

$$\left.\begin{aligned} &\mathbf{S} = \mathbf{C} + R\mathbf{J}, \quad \mathbf{J}^2 = 1, \quad \mathbf{J}.\mathbf{I}'' = 0, \\ &\mathbf{C} = \tfrac{1}{2}[\mathbf{S}' + \mathbf{I}''(\mathbf{S}'.\mathbf{I}'')], \quad 4R^2 = \mathbf{S}'^2 - (\mathbf{S}'.\mathbf{I}'')^2, \quad \mathbf{I}'' = \mathbf{S}''/|\mathbf{S}''|. \end{aligned}\right\} \quad (3·301)$$

Inequalities:

$$2\mathbf{S}'.\mathbf{S}''-\mathbf{S}''^2 \leqslant (\mathbf{S}'.\mathbf{I}'')^2 \leqslant \mathbf{S}^2 \leqslant \mathbf{S}'^2. \qquad (3\cdot302)$$

Let us translate these facts into analytical language. According to (3·216), we have

$$\left.\begin{aligned} &\text{Solution:}\quad \mathbf{S}\leftrightarrow p_i=u_{,i},\quad \Delta u=0,\quad (u)_B=f;\\ &L'\text{ (associated):}\quad \mathbf{S}'\leftrightarrow p'_i=u'_{,i},\quad (u')_B=f;\\ &L''\text{(complementary):}\quad \mathbf{S}''\leftrightarrow p''_i,\quad p''_{i,i}=0.\end{aligned}\right\} \qquad (3\cdot303)$$

That is all we need for the present, one point in L' and one in L'' (in addition to the origin); we shall suppose that suitable P-vector fields p'_i, p''_i have been chosen.

Some integrals have to be calculated:

$$\left.\begin{aligned} \mathbf{S}'^2&=\int p'_ip'_i\,dV=a^2,\quad\text{say,}\\ \mathbf{S}''^2&=\int p''_ip''_i\,dV=b^2,\quad\text{say,}\\ \mathbf{S}'.\mathbf{S}''&=\int p'_ip''_i\,dV=c,\quad\text{say;}\end{aligned}\right\} \qquad (3\cdot304)$$

then

$$\mathbf{I}''\leftrightarrow p''_i/b,\quad \mathbf{S}'.\mathbf{I}''=c/b. \qquad (3\cdot305)$$

By (3·301) the gradient of the solution u of our Dirichlet problem may be written in the form

$$u_{,i}=c_i+Rj_i, \qquad (3\cdot306)$$

where the P-vector field c_i and the scalar R are given by

$$\left.\begin{aligned} c_i&=\tfrac{1}{2}(p'_i+p''_ic/b^2),\\ 4R^2&=a^2-c^2/b^2,\end{aligned}\right\} \qquad (3\cdot307)$$

and j_i is some (unknown) P-vector field satisfying

$$\int j_ij_i\,dV=1,\quad \int j_ip''_i\,dV=0. \qquad (3\cdot308)$$

We note that by (3·306) and (3·308) we have

$$\int (u_{,i}-c_i)(u_{,i}-c_i)\,dV=R^2, \qquad (3\cdot309)$$

so that, if R is small, c_i is a good approximation to $u_{,i}$ in the mean square sense; c_i of course corresponds to the centre of the hypercircle and R is its radius.

In general the P-vector field c_i will not be a gradient. Since p'_i is, by definition, a gradient, it follows from (3·307) that c_i is a gradient if, and only if, p''_i is a gradient. The condition for this is

$$p''_{i,j}-p''_{j,i}=0; \qquad (3\cdot310)$$

thus c_i is a gradient if p''_i is chosen as in (3·222).

As remarked above, c_i may be a good approximation to $u_{,i}$, but the equation $u_{,i}=c_i$ is certainly false if $R \neq 0$; this is seen by (3·309). There are, however, other approximations to $u_{,i}$ which *might* give the correct value. Such approximations are given by any expression as on the right-hand side of (3·306), with j_i satisfying (3·308); the mean-square error then might (in the absence of further knowledge) be anything from zero to the square of the diameter of the hypercircle:

$$\int (u_{,i}-c_i-Rj_i)(u_{,i}-c_i-Rj_i)\,dV \leqslant 4R^2. \tag{3·311}$$

The formula (3·306) is of course exact for *some* j_i, but we do not know what that j_i is; all we know is that it satisfies (3·308).

The inequalities (3·302) give us

$$2c-b^2 \leqslant c^2/b^2 \leqslant \int u_{,i}u_{,i}\,dV \leqslant a^2, \tag{3·312}$$

or, more explicitly,

$$\begin{aligned} 2\int p'_i p''_i\,dV - \int p''_i p''_i\,dV &\leqslant \left(\int p'_i p''_i\,dV\right)^2 / \left(\int p''_i p''_i\,dV\right) \\ &\leqslant \int u_{,i}u_{,i}\,dV \leqslant \int p'_i p'_i\,dV. \end{aligned} \tag{3·313}$$

Of these inequalities, the last is very well known; it tells us that *the integral of the square of the gradient of the solution of the Dirichlet problem is less than the integral of the square of the gradient of any other function which satisfies the boundary condition* (Dirichlet (1), p. 127; Kellogg (1), p. 279; Sternberg and Smith (1), p. 187). Actually the inequality says 'cannot exceed', but the stronger statement is easily justified.

The other inequalities are more modern, and appear to be essentially due to Trefftz (1) (see Friedrichs (1)). Reference may also be made to Diaz and Weinstein (1), who, however, assume that p''_i is a gradient. It does not seem that the permissible discontinuities of (3·211) have been much considered (cf. Synge (4)).

Example: $(u)_B = \frac{1}{2}r^2$ *for a square.*

As a simple example to illustrate the above results, consider the Dirichlet problem

$$\Delta u = 0, \quad (u)_B = \tfrac{1}{2}r^2 = \tfrac{1}{2}(x_1^2+x_2^2), \tag{3·314}$$

where V is the interior of the square B bounded by the lines $x_1^2=1$, $x_2^2=1$. This is actually the torsion problem for a beam of square section, and the solution is known (Sokolnikoff (1), p. 128; Love (1), pp. 317, 323).

We take, as in (3·217),

$$u' = \tfrac{1}{2}(x_1^2+x_2^2), \quad \mathbf{S}' \leftrightarrow p'_i = u'_{,i} = (x_1, x_2), \tag{3·315}$$

and, as in (3·222),

$$\begin{aligned} \mathbf{S}'' \leftrightarrow p''_i &= \operatorname{grad} \tfrac{1}{4}(x_1^4 - 6x_1^2x_2^2 + x_2^4) \\ &= (x_1^3 - 3x_1x_2^2, \quad -3x_1^2x_2 + x_2^3). \end{aligned} \tag{3·316}$$

In selecting a harmonic function, we have gone to the fourth degree on account of the symmetry of the square. Then, as in (3·304),

$$\left.\begin{aligned} a^2 &= \mathbf{S}'^2 = \int (x_1^2+x_2^2)\,dV = \tfrac{8}{3}, \\ b^2 &= \mathbf{S}''^2 = \int [(x_1^3-3x_1x_2^2)^2+(-3x_1^2x_2+x_2^3)^2]\,dV = \tfrac{96}{35}, \\ c &= \mathbf{S}'.\mathbf{S}'' = \int [x_1(x_1^3-3x_1x_2^2)+x_2(-3x_1^2x_2+x_2^3)]\,dV = -\tfrac{16}{15}. \end{aligned}\right\} \tag{3·317}$$

Thus the hypercircle relation (3·306) reads

$$u_{,1} = c_1 + Rj_1, \quad u_{,2} = c_2 + Rj_2, \tag{3·318}$$

where

$$\left.\begin{aligned} c_1 &= \tfrac{1}{2}[x_1 + cb^{-2}(x_1^3-3x_1x_2^2)] = \tfrac{1}{2}x_1 - (\tfrac{7}{36})(x_1^3-3x_1x_2^2), \\ c_2 &= \tfrac{1}{2}[x_2 + cb^{-2}(-3x_1^2x_2+x_2^3)] = \tfrac{1}{2}x_2 + (\tfrac{7}{36})(3x_1^2x_2-x_2^3), \\ 4R^2 &= a^2 - c^2/b^2 = \tfrac{304}{135}, \end{aligned}\right\} \tag{3·319}$$

and j_i is some vector field satisfying

$$\int (j_1^2+j_2^2)\,dV = 1, \quad \int [j_1(x_1^3-3x_1x_2^2)+j_2(-3x_1^2x_2+x_2^3)]\,dV = 0. \tag{3·320}$$

We know that the gradient of the solution is certainly of the form (3·318), but we do not know what j_i is. We might take c_i as an approximation to the gradient of the solution, but it is a very poor one; the mean-square error, as in (3·309), is $R^2 = 76/135$.

The inequalities (3·312) give

$$-512/105 \leqslant 56/135 \leqslant \int u_{,i}u_{,i}\,dV \leqslant \tfrac{8}{3}, \tag{3·321}$$

or

$$0{\cdot}4148 \leqslant \int u_{,i}u_{,i}\,dV \leqslant 2{\cdot}6667. \tag{3·322}$$

The upper bound is very crude; the true value is 0·4175.

With such a simple approach close bounds are hardly to be expected, but they may be much improved by a new choice of $\mathbf{S}'$; instead of (3·315) let us take

$$u' = \tfrac{1}{2}(x_1^2+x_2^2) + \tfrac{1}{2}k(x_1^2-1)(x_2^2-1) \tag{3·323}$$

where k is a parameter to be chosen later. This is really the same thing as taking a second point on L'. We have then

$$\left.\begin{aligned} \mathbf{S}' \leftrightarrow p_i' &= [x_1 + kx_1(x_2^2-1),\, x_2 + kx_2(x_1^2-1)], \\ a^2 = \mathbf{S}'^2 &= \int [\{x_1 + kx_1(x_2^2-1)\}^2 + \{x_2 + kx_2(x_1^2-1)\}^2]\,dV \\ &= \tfrac{8}{3}(1 - \tfrac{4}{3}k + \tfrac{8}{15}k^2). \end{aligned}\right\} \tag{3·324}$$

We now choose $k = \tfrac{5}{4}$, which minimizes this expression, and gives

$$a^2 = \mathbf{S}'^2 = \tfrac{4}{9}. \tag{3·325}$$

Thus the bounds in (3·322) are improved to read

$$0{\cdot}4148 \leqslant \int u_{,i}u_{,i}\,dV \leqslant 0{\cdot}4445. \tag{3·326}$$

As the exact solution of this problem is known, and as we shall later develop more systematic ways for getting approximate solutions for such problems, we shall not here pursue any more refined calculations.

The vertices $\mathbf{V}'$, $\mathbf{V}''$ *and the hypercircle of class* $r+s$

Let us now see how the general method summarized in equations (2·773)–(2·782) is to be used in the approximate solution of the Dirichlet problem.

We take, in the linear subspaces L', L'' of (3·216), linear subspaces L'_r, L''_s of finite dimensionality. We write

$$\left.\begin{array}{l} \text{Solution:}\quad \mathbf{S} \leftrightarrow p_i = u_{,i},\quad \Delta u = 0,\quad (u)_B = f. \\ L'_r\text{:}\quad \text{associated:}\quad \mathbf{S}'_0 \leftrightarrow p'_i = u'_{,i},\quad (u')_B = f; \\ \qquad \text{homogeneous associated:}\quad \mathbf{T}'_\rho \leftrightarrow p'_{(\rho)i} = u'_{(\rho),i} \\ \qquad\qquad (u'_{(\rho)})_B = 0 \quad (\rho = 1, 2, \ldots, r). \\ L''_s\text{:}\quad \text{complementary:}\quad \mathbf{T}''_\sigma \leftrightarrow p''_{(\sigma)i},\ p''_{(\sigma)i,i} = 0 \\ \qquad\qquad (\sigma = 1, 2, \ldots, s). \end{array}\right\} \tag{3·327}$$

We note that u', $u'_{(\rho)}$, $p''_{(\sigma)i}$ may have the permissible discontinuities of (3·211). Let us suppose that these quantities have been chosen.

The next step is to solve the linear equations (2·776). In order to write these equations explicitly, the following integrals have to be evaluated:

$$\left.\begin{aligned} A'_{\mu\rho} &= \mathbf{T}'_\mu \cdot \mathbf{T}'_\rho = \int p'_{(\mu)i} p'_{(\rho)i}\, dV, \\ A'_\rho &= \mathbf{S}'_0 \cdot \mathbf{T}'_\rho = \int p'_i p'_{(\rho)i}\, dV, \\ A''_{\nu\sigma} &= \mathbf{T}''_\nu \cdot \mathbf{T}''_\sigma = \int p''_{(\nu)i} p''_{(\sigma)i}\, dV, \\ A''_\sigma &= \mathbf{S}'_0 \cdot \mathbf{T}''_\sigma = \int p'_i p''_{(\sigma)i}\, dV. \end{aligned}\right\} \tag{3·328}$$

In this notation (2·776) read

$$\left.\begin{aligned} \sum_{\mu=1}^{r} a'_\mu A'_{\mu\rho} + A'_\rho &= 0 \quad (\rho = 1, 2, \ldots, r), \\ \sum_{\nu=1}^{s} a''_\nu A''_{\nu\sigma} - A''_\sigma &= 0 \quad (\sigma = 1, 2, \ldots, s). \end{aligned}\right\} \tag{3·329}$$

If the vectors $\mathbf{T}'_\rho$, $\mathbf{T}''_\sigma$ are linearly independent (cf. footnotes to pp. 104, 108, 119), then $\det A'_{\mu\rho} \neq 0$, $\det A''_{\nu\sigma} \neq 0$, and there exist matrices $A'^{\alpha\rho}$, $A''^{\beta\sigma}$ such that, with Kronecker deltas as in (2·301),

$$\sum_{\rho=1}^{r} A'_{\mu\rho} A'^{\alpha\rho} = \delta^\alpha_\mu, \quad \sum_{\sigma=1}^{s} A''_{\nu\sigma} A''^{\beta\sigma} = \delta^\beta_\nu. \tag{3·330}$$

The solutions of (3·329) are then given by

$$a'_\alpha = -\sum_{\rho=1}^{r} A'_\rho A'^{\alpha\rho}, \quad a''_\beta = \sum_{\sigma=1}^{s} A''_\sigma A''^{\beta\sigma}, \tag{3·331}$$

and hence

$$\left.\begin{aligned}\sum_{\alpha=1}^{r} a'_\alpha A'_\alpha &= -\sum_{\alpha=1}^{r}\sum_{\rho=1}^{r} A'_\alpha A'^{\alpha\rho} A'_\rho,\\ \sum_{\beta=1}^{s} a''_\beta A''_\beta &= \sum_{\beta=1}^{s}\sum_{\sigma=1}^{s} A''_\beta A''^{\beta\sigma} A''_\sigma.\end{aligned}\right\} \tag{3·332}$$

This method of solving the linear equations (3·329) is systematic, but it is not the method we shall generally employ. If r and s are small (say, 2 or 3), we can solve the equations directly, and if r and s are large, it is usually more practical to solve by successive approximations, noting the remarks on p. 121 et seq.; we may also find help in the minimum principles given on p. 119.

Let us suppose that we have solved (3·329) in some way, so that the numbers a'_ρ, a''_σ are known. Then the vertices $\mathbf{V}'$, $\mathbf{V}''$ of L'_r, L''_s respectively are, by (2·775),

$$\left.\begin{aligned}\mathbf{V}' &= \mathbf{S}'_0 + \mathbf{T}', \quad &\mathbf{T}' &= \sum_{\rho=1}^{r} a'_\rho \mathbf{T}'_\rho \leftrightarrow \sum_{\rho=1}^{r} a'_\rho p'_{(\rho)i},\\ \mathbf{V}'' &= \mathbf{T}'', &\mathbf{T}'' &= \sum_{\sigma=1}^{s} a''_\sigma \mathbf{T}''_\sigma \leftrightarrow \sum_{\sigma=1}^{s} a''_\sigma p''_{(\sigma)i},\end{aligned}\right\} \tag{3·333}$$

and by (2·777)

$$\left.\begin{aligned}\mathbf{V}'^2 = \mathbf{S}'^2_0 + \mathbf{S}'_0 \,.\, \mathbf{T}' &= \mathbf{S}'^2_0 + \sum_{\rho=1}^{r} a'_\rho \mathbf{S}'_0 \,.\, \mathbf{T}'_\rho\\ &= \int p'_i p'_i \, dV + \sum_{\rho=1}^{r} a'_\rho A'_\rho,\\ \mathbf{V}''^2 = \mathbf{S}'_0 \,.\, \mathbf{T}'' &= \sum_{\sigma=1}^{s} a''_\sigma A''_\sigma.\end{aligned}\right\} \tag{3·334}$$

We note that expressions of the form (3·332) are needed for substitution here.

As an alternative to solving (3·329), we may try orthonormalization as in (2·778), but experience shows that this may be even more tedious than the solution of the linear equations. This is because a symmetry in the problem (as in the torsion of a square) may help us in solving (3·329), but this symmetry is lost in the process of orthonormalization, which requires us to arrange our vectors in some definite order; symmetry calls for an arrangement in groups, not in a sequence.

Having got the vertices as in (3·333), the hypercircle of class $r+s$ is given by (2·781); the formulae read

$$\left.\begin{aligned}&\mathbf{S} = \mathbf{C} + R\mathbf{J}, \quad \mathbf{J}^2 = 1,\\ &\mathbf{J}\,.\,\mathbf{T}'_\rho = 0, \quad \mathbf{J}\,.\,\mathbf{T}''_\sigma = 0 \quad (\rho = 1, 2, \ldots, r;\ \sigma = 1, 2, \ldots, s),\\ &\mathbf{C} = \tfrac{1}{2}(\mathbf{V}' + \mathbf{V}''), \quad 4\mathbf{C}^2 = \mathbf{V}'^2 + 3\mathbf{V}''^2,\\ &R = \tfrac{1}{2}\,|\,\mathbf{V}' - \mathbf{V}''\,|, \quad 4R^2 = \mathbf{V}'^2 - \mathbf{V}''^2.\end{aligned}\right\} \tag{3·335}$$

In the language of analysis, this means that the gradient of the solution is of the form

$$u_{,i} = c_i + Rj_i, \tag{3·336}$$

where

$$\left.\begin{aligned} \mathbf{C} \leftrightarrow c_i &= \tfrac{1}{2}\Big(p'_i + \sum_{\rho=1}^{r} a'_\rho p'_{(\rho)i} + \sum_{\sigma=1}^{s} a''_\sigma p''_{(\sigma)i}\Big), \\ 4R^2 = \mathbf{V}'^2 - \mathbf{V}''^2 &= \int p'_i p'_i \, dV + \sum_{\rho=1}^{r} a'_\rho A'_\rho - \sum_{\sigma=1}^{s} a''_\sigma A''_\sigma, \end{aligned}\right\} \tag{3·337}$$

and j_i is some (unknown) P-vector field satisfying

$$\left.\begin{aligned} &\int j_i j_i \, dV = 1, \\ &\int j_i p'_{(\rho)i} \, dV = 0 \quad (\rho = 1, 2, \ldots, r), \\ &\int j_i p''_{(\sigma)i} \, dV = 0 \quad (\sigma = 1, 2, \ldots, s). \end{aligned}\right\} \tag{3·338}$$

Although formulae have been written out here explicitly in order to show what they mean, it is generally wise to use the vectorial notation of function-space as far as possible; it is shorter and it helps one to have a geometrical realization of what is going on.

Bounds on $\mathbf{S}^2$ *and* $\mathbf{S}.\mathbf{G}$

In (2·782) we have the simple and important inequalities:

$$\mathbf{V}''^2 \leqslant \mathbf{S}^2 \leqslant \mathbf{V}'^2. \tag{3·339}$$

By (3·334) these may be written

$$\sum_{\sigma=1}^{s} a''_\sigma A''_\sigma \leqslant \int u_{,i} u_{,i} \, dV \leqslant \int p'_i p'_i \, dV + \sum_{\rho=1}^{r} a'_\rho A'_\rho, \tag{3·340}$$

where A'_ρ, A''_σ are as in (3·328) and a'_ρ, a''_σ are the solutions of (3·329); the sums occurring in (3·340) may be expressed as in (3·332).

A curious feature of (3·340) is that the bounds as written here do not seem to be necessarily positive. But of course they are, and their positive character depends on the way in which a'_ρ and a''_σ in (3·329) involve A'_μ and A''_ν.

If $\mathbf{G}$ is any F-vector, bounds on $\mathbf{S}.\mathbf{G}$ are given by (2·783), and their evaluation requires the solution of the linear equations (2·785). These equations are like (3·329), the same coefficients $A'_{\mu\rho}$, $A''_{\nu\sigma}$ occurring in both, but the other terms are different. Here we have an argument in favour of orthonormalization, for if we have once orthonormalized from $\mathbf{T}'_\rho$, $\mathbf{T}''_\rho$ to $\mathbf{I}'_\rho$, $\mathbf{I}''_\sigma$, both (3·329) and (2·785) are automatically solved. For the bounds on $\mathbf{S}.\mathbf{G}$ we would then simply use (2·786) in (2·783).

The method of the hypercircle will now be illustrated by two simple examples. The aim here is not to obtain highly accurate approximations, but rather to explore what can be done without heavy calculations by simple choices of associated and complementary F-vectors.

Example 1. $(u)_B = x^n$ *for a square*

We shall use here x and y for rectangular Cartesian coordinates. Let V be the interior of the square B formed by the lines $x=0$, $y=0$, $x=1$, $y=1$. Consider the Dirichlet problem:

$$\Delta u = 0, \quad (u)_B = x^n, \tag{3·341}$$

where $n \geqslant 1$, but it is not necessarily an integer. (The case $n=2$ is connected with the torsion problem; cf. p. 227.)

Let us first get a hypercircle of class 1, choosing

$$\mathbf{S}' \leftrightarrow \operatorname{grad} x^n = (nx^{n-1}, 0), \quad \mathbf{S}'' \leftrightarrow (1, 0), \tag{3·342}$$

the latter P-vector field having zero divergence, as required in (3·303). Then

$$\left.\begin{aligned} \mathbf{S}'^2 &= \iint n^2 x^{2n-2}\, dx\, dy = n^2/(2n-1), \\ \mathbf{S}''^2 &= \iint 1 \,.\, dx\, dy = 1, \\ \mathbf{S}'.\mathbf{S}'' &= \iint n x^{n-1}\, dx\, dy = 1, \\ \mathbf{I}'' &= \mathbf{S}'', \quad \mathbf{S}'.\mathbf{I}'' = 1. \end{aligned}\right\} \tag{3·343}$$

By (3·301) the centre of the hypercircle and the square of its radius are

$$\mathbf{C} \leftrightarrow [\tfrac{1}{2}(nx^{n-1}+1), 0], \quad R^2 = \tfrac{1}{4}(n-1)^2/(2n-1), \tag{3·344}$$

and (3·302) gives the bounds

$$1 \leqslant \mathbf{S}^2 \leqslant n^2/(2n-1). \tag{3·345}$$

The expression for $\mathbf{C}$ suggests that we regard the terms in the square brackets as an approximation to the components of the gradient of the solution, or, equivalently, that we accept $\frac{1}{2}(x^n + x)$ as an approximation to the solution u. But the mean-square error of this is R^2 and we see from (3·344) that in general R^2 is not small; hence we are not to regard the proposed approximate solution as a good one. But there is an exceptional case, namely, that in which the given exponent n is nearly equal to unity, for then R^2 is small. For example, if $n = 1{\cdot}01$, then $R^2 = 0{\cdot}00002451$, so that the mean-square error is small; the bounds on $\mathbf{S}^2$ are drawn close together and we have

$$1 \leqslant \mathbf{S}^2 \leqslant 1{\cdot}00009804. \tag{3·346}$$

The reason for this goodness of approximation is that if $n=1$ the solution of the problem (3·341) is $u=x$; then $\mathbf{S} \leftrightarrow (1, 0)$, which is precisely the $\mathbf{S}''$ we selected above. The result illustrates the application of the hypercircle method to a Dirichlet problem in which the boundary conditions differ only slightly from those of a problem for which the solution is known.

Let us return to the general problem (3·341), n being any number greater than unity, and proceed to the hypercircle of class $r+s$, following the method of (3·327).

We choose as associated F-vector

$$\mathbf{S}'_0 \leftrightarrow \operatorname{grad} x^n = (nx^{n-1}, 0). \tag{3·347}$$

In setting up homogeneous associated F-vectors $\mathbf{T}'_\rho$, we observe that the solution u of the problem must be symmetric with respect to the line $y=\frac{1}{2}$; thus u is an even function of $y-\frac{1}{2}$. If we restrict the functions $u'_{(\rho)}$ to such symmetry and remember that they are to vanish for $x=0$, $x=1$, $y=0$, $y=1$, the following functions suggest themselves:

$$u'_{(p,q)} = x^p(x-1)(y-\tfrac{1}{2})^{2q}y(y-1) \quad (p=1,2,\ldots;\ q=0,1,2,\ldots). \tag{3·348}$$

To keep the algebra from getting too complicated, we shall take only two such functions, writing

$$\left.\begin{aligned} \mathbf{T}'_1 &\leftrightarrow \operatorname{grad} x(x-1)\,y(y-1), \\ \mathbf{T}'_2 &\leftrightarrow \operatorname{grad} x^2(x-1)\,y(y-1). \end{aligned}\right\} \tag{3·349}$$

For the complementary F-vectors $\mathbf{T}''_\sigma$ we may use a function ψ as in (3·221). Symmetry suggests that ψ should be an *odd* function of $y-\frac{1}{2}$, and so we are led to consider functions of the form

$$\psi_{(p,q)} = x^p(y-\tfrac{1}{2})^{2q+1} \quad (p=0,1,\ldots;\ q=0,1,\ldots). \tag{3·350}$$

Again for simplicity we shall take only two such functions. Putting $p=0$, $q=0$ and $p=1$, $q=0$, and using the notation

$$\text{skewgrad} = (\partial/\partial y,\ -\partial/\partial x), \tag{3·351}$$

we obtain the two complementary F-vectors

$$\left.\begin{aligned} \mathbf{T}''_1 &\leftrightarrow \text{skewgrad}\,(y-\tfrac{1}{2}), \\ \mathbf{T}''_2 &\leftrightarrow \text{skewgrad}\,x(y-\tfrac{1}{2}). \end{aligned}\right\} \tag{3·352}$$

Now we have taken $r=2$, $s=2$, and so we shall obtain a hypercircle of class $2+2=4$.

Evaluation of the integrals in (3·328) gives

$$\left.\begin{aligned} &A'_{11} = \mathbf{T}'_1\cdot\mathbf{T}'_1 = \tfrac{1}{45}, \quad A'_{12} = \mathbf{T}'_1\cdot\mathbf{T}'_2 = \tfrac{1}{90}, \quad A'_{22} = \mathbf{T}'_2\cdot\mathbf{T}'_2 = \tfrac{4}{525}, \\ &A'_1 = \mathbf{S}'_0\cdot\mathbf{T}'_1 = -\frac{1}{6}\,\frac{n-1}{n+1}, \quad A'_2 = \mathbf{S}'_0\cdot\mathbf{T}'_2 = -\frac{1}{6}\,\frac{n(n-1)}{(n+2)(n+1)}; \\ &A''_{11} = \mathbf{T}''_1\cdot\mathbf{T}''_1 = 1, \quad A''_{12} = \mathbf{T}''_1\cdot\mathbf{T}''_2 = \tfrac{1}{2}, \quad A''_{22} = \mathbf{T}''_2\cdot\mathbf{T}''_2 = \tfrac{5}{12}, \\ &A''_1 = \mathbf{S}'_0\cdot\mathbf{T}''_1 = 1, \quad A''_2 = \mathbf{S}'_0\cdot\mathbf{T}''_2 = n/(n+1), \quad \mathbf{S}'^2_0 = n^2/(2n-1). \end{aligned}\right\} \tag{3·353}$$

Accordingly the equations (3·329) read

$$\left.\begin{aligned} &(\tfrac{1}{45})\,a'_1 + (\tfrac{1}{90})\,a'_2 = (n-1)/[6(n+1)], \\ &(\tfrac{1}{90})\,a'_1 + (\tfrac{4}{525})\,a'_2 = n(n-1)/[6(n+2)(n+1)]; \\ &\quad a''_1 + \tfrac{1}{2}a''_2 = 1, \\ &\tfrac{1}{2}a''_1 + (\tfrac{5}{12})\,a''_2 = n/(n+1), \end{aligned}\right\} \tag{3·354}$$

and the solutions are

$$\left.\begin{aligned}
a_1' &= \frac{15}{2}\frac{n-1}{n+1} - \frac{525}{26}\frac{(n-1)(n-2)}{(n+1)(n+2)},\\
a_2' &= \frac{525}{13}\frac{(n-1)(n-2)}{(n+1)(n+2)},\\
a_1'' &= -\frac{n-5}{2(n+1)},\\
a_2'' &= \frac{3(n-1)}{n+1}.
\end{aligned}\right\} \tag{3·355}$$

If we wish we can now write down the centre of the hypercircle by (3·337). But it is more interesting to get its radius R, which, by (3·337), is given by

$$\begin{aligned}
4R^2 &= \mathbf{S}_0'^2 + \sum_{\rho=1}^{2} a_\rho' A_\rho' - \sum_{\sigma=1}^{2} a_\sigma'' A_\sigma''\\
&= \frac{n^2}{2n-1} - \frac{5}{4}\left(\frac{n-1}{n+1}\right)^2 - \frac{175}{52}\left[\frac{(n-1)(n-2)}{(n+1)(n+2)}\right]^2 - \frac{5n^2-2n+5}{2(n+1)^2}.
\end{aligned} \tag{3·356}$$

By (3·339) the bounds for $\mathbf{S}^2$ are

$$\left.\begin{aligned}
\mathbf{V}''^2 &\leqslant \mathbf{S}^2 \leqslant \mathbf{V}'^2,\\
\mathbf{S}^2 &= \iint (u_x^2 + u_y^2)\,dx\,dy,\\
\mathbf{V}''^2 &= \frac{5n^2-2n+5}{2(n+1)^2},\\
\mathbf{V}'^2 &= \frac{n^2}{2n-1} - \frac{5}{4}\left(\frac{n-1}{n+1}\right)^2 - \frac{175}{52}\left[\frac{(n-1)(n-2)}{(n+1)(n+2)}\right]^2.
\end{aligned}\right\} \tag{3·357}$$

Note, as a check, that when $n=1$ these bounds coincide at the value 1.

It is interesting to see how close this hypercircle of class 4 brings us to the solution; this is shown in the following table:

Problem: $\Delta u = 0$, $(u)_B = x^n$ *on* $x=0$, $y=0$, $x=1$, $y=1$

n	$\mathbf{V}''^2$	$\mathbf{V}'^2$	$R^2 = (\mathbf{V}'^2 - \mathbf{V}''^2)/4$	$\mathbf{C}^2 = (\mathbf{V}'^2 + 3\mathbf{V}''^2)/4$	$R^2/\mathbf{C}^2$
1	1	1	0	1	0
2	1·1667	1·1944	0·0069	1·1736	0·0059
3	1·3750	1·4538	0·0197	1·3947	0·0141
4	1·5400	1·7011	0·0403	1·5903	0·0255
5	1·6667	1·9475	0·0702	1·7369	0·0404

The column R^2 gives the mean-square error if the centre $\mathbf{C}$ of the hypercircle is taken as an approximation to the gradient of the solution. The column $\mathbf{C}^2$ is calculated from (3·335) and (3·357). As an assessment of error, the ratio $R^2/\mathbf{C}^2$ is more significant than R^2; this may be called the 'relative mean-square error'. In view of the crudeness in the choices of $\mathbf{T}_\rho'$, $\mathbf{T}_\sigma''$, it is rather remarkable that this relative error does not amount to 5 % for $1 \leqslant n \leqslant 5$. If we wanted a closer approximation, we would use more homogeneous associated and complementary F-vectors generated from (3·348) and (3·350).

Example 2. *Square condenser*

Consider the square condenser shown in Fig. 3·32. The outer square B_0 of side $2(a+b)$ is charged to potential $u=1$ and the inner square B_1 of side $2a$ is earthed $(u=0)$. We have then the Dirichlet problem:

$$\Delta u=0, \quad (u)_{B_0}=1, \quad (u)_{B_1}=0. \tag{3·358}$$

P-space is the region between the two squares.

We shall use a very simple method, requiring only the most elementary calculations, but making use of the permissible discontinuities of (3·211). The solution **S** will be located on a hypercircle of class $(1+2)$.

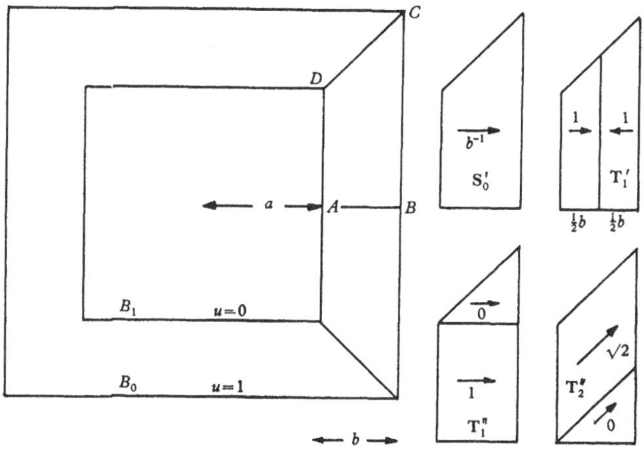

Fig. 3·32. Square condenser of inner side $2a$ and outer side $2(a+b)$. On the right are shown the P-vector fields in the octant $ABCD$ corresponding to $\mathbf{S}_0'$, $\mathbf{T}_1'$, $\mathbf{T}_1''$, $\mathbf{T}_2''$, respectively; the P-vector fields in the other octants are given by reflexions in the axes of symmetry of the squares.

The quadrilateral $ABCD$ in Fig. 3·32 represents one-eighth of the domain V. The four small diagrams on the right show the P-vector fields in $ABCD$ corresponding to $\mathbf{S}_0'$, $\mathbf{T}_1'$, $\mathbf{T}_1''$, $\mathbf{T}_2''$, respectively, the P-vector fields in the other seven octants being in each case obtained by reflexions in the axes of symmetry of the squares (the diagonals and the lines through the centre parallel to the sides).

The three-dimensional graph of u', the gradient of which corresponds to $\mathbf{S}_0'$, is like the surface of a picture frame, with $u'=0$ on B_1 and $u'=1$ on B_0. The graph of $\bar{u}_1'$, where grad $\bar{u}_1' \leftrightarrow \mathbf{T}_1'$, is like the roof of a hollow square building; there is a ridge, but this is a permissible discontinuity. The discontinuities for $\mathbf{T}_1''$ and $\mathbf{T}_2''$ are also permissible, since the normal components are continuous.

If the four small diagrams in Fig. 3·32 are superimposed, the nature of the overlappings of the subregions depends on whether a is greater than or less than b; we shall consider only the case where $b \leqslant a$.

The following results are immediate, and do not involve integrations on account of the constancy of the fields:

$$\left.\begin{aligned}&\mathbf{S}_0'^2 = 4 + 8a/b,\\ &A_{11}' = \mathbf{T}_1'.\mathbf{T}_1' = 4b^2 + 8ab, \quad A_1' = \mathbf{S}_0'.\mathbf{T}_1' = -2b,\\ &A_{11}'' = \mathbf{T}_1''.\mathbf{T}_1'' = 8ab, \quad A_{12}'' = \mathbf{T}_1''.\mathbf{T}_2'' = 8ab - 4b^2, \quad A_{22}'' = \mathbf{T}_2''.\mathbf{T}_2'' = 16ab,\\ &A_1'' = \mathbf{S}_0'.\mathbf{T}_1'' = 8a, \quad A_2'' = \mathbf{S}_0'.\mathbf{T}_2'' = 8a.\end{aligned}\right\} \tag{3·359}$$

Then the equations (3·329) read

$$\left.\begin{aligned}&a_1'(4b^2 + 8ab) - 2b = 0,\\ &a_1''.8ab + a_2''(8ab - 4b^2) - 8a = 0,\\ &a_1''(8ab - 4b^2) + a_2''.16ab - 8a = 0,\end{aligned}\right\} \tag{3·360}$$

and the solutions are

$$\left.\begin{aligned}&a_1' = \tfrac{1}{2}b^{-1}c(1+c)^{-1},\\ &a_1'' = b^{-1}(1+c)(1+2c-c^2)^{-1},\\ &a_2'' = b^{-1}c(1+2c-c^2)^{-1},\end{aligned}\right\} \tag{3·361}$$

where $c = \frac{1}{2}b/a \leqslant \frac{1}{2}$.

We have a linear 1-space L_1' and a linear 2-space L_2''; their vertices are

$$\mathbf{V}' = \mathbf{S}_0' + a_1'\mathbf{T}_1', \quad \mathbf{V}'' = a_1''\mathbf{T}_1'' + a_2''\mathbf{T}_2'', \tag{3·362}$$

and the centre of the hypercircle of class $(1+2)$ is

$$\mathbf{C} = \tfrac{1}{2}(\mathbf{V}' + \mathbf{V}''). \tag{3·363}$$

Fig. 3·33 shows the values of the components of the P-vector fields corresponding to $\mathbf{V}'$, $\mathbf{V}''$ and $\mathbf{C}$ in the octant $ABCD$ of Fig. 3·32, calculated for the case $b/a = \frac{2}{3}$, so that $c = \frac{1}{3}$. (The strength of each field is a function of a and b of the form $b^{-1}f(b/a)$.) The diagram for $\mathbf{C}$ shows the octant broken up into six parts, in each of which a constant P-vector field is assigned; these P-vector fields may be regarded as an approximation (a crude one) to the gradient of the solution of our Dirichlet problem.

Let us consider the error. We have

$$a_1'A_1' = -c(1+c)^{-1}, \quad a_1''A_1'' + a_2''A_2'' = 4c^{-1}(1+2c)(1+2c-c^2)^{-1}, \tag{3·364}$$

and so, by (3·334),

$$\left.\begin{aligned}&\mathbf{V}'^2 = \mathbf{S}_0'^2 + a_1'A_1' = 4 + 4c^{-1} - c(1+c)^{-1},\\ &\mathbf{V}''^2 = a_1''A_1'' + a_2''A_2'' = 4c^{-1}(1+2c)(1+2c-c^2)^{-1}.\end{aligned}\right\} \tag{3·365}$$

Hence

$$\left.\begin{aligned}&4R^2 = \mathbf{V}'^2 - \mathbf{V}''^2 = 4 + c(c^2 - 6c - 5)(1+c)^{-1}(1+2c-c^2)^{-1},\\ &4\mathbf{C}^2 = \mathbf{V}'^2 + 3\mathbf{V}''^2 = 4 - c(1+c)^{-1} + 4c^{-1}(4 + 8c - c^2)(1+2c-c^2)^{-1}.\end{aligned}\right\} \tag{3·366}$$

If we take the P-vector field corresponding to $\mathbf{C}$ as an approximation to the gradient of the solution, R^2 measures the mean-square error. Now R^2 as given by (3·366) is not small, and so we might be inclined to dismiss the suggested approximation as too poor. However, the thing we should consider is the *relative* mean-square error, $R^2/\mathbf{C}^2$. If we do this, we find a satisfactory approximation when c is small (thin condenser); for, expanding in power series, we find

$$\left.\begin{aligned} R^2 &= 1 - \tfrac{5}{4}c + \ldots, \\ \mathbf{C}^2 &= (4/c) + 1 + (\tfrac{11}{4})\, c + \ldots, \\ R^2/\mathbf{C}^2 &= (\tfrac{1}{4}c) - (\tfrac{21}{16})\, c^2 + \ldots . \end{aligned}\right\} \qquad (3{\cdot}367)$$

Thus, for a thin condenser (b/a small), $\mathbf{C}$ is a good approximation to the gradient of the solution in the sense of the relative mean square.

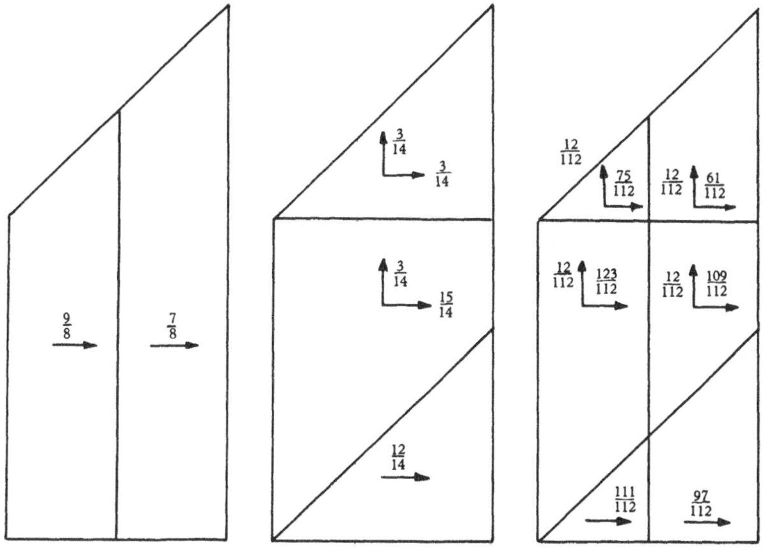

Fig. 3·33. Square condenser: P-vector fields (multiplied by b) for the vertices $\mathbf{V}'$, $\mathbf{V}''$ and the centre $\mathbf{C}$ of the hypercircle, for the case where $b = \frac{2}{3}a$, $c = \frac{1}{3}$.

The capacity of the condenser per unit length is $\mathbf{S}^2/2\pi$, and for it we have the bounds

$$\begin{aligned} \mathbf{V}''^2 &= 4c^{-1}(1+2c)(1+2c-c^2)^{-1} \\ &\leqslant \mathbf{S}^2 \leqslant 4 + 4c^{-1} - c(1+c)^{-1} = \mathbf{V}'^2. \end{aligned} \qquad (3{\cdot}368)$$

If $c \to 0$, then $c\mathbf{S}^2 \to 4$, so that for a thin condenser the capacity is approximately $2/\pi c = 4a/\pi b$, as indeed one would suppose from a rough argument which neglects the corners.

The following table gives some numerical values calculated from (3·365) and (3·366):

Square condenser inner side $2a$, outer side $2(a+b)$

$c = \frac{1}{2}b/a$	Lower bound for $2\pi \times$ capacity $\mathbf{V}''^2$	Upper bound for $2\pi \times$ capacity $\mathbf{V}'^2$	Mean square error of $\mathbf{C}$ R^2	Relative mean square error of $\mathbf{C}$ $R^2/\mathbf{C}^2$
0·01	400·039	403·990	0·9877	0·00246
0·1	40·336	43·909	0·8934	0·02167
0·2	20·588	23·833	0·8113	0·03791
0·3	14·128	17·102	0·7436	0·05000
0·4	10·976	13·714	0·6846	0·05871
0·5	9·143	11·667	0·6309	0·06454

For $c=0{\cdot}5$, we have $b/a=1$, and the condenser is far from thin; but the relative mean-square error is less than 7 %, which seems satisfactory in view of the great simplicity of the calculations.

Exercises

1. Extending the method of Example 2 above to three dimensions, establish the following bounds for the capacity $\mathbf{S}^2/4\pi$ of a cubical condenser, formed of two cubes of edges $2a$ and $2(a+b)$, the smaller situated symmetrically inside the larger:

$$24a^2b^{-1} \leqslant \mathbf{S}^2 \leqslant 24a^2b^{-1}(1+ba^{-1}+\tfrac{1}{3}b^2a^{-2}).$$

2. Suppose you have set up a hypercircle of class $r+s$ for a plane Dirichlet problem. Someone suggests a function v as a possible solution. Use the hypercircle to set upper and lower bounds for the mean-square error of the gradient of v, and verify that the lower bound is zero if the F-point corresponding to the gradient of v lies on the hypercircle.

3. Consider the Dirichlet problem

$$\Delta u = 0, \quad (u)_B = \tfrac{1}{2}x^2$$

for the square bounded by $x^2=1$, $y^2=1$. Take

$$\begin{aligned} &\mathbf{S}'_0 \leftrightarrow \operatorname{grad} \tfrac{1}{2}x^2 = (x, 0),\\ &\mathbf{T}'_1 \leftrightarrow \operatorname{grad} \cos \tfrac{1}{2}\pi x \cos \tfrac{1}{2}\pi y,\\ &\mathbf{T}''_1 \leftrightarrow \operatorname{grad} \cos \tfrac{1}{2}\pi x \cosh \tfrac{1}{2}\pi y. \end{aligned}$$

Show that $\quad \mathbf{V}'^2 = (\tfrac{4}{3}) - 8(2/\pi)^6, \quad \mathbf{V}''^2 = 8(2/\pi)^5 \tanh \tfrac{1}{2}\pi,$

and so locate the solution on a hypercircle of class $(1+1)$ with radius R where $R^2 = 0{\cdot}0083$.

4. Consider the Dirichlet problem

$$\Delta u = 0, \quad (u)_B = \tfrac{1}{2}(x^2+y^2),$$

for the right-angled triangle bounded by the lines

$$x=0, \quad y=0, \quad x+y=1.$$

Take
$$\begin{aligned} &\mathbf{S}'_0 \leftrightarrow \operatorname{grad} \tfrac{1}{2}(x^2+y^2), \quad &&\mathbf{T}'_1 \leftrightarrow \operatorname{grad} xy(x+y-1),\\ &\mathbf{T}''_1 \leftrightarrow \operatorname{grad}(x+y), &&\mathbf{T}''_2 \leftrightarrow \operatorname{grad} xy. \end{aligned}$$

Show that
$$\begin{aligned} &\mathbf{S}'^2_0 = \tfrac{1}{6}, \quad \mathbf{S}'_0 . \mathbf{T}'_1 = \tfrac{1}{60}, \quad \mathbf{S}'_0 . \mathbf{T}''_1 = \tfrac{1}{3}, \quad \mathbf{S}'_0 . \mathbf{T}''_2 = \tfrac{1}{12},\\ &\mathbf{T}'^2_1 = \tfrac{1}{90}, \quad \mathbf{T}''^2_1 = 1, \quad \mathbf{T}''_1 . \mathbf{T}''_2 = \tfrac{1}{3}, \quad \mathbf{T}''^2_2 = \tfrac{1}{6}, \end{aligned}$$

and hence obtain the bounds
$$\tfrac{15}{120} < \mathbf{S}^2 < \tfrac{17}{120}.$$

5. Consider the Dirichlet problem

$$\Delta u = 0, \quad (u)_B = (f)_B,$$

where the equation of B is $b(x_1, x_2) = 0$ and the functions $f(x_1, x_2)$ and $b(x_1, x_2)$ are continuous in the interior of B. Taking

$$\mathbf{S}'_0 \leftrightarrow f_{,i}, \quad \mathbf{T}'_1 \leftrightarrow b_{,i}, \quad \mathbf{T}''_1 \leftrightarrow (1,0), \quad \mathbf{T}''_2 \leftrightarrow (0,1),$$

establish the bounds

$$V^{-1}(\int_B f\,dx_1)^2 + V^{-1}(\int_B f\,dx_2)^2 \leqslant \mathbf{S}^2$$
$$\leqslant \int f_{,i} f_{,i}\,dV - (\int f_{,i} b_{,i}\,dV)^2 / \int b_{,i} b_{,i}\,dV,$$

where V is the area inside B.

6. Using polar coordinates, consider the Dirichlet problem

$$\Delta u = 0, \quad (u)_B = r,$$

where the domain is the interior of the quadrant bounded by $r=1$, $\theta=0$, $\theta=\frac{1}{2}\pi$. Take

$$\mathbf{S}_0' \leftrightarrow \operatorname{grad} r,$$
$$\mathbf{T}_1' \leftrightarrow \operatorname{grad} r(r-1)\sin 2\theta,$$
$$\mathbf{T}_1'' \leftrightarrow (\cos\theta, \ -\sin\theta),$$
$$\mathbf{T}_2'' \leftrightarrow (\sin\theta, \cos\theta),$$

the components shown being the (r,θ) components of the P-vector fields. Verify that the conditions (3·216) are satisfied, and establish the bounds

$$\pi\mathbf{V}''^2 = 2 \leqslant \pi\mathbf{S}^2 \leqslant \tfrac{1}{4}\pi^2 - \tfrac{2}{9} = \pi\mathbf{V}'^2,$$

or

$$2 \leqslant \pi\mathbf{S}^2 \leqslant 2{\cdot}2452.$$

3·4. BOUNDS FOR THE SOLUTION AND ITS DERIVATIVES AT AN INTERIOR POINT

By locating the solution of a Dirichlet problem on a hypercircle, we have succeeded in bounding the gradient of the solution in a mean-square sense. We now proceed to bound the solution and its derivatives *at a point*. This pointwise bounding starts only after the solution has been located on a hypercircle; we shall therefore suppose that a hypercircle of class $r+s$ has been obtained as in (3·335). The pointwise bounding is an application of (2·783), the trick being to choose $\mathbf{G}$ properly.*

The Green's F-vector

Fig. 3·41 shows a region V in the plane, simply or multiply connected. (In the multiply connected case, the holes are shaded.) We seek to bound the solution of the Dirichlet problem at any point P selected *inside* the region V. The method developed below is not applicable when P lies *on* the boundary B; that case requires special and more complicated techniques, which, indeed, work only when P is on a straight portion of B (Maple (1)). We shall confine ourselves here to the case where P lies inside V.

* The method followed here is that of Maple (1). For another method, see Diaz and Greenberg (1, 2) and Greenberg (1); for the connexion of their method with the hypercircle, see Synge (5). The second method involves the use of a Green's function in the usual sense, and it will be used in the discussion of pointwise bounds in elastic equilibrium (cf. p. 350).

With centre P and radius a describe a circle (interior v, circumference b). The radius a must be chosen sufficiently small, so that b does not cut B, although they may touch. There is no question of making a infinitesimal; indeed, for practical purposes it is best to make a as large as possible. We denote by $V-v$ the region obtained by removing the circle from the complete domain V.

Let x_i be any rectangular Cartesian coordinates and y_i rectangular Cartesian coordinates with origin at P, the axes of the two sets of coordinates being parallel. The comma notation for partial

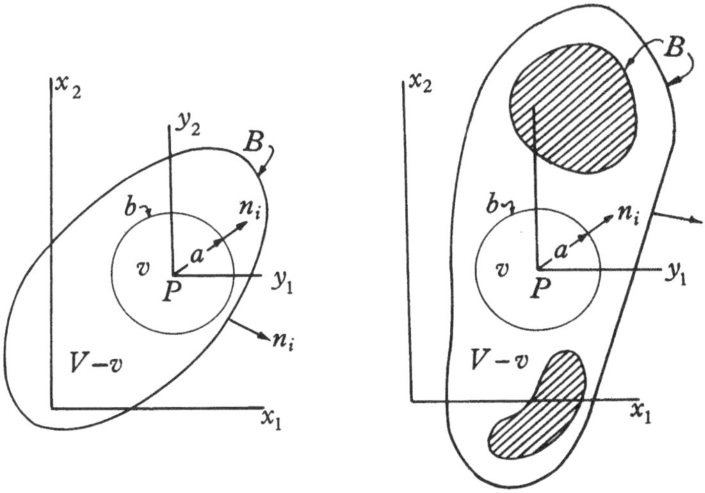

Simply connected Multiply connected

Fig. 3·41. P-space for the plane Dirichlet problem. The Green's vector vanishes inside the circle with centre P, radius a, area v and circumference b.

derivatives then applies equally to x_i and y_i ($\partial u/\partial x_i = \partial u/\partial y_i = u_{,i}$). We shall write r for distance from P, so that

$$r^2 = y_i y_i. \tag{3·401}$$

We now define a particular F-vector as follows:

$$\mathbf{G} \leftrightarrow g_i \begin{cases} g_i = y_i r^{-2} & \text{if} \quad r \geqslant a, \\ g_i = 0 & \text{if} \quad r < a. \end{cases} \tag{3·402}$$

We shall call **G** a *Green's vector*. The name is suggested by a rather loose analogy, because of course our Green's vector differs considerably from the Green's function usually associated with the Dirichlet problem; our Green's vector has no poles, but a discontinuity, and it satisfies no boundary conditions.

Calculating the square of $\mathbf{G}$, as in (3·203), we get

$$\mathbf{G}^2 = \int_V g_i g_i \, dV = \int_{V-v} r^{-2} \, dV. \tag{3·403}$$

Thus $\mathbf{G}^2$ tends to infinity if a tends to zero; but this is a limiting process we shall not use.

The field g_i has been chosen to satisfy

$$g_{i,i} = 0, \tag{3·404}$$

as is easily verified. This is essential for our purposes. For $r \geqslant a$, g_i is the gradient of the fundamental harmonic function $\log r$, and that is in fact the way in which it was generated. As we shall see, g_i serves to bound the solution of the Dirichlet problem at the point $y_i = 0$; the derivatives of g_i bound the derivatives of the solution.

Bounds for the solution

We suppose that the solution has been located on a hypercircle of class $r+s$ as in (3·335). The method works no matter what the length of the radius R of the hypercircle may be, but if close pointwise bounds are needed we must make R small, and this must be done before embarking on the technique described below.

In (2·783) we have bounds on $\mathbf{S}.\mathbf{G}$, where $\mathbf{S}$ is the solution and $\mathbf{G}$ any F-vector at all. These bounds may be written

$$|\mathbf{S}.\mathbf{G} - \mathbf{C}.\mathbf{G}| \leqslant RM, \tag{3·405}$$

where $\mathbf{C}$ is the centre of the hypercircle and M is given by

$$M^2 = \mathbf{G}^2 - \sum_{\rho=1}^{r} (\mathbf{G}.\mathbf{I}'_\rho)^2 - \sum_{\sigma=1}^{s} (\mathbf{G}.\mathbf{I}''_\sigma)^2 \quad (M \geqslant 0), \tag{3·406}$$

in terms of orthonormalized vectors as in (2·786), or more generally as in (2·784).

Let us calculate $\mathbf{S}.\mathbf{G}$ when $\mathbf{G}$ is the Green's vector (3·402). We have

$$\begin{aligned} \mathbf{S}.\mathbf{G} &= \int_V u_{,i} g_i \, dV = \int_{V-v} u_{,i} g_i \, dV \\ &= \int_B u g_i n_i \, dB - \int_b u g_i n_i \, db - \int_{V-v} u g_{i,i} \, dV, \end{aligned} \tag{3·407}$$

where n_i is the outward unit normal (see Fig. 3·41), dB, db are positive elements of arc and dV an element of area. The last integral vanishes, by (3·404). On b we have $g_i = y_i a^{-2}$, and so (3·407) may be written

$$\mathbf{S}.\mathbf{G} = \int_B u g_i n_i \, dB - a^{-2} \int_b u y_i n_i \, db. \tag{3·408}$$

But on b we have $y_i = an_i$, and so, by the well-known mean-value theorem for harmonic functions (u is the solution of the Dirichlet problem, and so is harmonic),

$$a^{-2}\int_b uy_i n_i db = a^{-1}\int_b u db = 2\pi(u)_P. \tag{3·409}$$

Thus (3·408) reads

$$\mathbf{S}.\mathbf{G} = \int_B ug_i n_i dB - 2\pi(u)_P. \tag{3·410}$$

When this is substituted in (3·405), we get bounds for $(u)_P$; *at any interior point P, the solution u of the Dirichlet problem is bounded above and below by the inequality*

$$\left| 2\pi(u)_P - \int_B ug_i n_i dB + \mathbf{C}.\mathbf{G} \right| \leqslant RM. \tag{3·411}$$

All the quantities here occurring, except $(u)_P$, are calculable in terms of the given boundary-value function f and the functions used in building up the hypercircle. Thus we have

$$\left.\begin{aligned} \int_B ug_i n_i dB &= \int_B fy_i n_i r^{-2} dB, \\ \mathbf{C}.\mathbf{G} = \int_V c_i g_i dV &= \int_{V-v} c_i y_i r^{-2} dV, \end{aligned}\right\} \tag{3·412}$$

where c_i is the P-vector field corresponding to the centre of the hypercircle, as in (3·337); R also is given by (3·337) and M is calculable by (3·406).

The inequality (3·411) means that, if RM is small, then a good approximation to the solution u at the point P is given by

$$(u)_P \sim (2\pi)^{-1}\left(\int_B ug_i n_i dB - \mathbf{C}.\mathbf{G}\right). \tag{3·413}$$

To make this a good approximation, the first requirement is to make R small. Once the vectors $\mathbf{S}_0'$, $\mathbf{I}_\rho'$, $\mathbf{I}_\sigma''$ (or equivalently $\mathbf{S}_0'$, $\mathbf{T}_\rho'$, $\mathbf{T}_\sigma''$) defining the hypercircle have been chosen, then R is fixed, and all we have under our control to make MR as small as possible is the radius a of the circle which we draw about the point P. We see from (3·406) and the remark after (3·403) that M is large if a is small; therefore we should make a as large as we can without letting the circle cut the boundary B. (We must have a complete circle b in V in order to apply the mean-value theorem as in (3·409).) The pointwise bounds obtained in the above manner are necessarily poor (unless R is very small indeed) if the point P lies near the boundary B, for then we can use only a small radius a.

Checks

Suppose that the radius of the hypercircle is zero ($R=0$). Then the centre **C** *is* the solution **S**. The right-hand side of (3·411) is zero, and the approximation (3·413) becomes exact. It takes only a moment to verify that the right-hand side of (3·413) does in fact represent $(u)_P$ accurately if we replace **C**.**G** by **S**.**G**.

Another interesting special case is that in which R is not zero, but where the boundary B is a circle and the point P is taken at its centre. We may then expand the circle b so that it coincides with B. Then the domain of integration involved in the calculation of $\mathbf{G}^2$, or of any scalar product involving **G**, disappears; hence $\mathbf{G}^2$ and all such scalar products vanish. Therefore $\mathbf{C}.\mathbf{G}=0$, $M=0$, and (3·411) becomes the equation

$$2\pi(u)_P=\int_B ug_i n_i dB. \tag{3·414}$$

The hypercircle has disappeared, and we are left with the mean-value theorem for a harmonic function! The result is of course trivial here, but the same idea (B a circle and P its centre) gives results which are not so trivial in other cases, as we shall see on pp. 160 and 162.

Bounds for the first derivatives

To obtain bounds for the first derivatives $(u_{,i})_P$, we need two Green's F-vectors $\mathbf{G}^{(p)}$ ($p=1,2$), defined as follows:

$$\mathbf{G}^{(p)}\leftrightarrow g_i^{(p)}\begin{cases} g_i^{(p)}=\delta_{ip}r^{-2}-2y_iy_pr^{-4} & \text{if} \quad r\geqslant a,\\ g_i^{(p)}=0 \quad \text{if} \quad r<a. \end{cases} \tag{3·415}$$

The secret of this formidable expression is that

$$g_i^{(p)}=g_{i,p}, \tag{3·416}$$

g_i being in fact $(\log r)_{,i}$ for $r\geqslant a$, as was already pointed out on p. 157. We note the essential equation

$$g_{i,i}^{(p)}=0, \tag{3·417}$$

which is indeed obvious since $g_i^{(p)}=(\log r)_{,ip}$.

Understanding that p is to take the values 1, 2, we have as in (3·405) the bounds

$$|\mathbf{S}.\mathbf{G}^{(p)}-\mathbf{C}.\mathbf{G}^{(p)}|\leqslant RM^{(p)}, \tag{3·418}$$

where
$$M^{(p)2}=\mathbf{G}^{(p)2}-\sum_{\rho=1}^{r}(\mathbf{G}^{(p)}.\mathbf{I}'_\rho)^2-\sum_{\sigma=1}^{s}(\mathbf{G}^{(p)}.\mathbf{I}''_\sigma)^2. \tag{3·419}$$

Now
$$\begin{aligned}\mathbf{S}.\mathbf{G}^{(p)}&=\int_V u_{,i}g_i^{(p)}dV=\int_{V-v}u_{,i}g_i^{(p)}dV\\ &=\int_B ug_i^{(p)}n_i dB-\int_b ug_i^{(p)}n_i db-\int_{V-v}ug_{i,i}^{(p)}dV,\end{aligned} \tag{3·420}$$

wherein the last integral vanishes by (3·417). Further

$$\int_b u g_i^{(p)} n_i \, db = \int_b u(\delta_{ip} a^{-2} - 2y_i y_p a^{-4}) \, n_i \, db$$

$$= -a^{-2} \int_b u n_p \, db = -a^{-2} \int_v u_{,p} \, dv = -\pi (u_{,p})_P. \quad (3·421)$$

The last step depends on the fact that $u_{,p}$, being the derivative of a harmonic function, is itself harmonic, and so its value at P is equal to the mean value in a circle with centre at P.* We have then

$$\mathbf{S}.\mathbf{G}^{(p)} = \int_B u g_i^{(p)} n_i \, dB + \pi (u_{,p})_P, \quad (3·422)$$

and (3·418) gives the following result: *at any interior point P, the partial derivatives $u_{,p}$ of the solution of the Dirichlet problem are bounded above and below by the inequality*

$$\left| \pi (u_{,p})_P + \int_B u g_i^{(p)} n_i \, dB - \mathbf{C}.\mathbf{G}^{(p)} \right| \leqslant RM^{(p)}. \quad (3·423)$$

Consequently if $RM^{(p)}$ is small, the following is a good approximation

$$(u_{,p})_P \sim \pi^{-1} \left(\mathbf{C}.\mathbf{G}^{(p)} - \int_B u g_i^{(p)} n_i \, dB \right). \quad (3·424)$$

Checks

Consider some special cases. First, suppose $R = 0$ and hence $\mathbf{C} = \mathbf{S}$. Then (3·424) becomes an exact expression for $(u_{,p})_P$; this is easy to verify directly.

As a second check, let B be a circle and P its centre. As on p. 159, we may expand b to coincide with B, and then (3·423) gives an equation analogous to (3·414):

$$\pi (u_{,p})_P = -\int_B u g_i^{(p)} n_i \, dB$$

$$= \int_B (2y_i y_p n_i r^{-4} - n_p r^{-2}) f \, dB = r^{-2} \int_B n_p f \, dB. \quad (3·425)$$

This is easy to verify directly, using Green's theorem.

A particular example is illustrated in Fig. 3·42. A circle of radius a is split into two semicircles, with $u = 1$ on the upper one and $u = 0$ on the lower one. Then (3·425) gives

$$(u_{,1})_P = 0, \quad (u_{,2})_P = 2/\pi a. \quad (3·426)$$

It is interesting that this result comes out so simply. On account of the discontinuity in the boundary condition, $\int_V u_{,i} u_{,i} \, dV$ does not exist, and we would expect that we would have to use the device of p. 137.

* If h is any harmonic function, differentiation with respect to the radius a of the circle gives

$$\frac{d}{da} \int_v h \, dv = \int_b h \, db = 2\pi a (h)_P,$$

and thus, on integration, $$\int_v h \, dv = \pi a^2 (h)_P.$$

Bounds for the second derivatives

To obtain bounds at a point for the second derivatives of the solution of the Dirichlet problem, we take a set of Green's F-vectors corresponding to P-vector fields obtained by differentiating (3·402) twice or (3·415) once:

$$\mathbf{G}^{(pq)} \leftrightarrow g_i^{(pq)} \begin{cases} g_i^{(pq)} = -2r^{-4}(\delta_{ip}y_q + \delta_{iq}y_p + \delta_{pq}y_i) \\ \qquad\qquad + 8r^{-6}y_iy_py_q \quad \text{if} \quad r \geqslant a, \\ g_i^{(pq)} = 0 \quad \text{if} \quad r < a. \end{cases} \tag{3·427}$$

We have of course $g_{i,i}^{(pq)} = 0$, since $g_{i,i} = 0$.

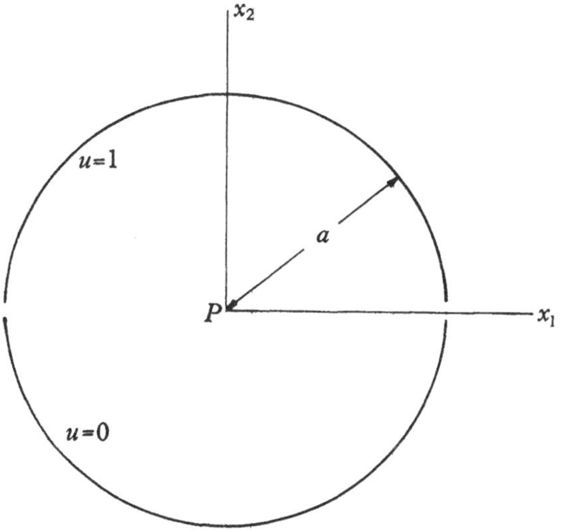

Fig. 3·42. A circular cylinder of radius a is split into two equal parts which are charged to potentials $u=1$ and $u=0$. The electric intensity at the centre is $2/(\pi a)$.

As in (3·405),
$$|\mathbf{S}.\mathbf{G}^{(pq)} - \mathbf{C}.\mathbf{G}^{(pq)}| \leqslant RM^{(pq)}, \tag{3·428}$$

where
$$M^{(pq)2} = \mathbf{G}^{(pq)2} - \sum_{\rho=1}^{r} (\mathbf{G}^{(pq)}.\mathbf{I}'_\rho)^2 - \sum_{\sigma=1}^{s} (\mathbf{G}^{(pq)}.\mathbf{I}''_\sigma)^2. \tag{3·429}$$

It is now a question of carrying out some calculations so as to express $\mathbf{S}.\mathbf{G}^{(pq)}$ in terms of a known integral and the value $(u_{,pq})_P$. We have

$$\mathbf{S}.\mathbf{G}^{(pq)} = \int_V u_{,i} g_i^{(pq)} dV = \int_{V-v} u_{,i} g_i^{(pq)} dV$$
$$= \int_B u g_i^{(pq)} n_i dB - \int_b u g_i^{(pq)} n_i db; \tag{3·430}$$

the integral over B is known in terms of the boundary values, and for the other integral we have, by (3·427),

$$\int_b ug_i^{(pq)} n_i\,db = 4a^{-4}\int_b uy_q n_p\,db - 2a^{-3}\delta_{pq}\int_b u\,db$$
$$= 4a^{-4}\int_v u_{,p} y_q\,dv, \tag{3·431}$$

by application of Green's theorem and the mean-value theorems for harmonic functions. Now differentiation with respect to a gives

$$\frac{d}{da}\int_v u_{,p} y_q\,dv = \int_b u_{,p} y_q\,db = a\int_b u_{,p} n_q\,db$$
$$= a\int_v u_{,pq}\,dv = \pi a^3 (u_{,pq})_P, \tag{3·432}$$

since $u_{,pq}$ is a harmonic function. Thus we get on integration

$$\int_v u_{,p} y_q\,dv = \tfrac{1}{4}\pi a^4 (u_{,pq})_P, \tag{3·433}$$

there being obviously no constant of integration, as we see on letting a tend to zero. Then (3·431) gives

$$\int_b ug_i^{(pq)} n_i\,db = \pi (u_{,pq})_P, \tag{3·434}$$

and so by (3·430)

$$\mathbf{S}\,.\,\mathbf{G}^{(pq)} = \int_B ug_i^{(pq)} n_i\,dB - \pi (u_{,pq})_P. \tag{3·435}$$

Then (3·428) gives this result: *at any interior point P, the second-order partial derivatives $u_{,pq}$ of the solution of the Dirichlet problem are bounded above and below by the inequality*

$$\left| \pi (u_{,pq})_P - \int_B ug_i^{(pq)} n_i\,dB + \mathbf{C}\,.\,\mathbf{G}^{(pq)} \right| \leqslant RM^{(pq)}. \tag{3·436}$$

If we take B to be a circle with P at its centre, and if we expand b into coincidence with B, then (3·436) becomes the equation

$$\pi (u_{,pq})_P = \int_B ug_i^{(pq)} n_i\,dB$$
$$= 2a^{-3}\int_B u(2n_p n_q - \delta_{pq})\,dB. \tag{3·437}$$

This formula expresses the values of the second derivatives of a harmonic function at the centre of a circle in terms of the values of the function on the circumference. It has emerged here as a by-product of the hypercircle method, but it can be easily verified directly. When applied to the example shown in Fig. 3·42, it gives zero values for all the second derivatives at P.

The nature of the calculations

It is hardly to be expected that the above method will yield close arithmetical values for pointwise bounds without considerable calculation. To see what calculation is involved, let us review the procedure, confining our attention to the bounds on $(u)_P$ as in (3·411), since the calculations for the pointwise bounds for derivatives are more complicated but not essentially different.

First, we must find a hypercircle. For use in (3·406) and (3·411), we need R, $\mathbf{C}$, $\mathbf{I}'_\rho$, $\mathbf{I}''_\sigma$, and it is desirable that orthonormalized vectors $\mathbf{I}'_\rho$, $\mathbf{I}''_\sigma$ should be used instead of the more general $\mathbf{T}'_\rho$, $\mathbf{T}''_\sigma$. To obtain close pointwise bounds, we need to make R small. The same hypercircle serves for pointwise bounds at all interior points P.

As in (3·411), we need for establishing bounds on $(u)_P$ the numerical values of the three quantities

$$\int_B u g_i n_i \, dB, \quad \mathbf{C}.\mathbf{G}, \quad M. \tag{3·438}$$

For the first we have

$$\int_B u g_i n_i \, dB = \int_B f y_i n_i r^{-2} dB = \int_B f \frac{\partial}{\partial n}(\log r)\, dB, \tag{3·439}$$

f being as usual the assigned boundary value of u. This integral is the logarithmic potential at P of a layer of doublets on B of density f.

As for the second quantity in (3·438), if we write $\mathbf{C} \leftrightarrow c_i$ we have

$$\begin{aligned} \mathbf{C}.\mathbf{G} &= \int_{V-v} c_i y_i r^{-2} dV \\ &= -\int_{V-v} c_{i,i} \log r \, dV + \int_B c_i n_i \log r \, dB \\ &\quad + \Sigma \int c_i n_i \log r \, ds - \log a \int_v c_{i,i} \, dv, \end{aligned} \tag{3·440}$$

where $\Sigma \int$ means a sum of integrals on those curves across which the normal component of c_i is discontinuous, the integrals to be taken over both sides of such curves, n_i being the unit normal pointing away from the side of integration. If $\mathbf{C}$ corresponds to a P-vector field of vanishing divergence (in general it does not), two of the above integrals disappear.

M is given by (3·406), assuming that we have orthonormalized to $\mathbf{I}'_\rho, \mathbf{I}''_\sigma$, and so we have to compute the quantities

$$\mathbf{G}^2, \quad \mathbf{G}.\mathbf{I}'_\rho, \quad \mathbf{G}.\mathbf{I}''_\sigma. \tag{3·441}$$

As in (3·403) we have $$\mathbf{G}^2 = \int_{V-v} r^{-2} dV. \tag{3·442}$$

For the other quantities in (3·441), let us write

$$\mathbf{I}'_\rho \leftrightarrow \overline{u}_{,i}', \quad (\overline{u}')_B = 0, \quad \mathbf{I}''_\sigma \leftrightarrow p''_i, \quad p''_{i,i} = 0.$$

Then
$$\left.\begin{aligned} \mathbf{G}.\mathbf{I}'_\rho &= \int_{V-v} \overline{u}_{,i}'(\log r)_{,i} dV = -a^{-1}\int_b u' db, \\ \mathbf{G}.\mathbf{I}''_\sigma &= \int_{V-v} p''_i (\log r)_{,i} dV = \int_B p''_i n_i \log r \, dB. \end{aligned}\right\} \tag{3·443}$$

The above reductions may serve to simplify calculations, but these are likely to remain formidable except in specially selected cases. For the torsion problem for a section bounded by two arcs of a circle and two chords equidistant from the centre, numerical bounds have been worked out by Maple (1).

Example: deformed tubular condenser

The following example has some interest for its own sake; it will also serve to illustrate the method of obtaining pointwise bounds, applied to a case in which P-space may differ only slightly from a domain for which we know the solution of the Dirichlet problem under the same boundary conditions.

Consider first a simple tubular condenser, the cross-section being the area between two concentric circles with centre O and radii a_1, a_2 $(a_1 < a_2)$. Let the inner circle be at potential $u = 1$ and the outer circle at potential $u = 0$. The solution of this Dirichlet problem is

$$u = -k \log(\rho/a_2), \quad k = [\log(a_2/a_1)]^{-1}, \tag{3·444}$$

where ρ is distance from the centre O.

Now let the outer circle be deformed into a curve B_0 lying wholly outside the circle $\rho = a_2$, the inner circle B_1 of radius a_1 remaining unaltered (Fig. 3·43); the potentials on B_0 and B_1 are respectively $u = 0$ and $u = 1$. At first we shall make no assumption regarding smallness of the deformation. Our object is to bound the solution of our new Dirichlet problem at a point P lying between B_0 and B_1.

As usual we denote by V the domain of the physical problem, i.e. the region between B_0 and B_1. We shall denote by $\tilde{V}$ the part of V exterior to the circle $\rho = a_2$. The complete boundary is $B = B_0 + B_1$.

First we must get a hypercircle. We let $\mathbf{S}'_0 \leftrightarrow \operatorname{grad} u'$, where

$$u' = \left\{\begin{matrix} -k\log(\rho/a_2) & \text{in} & V - \tilde{V}, \\ 0 \quad \text{in} \quad \tilde{V}. \end{matrix}\right\} \tag{3·445}$$

This function is continuous and satisfies the boundary conditions; the discontinuities in the derivatives are permissible. In polar coordinates (ρ, θ) we have

$$\mathbf{S}_0' \leftrightarrow \begin{cases} (-k/\rho, 0) & \text{in} \quad V-\tilde{V}, \\ (0,0) & \text{in} \quad \tilde{V}, \end{cases} \tag{3·446}$$

and we get

$$\mathbf{S}_0'^2 = 2\pi k. \tag{3·447}$$

Now let $\mathbf{T}_1'' \leftrightarrow \operatorname{grad} u''$, where

$$u'' = -k \log(\rho/a_2) \quad \text{in} \quad V. \tag{3·448}$$

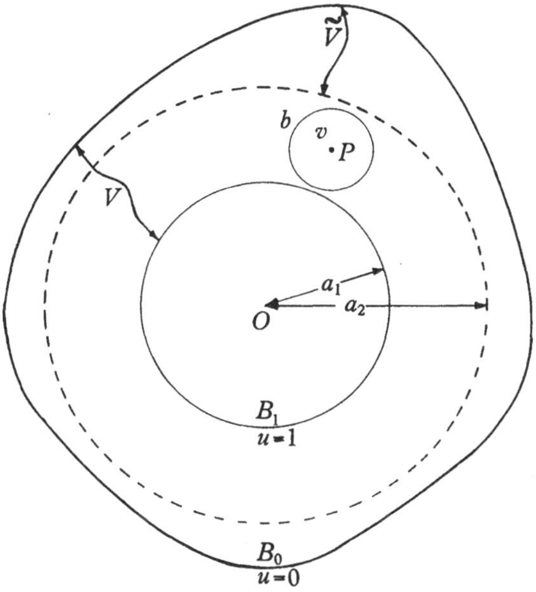

Fig. 3·43. Deformed tubular condenser; section bounded internally by the circle B_1 and externally by the curve B_0.

Note that u'' is u' continued into $\tilde{V}$. Obviously $\operatorname{div} \operatorname{grad} u''$ vanishes, as required for a complementary F-vector. Then

$$\left.\begin{aligned} \mathbf{T}_1'' &\leftrightarrow (-k/\rho, 0) \quad \text{in} \quad V, \\ \mathbf{T}_1''^2 &= \mathbf{S}_0'^2 + E = 2\pi k + E, \end{aligned}\right\} \tag{3·449}$$

where

$$E = k^2 \int_{\tilde{V}} \rho^{-1} d\rho\, d\theta. \tag{3·450}$$

If the deformation is small, then E is small, and in fact

$$E \sim k^2 a_2^{-2} \tilde{V}, \tag{3·451}$$

where $\tilde{V}$ is the area between B_0 and the circle $\rho = a_2$.

We see that

$$\mathbf{S}_0' \cdot \mathbf{T}_1'' = \mathbf{S}_0'^2 = 2\pi k. \tag{3·452}$$

We now have the elements of a hypercircle of class 1. There are no vectors of type $\mathbf{T}'_\rho$ and only one of type $\mathbf{T}''_\sigma$. We have $\mathbf{S}'_0$ as in (3·446) and the following expressions:

$$\left.\begin{aligned} \mathbf{I}''_1 = \mathbf{T}''_1(2\pi k + E)^{-\frac{1}{2}}, \quad \mathbf{S}'_0 . \mathbf{I}''_1 = 2\pi k(2\pi k + E)^{-\frac{1}{2}}, \\ \mathbf{V}' = \mathbf{S}'_0, \quad \mathbf{V}'' = \mathbf{I}''_1(\mathbf{S}'_0 . \mathbf{I}''_1) = \mathbf{T}''_1 . 2\pi k(2\pi k + E)^{-1}, \\ \mathbf{C} = \tfrac{1}{2}(\mathbf{V}' + \mathbf{V}'') = \tfrac{1}{2}[\mathbf{S}'_0 + \mathbf{T}''_1 . 2\pi k(2\pi k + E)^{-1}], \\ 4R^2 = \mathbf{V}'^2 - \mathbf{V}''^2 = 2\pi k E(2\pi k + E)^{-1}. \end{aligned}\right\} \tag{3·453}$$

The radius R is small if the deformation (and hence E) is small; this is true even if the deformation consists of small steep corrugations.

The capacity C of the condenser per unit length is such that $2\pi C = \mathbf{S}^2$, the square of the solution. As in (3·339), we have the following bounds for it:

$$(2\pi k)^2 (2\pi k + E)^{-1} = \mathbf{V}''^2 \leqslant \mathbf{S}^2 \leqslant \mathbf{V}'^2 = 2\pi k. \tag{3·454}$$

The capacity is reduced by the deformation, since k is the capacity for no deformation (put $E = 0$). For small E, the above inequalities are approximately

$$2\pi k - E \leqslant \mathbf{S}^2 \leqslant 2\pi k, \tag{3·455}$$

the approximate value of E being as in (3·451).

So much for the hypercircle. Now we seek bounds for the solution at a point P (Fig. 3·43), using (3·411):

$$\left.\begin{aligned} \left| 2\pi(u)_P - \int_B u g_i n_i dB + \mathbf{C} . \mathbf{G} \right| \leqslant RM, \\ M^2 = \mathbf{G}^2 - (\mathbf{G} . \mathbf{I}''_1)^2. \end{aligned}\right\} \tag{3·456}$$

By (3·439) and the boundary conditions on u, we have

$$\int_B u g_i n_i dB = \int_{B_1} \frac{\partial}{\partial n} (\log r)\, dB_1 = 0, \tag{3·457}$$

since $\log r$ is harmonic inside B_1. (Remember that ρ is measured from the centre O and r from the selected point P.)

To evaluate $\mathbf{C} . \mathbf{G}$ we need the following facts about the harmonic function $\log r$. The first is the well-known mean-value theorem for a function harmonic inside a circle. However, we shall prove the complete result, which reads:

The mean value of $\log PQ$, P being fixed and Q traversing the circumference of a circle with centre O and radius a, is

$$\left.\begin{aligned} &\log OP \text{ if } P \text{ lies outside the circle,} \\ &\log a \text{ if } P \text{ lies inside the circle.} \end{aligned}\right\} \tag{3·458}$$

To prove this, allow Q to traverse the boundary C of any region which contains neither O nor P. Then, since $\log OQ$ and $\log PQ$ are both harmonic in this region, we have

$$\int_C \left[\log PQ \frac{\partial}{\partial n} \log \frac{OQ}{a} - \log \frac{OQ}{a} \frac{\partial}{\partial n} \log PQ \right] ds = 0,$$

where $\partial/\partial n$ means differentiation along the normal drawn out of the region enclosed by C. Now let C consist of the circle C_0 with centre O and radius a, and of a little circle C_1 with centre O and radius ϵ, and also (if C_0 contains P) a little circle C_2 with centre P and radius ϵ. Then on C_0 we have

$$\frac{\partial}{\partial n}\log\frac{OQ}{a}=\frac{1}{a},\quad \log\frac{OQ}{a}=0,$$

and so we get

$$\frac{1}{a}\int_{C_0}\log PQ\,ds=\int_{C_1}\left[\frac{1}{\epsilon}\log PQ+\log\frac{\epsilon}{a}\frac{\partial}{\partial n}\log PQ\right]ds$$
$$-\int_{C_2}\left[\log\epsilon\frac{\partial}{\partial n}\log\frac{OQ}{a}+\frac{1}{\epsilon}\log\frac{OQ}{a}\right]ds,$$

the last integral being omitted if P is outside C_0. Since $ds=\epsilon\,d\theta$ on C_1 and on C_2, we get, in the limit as $\epsilon\to 0$,

$$\frac{1}{a}\int_{C_0}\log PQ\,ds=2\pi\log OP-2\pi\log\frac{OP}{a},$$

the last term being omitted if P is outside C_0. From this the two results contained in (3·458) follow at once.

Turning to (3·453), we note that $c_{i,i}=0$ since both $\mathbf{S}_0'$ and $\mathbf{T}_1''$ correspond to P-vector fields with vanishing divergence and $\mathbf{C}$ is a linear combination of them. Thus, using the second of (3·458) on the $\Sigma\int$ in (3·440) arising from $\rho=a_2$, we have

$$\mathbf{C}.\mathbf{G}=\int_B c_i n_i\log r\,dB-\pi k\log a_2. \tag{3·459}$$

The integral here breaks into two parts:

$$\left.\begin{aligned}\int_{B_1}c_i n_i\log r\,dB_1&=\pi k[1+2\pi k(2\pi k+E)^{-1}]\log(\rho)_P,\\ \int_{B_0}c_i n_i\log r\,dB_0&=-\pi k^2(2\pi k+E)^{-1}\int_{B_0}\rho^{-1}\cos\phi\log r\,dB_0,\end{aligned}\right\} \tag{3·460}$$

where ϕ is the angle between the radius vector from O and the normal to B_0.

As for M, we note that $\mathbf{G}^2$ is given by (3·442), and by (3·443) we have (with p_i'' corresponding to $\mathbf{I}_1''$, that is, $\mathbf{T}_1''/|\mathbf{T}_1''|$),

$$\begin{aligned}\mathbf{G}.\mathbf{I}_1''&=\int_B p_i'' n_i\log r\,dB\\ &=k(2\pi k+E)^{-\frac{1}{2}}\left[2\pi\log(\rho)_P-\int_{B_0}\rho^{-1}\cos\phi\log r\,dB_0\right].\end{aligned} \tag{3·461}$$

Thus (3·456) gives the following bounds for $(u)_P$, exact but rather clumsy:

$$\begin{aligned}&|\,2\pi(u)_P+\pi k[1+2\pi k(2\pi k+E)^{-1}]\log(\rho)_P\\ &\quad-\pi k^2(2\pi k+E)^{-1}\int_{B_0}\rho^{-1}\cos\phi\log r\,dB_0-\pi k\log a_2\,|\\ &\leqslant\tfrac{1}{2}(2\pi kE)^{\frac{1}{2}}(2\pi k+E)^{-\frac{1}{2}}\left[\int\!\!\int_{V-v}r^{-2}dV\right.\\ &\quad\left.-k^2(2\pi k+E)^{-1}\left\{2\pi\log(\rho)_P-\int_{B_0}\rho^{-1}\cos\phi\log r\,dB_0\right\}^2\right]^{\frac{1}{2}}.\end{aligned} \tag{3·462}$$

To simplify the result and make it more intelligible, let us now allow the outer boundary B_0 to approach the circle $\rho = a_2$, its tangent at each point at the same time approaching the tangent to the circle at the corresponding point, so that ϕ is small. Then, if we retain only principal parts, (3·462) becomes

$$2\pi \mid (u)_P + k \log(\rho/a_2)_P \mid \leqslant \tfrac{1}{2} k a_2^{-1} \tilde{V}^{\frac{1}{2}} \left[\iint_{V-\tilde{V}-v} r^{-2} dV - 2\pi k \{\log(\rho/a_2)_P\}^2 \right]^{\frac{1}{2}}. \tag{3·463}$$

This gives in approximate form an upper bound for the deviation of $(u)_P$ from the exact solution $-k \log(\rho/a_2)_P$ [cf. (3·444)] in the case of no deformation.

Exercises

1. Show from (3·403) that if the domain of the Dirichlet problem is bounded by the curve $r = f(\theta)$ in polar coordinates referred to P as origin, then the square of the Green's vector is

$$\mathbf{G}^2 = \int_0^{2\pi} \log f(\theta)\, d\theta - 2\pi \log a.$$

2. Show that $\mathbf{G}^2 \leqslant V/a^2 - \pi$, where V is the area of the domain of the Dirichlet problem.

3. If $\mathbf{I} \leftrightarrow (p_r, p_\theta)$ in polar coordinates referred to P as origin, show that

$$\mathbf{G}.\mathbf{I} = \iint_{V-v} p_r\, dr\, d\theta.$$

4. In the notation of (3·415), show that

$$\mathbf{G}^{(p)}.\mathbf{G}^{(q)} = \delta_{pq} \int_{V-v} r^{-4} dV,$$

so that the F-vectors $\mathbf{G}^{(1)}$ and $\mathbf{G}^{(2)}$ are orthogonal to one another.

5. Referring to (3·427), show that the three F-vectors $\mathbf{G}^{(11)}$, $\mathbf{G}^{(12)} = \mathbf{G}^{(21)}$, $\mathbf{G}^{(22)}$ all have the same magnitude (its square is $4\int_{V-v} r^{-6} dV$), and that $\mathbf{G}^{(11)}$ and $\mathbf{G}^{(22)}$ have opposite directions, with $\mathbf{G}^{(12)}$ orthogonal to them.

3·5. PYRAMID F-VECTORS

The method developed below gives systematically closer and closer approximations to the solution of a Dirichlet problem in the plane. It makes essential use of the permissible discontinuities of (3·211). The same idea may be extended to 3-space (see p. 301), and even to N-space [McMahon (1, 2)], but the mental difficulties of keeping track of the geometry increase with the dimensionality; for plane problems we can see what we are doing, and even construct graphs.

Pyramid functions

Consider a plane polygon, as in Fig. 3·51a, with a point A inside it. Now raise the point A out of the plane of the polygon, the displacement (of length h) to the new position A' being perpendicular to the plane of the polygon. Join A' to the corners and sides of the polygon. This creates a pyramid, having A' for vertex and the polygon for base (Fig. 3·51*b*).

Let x_i be rectangular Cartesian coordinates in the plane of the polygon, and let $x_i = a_i$ at A. Let $t(x_1, x_2)$ be the height of the point on the surface of the pyramid directly above the point x_i in the plane. It is an easy function to think about geometrically, for the pyramid is a three-dimensional graph of the function $t(x_1, x_2)$. We write $t = 0$ for points outside the polygon.

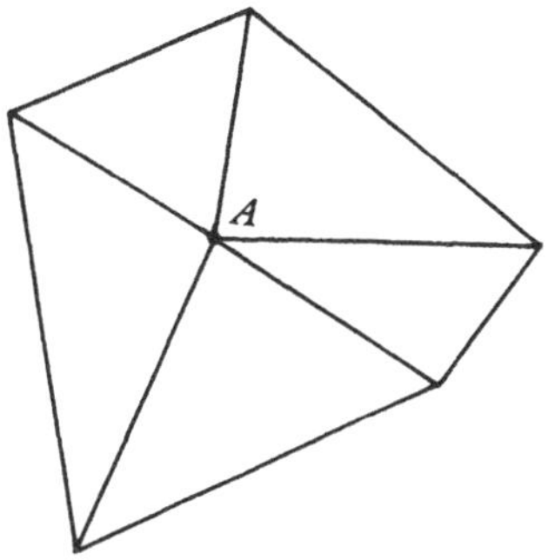

Fig. 3·51*a*. Polygonal base of a pyramid function.

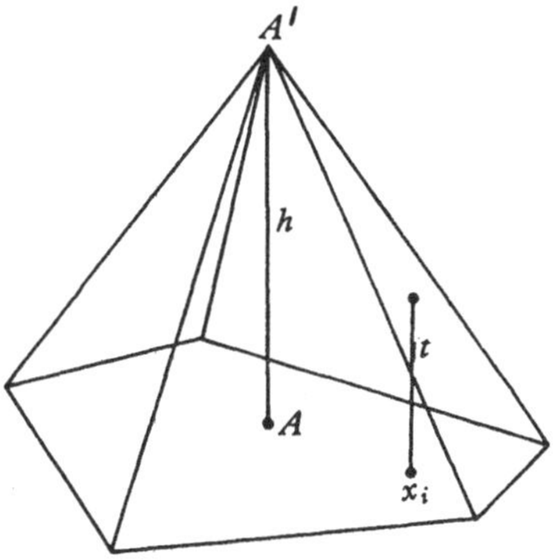

Fig. 3·51*b*. Three-dimensional graph of a pyramid function $t(x_1, x_2)$.

The following facts are obvious:

(i) $t(x_1, x_2)$ is a continuous function in the whole plane.

(ii) $t = 0$ on and outside the polygonal boundary.

(iii) $t = h$ when $x_i = a_i$.

(iv) The partial derivatives $t_{,i}$ are zero outside the polygon and have constant values in each of the triangles formed by joining A to the corners of the polygon (Fig. 3·51*a*).

(v) t is of the form

$$t(x_1, x_2) = h\tau(x_1, x_2), \qquad (3{\cdot}501)$$

where τ is independent of h, depending only on the geometry of the polygon and the position of A inside it.

We call $t(x_1, x_2)$ a *pyramid function*. One should not be confused by the three-dimensional graph as in Fig. 3·51*b*. It has been introduced only to facilitate thought; t is of course a function of position in the plane of the polygon and $t_{,i}$ a vector field in that plane.

Pyramid F-vectors of the first class

Suppose that we are dealing with a Dirichlet problem for P-space V, a region bounded by a curve B; V may be simply or multiply connected. Let us draw a polygon lying entirely in $V+B$ and generate in it a pyramid function $t(x_1, x_2)$. *Then* $t_{,i}$ *corresponds to a homogeneous associated F-vector* $\mathbf{T}'$, *for the conditions of* (3·216) *are satisfied and the discontinuities in* $t_{,i}$ *are permissible in accordance with* (3·211).

We call such an F-vector a *pyramid F-vector of the first class*. The corresponding P-vector field is as shown in Fig. 3·52; the P-vector

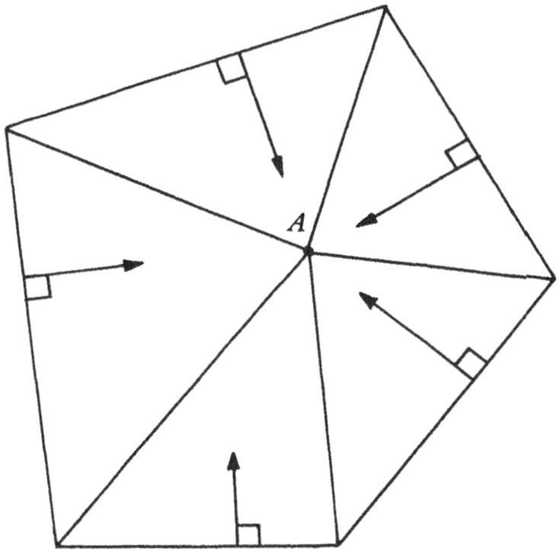

Fig. 3·52. P-vector field corresponding to a pyramid F-vector of the first class.

fields in the several triangles are perpendicular to the sides of the polygons, and their magnitudes are inversely proportional to the distances of the central point A from the sides.

Any linear combination of pyramid F-vectors of the first class is of course a homogeneous associated F-vector. The bases of the pyramids may or may not overlap. Suppose $\mathbf{T}'_1, \mathbf{T}'_2, \ldots$ are pyramid F-vectors of the first class, corresponding to the gradients of the pyramid functions $u'_1, u'_2, \ldots$. Then the linear combination

$$\mathbf{T}' = c_1 \mathbf{T}'_1 + c_2 \mathbf{T}'_2 + \ldots \tag{3·502}$$

corresponds to the gradient of the function

$$u' = c_1 u'_1 + c_2 u'_2 + \ldots. \tag{3·503}$$

The function u' is not itself a pyramid function, but it is a linear combination of them, and we may construct a polyhedral graph for u' by taking the pyramids of $u'_1, u'_2, \ldots$, pulling each one up by the appropriate factor $c_1, c_2, \ldots$, and then 'superimposing' them by adding together the several heights at any point. We call an F-vector constructed as in (3·502) out of pyramid functions a *polyhedral F-vector of the first class.*

If the bases of two pyramid F-vectors of the first class, say $\mathbf{T}'_1$, $\mathbf{T}'_2$, do not overlap, it is clear that they are orthogonal:

$$\mathbf{T}'_1 . \mathbf{T}'_2 = 0. \tag{3·504}$$

Under special circumstances they may be orthogonal even if their bases overlap.

Pyramid F-vectors of the second class

Consider the P-vector field of a pyramid F-vector of the first class as shown in Fig. 3·52, the gradient of a pyramid function

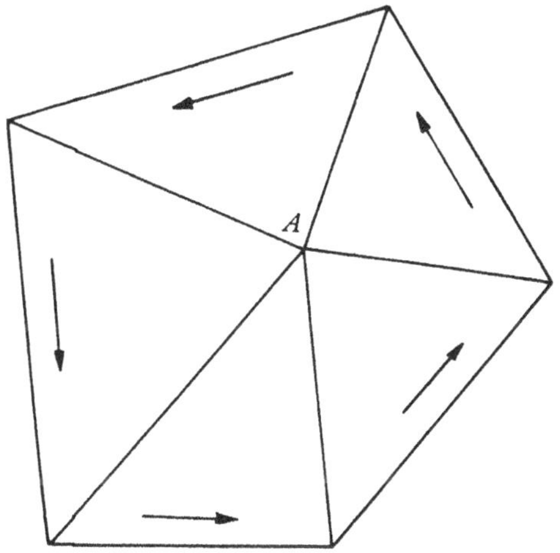

Fig. 3·53*a*. P-vector field corresponding to a pyramid F-vector of the second class.

$t(x_1, x_2)$. The normal component of this P-vector field is *not* continuous across the sides of the polygon (it vanishes outside), nor across the lines joining A to the vertices of the polygon. *But its tangential component is continuous across all these lines.*

Let us now rotate the P-vector field in each triangle through a right angle, all in the same sense (we shall choose the clockwise sense) and without altering the magnitudes of the P-vectors. The result is shown in Fig. 3·53*a*. By virtue of the *tangential* continuity of the old P-vector field, we know that the *normal* component of the new P-vector field is continuous. Moreover, its divergence vanishes inside each triangle, and of course outside the polygon where the field is zero, since the old P-vector field (and hence the new one) is of constant magnitude inside each triangle.

The F-vector corresponding to this new P-vector field we call a *pyramid F-vector of the second class.* It is clear that it is a *complementary F-vector,* since it satisfies the conditions (3·216) and the

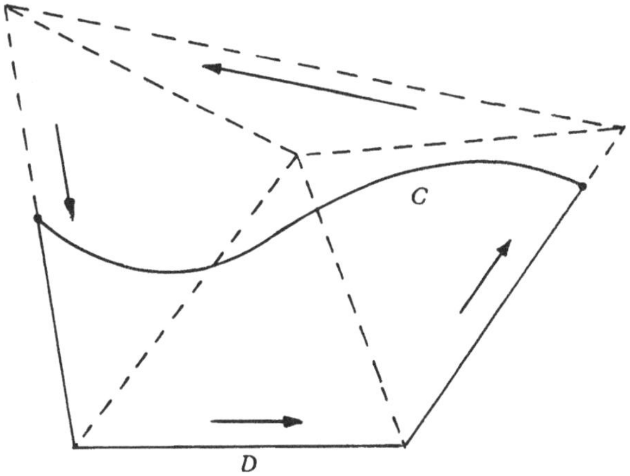

Fig. 3·53*b*. An integral property of pyramid F-vectors of the second class: $\int_C p''_i n_i ds = 0$.

discontinuities are permissible in the sense of (3·211).* If we write for the pyramid F-vector of the second class $\mathbf{T}'' \leftrightarrow p''_i$, then

$$p''_1 = t_{,2}, \quad p''_2 = -t_{,1} \quad \text{or} \quad p''_i = \epsilon_{ij} t_{,j}. \tag{3·505}$$

The following property of pyramid F-vectors of the second class is important. Let C be any curve drawn across the base (Fig. 3·53*b*). Let D be either of the portions into which C divides the boundary of the base. Then, since $p''_{i,i} = 0$, it follows that

$$\int_{C+D} p''_i n_i ds = 0,$$

the integral being taken round the loop formed by C and D, and n_i being the unit normal to the path of integration. But $p''_i n_i = 0$ on D, and therefore

$$\int_C p''_i n_i ds = 0. \tag{3·505a}$$

* Note that since no boundary condition is imposed on p''_i in (3·216), a pyramid F-vector of the second class with base overlapping the boundary B is a complementary F-vector.

In words, *the integral of the normal component of p''_i vanishes if the path of integration is any curve drawn across the base.*

The pyramid F-vector of the first class ($\mathbf{T}'$) is very closely associated with a pyramid graph. That of the second class ($\mathbf{T}''$) is not, at least not directly; it is not profitable to try to think of a three-dimensional graph having p''_i for slope. Indeed, if we want to think three-dimensionally of $\mathbf{T}''$, we have to fall back on the pyramid from which we generated $\mathbf{T}'$, recognizing that $\mathbf{T}''$ does not correspond to the slope of the pyramid, but to that slope turned through a right angle.

Triangulation and polyhedral graphs

The important fact which makes pyramid F-vectors useful is the following: Given any function u, continuous and with continuous

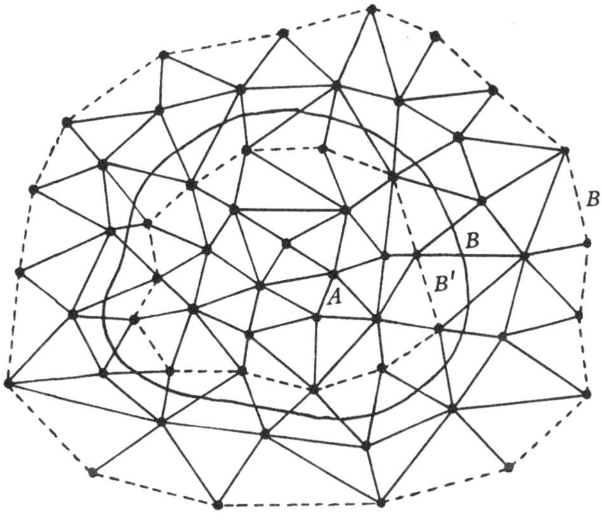

Fig. 3·54. Triangulation for a region V bounded by a curve B. The polygons B', B'' are shown by broken lines; B' is the outer boundary for the bases of pyramid F-vectors of the first class ($\mathbf{T}'_\rho$); B'' is the outer boundary for the bases of pyramid F-vectors of the second class ($\mathbf{T}''_\sigma$).

partial derivatives in a region V bounded by a curve B, we can approximate as closely as we like *to both u and its partial derivatives $u_{,i}$* by a linear combination of pyramid functions. This fact we have now to establish.

Let us *triangulate* the region V. This means that we draw any set of triangles, as in Fig. 3·54, such that every point of $V+B$ is a vertex of a triangle, a point on a side, or an interior point of a triangle.

In the applications the triangulation will be done in a systematic way, but for the general argument the triangulation may be quite irregular. We may suppose the net of triangles continued indefinitely outside B; all we actually need is the net inside and immediately outside B.

Suppose now that a three-dimensional graph of the function u is constructed; it is a smooth surface (say U) above (or perhaps below) the plane of the paper. U is terminated by a cylindrical wall rising from the curve B, for u is not supposed to be defined outside B.

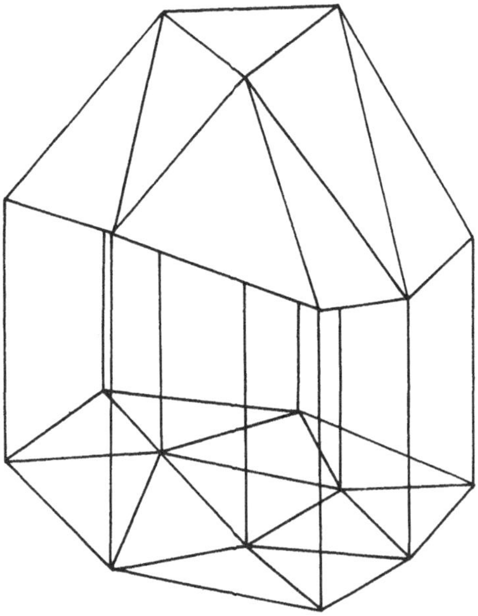

Fig. 3·55. Polyhedral graph.

The next step is to erect a perpendicular to the plane of the paper at each junction-point of the network of triangles, each perpendicular being terminated at its intersection with U. Perpendiculars erected at points outside B do not meet U; we terminate them arbitrarily.

The ends of the perpendiculars form a set of points in space, one above or below each junction-point. Let these points be joined by lines directly above the lines of the network, so that we have now a skew network in space. The triangles of this skew network are now filled with planes. It is in fact a roofing process, and the result is a polyhedral surface; a simple case is illustrated in Fig. 3·55.

It will now be assumed as intuitively obvious that it is possible to construct a polyhedral graph as above, based on a triangulation of fine enough mesh, so that if v is the function having this polyhedral graph, then v and its partial derivatives $v_{,i}$ approximate as closely as we like to u and $u_{,i}$ for all points inside the triangles ($v_{,i}$ does not exist at a vertex or on a side of a triangle of the network). In brief, *a given function and its first derivatives may be approximated as closely as we like by a polyhedral function based on a suitable triangulation.* (A rigorous proof of this is given on pp. 209–13.)

Approximation to a given function and its first derivatives by a linear combination of pyramid functions

Consider

(i) a given function u, with u and $u_{,i}$ continuous in $V+B$;

(ii) a polyhedral function v, constructed as described above and therefore such that v and $v_{,i}$ may approximate closely to u and $u_{,i}$;

(iii) a linear combination of pyramid functions, each having for central point A a junction-point of the triangulation underlying v, the base of the pyramid being bounded by the sides remote from A of those triangles that meet at A. (Thus in Fig. 3·54 the pyramid with A for central point has a hexagonal base.)

We shall now show that the coefficients in the linear combination (iii) may be chosen so that the polyhedral function defined by (iii) coincides with the polyhedral function v of (ii).

Fig. 3·56 shows part of the polyhedral graph of v, namely, the part above a base triangle $A_1A_2A_3$ of junction-points, these points being raised out of the plane through heights h_1, h_2, h_3 to positions A_1', A_2', A_3'. Let τ_1, τ_2, τ_3 denote pyramid functions, each of unit height, having A_1, A_2, A_3, respectively, for central points. Consider the function

$$w = h_1\tau_1 + h_2\tau_2 + h_3\tau_3. \tag{3·506}$$

Since each pyramid function is a linear function of the coordinates x_i over the triangle $A_1A_2A_3$, it follows that w is also a linear function of x_i. Further, w takes the values h_1, h_2, h_3 at the points A_1, A_2, A_3, respectively. But a linear function is determined by its values at three non-collinear points, and so, since $w=v$ at the vertices, it follows that $w=v$ over the triangle $A_1A_2A_3$.

If now we take not merely three pyramid functions, but a pyramid function for each junction-point of the net and write

$$w = h_1\tau_1 + h_2\tau_2 + h_3\tau_3 + h_4\tau_4 + \dots, \tag{3·507}$$

where h_1, h_2, h_3, h_4, ... are the values of v at the junction-points of the net, then $w=v$ all over the base of v. For the only pyramid functions which affect the value of w inside $A_1A_2A_3$ are τ_1, τ_2, τ_3, and so inside this triangle (3·507) reduces to (3·506); similarly (3·507) yields the correct value of w (i.e. $w=v$) inside all the other triangles.

This establishes what we sought to prove, viz. that we can create any given polyhedral function by a suitable linear combination of pyramid functions: symbolically, (ii) = (iii). Hence, since (ii) may approximate to (i), *we may approximate as closely as we like to a given function u and its first-order partial derivatives $u_{,i}$ by a linear combination of pyramid functions.*

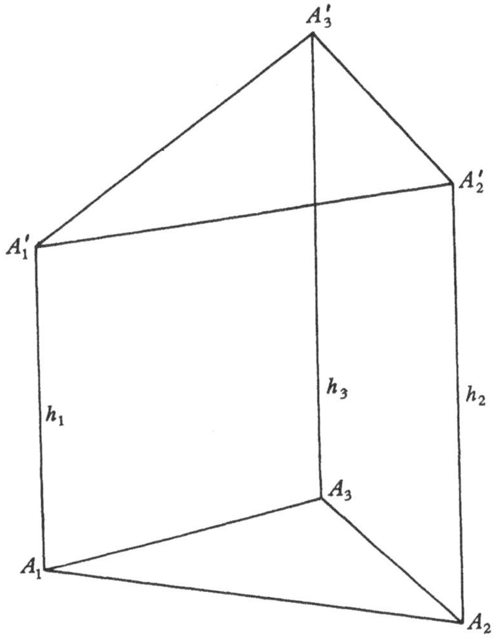

Fig. 3·56. The part of a polyhedral graph above the base triangle $A_1A_2A_3$.

We cannot approximate in this way to the *second* derivatives of u, because, for a polyhedral function, the second derivatives vanish in each triangle and do not exist on the sides.

Plan for the use of pyramid F-vectors in the hypercircle method

As we have already seen:

A pyramid F-vector of the first class (Fig. 3·52) is a homogeneous associated vector $(\mathbf{T}'_\rho)$, provided its base lies in $V+B$.

A pyramid F-vector of the second class (Fig. 3·53a) is a complementary vector $(\mathbf{T}''_\sigma)$.

The utility of these pyramid F-vectors in the hypercircle method is therefore obvious, but we need a systematic technique if we are to use them to obtain good approximations.

The first thing to decide on is a *network of triangulation*. The simplest is a pattern of equilateral triangles or one of right-angled isosceles triangles, leading respectively to *hexagonal F-vectors* and *square F-vectors*, to be discussed later (pp. 188 and 200). But of course we are by no means restricted to these simple triangulations.

Then the F-vectors should be *normalized*. This is a matter of convenience, and there are two rival plans: we might make them all of unit magnitude (in the sense of the F-metric) or we might make the height of every pyramid unity; the decision can be postponed until we come to special networks.

An important thing about a set of pyramid F-vectors is that each one has only a few neighbours, if we understand by *neighbours* those whose polygonal bases overlap. *This means that most of a set of pyramid F-vectors are orthogonal to one another*, a very useful fact for simplifying calculations.

Attention may be drawn again to the condition [cf. (3·216)] that the polygonal base of $\mathbf{T}'_\rho$ must not extend beyond the boundary B, since the function to the gradient of which $\mathbf{T}'_\rho$ corresponds is to vanish on B. Thus the polygon bounding all the pyramid F-vectors of the first class must not go outside B; this polygon (B') is shown by broken lines in Fig. 3·54. On the other hand, we need to use all those pyramid F-vectors of the second class ($\mathbf{T}''_\sigma$) with bases which lie in or overlap the interior of B; the polygon bounding the bases of all F-vectors of this type is marked B'' in Fig. 3·54.

The choice of the associated F-vector $\mathbf{S}'_0$ is important. If the boundary-value function f is given in the form $f(x_1, x_2)$, a function regular inside B, there is much to be said for choosing

$$\mathbf{S}'_0 \leftrightarrow f_{,i}, \tag{3·508}$$

as in (3·217). The only thing against this is the possible difficulty in calculating the scalar products $\mathbf{S}'_0 . \mathbf{T}'_\rho$, $\mathbf{S}'_0 . \mathbf{T}''_\sigma$. But this is easy if f is a simple function, like $\frac{1}{2}(x_1^2 + x_2^2)$ as in the torsion problem with which we shall deal in detail in Chapter 4.

Another plan is to use the network of triangulation to define $\mathbf{S}'_0$. The scheme is to represent $\mathbf{S}'_0$ by a polyhedral graph inside B' (see Fig. 3·54) with a suitable linkage of values between B' and B. In considering this, we must remember that we are thinking of a fine mesh, with many more triangles than are shown in Fig. 3·54, and we should bear in mind the validity of the plan in the limit when the triangulation becomes infinitely fine. We proceed as follows.

Our need is to find a function u', continuous inside and on B (although its derivatives need not be) and satisfying the boundary condition $(u')_B = f$.

We note that each junction-point on B' (Fig. 3·54) is joined to B by at least one line of the network. We use such lines (selecting one if there are several) to transmit to the junction-points on B' the values of f on B. Thus we assign u' at all junction-points on B'.

On the sides of B' we fill in values of u' according to the linear law (and so uniquely, once the values at the junctions have been assigned as above); we now have u' at all points of B'.

Next we assign u' arbitrarily at the junction-points inside B', preferably in some systematic way, and then fill the values linearly in the triangular meshes. Now we have for u', inside and on B', a function which can be represented by a polyhedral graph.

It remains to fill in u' between B' and B. This region is broken up by the network into a number of triangles and quadrilaterals, each of them having a curvilinear side (part of B). It is these curvilinear sides that cause trouble, and the method is much easier to use if B is a polygon, in which case we would incorporate it in the network and there would be no curvilinear parts; but let us consider the general case.

On the sides of these triangles and quadrilaterals belonging to B or to B', u' has already been assigned. Representing these assigned values as parts of a three-dimensional graph, it is easy to fill in a ruled three-dimensional graph for u' in each of these figures. This completes the description of u'; a polyhedral graph inside B' and a ruled graph between B and B'.

We define $\mathbf{S}'_0$ to be the F-vector corresponding to the gradient of u' as just determined; it is the sum of a linear combination of pyramid F-vectors of the first class with central points inside B' and a more awkward part corresponding to the region between B and B'. In computing the scalar products $\mathbf{S}'_0 . \mathbf{T}'_\rho$ this last awkward part does not come in, because the bases of $\mathbf{T}'_\rho$ do not extend outside B'; but in computing $\mathbf{S}'_0 . \mathbf{T}''_\sigma$ we cannot avoid it, since the bases of $\mathbf{T}''_\sigma$ run out to B'' (Fig. 3·54). In special cases, however, this difficulty may be circumvented (cf. p. 305).

Summary of plan for the use of pyramid F-vectors

(i) Triangulate the region V, carrying the triangulation out beyond the boundary B as in Fig. 3·54. A simple regular triangulation is to be preferred, incorporating B as part of it if B is polygonal.

(ii) Take pyramid F-vectors of both classes for this network ($\mathbf{T}'_\rho, \rho = 1, 2, \ldots, r$; $\mathbf{T}''_\sigma, \sigma = 1, 2, \ldots, s$). These should be normalized in some way, but not necessarily so as to give them unit magnitude. The set $\mathbf{T}'_\rho$ is to include all pyramids of which the bases do not go outside B (boundary B' in Fig. 3·54); the set $\mathbf{T}''_\sigma$ is to include all pyramids with bases wholly or partly inside B (boundary B'' in Fig. 3·54).

(iii) Choose $\mathbf{S}'_0$ as in (3·508) or in the other manner described above.

(iv) Calculate the scalar products as in (3·328):

$$\mathbf{T}'_\mu . \mathbf{T}'_\rho, \quad \mathbf{S}'_0 . \mathbf{T}'_\rho, \quad \mathbf{T}''_\nu . \mathbf{T}''_\sigma, \quad \mathbf{S}'_0 . \mathbf{T}''_\sigma.$$

(v) Solve the linear equations (3·329) to get a'_ρ, a''_σ.

(vi) Substitute in (3·333) and (3·334) to get

$$\mathbf{V}', \mathbf{V}'', \mathbf{V}'^2, \mathbf{V}''^2,$$

and so establish the bounds

$$\mathbf{V}''^2 \leqslant \mathbf{S}^2 \leqslant \mathbf{V}'^2.$$

(vii) Calculate $\mathbf{C} = \tfrac{1}{2}(\mathbf{V}' + \mathbf{V}'')$ as an approximation to the gradient of the solution, $4R^2 = \mathbf{V}'^2 - \mathbf{V}''^2$ as an estimate of error and $R^2/\mathbf{C}^2$ as an estimate of relative error. [$4\mathbf{C}^2 = \mathbf{V}'^2 + 3\mathbf{V}''^2$.]

The linear independence of $\mathbf{T}'_\rho$ and the linear dependence of $\mathbf{T}''_\sigma$

Referring to Fig. 3·54 and item (ii) of the preceding summary, we note that there is to be a $\mathbf{T}'_\rho$ centred at each junction-point inside B' (in Fig. 3·54 there are 11 such points). *These are linearly independent F-vectors.* To prove this, assume linear dependence, i.e. the existence of c_ρ (not all zero) such that $\sum_{\rho=1}^{r} c_\rho \mathbf{T}'_\rho = \mathbf{O}$. If u'_ρ is the pyramid function whose gradient corresponds to $\mathbf{T}'_\rho$, this means that $\sum_{\rho=1}^{r} c_\rho u'_\rho$ is constant inside B, and in fact this sum is zero since every u'_ρ vanishes on B'. But at the centre of $\mathbf{T}'_1$ we have

$$\sum_{\rho=1}^{r} c_\rho u'_\rho = c_1 u'_1,$$

and so $c_1 = 0$ because $u'_1 \neq 0$ at the point in question. Similarly all the c's vanish, and the linear independence is established.

Now consider $\mathbf{T}''_\sigma$. The centres of these F-vectors occupy all the junction-points inside the outer polygon B''. Although we do not actually need them in the hypercircle method (they violate the boundary condition for homogeneous associated vectors), let us consider the set $\mathbf{T}'_\rho$ extended to include all centres inside B''. For simplicity, suppose the normalization condition to be that all pyramids shall be of the same height.

Consider the F-vector
$$\sum_{\rho=1}^{s} \mathbf{T}'_{\rho}; \tag{3·509}$$
we note that the summation is from 1 to s, as for $\mathbf{T}''_{\sigma}$. This F-vector corresponds to the gradient of a polyhedral graph obtained by superimposing a number of pyramid graphs, all of the same height. This polyhedral graph is a flat table-land, flat, that is, all over the interior of B and out to the polygon between B and B'' (Fig. 3·54), and then falling down in slopes to B''. Thus, inside B, the P-vector field corresponding to (3·509) is *zero*.

Now turn the P-vector fields of the set $\mathbf{T}'_{\rho}$ $(\rho = 1, 2, \ldots, s)$ through a right angle, thus generating the set $\mathbf{T}''_{\sigma}$. It follows that the sum of these new P-vector fields is zero inside B, and so we have
$$\sum_{\sigma=1}^{s} \mathbf{T}''_{\sigma} = \mathbf{O}. \tag{3·510}$$
This, as a statement concerning F-vectors, concerns itself only with the domain of the problem, $V + B$; if this domain were changed so as to include B'', (3·510) would not be true for the set of pyramid F-vectors of the second class previously considered—this set would have to be extended out to a new polygonal boundary outside the new boundary of the domain of the problem.

The equation (3·510) is important; it tells us that *the set of pyramid F-vectors of the second class* $\mathbf{T}''_{\sigma}$, *as described in* (ii) *of the summary above, are linearly dependent.*

It will be recalled that in § 2·7 we regarded as standard the case where $\mathbf{T}'_{\rho}$ and $\mathbf{T}''_{\sigma}$ were linearly independent. The case of linear dependence was discussed on p. 103 and in footnotes to pp. 104, 108, 119; it was pointed out that linear dependence involved only minor modifications. In the present instance, the linear dependence (3·510) merely means that there is an indeterminacy when we come to solve (3·329); one of the a''_{σ} may be arbitrarily assigned.

The conjugate harmonic function

Let us first consider the Dirichlet problem
$$\Delta u = 0, \quad (u)_B = f, \tag{3·511}$$
for a *simply connected* region V.

Let us suppose that by means of pyramid F-vectors of both classes $(\mathbf{T}'_{\rho}, \mathbf{T}''_{\sigma})$ we have obtained vertices as in (3·333):
$$\mathbf{V}' = \mathbf{S}'_0 + \sum_{\rho=1}^{r} a'_{\rho} \mathbf{T}'_{\rho}, \quad \mathbf{V}'' = \sum_{\sigma=1}^{s} a''_{\sigma} \mathbf{T}''_{\sigma}. \tag{3·512}$$

Let us suppose that we have obtained a good approximation in the sense that $(\mathbf{V}' - \mathbf{V}'')^2$ is small. (We use the words 'good' and 'small' vaguely here, for we are at present seeking a general understanding.) Then if

$$\mathbf{S}_0' \leftrightarrow \operatorname{grad} u', \quad \mathbf{T}_\rho' \leftrightarrow \operatorname{grad} u'_{(\rho)}, \tag{3·513}$$

we have a good approximation to the solution u in the form

$$u \sim u' + \sum_{\rho=1}^{r} a_\rho' u'_{(\rho)}. \tag{3·514}$$

The gradient of the function on the right corresponds precisely to $\mathbf{V}'$.

Since $\mathbf{V}''$ is close to $\mathbf{V}'$ in F-space, the P-vector field to which $\mathbf{V}''$ corresponds is also a good approximation to the gradient of u. We recall that $\mathbf{T}'$ and $\mathbf{T}''$ are both generated from the same pyramid functions $u'_{(\mu)}$; we have in fact

$$\mathbf{T}_\mu' \leftrightarrow u'_{(\mu),i}, \quad \mathbf{T}_\mu'' \leftrightarrow \epsilon_{ij} u'_{(\mu),j}, \tag{3·515}$$

where ϵ_{ij} is as in (3·126). It is true that some of these pyramid functions are not used to form $\mathbf{T}'$ vectors because their bases cut B; but that does not affect the argument. We have accurately

$$\mathbf{V}'' \leftrightarrow \sum_{\sigma=1}^{s} a_\sigma'' \epsilon_{ij} u'_{(\sigma),j}, \tag{3·516}$$

and hence, since $\mathbf{V}''$ is a good approximation to the solution $\mathbf{S}$,

$$u_{,i} \sim \sum_{\sigma=1}^{s} a_\sigma'' \epsilon_{ij} u'_{(\sigma),j}. \tag{3·517}$$

Now let v be the harmonic function conjugate to u, so that as in (3·125) we have $u_{,i} = \epsilon_{ij} v_{,j}$; then (3·517) gives

$$v_{,j} \sim \sum_{\sigma=1}^{s} a_\sigma'' u'_{(\sigma),j}. \tag{3·518}$$

Integration gives

$$v \sim \sum_{\sigma=1}^{s} a_\sigma'' u'_{(\sigma)}, \tag{3·519}$$

if we omit an unimportant additive constant.

This is an interesting result: *we get by* (3·514) *a function approximating the solution u of the Dirichlet problem, and by* (3·519) *a polyhedral function approximating the conjugate harmonic function v.*

This result is quite satisfactory when V is simply connected. But if V is multiply connected, the conjugate harmonic function v will in general be multiple-valued and it then cannot possibly be approximated by a single-valued function as in (3·519). The argument leading to (3·519) must break down somewhere, and it breaks down in the assumption that we can succeed in drawing the vertices **V**′, **V**″ close together. *In a multiply connected region, we cannot hope (unless the boundary conditions of the Dirichlet problem are specially chosen) to draw* **V**′ *and* **V**″ *close together no matter how fine a mesh we use, if* **V**″ *is formed only of pyramid F-vectors of the second class.* We must do something to allow for the multiple-valuedness of v; this we shall now discuss.

Case of multiple connectivity

Suppose that the region V of the Dirichlet problem is multiply connected and the boundary condition $(u)_B = f$ such that the conjugate harmonic function v is multiple-valued. Let a triangulation be made and the junction points elevated so that they lie on the three-dimensional graph of v. On account of the multiple-valuedness of v we get an infinite number of points on the vertical erected at each junction-point, and no choice from these points can yield a smooth distribution of height. If we go round one of the irreducible circuits in V, we find on completing the circuit that v differs by a finite jump from what it was before.

To fix the ideas let us think of V as doubly connected, although the argument is essentially general. Fig. 3·57 will help to show in perspective what happens when we attempt to construct an approximate polyhedral graph for a multiple-valued function v. $ABCDEF$ is a broken line, part of the network in the plane; it serves as a cut to reduce connectivity. The heights AA'', BB'', ... represent the values of v on the far side of the cut, and AA', BB', ... the values on the near side of the cut, the differences $A'A''$, $B'B''$, ... being all equal to one another and to the jump in v (harmonic function). $GHIJK$ are other junction-points of the triangulation, on the near side of the cut, and $G'H'I'J'K'$ corresponding points on the polyhedral graph. This polyhedral graph is (for a fine mesh) a good approximation to both v and $v_{,i}$, but it cannot be constructed by superposition of pyramids.

If we forget about $A'B'C'D'E'F'$ and join $G'H'I'J'K'$ across to $A''B''C''D''E''F''$, filling in the new triangular faces (this is not shown in Fig. 3·57), we get a new polyhedral graph which *is* a superposition of pyramids, but gives a very poor approximation to $v_{,i}$ over the strip between $ABCDEF$ and $GHIJK$; if the mesh is taken very fine, the slope of this new polyhedral graph becomes very large. Let us for reference call this polyhedral graph Γ.

How can we remedy the defect of Γ, viz. its large slope on the near side of the cut? We do it as follows. We revise the points $G'H'I'J'K'$ by raising them through the v-jump, viz. the height

$A'A''$; we join the points so obtained across to $A''B''C''D''E''F''$. This gives us a set of triangular faces directly above the strip lying on the near side of the cut, and the gradient of these triangular faces approximates to $v_{,i}$. If we add to the gradient field of Γ the difference between the gradient field just described and the gradient field of Γ over the strip on this side of the cut, we get a gradient field which approximates $v_{,i}$ well everywhere for a fine mesh.

When we turn this gradient field through a right angle, we get a P-vector field corresponding to a linear combination of pyramid

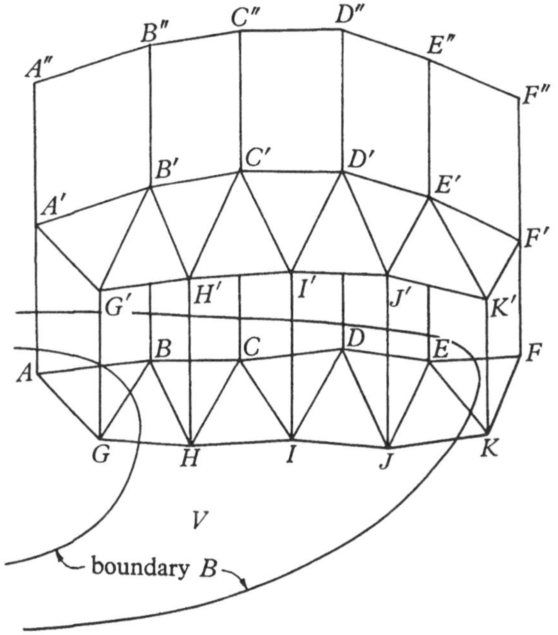

Fig. 3·57. The jump in a multiple-valued function v across the cut $ABCDEF$.

F-vectors of the second class, plus a P-vector field over the strip lying on this side of the cut; *this last corresponds to a complementary F-vector with only permissible discontinuities, since it is generated by rotation through a right angle from a gradient field and since (as is easily seen) it is parallel in each triangle to the edge of the triangle contained in $ABCDEF$ or $GHIJK$.*

This means that in the case of double connectivity we are to take as complementary vectors not only pyramid F-vectors of the second class, but also a new type which we shall call a *strip F-vector*; these are redefined more simply below. In cases of higher connectivity, we need a strip F-vector for each cut required to produce simple connectivity.

Strip F-vectors

Having seen above the purpose of a strip F-vector, let us give a simple direct definition.

Fig. 3·58 shows two broken lines, $ABCDEF$ and $GHIJK$, and lines joining them, forming a strip of triangles. Each triangle has one side on a boundary of the strip. In each of these triangles we assign a constant P-vector field parallel to the side of the triangle which lies in a boundary of the strip, *the strengths of these P-vector fields being chosen so that the normal component is continuous as we pass from triangle to triangle.* This ensures the satisfaction of the condition of permissible discontinuity for a complementary F-factor (3·211); the condition is of course also satisfied at the boundaries of the strip, the P-vector field being zero outside. The condition appears to be violated at the ends of the strip (AG, FK), *but we guard against this by letting the strip terminate on or outside the boundary B of the Dirichlet problem.*

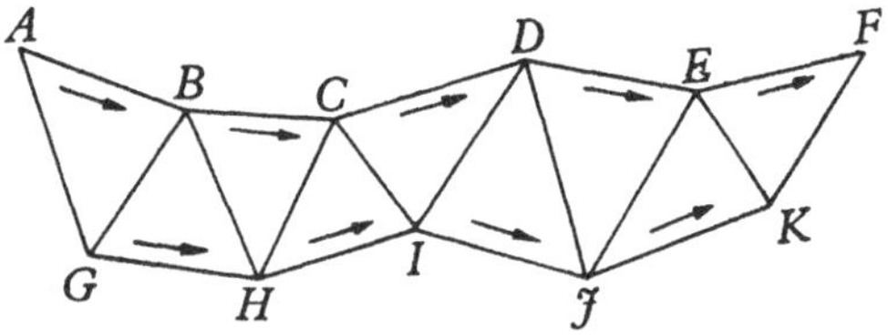

Fig. 3·58. P-vector field corresponding to a strip F-vector (complementary type $\mathbf{T}''_\sigma$); the normal components are continuous.

The above specification defines a *strip F-vector*; it is a complementary vector, and so takes its place with pyramid F-vectors of the second class.

For a given geometrical strip, there is only a single infinity of possible strip F-vectors, because the P-vector field in one of the triangles determines the P-vector field in all of them, by the condition of normal continuity. We might normalize a strip F-vector in some way, obtaining say $\mathbf{U}$. Then all strip F-vectors in the same geometrical strip are of the form $k\mathbf{U}$, where k is a constant.

Summary of procedure for the Dirichlet problem in a multiply connected plane region, using pyramid and strip F-vectors

(i) Triangulate and choose $\mathbf{S}'_0$.

(ii) Find the vertex $\mathbf{V}'$ as previously (multiple connectivity does not affect this).

(iii) Across each channel of the region take a strip of triangles of the network as in Fig. 3·58, thus making the region simply connected.

(iv) Include in the set of complementary F-vectors $\mathbf{T}''_\sigma$ not only the pyramid F-vectors of the second class as before, but also a strip vector (suitably normalized) on each of the strips of (iii) above.

(v) Proceed to find $\mathbf{V}''$ in the usual way.

Linear independence of strip F-vectors

It remains to discuss the possible linear dependence of the strip F-vectors with the pyramid F-vectors of the second class, i.e. the possibility of the existence of a relation of the form

$$\sum_{\sigma=1}^{s} c_\sigma \mathbf{P}_\sigma + \sum_{\tau=1}^{h} d_\tau \mathbf{U}_\tau = \mathbf{O}, \tag{3·520}$$

where $\mathbf{P}_\sigma$ are pyramid F-vectors of the second class and $\mathbf{U}_\tau$ strip F-vectors; h is the number of holes in the multiply connected region. We suppose $\mathbf{P}_\sigma$ normalized by the condition that all the generating pyramids are of the same height.

We consider the significance of (3·520) for each triangle of the network. If the triangle does not belong to a strip, only three $\mathbf{P}$'s are involved, and it is easily seen that (3·520) implies the equality of the corresponding c's. It follows that *all* the c's must be equal to one another, since the region with the strips removed is still connected and so we can work over it continuously, triangle by triangle. Thus, apart from the strips, the polyhedral graph from which $\sum_{\sigma=1}^{s} c_\sigma \mathbf{P}_\sigma$ is derived by turning the gradient through a right angle, is a flat table-land which comes up to each strip with the same height on both sides.

Now applying (3·520) to a triangle of a strip, we note that there occur three $\mathbf{P}$'s and only one $\mathbf{U}$. The three corresponding coefficients c are equal to one another, and the sum of the three $\mathbf{P}$-terms in (3·520) is zero. Hence the single d coefficient which occurs must be zero too. In fact, *the only relation of linear dependence of the form* (3·520) *is*

$$\sum_{\sigma=1}^{s} \mathbf{P}_\sigma = \mathbf{O}, \tag{3·521}$$

which is simply the relation of linear dependence already given in (3·510). *The strip F-vectors are linearly independent of one another and of the pyramid F-vectors of the second class.*

Bounds on the mean value of the gradient of the solution in a triangle of the network

Suppose we have carried out the calculations which locate the solution-point $\mathbf{S}$ on a hypercircle as in (3·335). Then the centre

C may be regarded as an approximation to **S**, the mean-square error being, from (3·336),

$$(\mathbf{S}-\mathbf{C})^2=\int(u_{,i}-c_i)(u_{,i}-c_i)\,dV=R^2, \tag{3·522}$$

where R is the radius of the hypercircle. Even though R may be small, the difference between $u_{,i}$ and c_i may be considerable in some parts of V and it is desirable to obtain local bounds. It is true that we have the method of § 3·4 for getting bounds at a point, but that method may prove formidable to apply. Here we shall be satisfied to get, not bounds at a point, but bounds on mean values taken over a mesh of the network.

Let $A_1A_2A_3$ be a triangle of the network (Fig. 3·59). Let **G** correspond to a P-vector field which is constant over this triangle and zero outside it (the discontinuity is no harm here). We shall get our bounds by using (2·783), which reads

$$|\,\mathbf{S}.\mathbf{G}-\mathbf{C}.\mathbf{G}\,|\leqslant R\,|\,\mathbf{G}_0\,|, \tag{3·523}$$

where $\mathbf{G}_0$ is given by (2·784).

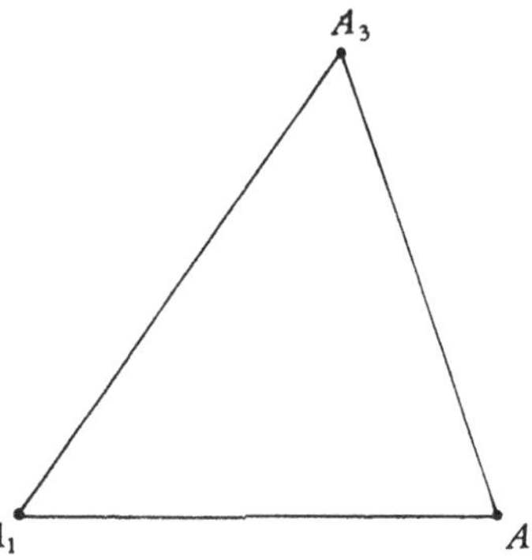

Fig. 3·59. A triangle of the network.

Now, if

$$\mathbf{G}\leftrightarrow m_i \tag{3·524}$$

over the triangle (a constant P-vector field), we have

$$\mathbf{S}.\mathbf{G}=m_i\int u_{,i}\,dA, \tag{3·525}$$

where the integration is taken over the triangle $A_1A_2A_3$. If we take in turn ($m_1=1, m_2=0$; $m_1=0, m_2=1$) it is clear that bounds on **S**.**G** will give bounds on the mean values of $u_{,1}$, $u_{,2}$ in the triangle.

As for **C**.**G**, we may turn to (3·335), or better to (2·781) and (2·775), so that orthonormalization is avoided. We note that $\mathbf{T}'_\rho$, $\mathbf{T}''_\sigma$ refer to pyramid F-vectors of the first and second classes respectively, including strip F-vectors in the case of multiple connectivity. We shall suppose in what follows (with very little loss of generality) that the triangle $A_1A_2A_3$ is not in a strip. The coefficients a'_ρ, a''_σ in (2·775) are supposed known.

Let the F-vectors be renumbered so that

$\mathbf{T}'_1$, $\mathbf{T}''_1$ are centred on A_1,
$\mathbf{T}'_2$, $\mathbf{T}''_2$ are centred on A_2,
$\mathbf{T}'_3$, $\mathbf{T}''_3$ are centred on A_3.

Then by (2·781) and (2·775)

$$\mathbf{C}\,.\,\mathbf{G}=\tfrac{1}{2}(\mathbf{V}'+\mathbf{V}'')\,.\,\mathbf{G}=\tfrac{1}{2}\mathbf{S}_0'\,.\,\mathbf{G}+\tfrac{1}{2}(a_1'\mathbf{T}_1'\,.\,\mathbf{G}+a_2'\mathbf{T}_2'\,.\,\mathbf{G}+a_3'\mathbf{T}_3'\,.\,\mathbf{G})$$
$$+\tfrac{1}{2}(a_1''\mathbf{T}_1''\,.\,\mathbf{G}+a_2''\mathbf{T}_2''\,.\,\mathbf{G}+a_3''\mathbf{T}_3''\,.\,\mathbf{G}), \qquad (3·526)$$

since $\mathbf{G}$ is orthogonal to all the pyramid F-vectors except those with bases overlapping the triangle $A_1A_2A_3$. Since $\mathbf{G}$ corresponds to a constant P-vector field inside the triangle, the last six of the scalar products are easy to evaluate; $\mathbf{S}_0'\,.\,\mathbf{G}$ is the only one that may be awkward.

We now turn to the right-hand side of (3·523) from which the required bounds are to be obtained. R is known by (3·335), but $|\,\mathbf{G}_0\,|$ presents difficulty. To determine it, we have to use (2·785); we have to solve the linear equations

$$\left.\begin{aligned}\sum_{\mu=1}^{r} c_\mu'\mathbf{T}_\mu'\,.\,\mathbf{T}_\rho'-\mathbf{G}\,.\,\mathbf{T}_\rho'=0 \quad (\rho=1,2,\ldots,r),\\ \sum_{\nu=1}^{s} c_\nu''\mathbf{T}_\nu''\,.\,\mathbf{T}_\sigma''-\mathbf{G}\,.\,\mathbf{T}_\sigma''=0 \quad (\sigma=1,2,\ldots,s).\end{aligned}\right\} \qquad (3·527)$$

In the notation of (3·328) these read

$$\left.\begin{aligned}\sum_{\mu=1}^{r} c_\mu' A_{\mu\rho}'-\mathbf{G}\,.\,\mathbf{T}_\rho'=0 \quad (\rho=1,2,\ldots,r),\\ \sum_{\nu=1}^{s} c_\nu'' A_{\nu\sigma}''-\mathbf{G}\,.\,\mathbf{T}_\sigma''=0 \quad (\sigma=1,2,\ldots,s).\end{aligned}\right\} \qquad (3·528)$$

Note that these equations do not involve $\mathbf{S}_0'$; otherwise they are rather like (3·329). Note also that the last terms vanish except for the values 1, 2, 3 of ρ and σ.

On account of the linear dependence relation (3·510) the solution of the second set of equations in (3·528) is not unique; one of the c'' coefficients may be chosen arbitrarily. When the equations (3·528) have been solved for c_ρ', c_σ'' (preferably by a method of successive approximations), we have, as in (2·784),

$$\mathbf{G}_0^2=\mathbf{G}^2-\sum_{\rho=1}^{3} c_\rho'\mathbf{G}\,.\,\mathbf{T}_\rho'-\sum_{\sigma=1}^{3} c_\sigma''\mathbf{G}\,.\,\mathbf{T}_\sigma'', \qquad (3·529)$$

$$\mathbf{G}^2=m_i m_i A=(m_1^2+m_2^2)\,A, \qquad (3·530)$$

A being the area of the triangle $A_1A_2A_3$.

To summarize, we bound the mean value of the gradient of the solution $(u_{,i})$ over a triangle of the network in the following way:

(i) A hypercircle is found as in (3·335) by means of a triangulation with use of pyramid F-vectors of both classes and strip F-vectors in the case of multiple connectivity. The coefficients a'_ρ, a''_σ of (2·775) are found and the radius R of the hypercircle.

(ii) We choose for **G** a constant P-vector field m_i in the mesh triangle we are interested in, and calculate **C**.**G** by (3·526).

(iii) We solve (3·528) for c'_ρ, c''_σ, choosing one of the latter arbitrarily. (This is the heaviest part of the work.)

(iv) We insert these values in (3·529). Note that we need only the six c's corresponding to the vertices of the triangle.

(v) We insert the values in (3·523) and so get the required bounds.

Exercises

1. A pyramid function is generated by raising a point P of a square to unit height above its plane. This function defines pyramid F-vectors of the first and second classes ($\mathbf{T}', \mathbf{T}''$). Show that, if the square lies wholly in the region V, then

$$\mathbf{T}'^2 = \mathbf{T}''^2 = a^2(A_1^{-1} + A_2^{-1} + A_3^{-1} + A_4^{-1}),$$

where $2a$ is a side of the square and A_1, A_2, A_3, A_4 the areas of the four triangles formed by joining P to the corners. (Note that $\mathbf{T}'^2 = \mathbf{T}''^2 = 4$ if P is at the centre of the square.)

2. Two squares of equal size are drawn, the sides of one square being parallel to the sides of the other and the centre of one square being at a corner of the other. By raising the centre of each square to unit height, we generate from each of them a pair of pyramid F-vectors (first and second classes). Show that all four F-vectors are orthogonal to one another, assuming that the region common to both squares lies wholly in the region V of P-space.

3. A plane is divided into parallelograms by two families of equidistant parallel lines. Then one family of diagonals is drawn, giving a triangulation of the plane. Show that the corresponding pyramid functions have hexagonal bases, and that if all the pyramids are of unit height and all domains of integration involved lie in V, then the scalar product of two neighbouring F-vectors of the same class is $-\cot\theta_1$ or $-\cot\theta_2$ or $-\cot\theta_3$, where $\theta_1, \theta_2, \theta_3$ are the angles of a mesh of the triangulation.

3·6. HEXAGONAL PYRAMID F-VECTORS

As remarked earlier, equilateral triangles and right-angled isosceles triangles give the simplest networks. We shall now discuss the former; the latter will be treated in § 3·7.

Definition and normalization

Fig. 3·61 shows a network of equilateral triangles. If we pull one of the junction-points vertically up out of the plane of the paper without disturbing the other junction-points, we generate a *hexagonal pyramid function* with level lines as shown. The hexagonal pyramid function gives rise to two F-vectors, *hexagonal pyramid F-vectors of the first and second classes*; their P-vector fields are shown in Fig. 3·62.

We shall write p for the magnitude of the P-vector field and $2a$ for the side of the hexagon, which is of course the same as the distance between junction-points in the triangular network. Also we shall denote the area of a mesh triangle by A, so that

$$A = a^2 . 3^{\frac{1}{2}}. \tag{3·601}$$

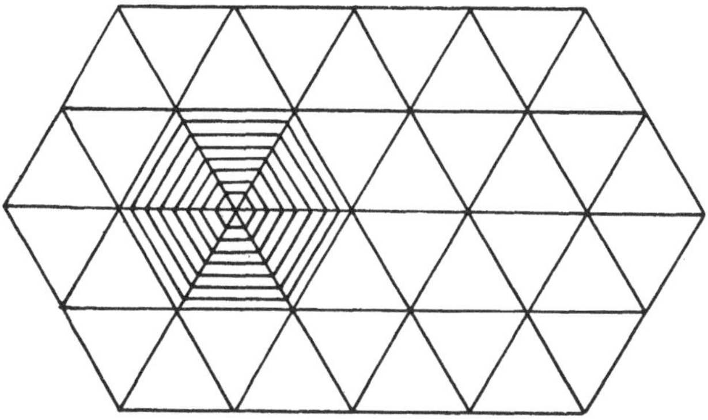

Fig. 3·61. Network of equilateral triangles and level lines of hexagonal pyramid function.

We shall normalize by making the height of each hexagonal pyramid unity. Then any two of p, a, A are expressible in terms of the third, and we have

$$pa = 3^{-\frac{1}{2}}, \quad p^2 A = 3^{-\frac{1}{2}}. \tag{3·602}$$

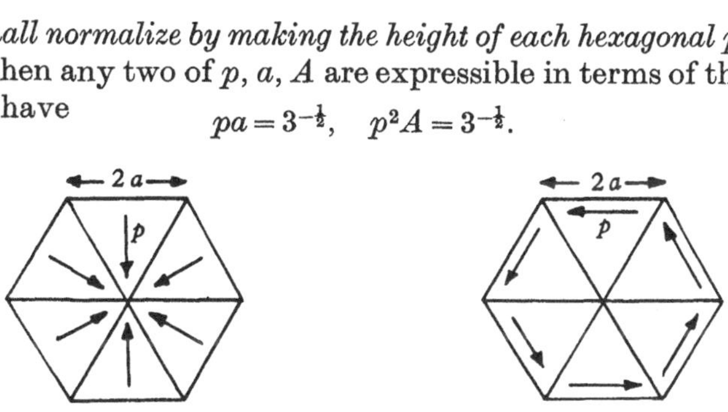

First class

Second class

Fig. 3·62. P-vector fields of hexagonal pyramid F-vectors.

The base of $\mathbf{T}'_\rho$ (hexagonal pyramid F-vector of the first class) is always contained in $V+B$; otherwise it would not be a homogeneous associated vector. Carrying out the integration prescribed in (3·203), we obtain

$$\mathbf{T}'^2_\rho = 6Ap^2 = 2 . 3^{\frac{1}{2}}. \tag{3·603}$$

On the other hand, the base of $\mathbf{T}''_\sigma$ (hexagonal pyramid F-vector of the second class) need not be entirely contained in $V+B$. If it is so contained, we have

$$\mathbf{T}''^2_\sigma = 6Ap^2 = 2 . 3^{\frac{1}{2}}. \tag{3·604}$$

But if only an area S of the base of $\mathbf{T}''_\sigma$ is contained in $V+B$, then (since the scalar product involves only integration over V)

$$\mathbf{T}''^2_\sigma = Sp^2. \tag{3·605}$$

Note that for such an 'incomplete' F-vector, we keep the same p as in (3·602).

Scalar products

From their property of lying in the orthogonal linear subspaces L', L'' [cf. (3·206) and (3·208)], it follows that *any hexagonal pyramid*

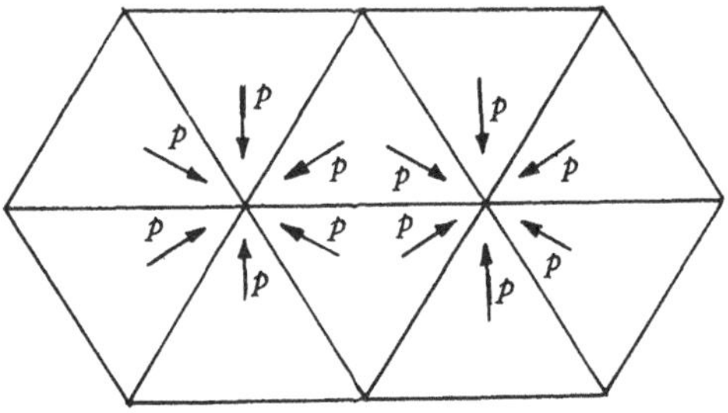

Neighbouring hexagonal pyramid F-vectors of the first class (calculation of $\mathbf{T}'_\mu . \mathbf{T}'_\rho$).

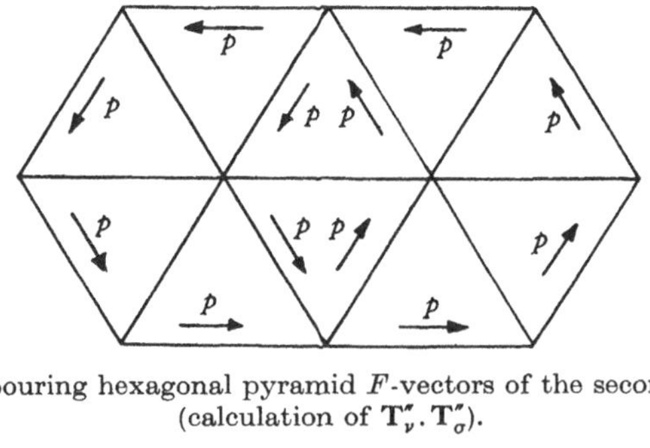

Neighbouring hexagonal pyramid F-vectors of the second class (calculation of $\mathbf{T}''_\nu . \mathbf{T}''_\sigma$).

Fig. 3·63

F-vector of the first class is orthogonal to any hexagonal pyramid F-vector of the second class.

As for two hexagonal pyramid F-vectors of the same class, we have to consider the scalar products $\mathbf{T}'_\mu . \mathbf{T}'_\rho$ and $\mathbf{T}''_\nu . \mathbf{T}''_\sigma$.

If the F-vectors are not neighbours (i.e. if their bases do not overlap), it is clear that they are orthogonal:

$$\mathbf{T}'_\mu . \mathbf{T}'_\rho = 0, \quad \mathbf{T}''_\nu . \mathbf{T}''_\sigma = 0. \tag{3·606}$$

If the F-vectors are neighbours (overlapping bases), we have the situations shown in Fig. 3·63.

In the case of F-vectors of the first class, the complete figure is contained in $V+B$, and since the overlapping P-vector fields are inclined to one another at 120°, we have

$$\mathbf{T}'_\mu . \mathbf{T}'_\rho = -p^2 A = -3^{-\frac{1}{2}}. \tag{3·607}$$

From (3·603) it follows then that *adjacent hexagonal pyramid F-vectors of the first class make with one another an angle θ where*

$$\cos\theta = -\tfrac{1}{6}. \tag{3·608}$$

Thus when we think of a fine mesh with many equilateral triangles and consequently many hexagonal pyramid F-vectors of the first class, the geometrical picture in F-space is simplified by the fact that most of these F-vectors are orthogonal to one another. Each F-vector is orthogonal to all the others except its neighbours, with each of which it makes the obtuse angle (3·608); in general there are six neighbours, but this number is reduced near the boundary B.

In the case of neighbouring hexagonal pyramid F-vectors of the second class, we look at the second diagram in Fig. 3·63. The only area that contributes to the scalar product is the overlapping rhombus, and if this is contained in $V+B$ we get, as in (3·607),

$$\mathbf{T}''_\nu . \mathbf{T}''_\sigma = -p^2 A = -3^{-\frac{1}{2}}, \tag{3·609}$$

and the same angle of inclination (3·608). If the overlapping rhombus is cut by B, we get

$$\mathbf{T}''_\nu . \mathbf{T}''_\sigma = -\tfrac{1}{2}p^2 S, \tag{3·610}$$

where S is the area of the portion of the rhombus contained in V.

Determination of the vertices $\mathbf{V}'$, $\mathbf{V}''$

To find the vertices $\mathbf{V}'$, $\mathbf{V}''$ we have to solve the linear equations (3·329). It is best to avoid orthonormalization and the matrix method of (3·331) and rely rather on successive approximations, making use of the orthogonalities which exist between the hexagonal pyramid F-vectors.

For $\mathbf{V}'$ we have to solve

$$\sum_{\mu=1}^{r} a'_\mu \mathbf{T}'_\mu . \mathbf{T}'_\rho + \mathbf{S}'_0 . \mathbf{T}'_\rho = 0 \quad (\rho = 1, 2, \ldots, r). \tag{3·611}$$

Here $\mathbf{T}'_\rho$ are hexagonal pyramid F-vectors of the first class, normalized by the condition of unit height. Hence, by (3·603), (3·607) and (3·606), we have

$$\left.\begin{aligned} &\mathbf{T}'_\mu . \mathbf{T}'_\rho = 2 . 3^{\frac{1}{2}} \text{ if } \mu = \rho, \\ &\mathbf{T}'_\mu . \mathbf{T}'_\rho = -3^{-\frac{1}{2}} \text{ if they are neighbours,} \\ &\mathbf{T}'_\mu . \mathbf{T}'_\rho = 0 \text{ if they are not neighbours.} \end{aligned}\right\} \tag{3·612}$$

Since $\mathbf{T}'_\rho$ has at most six neighbours, each summation in (3·611) contains at most seven terms. It is convenient to think of $\mathbf{T}'_\rho$ as *always* having six neighbours, and this can be done by the device of associating zero values of a' with fictitious neighbours. (These fictitious neighbours are like the extra $\mathbf{T}'$ included in (3·509).)

It is convenient to refer to the coefficients a'_μ as *weights* associated with the F-vectors.

Denoting by
$$\mathbf{T}'_{\rho 1}, \mathbf{T}'_{\rho 2}, \ldots, \mathbf{T}'_{\rho 6}; \quad a'_{\rho 1}, a'_{\rho 2}, \ldots, a'_{\rho 6} \tag{3·613}$$
the six neighbours of $\mathbf{T}'_\rho$ and the weights associated with them, we may write (3·611) in the following equivalent forms:
$$2 \,.\, 3^{\frac{1}{2}} a'_\rho - 3^{-\frac{1}{2}}(a'_{\rho 1} + a'_{\rho 2} + \ldots + a'_{\rho 6}) = -\mathbf{S}'_0 \,.\, \mathbf{T}'_\rho \quad (\rho = 1, 2, \ldots, r), \tag{3·614}$$

$$\boxed{a'_\rho = \tfrac{1}{6}(a'_{\rho 1} + a'_{\rho 2} + \ldots + a'_{\rho 6}) - \tfrac{1}{2} \,.\, 3^{-\frac{1}{2}} \mathbf{S}'_0 \,.\, \mathbf{T}'_\rho \quad (\rho = 1, 2, \ldots, r).} \tag{3·615}$$

It is understood that any one of $a'_{\rho 1}, a'_{\rho 2}, \ldots, a'_{\rho 6}$ is to be equated to zero if the corresponding $\mathbf{T}'$ is fictitious, in the sense that its base projects beyond the boundary B.

The formula (3·615) is very simple in structure and similar to a formula we would use if we were trying to solve a Poisson equation of the form (3·102) by finite differences.* That is, in fact, what we are doing, but in a refined way, since we have a precise knowledge of bounds in the mean-square sense once the equations have been solved. In words, (3·615) tells us that *each weight is the arithmetic mean of the six neighbouring weights, diminished by a certain known quantity*.

When the weights have been found, we get $\mathbf{V}'$ from (3·333) and $\mathbf{V}'^2$ from (3·334).

In the case of $\mathbf{V}''$, we have to find the weights a''_ν from the equations (3·329):
$$\sum_{\nu=1}^{s} a''_\nu \mathbf{T}''_\nu \,.\, \mathbf{T}''_\sigma - \mathbf{S}'_0 \,.\, \mathbf{T}''_\sigma = 0 \quad (\sigma = 1, 2, \ldots, s). \tag{3·616}$$
This looks like a repetition of what we have just done above, but there are differences.

Let us first suppose for simplicity that the region is simply connected, so that strip F-vectors do not occur; the $\mathbf{T}''$ are all hexagonal pyramid F-vectors of the second class.

* The difference equations usually employed for the Laplace or Poisson equation correspond to a square grid, not one based on equilateral triangles. We shall meet this more common type in (3·705) and (3·707). For a discussion of the convergence of iterative processes applied to these equations and of upper and lower bounds for their solution, see Diaz and Roberts (1, 2).

We recall (3·510), which tells us that the solution of (3·616) is not unique and that we can choose one of weights a'' arbitrarily. Probably we would give it the value zero.

That is not very important. The real difference between (3·616) and (3·611) arises from the raggedness of the set $\mathbf{T}''_\sigma$ near B; it is a raggedness in the matter of scalar products, for in calculating them we are to use only the parts of the bases lying inside B.

If the base of $\mathbf{T}''_\sigma$ does not go outside B, then the scalar product with a neighbour is as in (3·609). Hence we get a simple formula like (3·615):

$$a''_\sigma = \tfrac{1}{6}(a''_{\sigma1} + a''_{\sigma2} + \dots + a''_{\sigma6}) + \tfrac{1}{2} \,.\, 3^{-\frac{1}{2}} \mathbf{S}'_0 \,.\, \mathbf{T}''_\sigma. \qquad (3·617)$$

This holds for the weights of all hexagonal pyramid F-vectors of the second class $\mathbf{T}''_\sigma$ *whose bases do not go outside B.*

For other cases, the best we can do is to write

$$a''_\sigma \mathbf{T}''^2_\sigma = -(a''_{\sigma1} \mathbf{T}''_\sigma \,.\, \mathbf{T}''_{\sigma1} + a''_{\sigma2} \mathbf{T}''_\sigma \,.\, \mathbf{T}''_{\sigma2} + \dots + a''_{\sigma6} \mathbf{T}''_\sigma \,.\, \mathbf{T}''_{\sigma6}) + \mathbf{S}'_0 \,.\, \mathbf{T}''_\sigma. \qquad (3·618)$$

This formula applies to the case where the base of $\mathbf{T}''_\sigma$ *may go outside B* ((3·617) is a specialization when it does not). The notation is that

$$\mathbf{T}''_{\sigma1}, \mathbf{T}''_{\sigma2}, \dots, \mathbf{T}''_{\sigma6}; \quad a''_{\sigma1}, a''_{\sigma2}, \dots, a''_{\sigma6}$$

are the six neighbours of $\mathbf{T}''_\sigma$ and their weights. If we look at Fig. 3·54, we see that, for centres in the ring between B and B'', there will not be six neighbours; in such cases we proceed as we did for $\mathbf{T}'_\rho$, creating fictitious neighbours and giving them zero weight.

For multiply connected regions strip F-vectors occur in the set $\mathbf{T}''_\sigma$, and so there may be more than six neighbours; we can still use (3·618) provided we include in the parentheses on the right-hand side terms for *all* the neighbours of $\mathbf{T}''_\sigma$, including strip F-vectors.

When we have solved (3·617) and (3·618) for the weights a''_ν, we get $\mathbf{V}''$ from (3·333) and $\mathbf{V}''^2$ from (3·334).

In any piece of computation of this sort, we do not know whether we have done well until we have got R^2, where $4R^2 = \mathbf{V}'^2 - \mathbf{V}''^2$ as in (3·335). A small value of R^2 is what we need. If this has been achieved, we have not only the bounds $\mathbf{V}''^2$ and $\mathbf{V}'^2$ for $\mathbf{S}^2$; we have also a reasonable confidence (but not mathematical certainty) that the centre of the hypercircle, $\mathbf{C} = \tfrac{1}{2}(\mathbf{V}' + \mathbf{V}'')$, is a good approximation to the gradient of the solution of the Dirichlet problem all over the region.

The bisection process

In applying the method of hexagonal pyramid F-vectors, it is wise to start with a fairly coarse mesh and then reduce the size of the mesh by successive bisections.

Suppose we have found the weights a'_ρ, a''_σ (equivalently, the vertices $\mathbf{V}'$, $\mathbf{V}''$) using a network of triangles of which ABC (Fig. 3·64) is one. Now bisect the sides of this (and all other) triangles and join the points of bisection as shown, thus dividing the equilateral triangle ABC into four equilateral triangles. This gives us a new finer network with roughly four times the number of junction-

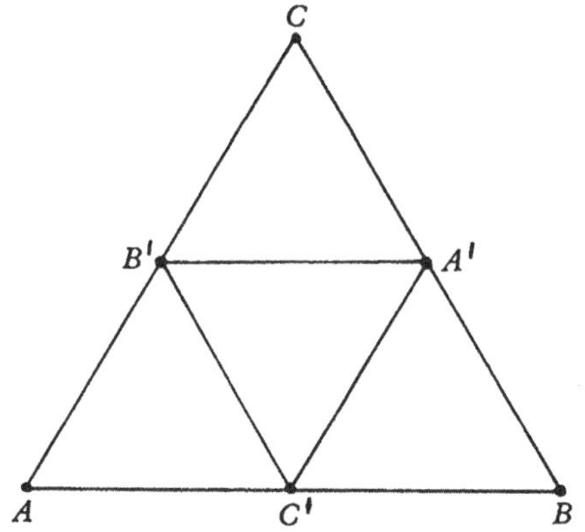

Fig. 3·64. The bisection process.

points and hence roughly four times the number of hexagonal pyramid F-vectors—'roughly', because of adjustments near the boundary.

Let us suppose that in the first stage with the coarser mesh we got the weights

$$a'_A,\quad a'_B,\quad a'_C,\quad a''_A,\quad a''_B,\quad a''_C. \tag{3·619}$$

In the next stage, after bisection, we have to solve again the linear equations (3·329), or the reduced forms (3·615), (3·617) and (3·618). This we do by successive approximations, inserting (as a basis) the old weights (3·619) at A, B, C, and at A', B', C' arithmetic means of the old weights, i.e. we would attach to A' the weights

$$\tfrac{1}{2}(a'_B + a'_C),\quad \tfrac{1}{2}(a''_B + a''_C).$$

This gives us a good start for the solution of the more numerous equations corresponding to the finer mesh.

In the case of multiple connectivity, a strip F-vector as in Fig. 3·58 would be similarly bisected; one-half would be rejected from the strip and absorbed into the general triangulation.

As far as $\mathbf{S}_0'$ is concerned, the bisection process is most easily carried out if $\mathbf{S}_0'$ is given as in (3·508); if it is given by a polyhedral graph, as described below (3·508), we have to make adjustments near B when we bisect the mesh.

However, it is not necessary to base higher approximations on lower ones; any network may be tackled independently, starting with zero weights.

In the above discussion, the method of hexagonal pyramid F-vectors has been treated as a general method for the approximate solution of the Dirichlet problem, applicable in the case of a general boundary and general boundary conditions. Naturally, it is far easier to apply when the boundary is simple (in particular, polygonal) and the boundary conditions simple also, as they are in condenser and torsion problems. To illustrate the method it seems best to use problems of the simpler types; but a curved boundary is treated in § 5·3.

The following example illustrates multiple connectivity, but not successive approximations by bisection; that will be demonstrated in Chapter 4 in connexion with the torsion of a regular hexagon.

Example: hexagonal condenser

Consider a hexagonal condenser as shown in Fig. 3·65, the inner hexagon B_0 having sides of length $\frac{1}{2}H$ and the outer hexagon B_1 sides of length H. The potentials are $u=0$ on B_0 and $u=1$ on B_1. Thus we have the Dirichlet problem:

$$\Delta u=0, \quad (u)_{B_0}=0, \quad (u)_{B_1}=1. \tag{3·620}$$

If the ratio of the outer and inner sides were any rational number, we could triangulate the region of the problem with equilateral triangles making the inner and outer hexagons part of the network. In our particular case this ratio is 2 and we can make the simple triangulation shown in Fig. 3·65; finer and finer meshes could be obtained from this by bisection as in Fig. 3·64. Naturally, we are not to expect very close bounds for the coarse mesh of Fig. 3·65, but in compensation the algebra is very simple.

For $\mathbf{S}_0'$ we take, in the quadrilateral $PQRS$, a P-vector field perpendicular to PQ and RS and of magnitude $(H\,.\,3^{\frac{1}{2}}/4)^{-1}$, pointing from PQ towards RS; in the other five similar quadrilaterals we take similar fields. The P-vector field, so defined over the whole region V, is the gradient of a continuous function u' which is 0 on B_0 and 1 on B_1.

We take eighteen homogeneous associated F-vectors $\mathbf{T}_\rho'$ $(\rho=1,2,\ldots,18)$, viz. pyramid F-vectors of the first class with centres at the points marked $1,2,\ldots,18$ in Fig. 3·65; these are normalized by the condition that the pyramids are of unit height. This makes the situation look much more forbidding than it really is, for it is evident from symmetry that the expression (3·333) for $\mathbf{V}'$ can involve only *two* weights, viz., a_1' and a_2'.

Using the formula (3·202) for the scalar product, and also (3·603) and (3·607), we obtain very easily

$$\left.\begin{aligned} &\mathbf{S}_0'^2 = 6.3^{\frac{1}{2}}, \quad \mathbf{S}_0'.\mathbf{T}_1' = -3^{-\frac{1}{2}}, \quad \mathbf{S}_0'.\mathbf{T}_2' = 0, \\ &\mathbf{T}_1'^2 = \mathbf{T}_2'^2 = 2.3^{\frac{1}{2}}, \quad \mathbf{T}_1'.\mathbf{T}_2' = -3^{-\frac{1}{2}}. \end{aligned}\right\} \tag{3·621}$$

From symmetry all other scalar products required follow from these.

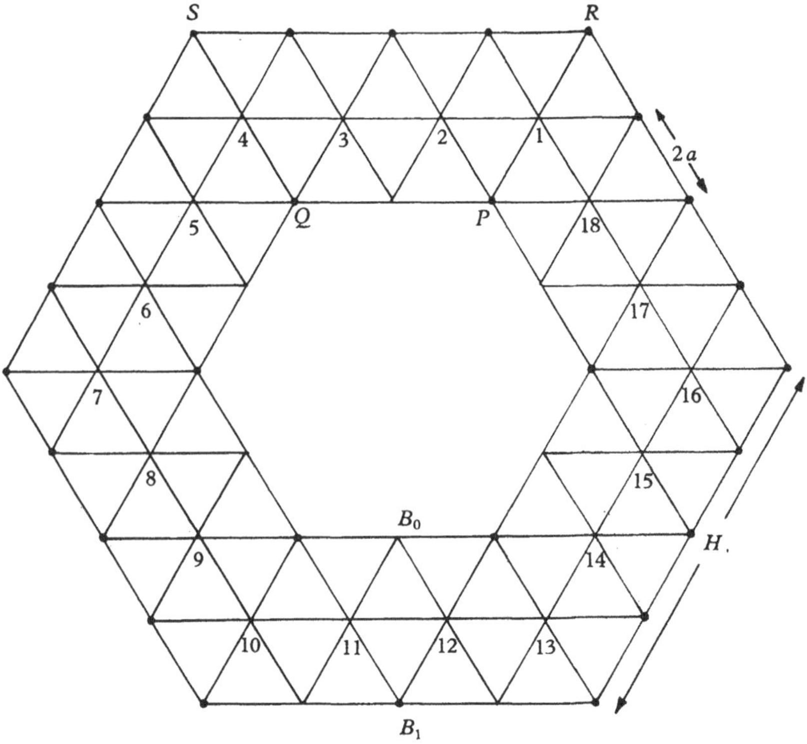

Fig. 3·65. Hexagonal condenser with triangulation and enumeration of hexagonal pyramid F-vectors of the first class.

The weights a_ρ' $(\rho = 1, 2, \ldots, 18)$ are to be found from (3·329) or from the handier equivalent formulae (3·615); we shall use the latter. From symmetry

$$a_3' = a_2', \quad a_{18}' = a_2',$$

and so application of (3·615) for $\rho = 1$ and $\rho = 2$ gives

$$a_1' = \tfrac{1}{6}.2a_2' + \tfrac{1}{6}, \quad a_2' = \tfrac{1}{6}(a_1' + a_2'), \tag{3·622}$$

and hence the weights $$a_1' = \tfrac{5}{28}, \quad a_2' = \tfrac{1}{28}. \tag{3·623}$$

All the weights follow by symmetry from these two.

We have now the vertex $\mathbf{V}'$ by (3·333):

$$\mathbf{V}' = \mathbf{S}_0' + \mathbf{T}', \quad \mathbf{T}' = \tfrac{1}{28}(5\Sigma\mathbf{T}_1' + \Sigma\mathbf{T}_2'), \tag{3·624}$$

where the summations are taken over symmetrically placed pyramids; the first summation has six terms (subscripts 1, 4, 7, 10, 13, 16) and the second has twelve terms.

For $\mathbf{V}'^2$ we have, by (3·334),

$$\mathbf{V}'^2 = \mathbf{S}_0'^2 + \mathbf{S}_0' . \mathbf{T}' = \mathbf{S}_0'^2 + 6a_1'\mathbf{S}_0' . \mathbf{T}_1' + 12a_2'\mathbf{S}_0' . \mathbf{T}_2'$$
$$= 6 . 3^{\frac{1}{2}} - (\tfrac{5}{14})\, 3^{\frac{1}{2}} = (\tfrac{79}{14})\, 3^{\frac{1}{2}} = 9·7737. \tag{3·625}$$

Hence we have, by (3·339), the following upper bound for the capacity C of the condenser per unit length ($2\pi C = \mathbf{S}^2$):

$$\mathbf{S}^2 \leqslant \mathbf{V}'^2 = 9·7737. \tag{3·626}$$

By way of comparison it is interesting to get another rough upper bound by a different method. Draw the circumscribed circle of the inner hexagon and the inscribed circle of the outer hexagon. Then the function

$$u' = \log(2r/H)/\log 3^{\frac{1}{2}}, \tag{3·627}$$

where r is distance from the centre, is a harmonic function taking the values 0, 1 for $r = \frac{1}{2}H$, $\frac{1}{2}H . 3^{\frac{1}{2}}$, i.e. on the inner and outer circle respectively. (It is in fact the potential in a circular condenser.) If we extend this function with value 0 back to the inner hexagon and value 1 out to the outer hexagon, we have a function of which the gradient corresponds to an associated F-vector, say $\mathbf{S}'$. It is easy to compute $\mathbf{S}'^2$ and so obtain from (3·302) the upper bound

$$\mathbf{S}^2 \leqslant \mathbf{S}'^2 = 11·4384. \tag{3·628}$$

Crude as the network is, our pyramid result (3·626) has bettered this.

Let us now get a lower bound on the capacity of the hexagonal condenser by means of complementary F-vectors. Since the region is doubly connected, a strip F-vector is required; one would suffice, but in the interests of symmetry it is better to take twelve.

Fig. 3·66 is the appropriate diagram. The triangulation of Fig. 3·65 is understood, but suppressed in order to show up the strip F-vectors.

Each junction point of Fig. 3·65 is now to be the centre of a hexagonal pyramid F-vector of the second class; these are normalized by the conditions that the height of the generating pyramid shall be unity and the circulation counter-clockwise as in Fig. 3·62. The centres are numbered as follows:

$a\,1, a\,2, \ldots, a\,24$ (outer ring),

$b\,1, b\,2, \ldots, b\,18$ (middle ring),

$c\,1, c\,2, \ldots, c\,12$ (inner ring).

Fig. 3·66 also shows twelve strip F-vectors at the corners, numbered $d\,1, d\,2, \ldots, d\,12$. These are normalized by the condition of giving unit rise outwards, so that the strength of each P-vector field is $2/H$ (H is the side of the outer hexagon).

It is easy to see that $\mathbf{S}'_0$ is orthogonal to all the hexagonal pyramid F-vectors of the second class. The only scalar products required (on account of symmetry, as explained below) are as follows, and they are easily calculated:

$$\left.\begin{aligned}
&\mathbf{S}'_0 . \mathbf{T}''_{d1} = 3^{\frac{1}{2}}/4, \\
&\mathbf{T}''^{2}_{a2} = 3^{\frac{1}{2}}, \quad \mathbf{T}''_{a2} . \mathbf{T}''_{b2} = -3^{-\frac{1}{2}}, \quad \mathbf{T}''_{a2} . \mathbf{T}''_{d1} = \tfrac{3}{8}, \\
&\mathbf{T}''^{2}_{b2} = 2 . 3^{\frac{1}{2}}, \quad \mathbf{T}''_{b2} . \mathbf{T}''_{a3} = -3^{-\frac{1}{2}}, \quad \mathbf{T}''_{b2} . \mathbf{T}''_{b3} = -3^{-\frac{1}{2}}, \quad \mathbf{T}''_{b2} . \mathbf{T}''_{d1} = \tfrac{1}{2}, \\
&\mathbf{T}''^{2}_{d1} = 3^{\frac{1}{2}}/4.
\end{aligned}\right\} \tag{3·629}$$

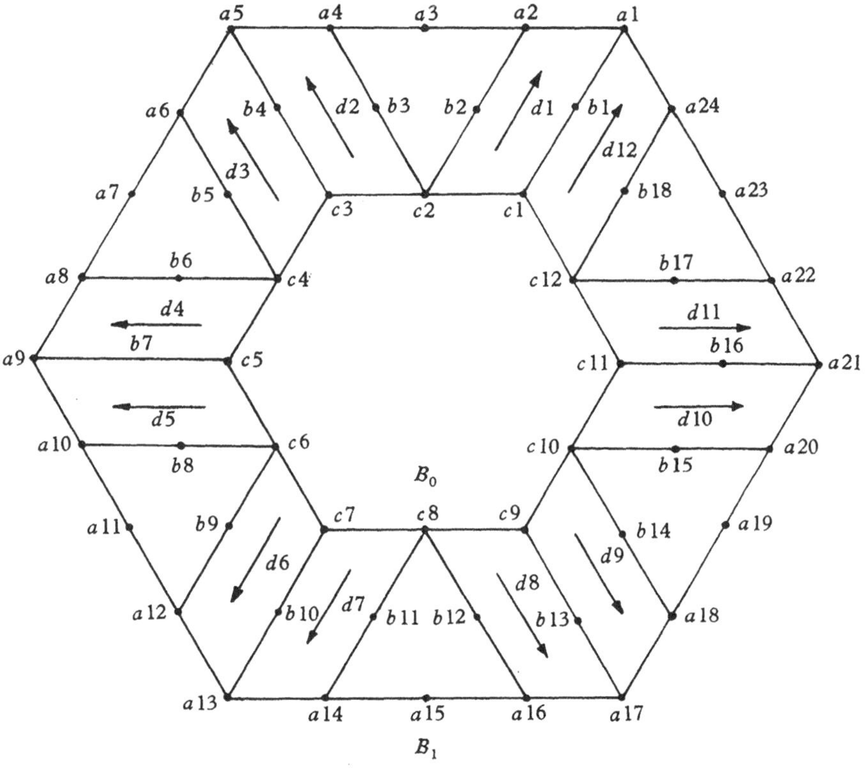

Fig. 3·66. Hexagonal condenser with enumeration of hexagonal pyramid F-vectors of the second class and strip F-vectors.

We have to solve (3·618). However, these equations are much simplified by symmetry considerations. It is evident that the P-vector field corresponding to the vertex $\mathbf{V}''$ will be symmetric with respect to all twelve lines of symmetry of the hexagons. This means that the P-vector field transforms into itself when we make a reflexion of it in any one of these lines of symmetry. Hence it is clear that:

(i) The weights of hexagonal pyramid F-vectors of the second class are equal and opposite if they occupy symmetric positions.

(ii) The weights of hexagonal F-vectors of the second class are zero if the centres lie on lines of symmetry.

(iii) The weights of the strip F-vectors are all the same.

Thus, changing the notation slightly to make it easier to write,

$$a''(a\,1) = a''(a\,3) = a''(b\,1) = a''(c\,1) = a''(c\,2) = 0, \tag{3·630}$$

and other equations of the same type. All weights vanish on the inner ring, and in fact there are only *three* weights to consider:

$$a''(a\,2), \quad a''(b\,2), \quad a''(d\,1). \tag{3·631}$$

For these three quantities (3·618) gives us the three equations

$$\left.\begin{aligned} a''(a\,2)\,\mathbf{T}''^2_{a2} &= -[a''(b\,2)\,\mathbf{T}''_{a2}\,.\,\mathbf{T}''_{b2} + a''(d\,1)\,\mathbf{T}''_{a2}\,.\,\mathbf{T}''_{d1}],\\ a''(b\,2)\,\mathbf{T}''^2_{b2} &= -[a''(a\,2)\,\mathbf{T}''_{b2}\,.\,\mathbf{T}''_{a2} - a''(b\,2)\,\mathbf{T}''_{b2}\,.\,\mathbf{T}''_{b3} + a''(d\,1)\,\mathbf{T}''_{b2}\,.\,\mathbf{T}''_{d1}],\\ a''(d\,1)\,\mathbf{T}''^2_{d1} &= -[a''(a\,2)\,\mathbf{T}''_{d1}\,.\,\mathbf{T}''_{a2} + a''(b\,2)\,\mathbf{T}''_{d1}\,.\,\mathbf{T}''_{b2}] + \mathbf{S}'_0\,.\,\mathbf{T}''_{d1}. \end{aligned}\right\} \tag{3·632}$$

We now substitute the numerical values for the squares and scalar products as given in (3·629), and solve the equations, obtaining

$$a''(a\,2) = -\tfrac{10}{37}\,.\,3^{\frac{1}{2}}, \quad a''(b\,2) = -\tfrac{6}{37}\,.\,3^{\frac{1}{2}}, \quad a''(d\,1) = \tfrac{64}{37}. \tag{3·633}$$

All the weights follow from these by symmetry.

We are now in a position to write down $\mathbf{V}''$ by (3·333), and we could plot its P-vector field if we so desired. But let us rather calculate $\mathbf{V}''^2$ from the same reference:

$$\mathbf{V}''^2 = \Sigma a''_\sigma\,\mathbf{S}'_0\,.\,\mathbf{T}''_\sigma = 12a''(d\,1)\,\mathbf{S}'_0\,.\,\mathbf{T}''_{d1}; \tag{3·634}$$

note that only the strip F-vectors contribute. We get

$$\mathbf{V}''^2 = \tfrac{192}{37}\,.\,3^{\frac{1}{2}} = 8·9879. \tag{3·635}$$

Thus, by (3·339) and the bound (3·626) already established, *we have the following bounds for the capacity per unit length* ($2\pi C = \mathbf{S}^2$) *of the hexagonal condenser shown in Figs.* 3·65 *and* 3·66:

$$8·9879 \leqslant \mathbf{S}^2 \leqslant 9·7737. \tag{3·636}$$

In view of the coarseness of the mesh and the simplicity of the calculations (which do not even require the integral calculus), the bounds must be regarded as reasonably close.

If we wished to push the bounds closer together, we would bisect the mesh and recalculate. But the number of linear equations to be solved would be greatly increased, and it would be advisable to resort to a method of solution by successive approximations. As this method will be sufficiently illustrated by later examples, we shall not pursue the question further here.

Exercises

1. Consider the Dirichlet problem $\Delta u = 0$, $(u)_B = \frac{1}{2}r^2$ for a regular hexagon of side H, r measuring distance from its centre. Take $\mathbf{S}'_0 \leftrightarrow \operatorname{grad} \frac{1}{2}r^2$ and for $\mathbf{T}'_1$ take a pyramid F-vector of the first class having the hexagon itself for base and normalized to unit height. Show that

$$\mathbf{S}'^2_0 = (\tfrac{5}{8})\,3^{\frac{1}{2}}H^4, \quad \mathbf{T}'^2_1 = 2\,.\,3^{\frac{1}{2}}, \quad \mathbf{S}'_0\,.\,\mathbf{T}'_1 = -3^{\frac{1}{2}}H^2,$$

and hence by (3·333) that the vertex $\mathbf{V}'$ is

$$\mathbf{V}' = \mathbf{S}_0' + \tfrac{1}{2}H^2\mathbf{T}_1', \quad \mathbf{V}'^2 = (\tfrac{1}{8})\,3^{\frac{1}{2}}H^4,$$

so that we have the upper bound

$$\mathbf{S}^2 \leqslant (\tfrac{1}{8})\,3^{\frac{1}{2}}H^4 = 0{\cdot}2165H^4.$$

Show that the three-dimensional graph of the function whose gradient corresponds to $\mathbf{V}'$ is cut by a vertical plane bisecting opposite sides of the hexagon in parabolas

$$z = \tfrac{1}{2}x^2 \pm 3^{-\frac{1}{2}}xH + \tfrac{1}{2}H^2,$$

where z is measured vertically and x is a coordinate on the line of section.

(This is the torsion problem for a regular hexagon and will be treated in detail in Chapter 4.)

2. Show that, irrespective of forms of polygonal bases, directions of sides and strengths of P-vector fields, *every* pyramid F-vector of the first class is orthogonal to *every* pyramid F-vector of the second class, provided the overlapping parts of their bases lie entirely in P-space (i.e. in the domain $V+B$).

3. A triangulation is made with equilateral triangles, as in Fig. 3·61. The centre of each triangle is then joined to the three vertices of that triangle, and a new network obtained in this way. Show that some of the bases of the corresponding pyramid functions will be triangular and some will have twelve sides.

4. For the hexagonal condenser of Fig. 3·65, bisect the mesh and obtain closer bounds for the capacity.

3·7. SQUARE PYRAMID F-VECTORS

Definition and normalization

The theory of square pyramid F-vectors follows the same general lines as that of hexagonal pyramid F-vectors, with one notable difference: all the hexagonal pyramid F-vectors of the same class are the same (except for the position of the vector fields in P-space), but there are two types of square pyramid F-vectors which have to be used in conjunction with one another.

Consider the network shown in Fig. 3·71, the mesh triangles being right-angled and isosceles. The junction points are of two types:

(i) four triangles meet at a junction point;

(ii) eight triangles meet at a junction point.

When we raise a junction point to generate a pyramid, we get square pyramids of two types, according as the junction point is of type (i) or type (ii). Level lines for the two types are shown in Fig. 3·71.

From these pyramids we generate *square pyramid F-vectors of the first and second classes*, as shown in Fig. 3·72.

It is useful to have a notation which distinguishes the two types. We shall continue to use $\mathbf{T}_\rho'$ for any pyramid F-vector of the first class and $\mathbf{T}''_\sigma$ for any pyramid F-vector of the second class or for a strip F-vector. For square pyramid F-vectors we shall use the following special notation (Fig. 3·72):

	First class	Second class
Small square (four triangles meet at centre)	$\mathbf{P}'$	$\mathbf{P}''$
Large square (eight triangles meet at centre)	$\mathbf{Q}'$	$\mathbf{Q}''$

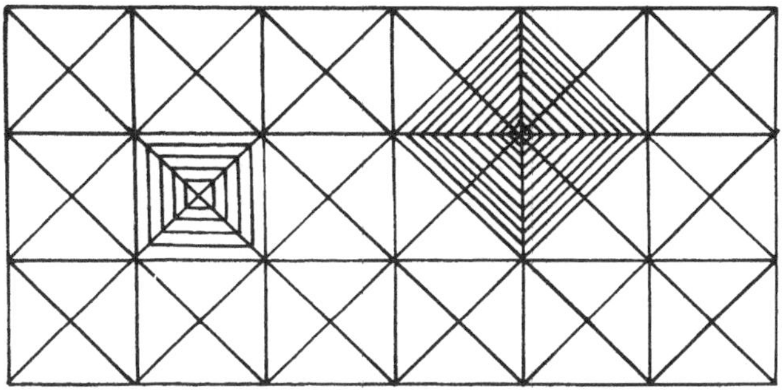

Fig. 3·71. Network of right angled isosceles triangles, giving two types of square pyramid function (level lines shown).

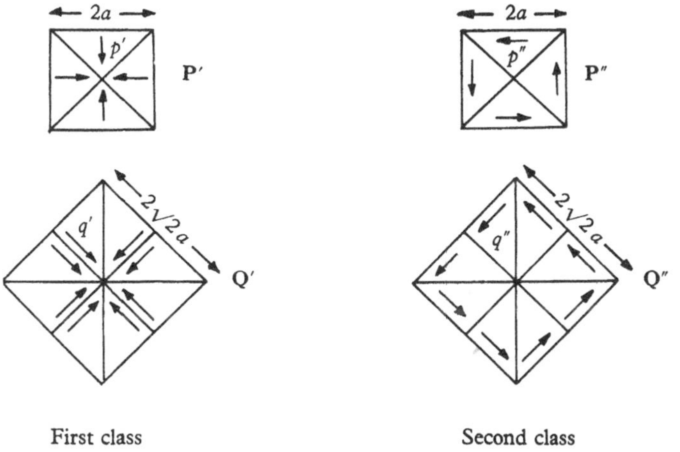

Fig. 3·72. P-vector fields of square pyramid F-vectors of both types (small square and large square) and of first and second classes.

We shall denote the side of the small square by $2a$, so that the side of the large square is $2 . 2^{\frac{1}{2}}a$. The strengths of the normalized P-vector fields will be denoted by p', p'', q', q''.

As a general rule we shall normalize by making the height of all generating pyramids equal to unity. Then

$$p' = p'' = a^{-1}, \quad q' = q'' = 2^{-\frac{1}{2}}a^{-1}. \tag{3·701}$$

Scalar products

In order that the square pyramid F-vectors of the first class ($\mathbf{P}'$ and $\mathbf{Q}'$) may be homogeneous associated vectors, we use only bases entirely inside $V+B$, i.e. inside the polygon B' of Fig. 3·54. But there is no such restriction on the bases of those of the second class ($\mathbf{P}''$ and $\mathbf{Q}''$).

If all bases involved are in $V+B$, then the basic formula (3·203) gives

$$\mathbf{P}'^2 = 4, \quad \mathbf{P}''^2 = 4, \quad \mathbf{Q}'^2 = 4, \quad \mathbf{Q}''^2 = 4. \tag{3·702}$$

$\mathbf{P}'$, $\mathbf{Q}'$ (being homogeneous associated F-vectors) are of course orthogonal to $\mathbf{P}''$, $\mathbf{Q}''$ (complementary F-vectors). All $\mathbf{P}'$ are orthogonal to one another, since their bases do not overlap; the same is true of all $\mathbf{P}''$. Although two $\mathbf{Q}'$ may overlap, their P-vector fields are P-orthogonal in the overlap, and so the F-vectors are orthogonal; this is true also of two $\mathbf{Q}''$.

In fact, for a network as in Fig. 3·71, if we consider only cases where the overlap if any is contained entirely in $V+B$, the only square pyramid F-vectors which are not orthogonal are either $\mathbf{P}'$ overlapping $\mathbf{Q}'$ or $\mathbf{P}''$ overlapping $\mathbf{Q}''$; for these it is easy to see that the scalar products are

$$\mathbf{P}'.\mathbf{Q}' = -1, \quad \mathbf{P}''.\mathbf{Q}'' = -1. \tag{3·703}$$

From (3·702) it follows then that the cosine of the F-angle is $-\frac{1}{4}$. To sum up, *the geometrical picture in F-space of a set of square pyramid F-vectors shows us a set of F-vectors which are of the same magnitude and mutually orthogonal, except that each has four neighbours with which it makes an obtuse angle with cosine* $-\frac{1}{4}$. This simple statement must be modified near the boundary B. Since, in calculating scalar products, we integrate over $V+B$, the values for $\mathbf{P}''^2$ and $\mathbf{Q}''^2$ shown in (3·702) do not hold if the base goes outside $V+B$; further, the value for $\mathbf{P}''.\mathbf{Q}''$ in (3·703) does not hold if the region of overlap goes outside $V+B$.

Determination of the vertices $\mathbf{V}'$, $\mathbf{V}''$

To determine $\mathbf{V}'$, we have to solve (3·329):

$$\sum_{\mu=1}^{r} a'_\mu \mathbf{T}'_\mu . \mathbf{T}'_\rho + \mathbf{S}'_0 . \mathbf{T}'_\rho = 0 \quad (\rho = 1, 2, \ldots, r). \tag{3·704}$$

By means of (3·702) and (3·703) we reduce these equations to a form like (3·615) for hexagons:

$$\boxed{a'_\rho = \tfrac{1}{4}(a'_{\rho 1} + a'_{\rho 2} + a'_{\rho 3} + a'_{\rho 4}) - \tfrac{1}{4}\mathbf{S}'_0 . \mathbf{T}'_\rho \quad (\rho = 1, 2, \ldots, r).} \tag{3·705}$$

Here a'_ρ is the weight of the square pyramid F-vector $\mathbf{T}'_\rho$ of the first class (small square or large square) and $a'_{\rho 1}, a'_{\rho 2}, a'_{\rho 3}, a'_{\rho 4}$ the weights of its four neighbours (in diagonal directions); near the boundary we supply fictitious neighbours and give them zero weight, as described on p. 192 in the case of hexagons. Thus *each weight is the arithmetic mean of the four neighbouring weights, reduced by the last term in* (3·705). The formal symmetry as between the two types ($\mathbf{P}'$ and $\mathbf{Q}'$) is interesting.

For $\mathbf{V}''$ we have to solve, as in (3·329),

$$\sum_{\nu=1}^{s} a''_\nu \mathbf{T}''_\nu . \mathbf{T}''_\sigma - \mathbf{S}'_0 . \mathbf{T}''_\sigma = 0 \quad (\sigma = 1, 2, \ldots, s). \tag{3·706}$$

Considering first square pyramid F-vectors of the second class, we see that *if the base of* $\mathbf{T}''_\sigma$ *does not go outside* B, *then*

$$a''_\sigma = \tfrac{1}{4}(a''_{\sigma 1} + a''_{\sigma 2} + a''_{\sigma 3} + a''_{\sigma 4}) + \tfrac{1}{4}\mathbf{S}'_0 . \mathbf{T}''_\sigma, \tag{3·707}$$

with a notation for neighbours as in (3·705); *but if the base of* $\mathbf{T}''_\sigma$ *does go outside* B, *we must use the more general formula*

$$\begin{aligned} a''_\sigma \mathbf{T}''^2_\sigma = -(a''_{\sigma 1} \mathbf{T}''_\sigma . \mathbf{T}''_{\sigma 1} + a''_{\sigma 2} \mathbf{T}''_\sigma . \mathbf{T}''_{\sigma 2} + a''_{\sigma 3} \mathbf{T}''_\sigma . \mathbf{T}''_{\sigma 3} \\ + a''_{\sigma 4} \mathbf{T}''_\sigma . \mathbf{T}''_{\sigma 4}) + \mathbf{S}'_0 . \mathbf{T}''_\sigma. \end{aligned} \tag{3·708}$$

In the case of a multiply connected region strip F-vectors will occur, and then an F-vector may have more than four neighbours. The formulae (3·707) and (3·708) must be modified when an overlapping strip F-vector is involved, by the inclusion of an additional term in the parentheses. In any case of doubt as to the proper procedure, reference may be made to the basic equations (3·329).

When the equations (3·705) and (3·707) or (3·708), or a modification of them, have been solved, the values of $\mathbf{V}'^2$ and $\mathbf{V}''^2$ are given by (3·334), and hence bounds on $\mathbf{S}^2$ obtained from (3·339).

The bisection process

Fig. 3·73 shows a network in full lines. When the broken lines are added, we obtain a finer mesh. This bisection process may be used to get closer and closer approximations. On account of the occurrence of the two types of square pyramid F-vector (small square and large square) the bisection process is not quite as simple as in the case of hexagonal pyramid F-vectors (cf. Fig. 3·64, p. 194). The following facts may be noted:

(i) A junction point A (Fig. 3·73) which was the centre of a large square retains that status.

(ii) A junction point B which was the centre of a small square becomes the centre of a large square.

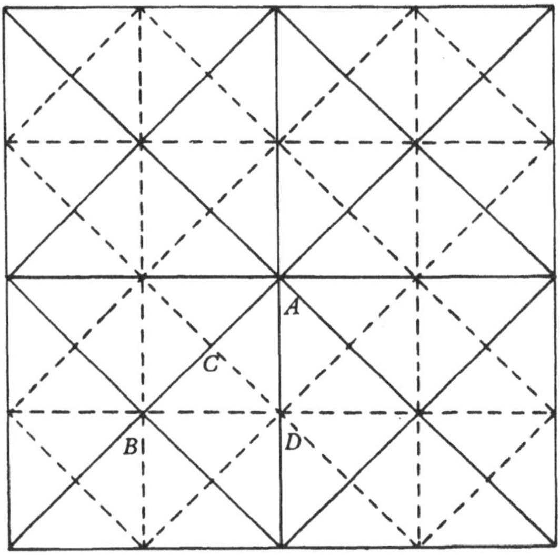

Fig. 3·73. Bisection of a square network. The old network is shown in full lines; the new network is formed by adding the broken lines.

(iii) A point C, which was not a junction point in the old network, becomes the centre of a small square.

(iv) A point D, which was not a junction point in the old net-network, becomes the centre of a large square.

(v) Bisection of a fine triangulation roughly quadruples the number of junction points and the number of triangles.

Example: square condenser

We shall now carry out for a square condenser calculations similar to those already made for the hexagonal condenser shown in Figs. 3·65 and 3·66.

Fig. 3·74 shows a square condenser of outer side s and inner side $\frac{1}{3}s$; this shape is selected because it gives an example of the use of square pyramid F-vectors without undue complications. The inner square B_0 is at potential $u=0$ and the outer square B_1 at potential $u=1$.

We shall use the triangulation shown in Fig. 3·74 to obtain the vertex $\mathbf{V}'$, taking all junction points not on the boundary as centres of square pyramid F-vectors of the first class. These are enumerated as follows:

Small squares ($\mathbf{P}'$):
  Outer ring: $a\,1, a\,2, \ldots, a\,20$.
  Inner ring: $b\,1, b\,2, \ldots, b\,12$.

Large squares ($\mathbf{Q}'$):
  $1, 2, \ldots, 16$.

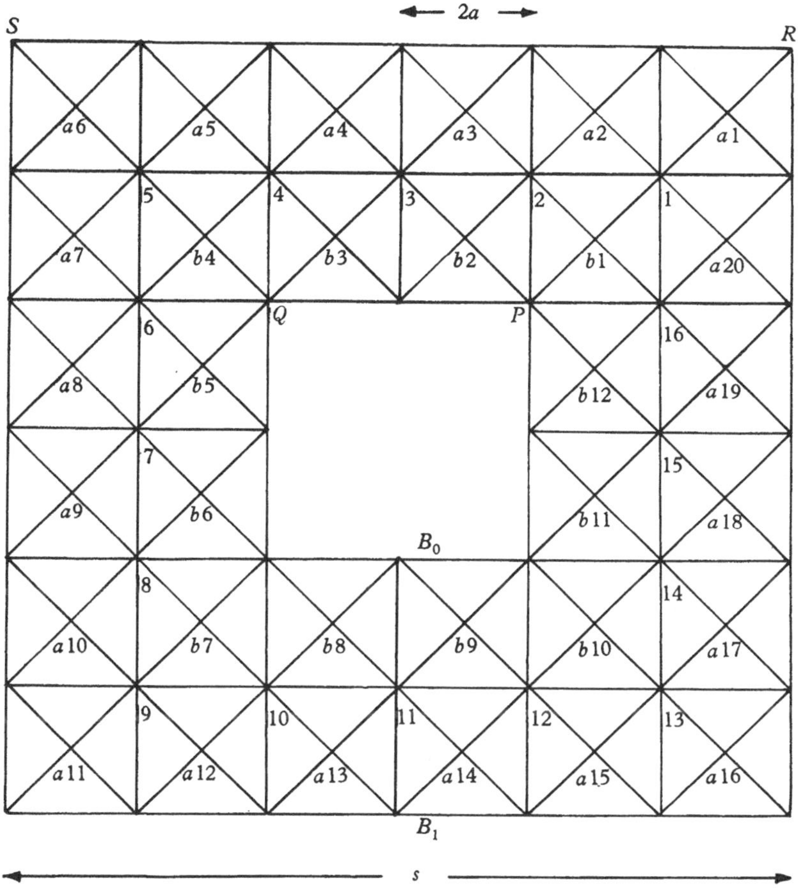

Fig. 3·74. Square condenser with triangulation and enumeration of square pyramid F-vectors of the first class.

For $\mathbf{S}'_0$ we choose a P-vector field of magnitude $3s^{-1}$ pointing up the page in the quadrilateral $PQRS$, the field in the rest of the region between B_0 and B_1 being given by reflexions in the diagonals.

Remembering the normalization (3·701), we find

$$\mathbf{S}'_0 . \mathbf{T}'_1 = \mathbf{S}'_0 . \mathbf{T}'_{a1} = \mathbf{S}'_0 . \mathbf{T}'_{b1} = -\tfrac{1}{2}, \tag{3·709}$$

with the same equations for the centres on the other three diagonals; $\mathbf{S}_0'$ is orthogonal to all the rest of the square pyramid F-vectors of the first class.

There are eight independent weights to be found, namely, those corresponding to the centres

$$1, 2, 3; \quad a\,1, a\,2, a\,3; \quad b\,1, b\,2;$$

to find these weights we use (3·705). Taking account of symmetry, we have the following eight equations:

$$\left.\begin{aligned}
a'(1) &= \tfrac{1}{4}[a'(a\,1) + 2a'(a\,2) + a'(b\,1)] + \tfrac{1}{8},\\
a'(2) &= \tfrac{1}{4}[a'(a\,2) + a'(a\,3) + a'(b\,2) + a'(b\,1)],\\
a'(3) &= \tfrac{1}{2}[a'(a\,3) + a'(b\,2)],\\
a'(a\,1) &= \tfrac{1}{4}a'(1) + \tfrac{1}{8},\\
a'(a\,2) &= \tfrac{1}{4}[a'(1) + a'(2)],\\
a'(a\,3) &= \tfrac{1}{4}[a'(2) + a'(3)];\\
a'(b\,1) &= \tfrac{1}{4}[a'(1) + 2a'(2)] + \tfrac{1}{8},\\
a'(b\,2) &= \tfrac{1}{4}[a'(2) + a'(3)].
\end{aligned}\right\} \tag{3·710}$$

These equations are easy to solve; we get

$$\left.\begin{aligned}
&a'(1) = \tfrac{33}{116}, && a'(2) = \tfrac{3}{29}, && a'(3) = \tfrac{1}{29},\\
&a'(a\,1) = \tfrac{91}{464}, && a'(a\,2) = \tfrac{45}{464}, && a'(a\,3) = \tfrac{1}{29},\\
&a'(b\,1) = \tfrac{115}{464}, && a'(b\,2) = \tfrac{1}{29}.
\end{aligned}\right\} \tag{3·711}$$

With these values, we can get $\mathbf{V}'$ from (3·333) if we wish. But let us go on to $\mathbf{V}'^2$ as in (3·334). In view of (3·709), we get

$$\mathbf{V}'^2 = \mathbf{S}_0'^2 + 4(-\tfrac{1}{2})\,[a'(1) + a'(a\,1) + a'(b\,1)]. \tag{3·712}$$

By easy calculation we have
$$\mathbf{S}_0'^2 = 8, \tag{3·713}$$

and so
$$\mathbf{V}'^2 = \tfrac{759}{116}. \tag{3·714}$$

Thus for the capacity $C = \mathbf{S}^2/2\pi$ of the condenser (per unit length) we have the upper bound

$$\mathbf{S}^2 \leqslant \mathbf{V}'^2 = \tfrac{759}{116} = 6{\cdot}5431. \tag{3·715}$$

Let us now get a lower bound through $\mathbf{V}''$. For this we use square pyramid F-vectors of the second class with centres at points shown in Fig. 3·75, and also strip F-vectors at the corners. The square F-vectors are enumerated as follows:

Small squares ($\mathbf{P}''$):
- Outer ring: $a\,1, a\,2, \ldots, a\,20$.
- Inner ring: $b\,1, b\,2, \ldots, b\,12$.

Large squares ($\mathbf{Q}''$):
- Outer ring: $A\,1, A\,2, \ldots, A\,24$.
- Middle ring: $B\,1, B\,2, \ldots, B\,16$.
- Inner ring: $C\,1, C\,2, \ldots, C\,8$.

The strip F-vectors are normalized to P-vector strength $2^{-\frac{1}{2}}.3s^{-1}$; they are designated

$$1, 2, \ldots, 8.$$

Their P-vector fields point outwards.

The symmetry argument of pp. 198–9 may be used here also; it tells us that the weights of the square pyramid F-vectors vanish if their centres are on lines of symmetry of the squares, and so we can simplify at once by

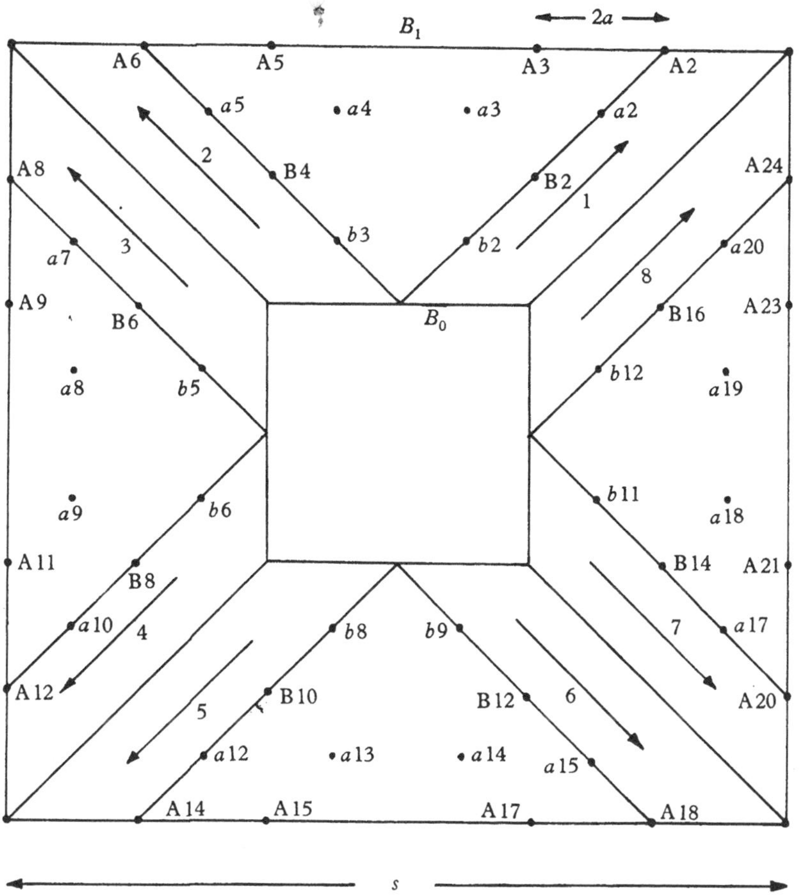

Fig. 3·75. Square condenser with enumeration of square pyramid F-vectors of the second class and strip F-vectors.

omitting them. This is done in Fig. 3·75; there the inner ring (designated C above) does not appear at all for this reason. The weights of all eight strip F-vectors are the same.

We have now to find the weights. The 'mean-value' equation (3·707) is not of much use here, for the mesh is too coarse. It is useful only when working in the interior of a fine mesh. Instead we shall use (3·708), or equivalently (3·329); if the former, we are to remember that a strip F-vector may also be a neighbour.

We need some scalar products, all of which are easy to calculate. We note that $\mathbf{S}_0'$ is orthogonal to all the square pyramid F-vectors of the second class and that

$$\mathbf{S}_0'.\mathbf{T}_1''=\tfrac{1}{4} \tag{3·716}$$

($\mathbf{T}_1''$ is of course the strip F-vector according to our enumeration above). Further

$$\mathbf{T}_1''^2=\mathbf{T}_1''.\mathbf{T}_{A2}''=\mathbf{T}_1''.\mathbf{T}_{B2}''=\mathbf{T}_1''.\mathbf{T}_{a2}''=\mathbf{T}_1''.\mathbf{T}_{b2}''=\tfrac{1}{4},\quad \mathbf{T}_1''.\mathbf{T}_{A3}''=0, \tag{3·717}$$

and by (3·702), combined with the fact that only half the bases of $\mathbf{T}_{A2}''$ and $\mathbf{T}_{A3}''$ lie in V,

$$\mathbf{T}_{A2}''^2=\mathbf{T}_{A3}''^2=2,\quad \mathbf{T}_{B2}''^2=\mathbf{T}_{a2}''^2=\mathbf{T}_{a3}''^2=\mathbf{T}_{b2}''^2=4; \tag{3·718}$$

by (3·703) the scalar product of (diagonal) neighbours is -1.

There are seven independent weights to be found,

$$1,\quad A2,\quad A3,\quad B2,\quad a2,\quad a3,\quad b2,$$

and for them, by (3·708) or equivalently (3·329), we have the following seven equations, which will be written down here as they come from (3·708) to make the work easier to follow:

$$\left.\begin{aligned}
&a''(1).\tfrac{1}{4}=-[a''(A2).\tfrac{1}{4}+a''(B2).\tfrac{1}{4}+a''(a2).\tfrac{1}{4}+a''(b2).\tfrac{1}{4}]+\tfrac{1}{4},\\
&a''(A2).2=-[a''(1).\tfrac{1}{4}+a''(a2)(-1)],\\
&a''(A3).2=-[a''(a2)(-1)+a''(a3)(-1)],\\
&a''(B2).4=-[a''(1).\tfrac{1}{4}+a''(a2)(-1)+a''(a3)(-1)+a''(b2)(-1)],\\
&a''(a2).4=-[a''(1).\tfrac{1}{4}+a''(A2)(-1)+a''(A3)(-1)+a''(B2)(-1)],\\
&a''(a3).4=-[a''(A3)(-1)+a''(B2)(-1)],\\
&a''(b2).4=-[a''(1).\tfrac{1}{4}+a''(B2)(-1)].
\end{aligned}\right\} \tag{3·719}$$

The matrix of these equations is simple and they are easy to solve. Using a common denominator 676, the solutions are compactly written

$$\begin{aligned}
&a''(1, && =(1980,\\
&\quad A2,\quad A3,\quad B2, && \quad -433,\,-255,\,-301,\\
&\quad a2,\quad a3,\quad b2) && \quad -371,\,-139,\,-199)/676.
\end{aligned} \tag{3·720}$$

Now we have $\mathbf{V}''$ by (3·333) if we want it, and by (3·334)

$$\mathbf{V}''^2=8a''(1)\,\mathbf{S}_0'.\mathbf{T}_1''=8.\tfrac{1980}{676}.\tfrac{1}{4}=\tfrac{990}{169}=5·8580. \tag{3·721}$$

Combining this with (3·715), we have, by (3·339), *the following bounds for the capacity* $C=\mathbf{S}^2/2\pi$ *of the square condenser (per unit length) shown in Figs.* 3·74 *and* 3·75*:*

$$5·8580=\tfrac{990}{169}=\mathbf{V}''^2\leqslant\mathbf{S}^2\leqslant\mathbf{V}'^2=\tfrac{759}{116}=6·5431. \tag{3·722}$$

If we wished to get closer bounds, we would apply the bisection process of Fig. 3·73; then the number of equations to be solved would be greatly increased and we would have to solve them by successive approximations, as we shall do in later examples.

Exercises

[The first four of the exercises which follow deal with the torsion problem for a square. The exact solution (for which $\mathbf{S}^2 = 0{\cdot}02609s^4$) is well known, and will be worked out following the hypercircle plan in § 4·2. The exercises are intended to familiarize the reader with square pyramid F-vectors; the bounds are crude but the calculations very simple.]

1. Consider the Dirichlet problem $\Delta u = 0$, $(u)_B = \frac{1}{2}r^2$ for a square of side s, the origin being at the centre. By dividing the square into four equal triangles and using one square pyramid F-vector of the first class, locate the solution on a hypercircle of class $1+0$ and prove that

$$\mathbf{S}^2 \leqslant \tfrac{1}{18}s^4 = 0{\cdot}05556s^4.$$

2. In the above problem, divide the square into four squares and triangulate by drawing the diagonals. Using five square pyramid F-vectors of the first class (four small and one large), locate the solution on a hypercircle of class $5+0$ and show that $\mathbf{V}'^2 = \frac{1}{18}s^4$, so that the bound obtained in Example 1 is not improved.

3. Show that, on account of symmetry, square pyramid F-vectors of the second class give no lower bounds (except a trivial zero) for the triangulations described above. But if we triangulate by drawing nine squares and their diagonals, we may obtain

$$\mathbf{S}^2 \geqslant (\tfrac{1}{3}s)^4 = 0{\cdot}01235s^4$$

by a very simple calculation in which only one weight has to be found in order to get the vertex $\mathbf{V}''$.

4. With the triangulation of Example 3, obtain the upper bound

$$\mathbf{S}^2 \leqslant \mathbf{V}'^2 = 47s^4/1134 = 0{\cdot}04145s^4$$

by solving four linear equations (use symmetry).

5. Find bounds for the capacity per unit length of a square condenser for which the side of the inner square is *half* that of the outer square, following the same general plan as in Figs. 3·74 and 3·75.

3·8. APPROXIMATION BY LINEAR INTERPOLATION

We shall now give mathematical precision to the statement made on p. 175 that a given function and its first derivatives may be approximated as closely as we like by a polyhedral function based on a suitable triangulation.

Let a function u be given in a plane region V. Let V be triangulated, and in each triangle $P^{(1)}P^{(2)}P^{(3)}$ of the triangulation (Fig. 3·81) let a function v be defined by the determinantal equation

$$\begin{vmatrix} v & u^{(1)} & u^{(2)} & u^{(3)} \\ x_1 & x_1^{(1)} & x_1^{(2)} & x_1^{(3)} \\ x_2 & x_2^{(1)} & x_2^{(2)} & x_2^{(3)} \\ 1 & 1 & 1 & 1 \end{vmatrix} = 0, \qquad (3{\cdot}801)$$

where $u^{(1)}$, $u^{(2)}$, $u^{(3)}$ are the values of u at the vertices and $x_i^{(1)}$, $x_i^{(2)}$, $x_i^{(3)}$ the coordinates of the vertices. It is clear that (3·801) defines v as a linear function of x_i, with $v=u$ at the vertices; we call it a *linear interpolation*.

The solution of (3·801) may be written in the form

$$v = u^{(1)}\tau^{(1)} + u^{(2)}\tau^{(2)} + u^{(3)}\tau^{(3)}, \tag{3·802}$$

where $\tau^{(1)}$, $\tau^{(2)}$, $\tau^{(3)}$ are given by

$$\begin{vmatrix} \tau^{(1)} & 1 & 0 & 0 \\ x_1 & x_1^{(1)} & x_1^{(2)} & x_1^{(3)} \\ x_2 & x_2^{(1)} & x_2^{(2)} & x_2^{(3)} \\ 1 & 1 & 1 & 1 \end{vmatrix} = 0 \tag{3·803}$$

and two similar equations. These are linear interpolations with the values (1, 0, 0), (0, 1, 0), (0, 0, 1) at the vertices of the triangle, and

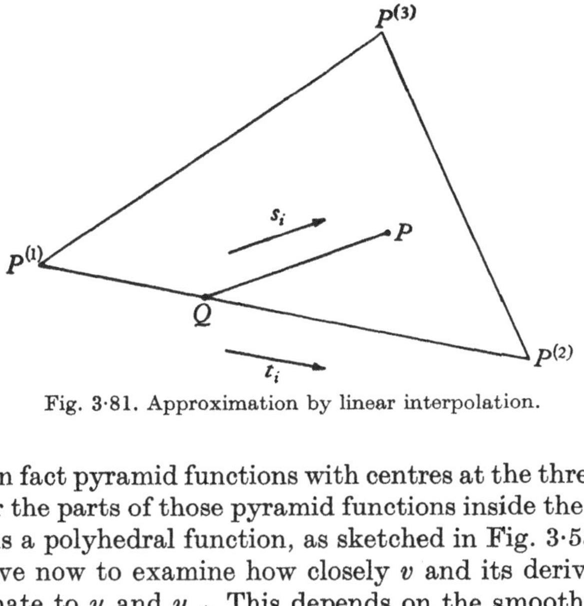

Fig. 3·81. Approximation by linear interpolation.

thus are in fact pyramid functions with centres at the three vertices (or rather the parts of those pyramid functions inside the triangle). Hence v is a polyhedral function, as sketched in Fig. 3·55, p. 174.

We have now to examine how closely v and its derivatives $v_{,i}$ approximate to u and $u_{,i}$. This depends on the smoothness of u, and we shall assume $u, u_{,i}$ continuous, with the second derivatives bounded by the condition

$$\left|\frac{d^2u}{ds^2}\right| \leqslant M, \tag{3·804}$$

where M is some positive number and the derivative is calculated at any point along any straight line with line element ds. Equivalently

$$|u_{,ij}t_i t_j| \leqslant M, \tag{3·805}$$

where t_i is any unit vector. This implies

$$|u_{,ij}s_i t_j| \leqslant M, \tag{3·806}$$

where s_i and t_i are any two unit vectors (this is easily proved by using those axes for which $u_{,12}=0$ at the point in question).

Let ϕ be defined by

$$\phi = u - v. \tag{3·807}$$

Then $\phi = 0$ at the vertices of the triangle, and $\phi_{,ij}=u_{,ij}$, since v is a linear function; therefore by (3·806)

$$|\phi_{,ij}s_i t_j| \leqslant M. \tag{3·808}$$

Since $\phi = 0$ at $P^{(1)}$ and $P^{(2)}$, there exists some point Q on the segment $P^{(1)}P^{(2)}$ such that

$$\phi_{,i}t_i = 0 \tag{3·809}$$

at Q, where t_i is a unit vector along $P^{(1)}P^{(2)}$. Let P be any point in the triangle and s_i a unit vector along QP. Then by (3·809)

$$(\phi_{,i}t_i)_P = \int_Q^P \frac{d}{ds}(\phi_{,i}t_i)\,ds = \int_Q^P \phi_{,ij}t_i s_j\,ds, \tag{3·810}$$

and so, by (3·808),

$$|\phi_{,i}t_i|_P \leqslant Ml, \tag{3·811}$$

where l is the greatest side of the triangle.

This tells us that the orthogonal projection of the vector grad ϕ on any one of the three sides of the triangle cannot exceed Ml. Let us then take a circle with radius Ml and draw the six lines which are tangent to this circle and parallel to the sides of the triangle. These six lines bound a hexagon, and the information we have about grad ϕ tells us that, if we draw the vector grad ϕ from the centre of the circle, the extremity of this vector cannot lie outside the hexagon. Now the vertices of the hexagon are at distances

$$Ml \sec \tfrac{1}{2}A, \quad Ml \sec \tfrac{1}{2}B, \quad Ml \sec \tfrac{1}{2}C$$

from the centre of the circle, A, B and C being the angles of the triangle, and we conclude that

$$|\operatorname{grad} \phi| \leqslant kMl, \tag{3·812}$$

where

$$k = \sec \tfrac{1}{2}\theta, \tag{3·813}$$

θ being the greatest of the three angles A, B, C.

Since

$$|\phi_{,1}| \leqslant |\operatorname{grad} \phi|, \quad |\phi_{,2}| \leqslant |\operatorname{grad} \phi|, \tag{3·814}$$

we see that $v_{,i}$ approximates to $u_{,i}$ in the sense that

$$|u_{,1}-v_{,1}| \leqslant kMl, \quad |u_{,2}-v_{,2}| \leqslant kMl, \tag{3·815}$$

l being the greatest side of the triangle and k as in (3·813). For any triangulation a bisection process as shown in Figs. 3·64 and 3·73 may be applied to give a finer triangulation, each triangle being cut into four triangles, and this process may be continued indefinitely; it does not change angles, but it makes l as small as we please. Thus (3·815) tells us that *we can make the derivatives* $v_{,i}$ *of the linear interpolation approximate to* $u_{,i}$ *as closely as we like by taking a sufficiently fine triangulation.*

We note that flat triangles give a poor approximation, since k tends to infinity as θ tends to 180°. For an equilateral triangle

$$k = \sec 30^\circ = 2.3^{-\frac{1}{2}}, \tag{3·816}$$

and for a right-angled isosceles triangle

$$k = \sec 45^\circ = 2^{\frac{1}{2}}. \tag{3·817}$$

To get a bound for $|u-v|$, we integrate along a straight line from a vertex of the triangle to any point P in it; then

$$(\phi)_P = \int \phi_{,i} dx_i, \quad |\phi|_P \leqslant |\operatorname{grad} \phi|_{\max.} l, \tag{3·818}$$

and so

$$|u-v| \leqslant kMl^2. \tag{3·819}$$

Thus the linear interpolation v approximates to u as closely as we like if we make the triangulation fine enough. This approximation, being of order l^2, is better than the approximation for the derivatives, which is of order l.

A different method of interpolation has been given by Pólya (1). He divides the plane into squares of side h by the lines $x = ih$, $y = jh$, where i and j are integers, and in the square with corners $[ih, jh]$, $[(i+1)h, jh]$, $[ih, (j+1)h]$, $[(i+1)h, (j+1)h]$ takes the bilinear interpolation

$$v = u_{ij}(1-\xi)(1-\eta) + u_{i+1,j}\xi(1-\eta) + u_{i,j+1}(1-\xi)\eta + u_{i+1,j+1}\xi\eta, \tag{3·820}$$

where ξ and η are given by

$$x = (i+\xi)h, \quad y = (j+\eta)h, \tag{3·821}$$

and u_{ij}, $u_{i+1,j}$, $u_{i,j+1}$, $u_{i+1,j+1}$ are the values of the given function at the corners of the square. If we take a set of four such squares forming a square of side $2h$, and set zero values on the boundary and unity at the centre, we have a type of pyramid function on a square base; from it pyramid F-vectors of the two classes can be generated by differentiation and rotation of the gradient field.

Since a network of squares is usually the easiest to work with, and since Pólya's method involves only one type instead of the two types shown in Fig. 3·72, it may well be that his method offers advantages over interpolation based on a triangular network.

Exercise

A square with sides $x_1 = \pm 1$, $x_2 = \pm 1$ is divided into n^2 small squares, and each of these small squares is divided into four triangles by its diagonals. Consider the function $u = \frac{1}{2}(x_1^2 + x_2^2)$ and the polyhedral function v such that $v = u$ at the vertices of the triangulation. Show that

$$M = 1, \quad l = 2/n, \quad k = 2^{\frac{1}{2}},$$

and hence that all over the square

$$|u_{,i} - v_{,i}| \leqslant 2.2^{\frac{1}{2}}/n, \quad |u - v| \leqslant 4.2^{\frac{1}{2}}/n^2.$$

CHAPTER 4

THE TORSION PROBLEM

4·1. THE TORSION PROBLEM AS A NEUMANN PROBLEM AND AS AN EXTENDED DIRICHLET PROBLEM

The problem of the twisting or torsion of an elastic cylinder is a standard problem in the theory of elasticity [cf. Love(1), p. 310; Sokolnikoff(1), p. 109]. It was mentioned on p. 130 with the remark that, for a simply connected cross-section, it is mathematically identical with other physical problems—fluid in a rotating container, viscous flow through a pipe, and the equilibrium of an elastic membrane. If we solve one such problem for a certain boundary, we solve all.

But for a multiply connected cross-section the torsion problem no longer coincides mathematically with the other problems mentioned. It is no longer a Dirichlet problem in the ordinary sense, and the methods of Chapter 3 cannot be applied directly to it. The torsion problem is of course a problem of elastic equilibrium, and as such comes under the theory of § 5·4; but, relative to most elastic problems, it is simpler and richer in results, and we shall devote the present chapter to it, paying particular attention to multiply connected cross-sections (the simply connected cross-section is a special case).

The torsion problem stated

We consider a cylinder of homogeneous isotropic elastic material (rigidity μ) and take coordinates x_1, x_2, x_3 with the axis of x_3 parallel to the generators of the cylinder. The following system of displacements is assumed:

$$u_1 = -\alpha x_2 x_3, \quad u_2 = \alpha x_1 x_3, \quad u_3 = \alpha\phi(x_1, x_2). \tag{4·101}$$

Here α is a constant (angle of twist per unit length). The function ϕ is called the *warping function*; the torsion problem consists in finding it, or an equivalent function.

These displacements give only two components of stress,

$$E_{13} = \mu\alpha(\phi_{,1} - x_2), \quad E_{23} = \mu\alpha(\phi_{,2} + x_1) \tag{4·102}$$

(see § 5·4 for notation), and the equations of equilibrium reduce to one:

$$\Delta\phi = 0. \tag{4·103}$$

There is to be no traction across the surface of the cylinder; this gives

$$\frac{\partial \phi}{\partial n} = n_1 x_2 - n_2 x_1, \tag{4·104}$$

where n_i $(i = 1, 2)$ is the unit normal to the surface and ∂n an element of the normal in the sense of n_i.

Fig. 4·11 shows a simply connected cross-section V with boundary B, and Fig. 4·12 shows a multiply connected cross-section V with boundary

$$B = B_0 + B_1 + B_2 + \dots;$$

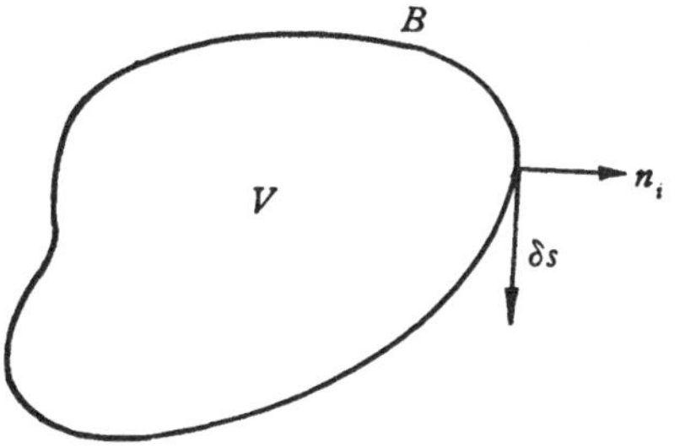

Fig. 4·11. Simply connected section.

B_0 is the outer boundary and $B_1, B_2, \dots$ the boundaries of the holes. We shall assume h holes in general, so that $h = 0$ for simple connectivity and $h = 2$ in Fig. 4·12. In (4·104) n_i may be taken in either sense; we shall take it pointing out of V, and when

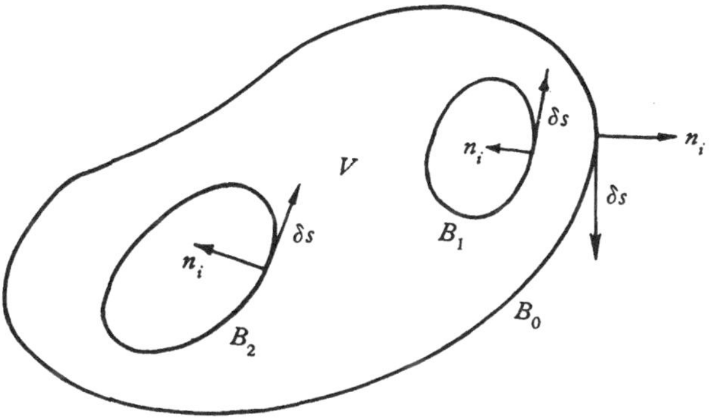

Fig. 4·12. Multiply connected section.

we have occasion to use an element ∂s of B, it will be taken in the sense shown.

Let r denote distance from the origin, so that

$$r^2 = x_1^2 + x_2^2. \tag{4·105}$$

Then on B

$$\frac{\partial x_1}{\partial s} = n_2, \quad \frac{\partial x_2}{\partial s} = -n_1, \tag{4·106}$$

and so

$$\frac{\partial}{\partial s}(\tfrac{1}{2}r^2) = n_2 x_1 - n_1 x_2. \tag{4·107}$$

Thus, by (4·103) and (4·104), *the torsion problem is the Neumann problem*

$$\Delta\phi=0, \quad \left(\frac{\partial\phi}{\partial n}\right)_B=(n_1x_2-n_2x_1)_B=-\left[\frac{\partial}{\partial s}(\tfrac{1}{2}r^2)\right]_B. \qquad (4\cdot108)$$

It is clear that $$\int_{B_\lambda}\frac{\partial\phi}{\partial n}dB_\lambda=0 \quad (\lambda=1,2,\ldots,h), \qquad (4\cdot109)$$

and so by (3·128) ϕ possesses a single-valued harmonic conjugate ψ such that $\phi+i\psi$ is an analytic function of x_1+ix_2 and

$$\phi_{,1}=\psi_{,2}, \quad \phi_{,2}=-\psi_{,1}. \qquad (4\cdot110)$$

Moreover [cf. (3·134)] the Neumann problem (4·108) is equivalent to *the extended Dirichlet problem*

$$\Delta\psi=0, \quad (\psi)_{B_0}=\tfrac{1}{2}r^2, \quad (\psi)_{B_\lambda}=\tfrac{1}{2}r^2+C_\lambda \quad (\lambda=1,2,\ldots,h), \quad (4\cdot111)$$

where the constants C_λ are not given a priori, but must be chosen to satisfy the h conditions

$$\int_{B_\lambda}\frac{\partial\psi}{\partial n}dB_\lambda=0 \quad (\lambda=1,2,\ldots,h). \qquad (4\cdot112)$$

We have thus two alternative ways of looking at the torsion problem: as a Neumann problem or as an extended Dirichlet problem (an ordinary Dirichlet problem in the case of simple connectivity).

We note that by (4·102) we have the following expressions for the stress:

$$\left.\begin{aligned}E_{13}&=\mu\alpha(\phi_{,1}-x_2)=\mu\alpha(\psi_{,2}-x_2)=\mu\alpha(\psi-\tfrac{1}{2}r^2)_{,2},\\ E_{23}&=\mu\alpha(\phi_{,2}+x_1)=\mu\alpha(-\psi_{,1}+x_1)=-\mu\alpha(\psi-\tfrac{1}{2}r^2)_{,1}.\end{aligned}\right\} \qquad (4\cdot113)$$

The *lines of stress* are those curves to which the vector (E_{13}, E_{23}) is tangent; they satisfy $dx_1:dx_2=E_{13}:E_{23}$, and so have the equations

$$\psi-\tfrac{1}{2}r^2=\text{const.} \qquad (4\cdot114)$$

This family of curves includes $B_0, B_1, \ldots, B_h$. The *level lines*, or lines of constant warping, are given by $\phi=\text{const.}$

Torsional rigidity

It follows from (4·113) that

$$\int_V E_{13}dV=0, \quad \int_V E_{23}dV=0. \qquad (4\cdot115)$$

There is therefore no resultant force on the cross-section, and the total reaction on it is a couple or torque which may be written

$$\mu\alpha\Gamma=\int_V(x_1E_{23}-x_2E_{13})\,dV. \qquad (4\cdot116)$$

It is convenient to divide by $\mu\alpha$ and refer to Γ as the *torsional rigidity* (it is more usual but less convenient to include the factor μ in the definition of torsional rigidity). Thus

$$\begin{aligned}\Gamma &= I + \int_V (x_1\phi_{,2} - x_2\phi_{,1})\,dV \\ &= I - \int_V (x_1\psi_{,1} + x_2\psi_{,2})\,dV, \end{aligned} \tag{4·117}$$

where I is the (polar) moment of inertia of the cross-section,

$$I = \int_V r^2 dV. \tag{4·118}$$

The formulae (4·117) can be written in other forms. Using Green's theorem, with (4·111) and (4·112), we get

$$\begin{aligned}\int_V (x_1\psi_{,1} + x_2\psi_{,2})\,dV &= \int_V [\psi_{,1}(\tfrac{1}{2}r^2)_{,1} + \psi_{,2}(\tfrac{1}{2}r^2)_{,2}]\,dV \\ &= \int_B \tfrac{1}{2}r^2 \frac{\partial\psi}{\partial n}\,dB \\ &= \int_B \psi\frac{\partial\psi}{\partial n}\,dB \\ &= \int_V \psi_{,i}\psi_{,i}\,dV, \end{aligned} \tag{4·119}$$

and so we have formulae due to Diaz and Weinstein (2):

$$\Gamma = I - \int_V \psi_{,i}\psi_{,i}\,dV = I - \int_V \phi_{,i}\phi_{,i}\,dV. \tag{4·120}$$

The goal in the torsion problem is to find ϕ or ψ; but as a more limited objective we may seek the single number Γ.

Transformation of axes

It is easy to get confused about the transformation of axes in the torsion problem, and to forestall such confusion we shall go into the matter here. It is often convenient to choose the origin at the centroid of V and the axes in the principal directions with respect to moments of inertia; but it is by no means necessary to do so. The origin and axes have so far been quite arbitrary and will remain so.

We need to use the word *invariant* in two different applications: first, as applied to a function of position in the plane of the section, it means that the *value* at each *point* is independent of the axes used; secondly, as applied to a single number like Γ, it means that the number is independent of the choice of axes.

We consider

(i) rotation of the axes about O,

(ii) translation of axes.

Rotation is easy to discuss. The conditions (4·108) are invariant in form and the function ϕ is invariant in the first sense (we neglect a trivial arbitrary additive constant). Further, (4·111) and (4·112) are invariant in form and the function ψ is invariant. The stress components transform like the components of a vector and the torsional rigidity Γ is an invariant number.

Translation of axes is a little trickier. Let us change the origin to the point $x_i = c_i$, so that the new coordinates are $x'_i = x_i - c_i$. Let ϕ and ϕ' correspond to the two sets of axes, both functions evaluated at the same point in the plane. Then by (4·108)

$$\Delta(\phi' - \phi) = 0, \quad [\partial(\phi' - \phi)/\partial n]_B = n_2 c_1 - n_1 c_2. \tag{4·121}$$

Dropping a trivial additive constant, the solution of this simple Neumann problem is

$$\phi' - \phi = c_1 x_2 - c_2 x_1. \tag{4·122}$$

This is the law of transformation of ϕ under translation of axes.

We now consider translation in connexion with (4·111). We have

$$\tfrac{1}{2}r'^2 = \tfrac{1}{2}x'_i x'_i = \tfrac{1}{2}(x_i - c_i)(x_i - c_i) = \tfrac{1}{2}r^2 - c_i x_i + \tfrac{1}{2}c_i c_i. \tag{4·123}$$

Then (4·111) gives

$$\left.\begin{aligned} &\Delta(\psi' - \psi) = 0, \quad (\psi' - \psi)_{B_0} = -c_i x_i + \tfrac{1}{2}c_i c_i, \\ &(\psi' - \psi)_{B_\lambda} = -c_i x_i + \tfrac{1}{2}c_i c_i + C'_\lambda - C_\lambda \quad (\lambda = 1, 2, \ldots, h). \end{aligned}\right\} \tag{4·124}$$

The solution for ψ' and C'_λ is given by

$$\psi' - \psi = -c_i x_i + \tfrac{1}{2}c_i c_i, \quad C'_\lambda = C_\lambda \quad (\lambda = 1, 2, \ldots, h), \tag{4·125}$$

satisfying also (4·112) because

$$\int_{B_\lambda} \frac{\partial \psi'}{\partial n} dB_\lambda = \int_{B_\lambda} \frac{\partial \psi}{\partial n} dB_\lambda - \int_{B_\lambda} c_i n_i \, dB_\lambda = 0. \tag{4·126}$$

Equation (4·125) *is the law of transformation of ψ under translation of axes; we note that the numbers C_λ are invariants.*

We verify from (4·122) or (4·125) that (4·113) gives, under translation, $E'_{13} = E_{13}$, $E'_{23} = E_{23}$, so that stress is invariant under translation of axes, as of course it must be by its physical nature. As for the torsional rigidity, we establish its invariance under translation of axes by use of the equations (4·115); we get from (4·116)

$$\mu\alpha(\Gamma' - \Gamma) = \int_V (-c_1 E_{23} + c_2 E_{13})\, dV = 0. \tag{4·127}$$

Thus the torsional rigidity Γ *is invariant under both rotation and translation of axes.*

Exercises

1. Show that for a circular section bounded by $(x_i - c_i)(x_i - c_i) = a^2$ the solution of the torsion problem and the torsional rigidity are

$$\psi = c_i x_i - \tfrac{1}{2} c_i c_i + \tfrac{1}{2} a^2, \quad \Gamma = \tfrac{1}{2}\pi a^4. \tag{4·128}$$

Note that Γ is the moment of inertia about the centre of the circle.

2. For an elliptical boundary with equation $x_1^2/a_1^2 + x_2^2/a_2^2 = 1$ the solution is [cf. Sokolnikoff(1), p. 120]

$$\psi = \tfrac{1}{2} c(x_1^2 - x_2^2) + k, \quad c = \frac{a_1^2 - a_2^2}{a_1^2 + a_2^2}, \quad k = \frac{a_1^2 a_2^2}{a_1^2 + a_2^2}.$$

3. For an equilateral triangle show that the solution may be written [cf. Sokolnikoff(1), p. 125]

$$\operatorname{grad} \psi = [c(x_1^2 - x_2^2), \quad -2c x_1 x_2],$$

where c is a constant.

4. Show that
$$\psi = c r^n \cos n\theta - A$$

(c, n, A constants) is the solution of the torsion problem for the section with boundary, in polar coordinates,

$$c r^n \cos n\theta = \tfrac{1}{2} r^2 + A.$$

4·2. THE HYPERCIRCLE METHOD APPLIED TO THE TORSION PROBLEM

In applying the hypercircle method to the torsion problem, P-space is the cross-section V and an F-vector corresponds to a P-vector field in V:

$$\mathbf{S} \leftrightarrow p_i. \tag{4·201}$$

The scalar product is
$$\mathbf{S}.\mathbf{S}' = \int p_i p_i' \, dV. \tag{4·202}$$

The moment of inertia F-vector and the spin F-vector

An important role is played by the *moment of inertia F-vector* $\mathbf{S}_0'$, defined by

$$\mathbf{S}_0' \leftrightarrow (x_1, x_2) = \operatorname{grad} \tfrac{1}{2} r^2, \tag{4·203}$$

so called because

$$\mathbf{S}_0'^2 = \int (x_1^2 + x_2^2)\, dV = \int r^2 dV = I. \tag{4·204}$$

This F-vector is used when we treat the torsion problem as an extended Dirichlet problem.

When we treat it as a Neumann problem, we use instead the *spin F-vector*

$$\mathbf{S}_0'' \leftrightarrow (x_2, -x_1), \tag{4·205}$$

so called because this is the velocity field for a disk spinning with unit angular velocity. Note that the P-vector field of $\mathbf{S}_0''$ is simply that of $\mathbf{S}_0'$ turned through a right angle; we have

$$\mathbf{S}_0' . \mathbf{S}_0'' = 0, \quad \mathbf{S}_0''^2 = I. \tag{4·206}$$

Splitting the Neumann problem

Consider the torsion problem in the Neumann form (4·108). For the solution we set

$$\mathbf{S} \leftrightarrow \phi_{,i}. \tag{4·207}$$

We define two linear subspaces as follows:

$$\left.\begin{aligned} L'\colon\quad & \mathbf{S}' \leftrightarrow p_i', \quad p_i' = u_{,i}'; \\ L''\colon\quad & \mathbf{S}'' \leftrightarrow p_i'', \quad p_{i,i}'' = 0, \quad (p_i'' n_i)_B = (n_1 x_2 - n_2 x_1)_B. \end{aligned}\right\} \tag{4·208}$$

The solution $\mathbf{S}$ is on L' and on L''. Taking F-vectors $\mathbf{T}'$, $\mathbf{T}''$ lying in L' and L'' respectively, we have

$$\mathbf{T}' . \mathbf{T}'' = \int p_i' p_i'' dV = \int u_{,i}' p_i'' dV = \int u' p_i'' n_i dB = 0, \tag{4·209}$$

since for $\mathbf{T}''$ we have $p_i'' n_i = 0$ on B. Thus L' and L'' are mutually orthogonal, and the problem is that of finding the intersection of two orthogonal linear subspaces. The hypercircle method is applicable. Note that L' (but not L'') contains the origin of F-space.

Certain discontinuities are permissible, as we see on inspecting (4·209):

$$\left.\begin{aligned} & u' \text{ must be continuous, but } u_{,i}' \text{ need only be piecewise continuous;} \\ & p_i'' n_i \text{ (across any line) must be continuous, but } p_i'' \text{ need only be piecewise continuous.} \end{aligned}\right\} \tag{4·210}$$

Note that the F-point $\mathbf{S}_0''$ is on L''.

Some inequalities for torsional rigidity

Although we shall in the main treat the torsion problem as an extended Dirichlet problem (see below), some inequalities are slightly easier to obtain from the Neumann treatment. We note the following formulae, obtained from (2·771) and (2·772) by interchanging $\mathbf{S}'$ and $\mathbf{S}''$, because L' (not L'') contains $\mathbf{O}$:

$$\mathbf{S}^2 = \mathbf{S} . \mathbf{S}'', \tag{4·211}$$

$$2\mathbf{S}'' . \mathbf{S}' - \mathbf{S}'^2 \leqslant (\mathbf{S}'' . \mathbf{I}')^2 \leqslant \mathbf{S}^2 \leqslant \mathbf{S}''^2, \quad \mathbf{I}' = \mathbf{S}' / |\mathbf{S}'|. \tag{4·212}$$

By (4·120) and (4·206) we have

$$\Gamma = I - \mathbf{S}^2 = \mathbf{S}_0''^2 - \mathbf{S}^2, \tag{4·213}$$

and hence, since we can put $\mathbf{S}'' = \mathbf{S}_0''$ in (4·212),

$$0 \leqslant \Gamma \leqslant I. \tag{4·214}$$

It is easy to show that $\Gamma = 0$ cannot occur [use (4·215) below] and that $\Gamma = I$ if, and only if, V is a circle or a ring bounded by two concentric circles; thus Γ *is always positive and* (*with these special cases excepted*) *less than the moment of inertia about the centroid.**

By (4·211) we can write Γ in another form:

$$\Gamma = \mathbf{S}_0''^2 - \mathbf{S}^2 = \mathbf{S}_0''^2 - 2\mathbf{S}^2 + \mathbf{S}^2 = (\mathbf{S}_0'' - \mathbf{S})^2. \tag{4·215}$$

Consider now a section with h holes (Fig. 4·21), and the following $h+2$ torsion problems:

P: the problem with the h holes,

P_0: the problem with all the holes filled in,

P_λ: the problem for a cylinder bounded externally by B_λ $(\lambda = 1, 2, \ldots, h)$.

Write Γ, Γ_0, Γ_λ for the corresponding torsional rigidities. Let $\mathbf{S}_0$ be the solution of problem P_0; the domain is V_0, the interior of B_0. Consider the P-vector fields which are the gradients of the solutions of the problems P, P_λ, each of which has its own domain (not overlapping); from (4·108) it is clear that they give a P-vector field in V_0 for which $p_i n_i$ is continuous across B_λ and satisfies the boundary condition on B_0 for the problem P_0. If $\mathbf{S}''$ corresponds to this set of P-vector fields, then $\mathbf{S}''$ is a point on L'' for the problem P_0, and so, by (4·212),

$$\mathbf{S}_0^2 \leqslant \mathbf{S}''^2. \tag{4·216}$$

Add equations of the form (4·213) for the problems P, P_λ and obtain

$$\Gamma + \sum_{\lambda=1}^{h} \Gamma_\lambda = I_0 - \mathbf{S}''^2, \tag{4·217}$$

where I_0 is the moment of inertia of V_0; for the problem P_0 (4·213) gives

$$\Gamma_0 = I_0 - \mathbf{S}_0^2. \tag{4·218}$$

It follows from (4·216), (4·217) and (4·218) that

$$\Gamma_0 \geqslant \Gamma + \sum_{\lambda=1}^{h} \Gamma_\lambda. \tag{4·219}$$

We may state that *the torsional rigidity of any cross-section exceeds* (*or is equal to*) *the sum of the torsional rigidities of any set of parts into*

* Γ is invariant under change of origin but I is not, being a minimum with respect to the centroid of V.

which it may be decomposed [Weinberger (1)]. Actually this statement is a little more general than what we proved above, but the same form of proof holds when some of the holes are left open in problem P_0.

We can also write (4·219) in the form

$$\Gamma \leqslant \Gamma_0 - \sum_{\lambda=1}^{h} \Gamma_\lambda, \tag{4·220}$$

and say that *the torsional rigidity of any multiply connected cross-section is less than (or equal to) the torsional rigidity of the cross-section with holes filled in, minus the sum of the torsional rigidities of cylinders filling the holes.*

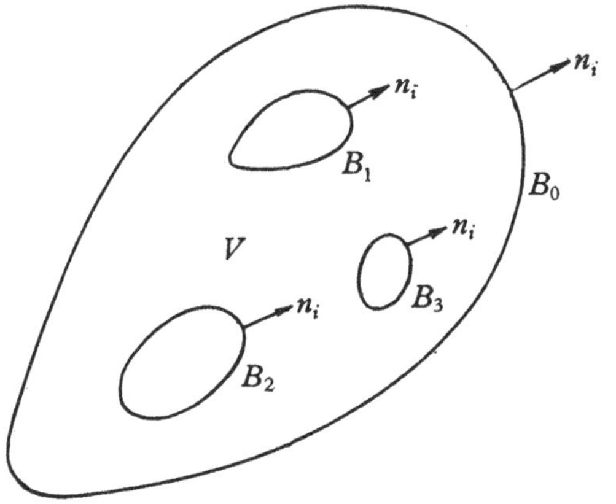

Fig. 4·21. The torsion problem for a multiply connected section ($h=3$).

For example, by (4·128), the torsional rigidity of a circular cylinder of radius a with n circular holes of radius b drilled in it satisfies

$$\Gamma \leqslant \tfrac{1}{2}\pi(a^4 - nb^4). \tag{4·221}$$

Should we wish to apply the hypercircle method to the torsion problem in Neumann form, we would take linear subspaces of finite dimensionality immersed in L' and L'' and find their vertices $\mathbf{V}'$, $\mathbf{V}''$. Then by (2·782) (interchanging $\mathbf{V}'$ and $\mathbf{V}''$, since L' contains $\mathbf{O}$)

$$\mathbf{V}'^2 \leqslant \mathbf{S}^2 \leqslant \mathbf{V}''^2, \tag{4·222}$$

and so, by (4·213), the torsional rigidity is bounded by

$$I - \mathbf{V}''^2 \leqslant \Gamma \leqslant I - \mathbf{V}'^2. \tag{4·223}$$

Splitting the extended Dirichlet problem

Fig. 4·21 shows a multiply connected section with h holes; the outer boundary is B_0 and the boundaries of the holes are

$B_1, B_2, \ldots, B_h$. Then, as in (4·111) and (4·112), our problem is to find a function ψ and h constants $C_1, C_2, \ldots, C_h$ to satisfy the following conditions:*

$$\Delta\psi = 0, \quad (\psi)_{B_0} = \tfrac{1}{2}r^2, \quad (\psi)_{B_\lambda} = \tfrac{1}{2}r^2 + C_\lambda,$$
$$\int_{B_\lambda} (\partial\psi/\partial n)\, dB_\lambda = 0 \quad (\lambda = 1, 2, \ldots, h). \tag{4·224}$$

In (3·216) we defined, for the Dirichlet problem, associated F-vectors $\mathbf{S}'$, homogeneous associated F-vectors $\mathbf{T}'$ and complementary F-vectors $\mathbf{T}''(=\mathbf{S}'')$. To meet our present needs we shall modify these definitions but keep the old names, as follows:

$$\left.\begin{array}{ll}
\text{Solution: } \mathbf{S} \leftrightarrow p_i = \psi_{,i}, \text{ where } \psi \text{ satisfies (4·224).} & \\
L': \text{ associated:} & \\
\quad \mathbf{S}' \leftrightarrow p'_i = u'_{,i}, \quad (u')_{B_0} = \tfrac{1}{2}r^2, \quad (u')_{B_\lambda} = \tfrac{1}{2}r^2 + C_\lambda & \\
\quad\quad (\lambda = 1, 2, \ldots, h), & \\
\quad \text{the constants } C_\lambda \text{ having any values at all.} & \\
\quad \text{homogeneous associated:} & \\
\quad \mathbf{T}' \leftrightarrow \bar{p}'_i = \bar{u}'_{,i}, \quad (\bar{u}')_{B_0} = 0, \quad (\bar{u}')_{B_\lambda} = D_\lambda & \\
\quad\quad (\lambda = 1, 2, \ldots, h), & \\
\quad \text{the constants } D_\lambda \text{ having any values at all.} & \\
L'': \text{ complementary:} & \\
\quad \mathbf{T}'' \leftrightarrow p''_i, \quad p''_{i,i} = 0, \quad \displaystyle\int_{B_\lambda} p''_i n_i \, dB_\lambda = 0 \quad (\lambda = 1, 2, \ldots, h). &
\end{array}\right\} \tag{4·225}$$

The permissible discontinuities are as before, as in (3·211): u' and $\bar{u}'$ are to be continuous, but their partial derivatives need only be piecewise continuous; p''_i need only be piecewise continuous, but across any curve there must be continuity of $p''_i n_i$. These conditions ensure that, in the evaluation of scalar products as in (4·226), no line integrals come in except those on B_0 and B_λ.

The points $\mathbf{S}'$ in F-space define a linear subspace L', in which the F-vectors $\mathbf{T}'$ lie. The points $\mathbf{T}''$ define a linear subspace L'' and they lie in it; it contains the origin $\mathbf{O}$. We have now to show that L' and L'' are orthogonal; if we can show that, then the hypercircle theory of § 2·7 applies, since it is clear from (4·225) that the solution $\mathbf{S}$ is the intersection of L' and L''.

To prove this orthogonality, we have (with the unit normal drawn out from each curve as in Fig. 4·21)

$$\begin{aligned}\mathbf{T}'.\mathbf{T}'' &= \int \bar{u}'_{,i} p''_i \, dV \\ &= -\int \bar{u}' p''_{i,i} \, dV + \int \bar{u}' p''_i n_i \, dB_0 - \sum_{\lambda=1}^{h} \int \bar{u}' p''_i n_i \, dB_\lambda,\end{aligned} \tag{4·226}$$

* For a simply connected domain we have only $\Delta\psi = 0$, $(\psi)_B = \tfrac{1}{2}r^2$.

and each of these integrals vanishes by (4·225); thus the orthogonality of L' and L'' is established, and we can use the hypercircle theory of § 2·7 for the case where L'' contains $\mathbf{O}$.

We note that the moment of inertia F-vector $\mathbf{S}_0'$ of (4·203) gives a point on L':

$$\mathbf{S}_0' \leftrightarrow (x_1, x_2) = \operatorname{grad} \tfrac{1}{2}r^2. \tag{4·227}$$

Summary of procedure for the torsion problem, applicable to a multiply connected section

We have just seen that the torsion problem for a multiply connected section reduces to finding the intersection of two orthogonal linear subspaces, L', L''. But we do not have a Dirichlet problem in the sense of Chapter 3, and so the formulae of that chapter have to be used with a little caution. It will be useful to summarize here a standard procedure for the torsion problem, applicable to a multiply connected section, but of course covering the special case of simple connectivity. The reader will have no difficulty in verifying the validity of the steps.

(i) Choose $\mathbf{S}_0' \leftrightarrow (x_1, x_2)$, the moment of inertia F-vector as in (4·227); calculate $\mathbf{S}_0'^2 = I$, the moment of inertia of the section.

(ii) Choose r homogeneous associated F-vectors $\mathbf{T}_\rho'$ $(\rho = 1, 2, \ldots, r)$ as in (4·225) and calculate the scalar products

$$A_\rho' = \mathbf{S}_0' \, . \, \mathbf{T}_\rho', \quad A_{\mu\rho}' = \mathbf{T}_\mu' \, . \, \mathbf{T}_\rho'.$$

(iii) Choose s complementary F-vectors $\mathbf{T}_\sigma''$ $(\sigma = 1, 2, \ldots, s)$ as in (4·225) and calculate $A_\sigma'' = \mathbf{S}_0' \, . \, \mathbf{T}_\sigma''$, $A_{\nu\sigma}'' = \mathbf{T}_\nu'' \, . \, \mathbf{T}_\sigma''$.

(iv) Solve the equations (2·776), viz.

$$\left.\begin{aligned} \sum_{\mu=1}^{r} a_\mu' A_{\mu\rho}' + A_\rho' = 0 \quad (\rho = 1, 2, \ldots, r), \\ \sum_{\nu=1}^{s} a_\nu'' A_{\nu\sigma}'' - A_\sigma'' = 0 \quad (\sigma = 1, 2, \ldots, s). \end{aligned}\right\} \tag{4·228}$$

(v) Obtain the vertices $\mathbf{V}'$, $\mathbf{V}''$ by

$$\mathbf{V}' = \mathbf{S}_0' + \sum_{\rho=1}^{r} a_\rho' \mathbf{T}_\rho', \quad \mathbf{V}'' = \sum_{\sigma=1}^{s} a_\sigma'' \mathbf{T}_\sigma'', \tag{4·229}$$

and their squares by

$$\mathbf{V}'^2 = \mathbf{S}_0'^2 + \sum_{\rho=1}^{r} a_\rho' A_\rho', \quad \mathbf{V}''^2 = \sum_{\sigma=1}^{s} a_\sigma'' A_\sigma'', \tag{4·230}$$

as in (2·775) and (2·777).

(vi) The hypercircle of class $r+s$ is then as in (2·781), and the basic inequalities are

$$\left.\begin{aligned} \mathbf{V}''^2 \leqslant \mathbf{S}^2 \leqslant \mathbf{V}'^2, \\ \mathbf{S}_0'^2 - \mathbf{V}'^2 = -\sum_{\rho=1}^{r} a_\rho' A_\rho' \leqslant \Gamma \leqslant \mathbf{S}_0'^2 - \mathbf{V}''^2 = \mathbf{S}_0'^2 - \sum_{\sigma=1}^{s} a_\sigma'' A_\sigma''. \end{aligned}\right\} \tag{4·231}$$

(vii) In the case of orthonormal vectors $\mathbf{I}'_\rho$, $\mathbf{I}''_\sigma$ (instead of $\mathbf{T}'_\rho$, $\mathbf{T}''_\sigma$) the above formulae simplify to

$$\left.\begin{aligned}
&\mathbf{V}' = \mathbf{S}'_0 - \sum_{\rho=1}^{r} \mathbf{I}'_\rho(\mathbf{S}'_0 . \mathbf{I}'_\rho), \quad \mathbf{V}'' = \sum_{\sigma=1}^{s} \mathbf{I}''_\sigma(\mathbf{S}'_0 . \mathbf{I}''_\sigma),\\
&\mathbf{V}'^2 = \mathbf{S}'^2_0 - \sum_{\rho=1}^{r} (\mathbf{S}'_0 . \mathbf{I}'_\rho)^2, \quad \mathbf{V}''^2 = \sum_{\sigma=1}^{s} (\mathbf{S}'_0 . \mathbf{I}''_\sigma)^2,\\
&\sum_{\rho=1}^{r} (\mathbf{S}'_0 . \mathbf{I}'_\rho)^2 \leqslant \Gamma \leqslant \mathbf{S}'^2_0 - \sum_{\sigma=1}^{s} (\mathbf{S}'_0 . \mathbf{I}''_\sigma)^2.
\end{aligned}\right\} \tag{4·232}$$

Approximation to stress and warping

What follows applies to any section, simply or multiply connected. We introduce the following notation. Let $\mathbf{U}$ be any F-vector and q_i the corresponding P-vector field. We shall use the symbol $\tilde{\mathbf{U}}$ to denote the F-vector corresponding to the P-vector field obtained by rotating q_i through a right angle, counter-clockwise. Thus

$$\left.\begin{aligned}
&\mathbf{U} \leftrightarrow (q_1, q_2), \quad \tilde{\mathbf{U}} \leftrightarrow (-q_2, q_1),\\
&\mathbf{U} \leftrightarrow q_i, \quad \tilde{\mathbf{U}} \leftrightarrow -\epsilon_{ij} q_j,
\end{aligned}\right\} \tag{4·233}$$

to write the correspondences in two slightly different but equivalent ways. This type of transition has already been used (p. 181) in the connexion between pyramid F-vectors of the two classes.

Consider now the torsion problem and the P-vector field of stress, as given by (4·113). Let $\mathbf{E}$ denote the corresponding F-vector, so that

$$\mathbf{E} \leftrightarrow \mu\alpha(\psi_{,2} - x_2, \ -\psi_{,1} + x_1). \tag{4·234}$$

Then

$$\tilde{\mathbf{E}} \leftrightarrow \mu\alpha(\psi_{,1} - x_1, \psi_{,2} - x_2). \tag{4·235}$$

Now $\psi_{,i} \leftrightarrow \mathbf{S}$ (the solution) and $x_i \leftrightarrow \mathbf{S}'_0$ (the moment of inertia F-vector), and so

$$\tilde{\mathbf{E}} = \mu\alpha(\mathbf{S} - \mathbf{S}'_0). \tag{4·236}$$

If, as in (4·229), we have obtained the vertices $\mathbf{V}'$, $\mathbf{V}''$, and if these are close together (equivalently, $\mathbf{V}'^2 - \mathbf{V}''^2$ small), then either is a good approximation to $\mathbf{S}$, and so by (4·236) we have the following good approximations for $\tilde{\mathbf{E}}$ (and hence by (4·233) for the stress $\mathbf{E}$):

$$\left.\begin{aligned}
&\tilde{\mathbf{E}}/\mu\alpha \sim \mathbf{V}' - \mathbf{S}'_0 = \sum_{\rho=1}^{r} a'_\rho \mathbf{T}'_\rho,\\
&\tilde{\mathbf{E}}/\mu\alpha \sim \mathbf{V}'' - \mathbf{S}'_0 = \sum_{\sigma=1}^{s} a''_\sigma \mathbf{T}''_\sigma - \mathbf{S}'_0.
\end{aligned}\right\} \tag{4·237}$$

In the case of orthonormal F-vectors $\mathbf{I}'_\rho$, $\mathbf{I}''_\sigma$ (instead of $\mathbf{T}'_\rho$, $\mathbf{T}''_\sigma$) we have the approximations

$$\left.\begin{aligned}\tilde{\mathbf{E}}/\mu\alpha &\sim -\sum_{\rho=1}^{r}\mathbf{I}'_\rho(\mathbf{S}'_0\,.\,\mathbf{I}'_\rho),\\ \tilde{\mathbf{E}}/\mu\alpha &\sim \sum_{\sigma=1}^{s}\mathbf{I}''_\sigma(\mathbf{S}'_0\,.\,\mathbf{I}''_\sigma)-\mathbf{S}'_0.\end{aligned}\right\} \tag{4·238}$$

If u'_ρ is the function whose gradient corresponds to $\mathbf{T}'_\rho$, then, as in (4·237),

$$\mathbf{V}'-\mathbf{S}'_0 \leftrightarrow \operatorname{grad}\sum_{\rho=1}^{r} a'_\rho u'_\rho, \tag{4·239}$$

and it follows from (4·237) that *the lines of stress* (i.e. the lines along which the stress P-vector points) *are approximately the level lines of the function* $\sum_{\rho=1}^{r} a'_\rho u'_\rho$ *and so are represented approximately by the equations*

$$\sum_{\rho=1}^{r} a'_\rho u'_\rho = \text{const.} \tag{4·240}$$

We can also get approximate information about the warping function ϕ. As in (4·110), we have $\phi_{,1}=\psi_{,2}$, $\phi_{,2}=-\psi_{,1}$, and so the lines $\phi=\text{const.}$ are the lines along which $\operatorname{grad}\psi$ points. Now the vertex $\mathbf{V}''$, as an approximation to $\mathbf{S}$, corresponds approximately to $\operatorname{grad}\psi$. If then the complementary F-vectors are given in the form

$$\mathbf{T}''_\sigma \leftrightarrow \epsilon_{ij}u''_{\sigma,j} \tag{4·241}$$

(this P-vector field having zero divergence as required by (4·225)), it is easy to see that *the lines of constant warping* ($\phi=\text{const.}$) *are approximately the level lines of the function* $\sum_{\sigma=1}^{s} a''_\sigma u''_\sigma$ *and so are represented approximately by the equations*

$$\sum_{\sigma=1}^{s} a''_\sigma u''_\sigma = \text{const.} \tag{4·242}$$

Torsion of a rectangle

In this book the method of the hypercircle is for the most part used to get approximations, with upper and lower bounds in the mean-square sense, but here, by taking infinite sequences of F-vectors, we shall get an exact solution. Part of the work really follows the usual method [Love(1), pp. 317, 323; Sokolnikoff(1), p. 128], although it looks different on account of the F-space notation; in this part we deal with an infinite sequence of F-points in the linear subspace L'', corresponding to the gradients of harmonic functions. But we also use an infinite sequence of F-vectors lying in the other linear subspace L'. In this way we reduce the radius of the hypercircle to zero and so get the exact solution. Since the domain is simply connected, we have an ordinary Dirichlet problem.

Using x and y for rectangular Cartesian coordinates, we take for P-space the rectangular domain bounded by the lines $x = \pm a$, $y = \pm b$. For this domain we consider *three* Dirichlet problems, (a), (b), (c), corresponding respectively to the three boundary conditions:

$$(a) \quad (u)_B = \tfrac{1}{2}x^2, \qquad (b) \quad (u)_B = \tfrac{1}{2}y^2, \qquad (c) \quad (u)_B = \tfrac{1}{2}(x^2+y^2).$$

Problem (c) is the torsion problem; the others are closely connected with it. If $\mathbf{S}_a$, $\mathbf{S}_b$, $\mathbf{S}_c$ are the respective solutions, then obviously

$$\mathbf{S}_c = \mathbf{S}_a + \mathbf{S}_b. \tag{4·243}$$

We consider yet a fourth Dirichlet problem, viz. $(u)_B = \tfrac{1}{2}(x^2 - y^2)$. Since the right-hand side is a harmonic function, the solution is $u = \tfrac{1}{2}(x^2 - y^2)$; for the corresponding F-vector we write $\mathbf{H} \leftrightarrow (x, -y)$. It follows that

$$\mathbf{S}_a - \mathbf{S}_b = \mathbf{H} \leftrightarrow (x, -y), \tag{4·244}$$

and then (4·243) gives

$$\mathbf{S}_c = 2\mathbf{S}_a - \mathbf{H}. \tag{4·245}$$

Thus the solution of the torsion problem (c) is given in terms of the solution of (a).

As associated F-vectors for the three problems we shall take respectively the gradients of $\tfrac{1}{2}x^2$, $\tfrac{1}{2}y^2$, $\tfrac{1}{2}(x^2+y^2)$:

$$\left.\begin{aligned} &\mathbf{S}'_a \leftrightarrow (x, 0), \quad \mathbf{S}'_b \leftrightarrow (0, y), \quad \mathbf{S}'_c = \mathbf{S}'_a + \mathbf{S}'_b \leftrightarrow (x, y), \\ &\mathbf{S}'^2_a = \tfrac{4}{3}a^3 b, \quad \mathbf{S}'^2_b = \tfrac{4}{3}ab^3, \quad \mathbf{S}'^2_c = \mathbf{S}'^2_a + \mathbf{S}'^2_b = \tfrac{4}{3}ab(a^2+b^2). \end{aligned}\right\} \tag{4·246}$$

If $k_r = \tfrac{1}{2}\pi(2r+1)$, the functions

$$\cos(k_p x/a)\cos(k_q y/b) \quad (p, q = 0, 1, 2, \ldots)$$

all vanish on the boundary B. Therefore the gradient of any one of these functions corresponds to a homogeneous associated F-vector

$$\begin{aligned} \mathbf{T}'_{pq} \leftrightarrow [&-(k_p/a)\sin(k_p x/a)\cos(k_q y/b), \\ &-(k_q/b)\cos(k_p x/a)\sin(k_q y/b)]. \end{aligned} \tag{4·247}$$

A simple calculation gives

$$\mathbf{T}'^2_{pq} = ab[(k_p/a)^2 + (k_q/b)^2]. \tag{4·248}$$

The F-vectors $\mathbf{T}'_{pq}$ are easily seen to be orthogonal, and so, if we define

$$\mathbf{I}'_{pq} = \mathbf{T}'_{pq}/|\mathbf{T}'_{pq}| \quad (p, q = 0, 1, 2, \ldots), \tag{4·249}$$

we have an infinite orthonormal set of homogeneous associated F-vectors.

Another simple calculation gives

$$\mathbf{S}'_a \,.\, \mathbf{I}'_{pq} = A_{pq}, \tag{4·250}$$

where $A_{pq} = -4(-1)^{p+q}(ab)^{\frac{1}{2}}(k_p k_q)^{-1}[(k_p/a)^2 + (k_q/b)^2]^{-\frac{1}{2}}$. (4·251)

From symmetry it follows that

$$\mathbf{S}'_b \cdot \mathbf{I}'_{pq} = A_{pq}, \tag{4·252}$$

and so, by (4·246),

$$\mathbf{S}'_c \cdot \mathbf{I}'_{pq} = 2A_{pq}. \tag{4·253}$$

We have now completed the computations necessary to give the vertices $\mathbf{V}'$ for our three problems in accordance with (2·779), or strictly the limits of these vertices as the ranges of p and q tend to infinity. If Σ denotes the infinite double summation for $p, q = 0, 1, 2, \ldots$, we obtain from (2·779)

$$\left.\begin{aligned} \mathbf{V}'_a &= \mathbf{S}'_a - \Sigma \mathbf{I}'_{pq} A_{pq}, \\ \mathbf{V}'_b &= \mathbf{S}'_b - \Sigma \mathbf{I}'_{pq} A_{pq}, \\ \mathbf{V}'_c &= \mathbf{V}'_a + \mathbf{V}'_b = \mathbf{S}'_c - 2\Sigma \mathbf{I}'_{pq} A_{pq}. \end{aligned}\right\} \tag{4·254}$$

Also, by (2·780),

$$\left.\begin{aligned} \mathbf{V}'^2_a &= \mathbf{S}'^2_a - \Sigma A^2_{pq}, \\ \mathbf{V}'^2_b &= \mathbf{S}'^2_b - \Sigma A^2_{pq}, \\ \mathbf{V}'^2_c &= \mathbf{S}'^2_c - 4\Sigma A^2_{pq}. \end{aligned}\right\} \tag{4·255}$$

Let us evaluate the sum occurring here. We note [Hobson (1), p. 345] that

$$\sum_{r=0}^{\infty} (2r+1)^{-2} = \tfrac{1}{8}\pi^2, \quad \sum_{r=0}^{\infty} (2r+1)^{-4} = \tfrac{1}{96}\pi^4, \tag{4·256}$$

$$\sum_{r=0}^{\infty} [(2r+1)^2 + x^2]^{-1} = \tfrac{1}{4}\pi x^{-1} \tanh \tfrac{1}{2}\pi x, \tag{4·257}$$

so that

$$\sum_{r=0}^{\infty} k_r^{-2} = \tfrac{1}{2}, \quad \sum_{r=0}^{\infty} k_r^{-4} = \tfrac{1}{6}, \quad \sum_{r=0}^{\infty} (k_r^2 + X^2)^{-1} = \tfrac{1}{2} X^{-1} \tanh X. \tag{4·258}$$

Hence, carrying out the double summation first in one order and then in the other, we find

$$\begin{aligned} \Sigma A^2_{pq} &= \tfrac{4}{3}ab^3 - 8b^4 \sum_{q=0}^{\infty} k_q^{-5} \tanh(k_q a/b) \\ &= \tfrac{4}{3}a^3 b - 8a^4 \sum_{p=0}^{\infty} k_p^{-5} \tanh(k_p b/a). \end{aligned} \tag{4·259}$$

Substitution from (4·246) and (4·259) in (4·255) gives

$$\mathbf{V}'^2_a = 8a^4 \sum_{p=0}^{\infty} k_p^{-5} \tanh(k_p b/a). \tag{4·260}$$

We have actually solved the torsion problem in (4·254), but we do not know that yet, for the mere fact that the number of F-vectors used is infinite is no assurance that $\mathbf{V}'_c = \mathbf{S}_c$. That it is so we shall now establish by working in the other linear subspace L''.

The functions $\cos(k_r x/a)\cosh(k_r y/a)$ $(r = 0, 1, 2, \ldots)$ are harmonic; the gradient of each of them has therefore zero divergence and so corresponds by (4·225) to a complementary F-vector, and we write

$$\mathbf{T}''_r \leftrightarrow [-(k_r/a)\sin(k_r x/a)\cosh(k_r y/a),\\ (k_r/a)\cos(k_r x/a)\sinh(k_r y/a)]. \quad (4{\cdot}261)$$

A simple calculation gives

$$\mathbf{T}''^2_r = k_r \sinh(2k_r b/a), \quad (4{\cdot}262)$$

and the square root gives $|\mathbf{T}''_r|$; since the vectors $\mathbf{T}''_r$ are orthogonal, as is easily seen, we have then an infinite orthonormal set of complementary F-vectors

$$\mathbf{I}''_r = \mathbf{T}''_r / |\mathbf{T}''_r| \quad (r = 0, 1, 2, \ldots). \quad (4{\cdot}263)$$

These F-vectors are particularly suited to the problem (a). By direct calculation we get

$$\mathbf{S}'_a \cdot \mathbf{I}''_r = A_r, \quad (4{\cdot}264)$$

where $A_r = -4(-1)^r a^2 k_r^{-2} \sinh(k_r b/a)\,[k_r \sinh(2k_r b/a)]^{-\frac{1}{2}}$. (4·265)

The vertex $\mathbf{V}''_a$ is then, by (2·779),

$$\mathbf{V}''_a = \sum_{r=0}^{\infty} \mathbf{I}''_r (\mathbf{S}'_a \cdot \mathbf{I}''_r) = \sum_{r=0}^{\infty} A_r \mathbf{I}''_r, \quad (4{\cdot}266)$$

and by (2·780)

$$\mathbf{V}''^2_a = \sum_{r=0}^{\infty} A_r^2 = 8a^4 \sum_{r=0}^{\infty} k_r^{-5} \tanh(k_r b/a). \quad (4{\cdot}267)$$

But this is precisely the expression in (4·260), and so $\mathbf{V}'^2_a = \mathbf{V}''^2_a$. By (2·781) this implies the vanishing of the radius of the hypercircle, and so $\mathbf{V}'_a = \mathbf{V}''_a = \mathbf{S}_a$; hence by (4·254) the solution of problem (a) is

$$\mathbf{S}_a = \mathbf{S}'_a - \Sigma A_{pq} \mathbf{I}'_{pq} = \sum_{r=0}^{\infty} A_r \mathbf{I}''_r. \quad (4{\cdot}268)$$

But if $\mathbf{S}_a = \mathbf{V}'_a$, then similarly $\mathbf{S}_b = \mathbf{V}'_b$, and addition gives, by (4·243) and (4·254),

$$\mathbf{S}_c = \mathbf{V}'_c = \mathbf{S}'_c - 2\Sigma A_{pq} \mathbf{I}'_{pq}. \quad (4{\cdot}269)$$

This is the solution of the torsion problem for a rectangle, with $\mathbf{S}'_c$ as in (4·246), $\mathbf{I}'_{pq}$ as in (4·249) and A_{pq} as in (4·251).

It is more usual to express the solution in terms of hyperbolic functions, and this we can do by means of (4·245) and (4·268):

$$\mathbf{S}_c = -\mathbf{H} + 2\sum_{r=0}^{\infty} A_r \mathbf{I}''_r, \quad (4{\cdot}270)$$

where $\mathbf{H}$ is as in (4·244), $\mathbf{I}''_r$ as in (4·263) and A_r as in (4·265).

The last of (4·255) gives

$$\mathbf{S}_c^2 = \mathbf{V}_c'^2 = \mathbf{S}_c'^2 - 4\Sigma A_{pq}^2, \tag{4·271}$$

and so, by (4·120) and (4·259), the *torsional rigidity* of the rectangle is

$$\begin{aligned}\Gamma &= \mathbf{S}_c'^2 - \mathbf{S}_c^2 = 4\Sigma A_{pq}^2 \\ &= \tfrac{16}{3}a^3 b - 32a^4 \sum_{p=0}^{\infty} k_p^{-5} \tanh(k_p b/a) \\ &= \tfrac{16}{3}ab^3 - 32b^4 \sum_{q=0}^{\infty} k_q^{-5} \tanh(k_q a/b)\end{aligned} \tag{4·272}$$

[Sokolnikoff (1), p. 132].

In the case of a square ($b = a$, side $= 2a$), we have [Love (1), p. 324]

$$\begin{aligned}\Gamma &= 32a^4\left(\tfrac{1}{6} - \sum_{p=0}^{\infty} k_p^{-5} \tanh k_p\right) \\ &= 2{\cdot}2492\,a^4.\end{aligned} \tag{4·273}$$

Example: torsion of a hollow square

In §4·5 we shall deal systematically with the torsion of a certain hollow square, using pyramid F-vectors. Here we shall be content to establish rough bounds for the torsional rigidity, but we shall retain general values for the sides of the inner and outer squares, so that the case of a thin square will be included as a limit, and as we approach that limit the bounds become more significant.

Fig. 4·22 shows a hollow square with outer side $2a$ and inner side $2b$. We take the origin at the centre and axes Oxy as shown. We choose

$$\mathbf{S}_0' \leftrightarrow (x, y) \tag{4·274}$$

as usual.

To get a homogeneous associated F-vector in accordance with (4·225), we need a function $\bar{u}'$ which vanishes on the outer square and is constant on the inner square. We define such a function by $\bar{u}' = x - a$ in the octant $ABCD$ of Fig. 4·22, the values over the rest of the section being defined by reflexions in the lines of symmetry, equal values being associated with symmetric points. The function is continuous and the discontinuities in its derivatives are permissible. For homogeneous associated F-vector (we shall be content with one) we put then

$$\mathbf{T}_1' \leftrightarrow \operatorname{grad} \bar{u}' \; [= (1, 0) \text{ in } ABCD]. \tag{4·275}$$

To get a complementary F-vector in accordance with (4·225), we need a P-vector field with vanishing divergence, with continuous normal component, and with the integral of the normal component around B_1 vanishing. Moreover, the P-vector field should possess the symmetry of the problem, i.e. it should reflect into itself when reflected in any one of the four lines of symmetry.

It is easy to find a simple P-vector field satisfying these requirements. It is shown in Fig. 4·23 and the corresponding complementary F-vector may be described by

$$\mathbf{T}_1'' \leftrightarrow \begin{cases} (-x+b, y) \text{ in rectangle } ABED, \\ (b, b) \text{ in triangle } CDE, \end{cases} \tag{4·276}$$

the rest of the P-vector field being obtained by symmetry. We shall content ourselves with one complementary F-vector, so that the hypercircle will be of class $1+1$.

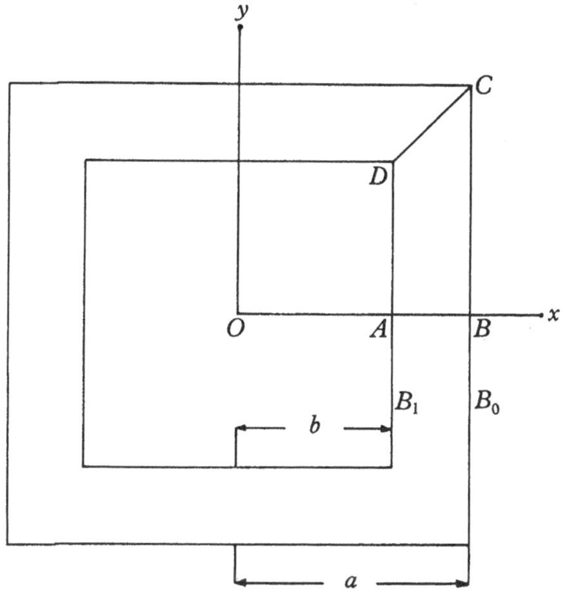

Fig. 4·22. Torsion of a hollow square.

By simple integrations we find

$$\left.\begin{aligned} &\mathbf{S}_0'^2 = I = \tfrac{8}{3}(a^4-b^4), \\ &A_1' = \mathbf{S}_0' . \mathbf{T}_1' = \tfrac{8}{3}(a^3-b^3), \quad A_{11}' = \mathbf{T}_1'^2 = 4(a^2-b^2), \\ &A_1'' = \mathbf{S}_0' . \mathbf{T}_1'' = \tfrac{4}{3}ab(a^2-b^2), \quad A_{11}'' = \mathbf{T}_1''^2 = \tfrac{8}{3}b(a-b)(a^2+ab-b^2). \end{aligned}\right\} \tag{4·277}$$

Then by (4·228)

$$\left.\begin{aligned} a_1' &= -A_1'/A_{11}' = -\tfrac{2}{3}(a+b)^{-1}(a^2+ab+b^2), \\ a_1'' &= A_1''/A_{11}'' = \tfrac{1}{2}a(a+b)(a^2+ab-b^2)^{-1}. \end{aligned}\right\} \tag{4·278}$$

With these values of the constants the vertices are, by (4·229),

$$\mathbf{V}' = \mathbf{S}_0' + a_1'\mathbf{T}_1', \quad \mathbf{V}'' = a_1''\mathbf{T}_1'', \tag{4·279}$$

and we get (crude) alternative approximations to the stress distribution from (4·237):

$$\tilde{\mathbf{E}}/\mu\alpha \sim a_1'\mathbf{T}_1', \quad \tilde{\mathbf{E}}/\mu\alpha \sim a_1''\mathbf{T}_1'' - \mathbf{S}_0'. \tag{4·280}$$

By (4·231) we have as bounds on the torsional rigidity

$$-a_1' A_1' \leqslant \Gamma \leqslant \mathbf{S}_0'^2 - a_1'' A_1'', \tag{4·281}$$

or explicitly

$$\frac{16}{9}\frac{a-b}{a+b}(a^2+ab+b^2)^2 \leqslant \Gamma \leqslant \tfrac{2}{3}(a^2-b^2)\left[4(a^2+b^2) - \frac{a^2b(a+b)}{a^2+ab-b^2}\right]. \tag{4·282}$$

In general these bounds are not close together (see p. 270 for the case $a=2b$), but the result is of interest as a simple illustration of the application of the hypercircle method to a problem of torsion for a doubly

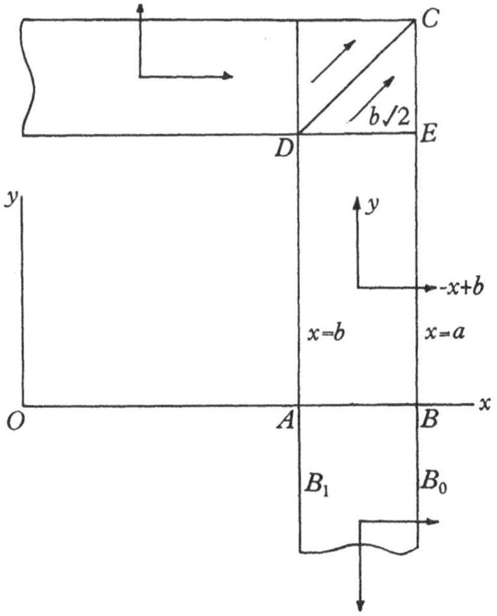

Fig. 4·23. Torsion of a hollow square: P-vector field of the complementary F-vector $\mathbf{T}_1''$.

connected section. It becomes more interesting in the case where $a-b$ is small, so that the square is thin. Then (4·282) yields the precise result

$$8 \leqslant \lim_{b\to a} \Gamma a^{-3}(a-b)^{-1} \leqslant 8, \tag{4·283}$$

so that the limit in question is in fact 8. This confirms the result given by the less mathematically precise theory of the torsion of thin tubes [Sokolnikoff (1), p. 174].

Without proceeding to the limit, we can get from (4·282) approximate bounds for a thin square of thickness $t=a-b$. Expanding both sides in powers of t and retaining only the first and second powers, we get

$$8a^3t\left(1-\frac{3}{2}\frac{t}{a}\right) \leqslant \Gamma \leqslant 8a^3t\left(1-\frac{t}{a}\right). \tag{4·284}$$

If we go to the other limit in (4·282), making the size of the hole tend to zero, we get the bounds

$$1\cdot778 = \tfrac{16}{9} \leqslant \lim_{b\to 0} \Gamma a^{-4} \leqslant \tfrac{8}{3} = 2\cdot6667. \tag{4·285}$$

For no hole at all the value is 2·2492, as in (4·273).

Let us recall the inequalities (2·772):

$$2\mathbf{S}'.\mathbf{S}'' - \mathbf{S}''^2 \leqslant (\mathbf{S}'.\mathbf{I}'')^2 \leqslant \mathbf{S}^2 \leqslant \mathbf{S}'^2. \tag{4·286}$$

By the last of these and (4·120)

$$\Gamma = I - \mathbf{S}^2 \geqslant \mathbf{S}_0'^2 - \mathbf{S}'^2, \quad \Gamma \leqslant I = \mathbf{S}_0'^2, \tag{4·287}$$

and so the torsional rigidity is, quite generally, bounded by

$$\mathbf{S}_0'^2 - \mathbf{S}'^2 \leqslant \Gamma \leqslant \mathbf{S}_0'^2, \tag{4·288}$$

where $\mathbf{S}_0'$ is the moment of inertia F-vector and $\mathbf{S}'$ any point on L'.

By taking $\mathbf{S}' \leftrightarrow \operatorname{grad} u'$, where $u' = \frac{1}{2}(y^2 + a^2)$ in the octant $ABCD$ of Fig. 4·22, with values elsewhere given by symmetry, Weinstein (2) gave the following simple bounds for the torsional rigidity of the hollow square:

$$2(a^4 - b^4) < \Gamma < \tfrac{8}{3}(a^4 - b^4). \tag{4·289}$$

By (4·220) Weinberger (1) improved this upper bound to

$$\Gamma < 2\cdot2492(a^4 - b^4); \tag{4·290}$$

he considered also hollow sections, including a square with four holes in it, a problem discussed by Courant (1).

The interest in these results lies in the ease with which they are obtained. For other work on torsional rigidity, see Pólya and Szegö (1), Pólya and Weinstein (1) and Barta (1).

Exercises

1. Show that the torsional rigidity of a semicircular area of radius a satisfies

$$\tfrac{1}{32}\pi a^4 < \Gamma < \tfrac{1}{4}\pi a^4.$$

2. A section consists of a square of side $2a$ with semicircles described on two opposite sides. Noting that the torsional rigidity of an ellipse with semi-axes a, b is $\pi a^3 b^3 (a^2 + b^2)^{-1}$, establish the inequalities

$$\tfrac{8}{5}\pi < \Gamma/a^4 < \tfrac{16}{3} + \tfrac{3}{2}\pi,$$

or

$$5\cdot0265 < \Gamma/a^4 < 10\cdot0457.$$

3. Modify the preceding exercise by taking, instead of a square, a rectangle with sides $(2a, 2b)$ with semicircles described on the sides $2a$, and investigate bounds on the torsional rigidity for the case where the ratio b/a is small. Retaining only the most important small terms occurring, prove that

$$\tfrac{1}{2}\pi + \pi b/a < \Gamma/a^4 < \tfrac{1}{2}\pi + 4b/a.$$

4. For a hollow square as in Fig. 4·22 take $b = \frac{9}{10}a$ and plot the approximate lines of stress as given by the two approximations in (4·280). (In view of (4·283) we may expect the approximations to be fairly good.)

5. Consider the torsion of any thin-walled polygon of uniform thickness t. Following the suggestion of (4·275) and (4·276), devise a simple homogeneous associated F-vector $\mathbf{T}_1'$ and a simple complementary F-vector $\mathbf{T}_1''$ and use them to obtain bounds on the torsional rigidity, approximating as in (4·284) with rejection of cubes and higher powers of t.

6. For the hollow square of Fig. 4·22 improve the bounds (4·282) by taking a second complementary F-vector $\mathbf{T}_2''$ corresponding to a P-vector field as indicated in Fig. 4·24.

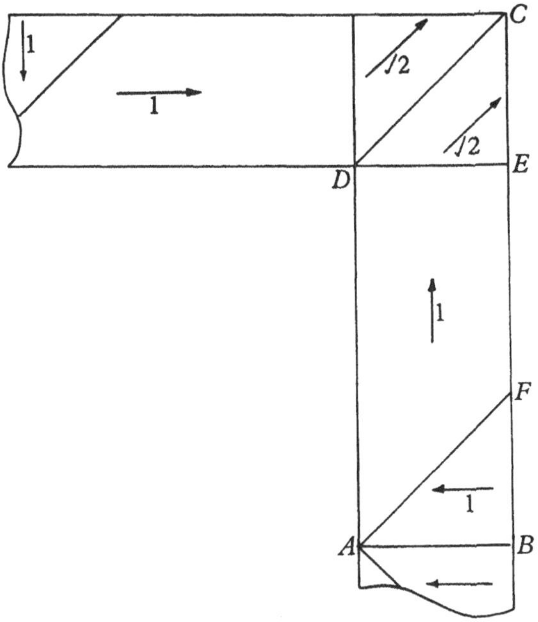

Fig. 4·24. Torsion of a hollow square: P-vector field for F-vector $\mathbf{T}_2''$ in Exercise 6.

4·3. PYRAMID F-VECTORS IN THE TORSION PROBLEM

When the section is simply connected, the torsion problem is a Dirichlet problem and the pyramid F-vectors of §§ 3·5, 3·6 and 3·7 are consequently available for use. They may also be used in the case of a multiply connected section, but this requires justification and comment.

We have to compare (3·216), which summarizes the splitting for the Dirichlet problem, with (4·225), which summarizes it for the torsion problem, including the case of multiple connectivity. It is clear that in both (3·216) and (4·225) the requirements for a homogeneous associated F-vector are met by a pyramid F-vector of the first class (Fig. 3·52), provided that its base lies in the section. But in (4·225) these will not suffice, for we need a function $\bar{u}'$ which is constant (not zero, in general) on the boundaries of the holes. Thus we supplement the pyramid F-vectors of the first class, as previously considered, with *ring F-vectors*, which are described below.

In the Dirichlet problem for a multiply connected region, we found it necessary to introduce strip F-vectors (Fig. 3·58) to take care of the multiple-valued character of the conjugate harmonic function. In the torsion problem the conjugate harmonic function (the warping function ϕ) is single-valued, and so *strip F-vectors will not occur.*

Ring F-vectors

Fig. 4·31 illustrates the P-vector field corresponding to a ring F-vector. A doubly connected region V is shown for simplicity; in general we need a ring F-vector surrounding each hole. There is no particular virtue in the triangulation used in Fig. 4·31; some triangulation is needed for purposes of illustration, and that chosen consists of isosceles right-angled triangles, as shown at the left-hand top corner of the diagram. The pyramid F-vectors all have hexagonal bases. To avoid confusion, only a few important junction-points are filled in in the main part of the diagram.

In general terms, the P-vector field of the ring F-vector is the gradient of a hill which slopes up from the plane of the paper to a plateau which contains the inner boundary B_1. It is constructed as follows. Imagine the network extended over the whole plane. Draw a closed polygon consisting entirely of network edges, contained entirely in $V+B_1$. (This is the inner closed polygon in Fig. 4·31.) Raise all junction points on and inside this closed polygon, all to the same height. *The resulting gradient field is the P-vector field of a ring F-vector, provided the outer closed polygon in Fig.* 4·31 *is contained in* $V+B_0$. The P-vector field vanishes everywhere except in the ring between the two closed polygons, and it is the gradient of a polyhedral function which is zero on B_0 and constant on B_1; the requirements of (4·225) are thus fulfilled.

Another way of looking at a ring F-vector is to think of it as the sum of pyramid F-vectors of the first class, all of the same height, with centres inside and on the inner closed polygon of Fig. 4·31.

Thus in the torsion problem for a multiply connected section with h holes, we are to include in the set of homogeneous associated F-vectors not only all pyramid F-vectors of the first class with bases in $V+B$, but also h ring F-vectors (suitably normalized). The vertex $\mathbf{V}'$ is then to be found from (4·228) and (4·229), the set $\mathbf{T}'_\rho$ including the ring F-vectors.

Pyramid F-vectors of the second class in the case of multiple connectivity

In the torsion problem for a multiply connected section, a complementary F-vector $\mathbf{T}''$ corresponds [cf. (4·225)] to a vector field p''_i which has vanishing divergence and satisfies

$$\int_{B_\lambda} p''_i n_i \, dB_\lambda = 0$$

for each hole. This integral condition is of course satisfied by a pyramid F-vector of the second class if its base does not intersect the boundary of a hole, and it follows from (3·505a) that it is also

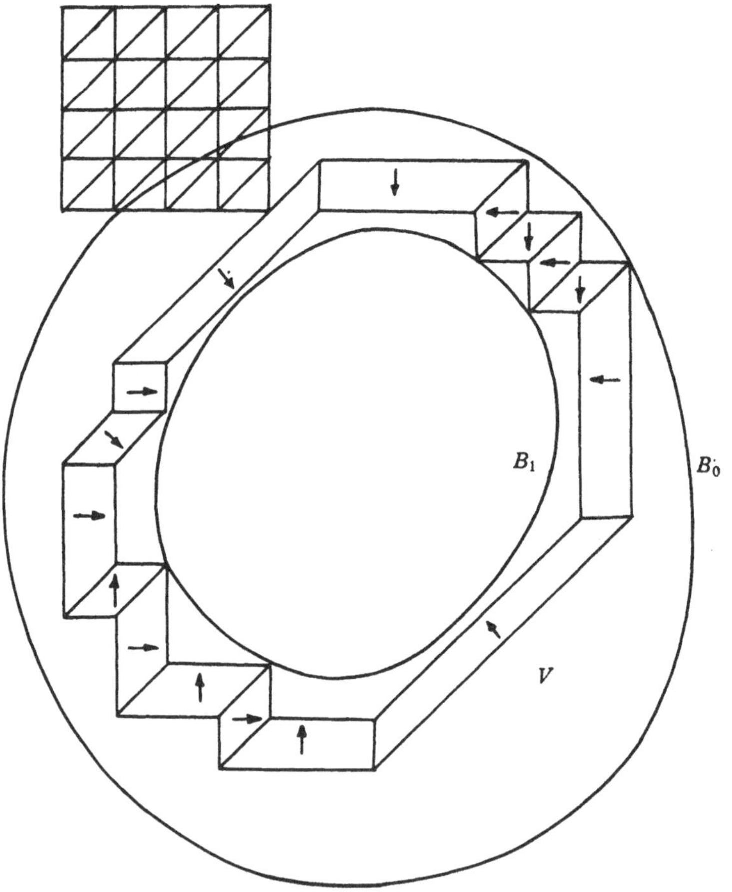

Fig. 4·31. Ring F-vector for the torsion problem for a doubly connected section; the P-vector field is indicated by arrows. The general mesh pattern is shown at the left top corner.

satisfied even if the base does intersect the boundary of a hole. Therefore, for the determination of the vertex $\mathbf{V}''$ by (4·228) and (4·229), we should use as complementary F-vectors all pyramid F-vectors of the second class given by the triangulation, including those with bases lying wholly in the domain of the section and those with bases intersecting the outer boundary and the boundaries of the holes.

Scalar products with $\mathbf{S}_0'$

Here, as throughout the whole of the discussion of the torsion problem, $\mathbf{S}_0'$ is the moment of inertia F-vector of the section [$\mathbf{S}_0' \leftrightarrow x_i$ as in (4·203)].

Consider an F-vector $\mathbf{U}$ corresponding to a P-vector field q_i which is constant inside an area A of the section V and zero outside that area. Then

$$\mathbf{S}_0' . \mathbf{U} = \int_A x_i q_i dV = \bar{x}_i q_i A, \qquad (4·301)$$

where $\bar{x}_i$ is the centroid of A. We recognize here the ordinary scalar product of the position-vector of the centroid and the P-vector field; this is of course independent of the directions of the axes, and we can sometimes simplify calculations by special choice of these directions.

Consider now a *hexagonal pyramid F-vector of the first class* $\mathbf{T}'$, as shown in Fig. 3·62 (p. 189); we suppose it normalized by the condition of unit height as in (3·602), and we shall further suppose its base contained entirely in $V+B$. Then application of (4·301) to each of the six equilateral triangles forming the base, with combination of opposite pairs, gives

$$\mathbf{S}_0' . \mathbf{T}' = -4A = -4a^2 . 3^{\frac{1}{2}}, \qquad (4·302)$$

where A is the area of each of the six equilateral triangles and $2a$ is its side.

Similarly, for a *hexagonal pyramid F-vector of the second class* $\mathbf{T}''$, as shown in Fig. 3·62 also, with base included in $V+B$, we find

$$\mathbf{S}_0' . \mathbf{T}'' = 0. \qquad (4·303)$$

These results are of course independent of the orientations of the hexagons relative to the coordinate axes. They are also independent of the position of the origin. In the case of incomplete pyramids we do not have this independence (see Figs. 4·32 and 4·33).

If the hexagonal bases extend outside V, we must go back to (4·301) to compute the scalar products.

We pass now to *square pyramid F-vectors*, as shown in Fig. 3·72 (p. 201). Let all bases be included in $V+B$ and let the normalization be as in (3·701), so that all pyramids are of unit height. Here are the results, obtained from (4·301), $2a$ being the side of the small square:

$$\left.\begin{array}{l}\text{Square pyramid } F\text{-vectors of the first class } (\mathbf{T}'):\\ \quad\text{Small square: } \mathbf{S}'_0\,.\,\mathbf{P}' = -\tfrac{8}{3}a^2,\\ \quad\text{Large square: } \mathbf{S}'_0\,.\,\mathbf{Q}' = -\tfrac{16}{3}a^2.\end{array}\right\} \tag{4·304}$$

$$\left.\begin{array}{l}\text{Square pyramid } F\text{-vectors of the second class } (\mathbf{T}''):\\ \quad\text{Small square: } \mathbf{S}'_0\,.\,\mathbf{P}'' = 0,\\ \quad\text{Large square: } \mathbf{S}'_0\,.\,\mathbf{Q}'' = 0.\end{array}\right\} \tag{4·305}$$

As before, these results are independent of the choice of axes. If the square bases extend outside V, we must go back to (4·301).

If $\mathbf{U}$ is a strip F-vector or a ring F-vector, the scalar product $\mathbf{S}'_0\,.\,\mathbf{U}$ can be easily calculated from (4·301), since the strip or ring can be split into triangles in each of which the P-vector field is constant. However, as remarked earlier, strip F-vectors do not occur in torsion problems.

Cases of broken pyramids

Let us now consider the evaluation of $\mathbf{S}'_0\,.\,\mathbf{U}$, where $\mathbf{U}$ is a pyramid F-vector with a base broken by the boundary B, so that part of the base extends outside V. In general the evaluation involves the form of B, but there are simple and useful cases where B consists of straight portions so that the portion of the base inside V consists of complete mesh triangles.

Some scalar products of this sort are shown in Fig. 4·32 for hexagonal pyramid F-vectors with broken bases, including both first class ($\mathbf{T}'$) and second class ($\mathbf{T}''$). Fig. 4·33 shows some similar results for square pyramid F-vectors of both classes, but only those of the $\mathbf{Q}$-type (large square, Fig. 3·72). In each case x_i are the coordinates of the central point of the pyramid.

In Figs. 4·32 and 4·33 the triangles drawn represent the parts of the bases inside V.

Exercises

1. Consider a cross-section V bounded externally by a polygon B_0 and internally by a polygon B_1, neither polygon being in general a regular one. Let B_1 have n sides. Make a simple triangulation of V by joining the vertices of B_0 and B_1 without intersections of the joins. Let $\mathbf{T}'_1, \mathbf{T}'_2, \ldots, \mathbf{T}'_n$ be pyramid F-vectors of the first class with centres at the vertices of B_1, normalized by

$$\text{First class}$$

$$\mathbf{S}_0'.\,\mathbf{T}' = 2a\,(x_2 - 3^{\frac{1}{2}}a)$$

$$\mathbf{S}_0'.\,\mathbf{T}' = -2a\,(x_2 + 3^{\frac{1}{2}}a)$$

$$\mathbf{S}_0'.\,\mathbf{T}' = a\,(\tfrac{1}{2}.3^{\frac{1}{2}}x_1 + \tfrac{3}{2}x_2 - \tfrac{4}{3}.3^{\frac{1}{2}}a)$$

$$\mathbf{S}_0'.\,\mathbf{T}' = -a\,(\tfrac{1}{2}.3^{\frac{1}{2}}x_1 + \tfrac{3}{2}x_2 + \tfrac{8}{3}.3^{\frac{1}{2}}a)$$

$$\text{Second class}$$

$$\mathbf{S}_0'.\,\mathbf{T}'' = 2ax_1$$

$$\mathbf{S}_0'.\,\mathbf{T}'' = -2ax_1$$

$$\mathbf{S}_0'.\,\mathbf{T}'' = a\,(\tfrac{3}{2}x_1 - \tfrac{1}{2}.3^{\frac{1}{2}}x_2)$$

$$\mathbf{S}_0'.\,\mathbf{T}'' = a\,(-\tfrac{3}{2}x_1 + \tfrac{1}{2}.3^{\frac{1}{2}}x_2)$$

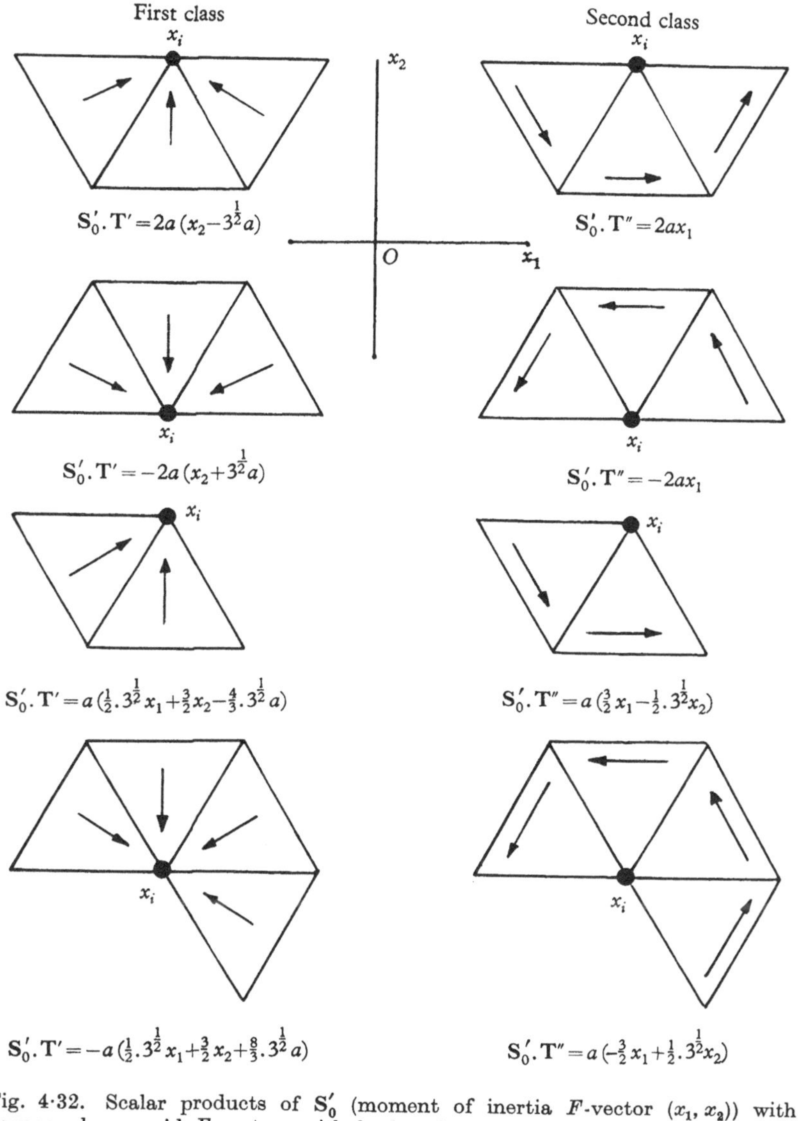

Fig. 4·32. Scalar products of $\mathbf{S}_0'$ (moment of inertia F-vector (x_1, x_2)) with hexagonal pyramid F-vectors with broken bases; $2a$ is the side of the mesh triangle; the directions of the axes are shown; the point x_i in each case is the centre of the completed hexagon. The F-vectors are normalized by (3·602).

making the heights of the corresponding pyramid functions unity. Taking
the ring F-vector $\mathbf{R} = \mathbf{T}'_1 + \mathbf{T}'_2 + \ldots + \mathbf{T}'_n$, show that

$$\Gamma \geqslant (\mathbf{S}'_0 \cdot \mathbf{R})^2 / \mathbf{R}^2,$$

where $\mathbf{S}'_0$ is the moment of inertia F-vector (4·227).

First class

$\mathbf{S}'_0 \cdot \mathbf{Q}' = a(2x_2 - \tfrac{8}{3}a)$

$\mathbf{S}'_0 \cdot \mathbf{Q}' = -a(2x_2 + \tfrac{8}{3}a)$

$\mathbf{S}'_0 \cdot \mathbf{Q}' = a(x_1 + x_2 - \tfrac{4}{3}a)$

$\mathbf{S}'_0 \cdot \mathbf{Q}' = -a(x_1 + x_2 + 4a)$

Second class

$\mathbf{S}'_0 \cdot \mathbf{Q}'' = 2ax_1$

$\mathbf{S}'_0 \cdot \mathbf{Q}'' = -2ax_1$

$\mathbf{S}'_0 \cdot \mathbf{Q}'' = a(x_1 - x_2)$

$\mathbf{S}'_0 \cdot \mathbf{Q}'' = a(x_2 - x_1)$

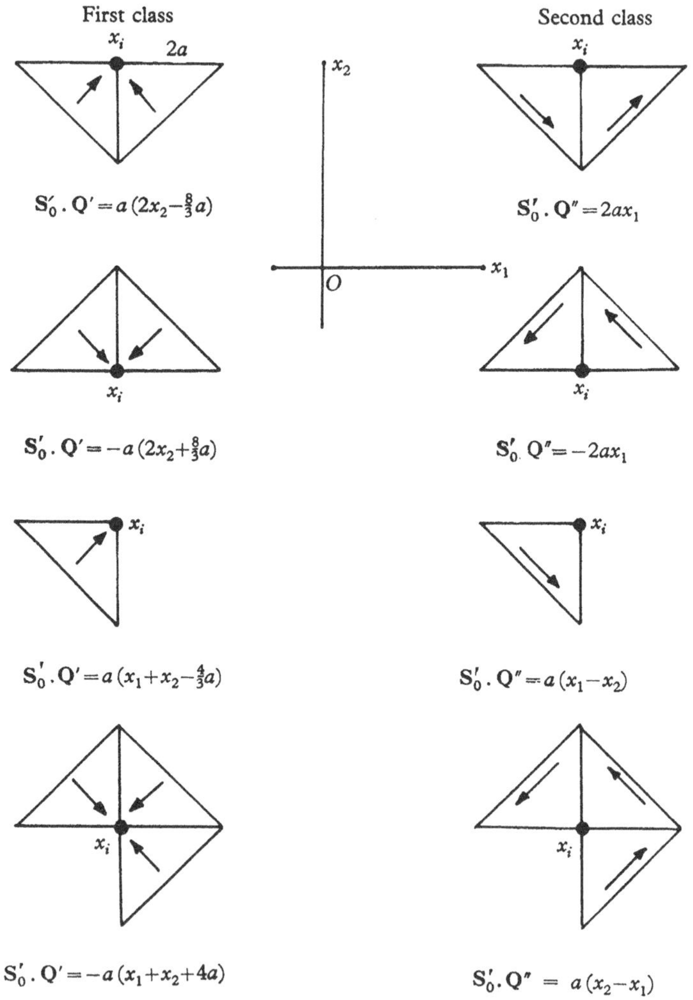

Fig. 4·33. Scalar products of $\mathbf{S}'_0$ (moment of inertia F-vector (x_1, x_2)) with square
pyramid F-vectors with broken bases (**Q**-type, large square, Fig. 3·72); the sides
of a mesh triangle are $2a$, $2a$, $2a\sqrt{2}$; the directions of the axes are shown; the point
x_i in each case is the centre of the completed square. The F-vectors are normalized
by (3·701).

2. As a particular case of Example 1, consider the hollow square of Fig. 4·22, and make a triangulation in which each vertex of the inner square is joined to two vertices of the outer square. Show that in this case

$$\mathbf{R}^2 = 4(a+b)/(a-b), \quad \mathbf{S}_0'.\mathbf{R} = -\tfrac{8}{3}(a^2+ab+b^2),$$

and hence that

$$\Gamma \geqslant \frac{16}{9}\frac{(a^3-b^3)^2}{a^2-b^2},$$

the same lower bound as in (4·282).

4·4. THE TORSION OF A REGULAR HEXAGON

The torsion problem for a beam of regular hexagonal cross-section lends itself particularly well to treatment by hexagonal pyramid F-vectors. In fact, the method is here so closely wedded to the problem that the reader may justly complain that it is not a fair test of the method! That is true; but if we have to select a problem to illustrate a method, it seems most natural to select one to which the method can be applied with the greatest ease. For an application of hexagonal pyramid F-vectors to a problem with a curved boundary, see pp. 316–29.

General plan

We shall denote the side of the hexagonal cross-section by H. It is clear that we can triangulate the cross-section with equilateral triangles of side H/n, where n is any integer. We shall, however, use only $n=1, 2, 4, 8, \ldots$, these being the values which occur naturally in the bisection process described on p. 194.

The triangulations for $n=1, 2, 4, 8$ are shown in Fig. 4·41. But in most of the arithmetical calculations it is really unnecessary to use the complete hexagon; on account of symmetry, a twelfth part of it suffices. It is by no means essential that we should start with $n=1$ and build up a closer mesh by bisections; but we shall do so because it is interesting to see how the bounds on the torsional rigidity converge.

Since the region of the problem is simply connected, the torsion problem is a Dirichlet problem, and so we can use all results of Chapter 3 if we want to. However, it is unnecessary to do this, since the essential steps have been summarized on p. 224. To simplify the computations, we introduce special notation appropriate to the problem. In what follows, n has any integer value.

Consider first hexagonal pyramid F-vectors of the first class ($\mathbf{T}_\rho'$). Since the region is simply connected, we use only those with bases contained in it. Thus there is only one of them for $n=1$ and there are seven of them for $n = 2$ (Fig. 4·41). The weights a' are to be found from (4·228), which are equivalent to (3·615) in the present case; let us insert for the scalar product in (3·615) the value (4·302), thus obtaining for each weight a' the following expression in terms of the weights of its six neighbours:

$$a'_\rho = \tfrac{1}{6}(a'_{\rho 1} + a'_{\rho 2} + \ldots + a'_{\rho 6}) + 2a^2, \tag{4·401}$$

where $2a = H/n$, the side of a mesh triangle.

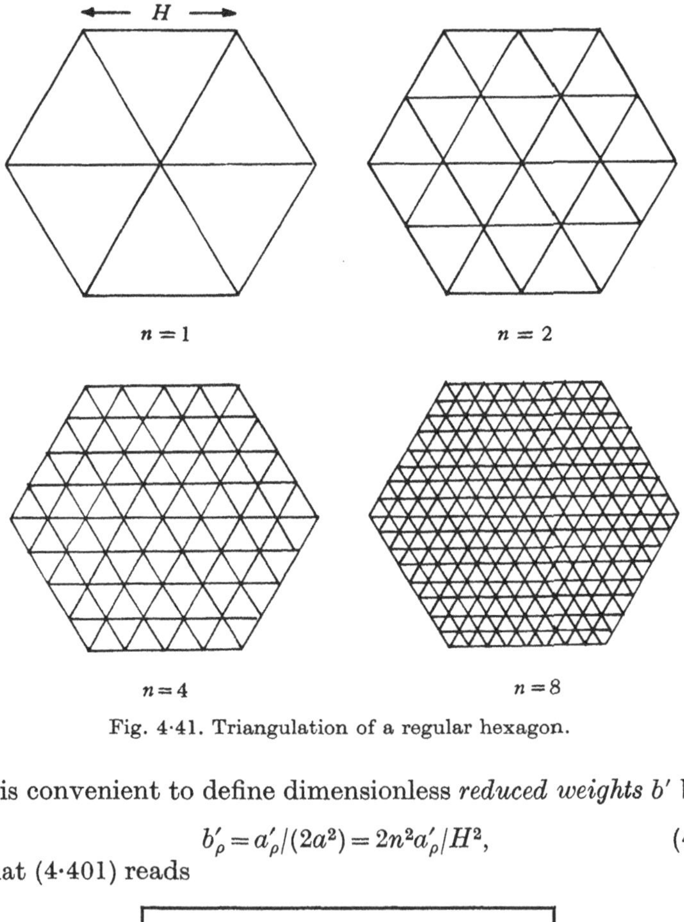

Fig. 4·41. Triangulation of a regular hexagon.

It is convenient to define dimensionless *reduced weights* b' by

$$b'_\rho = a'_\rho/(2a^2) = 2n^2 a'_\rho/H^2, \tag{4·402}$$

so that (4·401) reads

$$\boxed{b'_\rho = \tfrac{1}{6}(b'_{\rho 1} + b'_{\rho 2} + \ldots + b'_{\rho 6}) + 1.} \tag{4·403}$$

This is the basic formula for the computation of the vertex $\mathbf{V}'$; it gives the weight of each hexagonal pyramid F-vector of the first class

($\mathbf{T}'_\rho$) in terms of the reduced weights of its six neighbours, including the case where a neighbour is fictitious in the sense that its centre is on the boundary B; then its weight is to be taken equal to zero.

When the reduced weights b'_ρ have been found from (4·403), (4·229) gives the vertex $\mathbf{V}'$ in the form

$$\mathbf{V}' = \mathbf{S}'_0 + \tfrac{1}{2}(H^2/n^2) \sum_{\rho=1}^{r} b'_\rho \mathbf{T}'_\rho, \tag{4·404}$$

where $\mathbf{S}'_0$ is the moment of inertia F-vector, $\mathbf{S}'_0 \leftrightarrow (x, y)$. By (4·230) and (4·302), we have

$$\begin{aligned}\mathbf{V}'^2 &= \mathbf{S}'^2_0 + \tfrac{1}{2}(H^2/n^2) \sum_{\rho=1}^{r} b'_\rho \mathbf{S}'_0 . \mathbf{T}'_\rho \\ &= \mathbf{S}'^2_0 - \tfrac{1}{2} . 3^{\frac{1}{2}} . (H/n)^4 \sum_{\rho=1}^{r} b'_\rho .\end{aligned} \tag{4·405}$$

Then (4·231) gives as lower bound for the torsional rigidity

$$\boxed{\mathbf{S}'^2_0 - \mathbf{V}'^2 = \tfrac{1}{2} . 3^{\frac{1}{2}}(H/n)^4 \sum_{\rho=1}^{r} b'_\rho \leqslant \Gamma.} \tag{4·406}$$

$$(\tfrac{1}{2} . 3^{\frac{1}{2}} = 0·866025).$$

Next we consider hexagonal pyramid F-vectors of the second class ($\mathbf{T}''_\sigma$). We have to include not only hexagonal bases lying inside the cross-section but also those centred on the boundary. Symmetry tells us, as on pp. 198–9, that we are to give zero weight to those with centres on lines of symmetry and equal and opposite weights to those with centres symmetrically placed. Thus we need use only one-twelfth of the whole cross-section, as shown in Fig. 4·42, where one of the bases centred on the half-edge PQ is sketched.

Turning to (3·617) and noting (4·303), we see that if the centre of $\mathbf{T}''_\sigma$ lies *inside* the cross-section, then

$$a''_\sigma = \tfrac{1}{6}(a''_{\sigma 1} + a''_{\sigma 2} + \ldots + a''_{\sigma 6}); \tag{4·407}$$

the weight is the arithmetic mean of the weights of the six neighbours, zero weights being assigned on the lines of symmetry.

If $\mathbf{T}''_\sigma$ is centred on the edge, as in Fig. 4·42, we use (3·618), inserting the following values:

$$\left.\begin{aligned} &\mathbf{T}''^2_\sigma = 3^{\frac{1}{2}}, \text{ by (3·604), half the base being included;} \\ &\mathbf{T}''_\sigma . \mathbf{T}''_{\sigma 1} = -3^{-\frac{1}{2}} \text{ if } \mathbf{T}''_{\sigma 1} \text{ is centred inside;} \\ &\mathbf{T}''_\sigma . \mathbf{T}''_{\sigma 3} = -\tfrac{1}{2} . 3^{-\frac{1}{2}} \text{ if } \mathbf{T}''_{\sigma 3} \text{ is also centred on the edge} \\ &\qquad \text{(halving the above value).} \end{aligned}\right\} \tag{4·408}$$

We note that, of the six possible neighbours, two lie inside, two lie on the edge, and two are lacking. As for the last term in (3·618), the top right-hand diagram in Fig. 4·32 gives $\mathbf{S}_0'.\mathbf{T}_\sigma'' = 2ax_\sigma$, where x_σ is

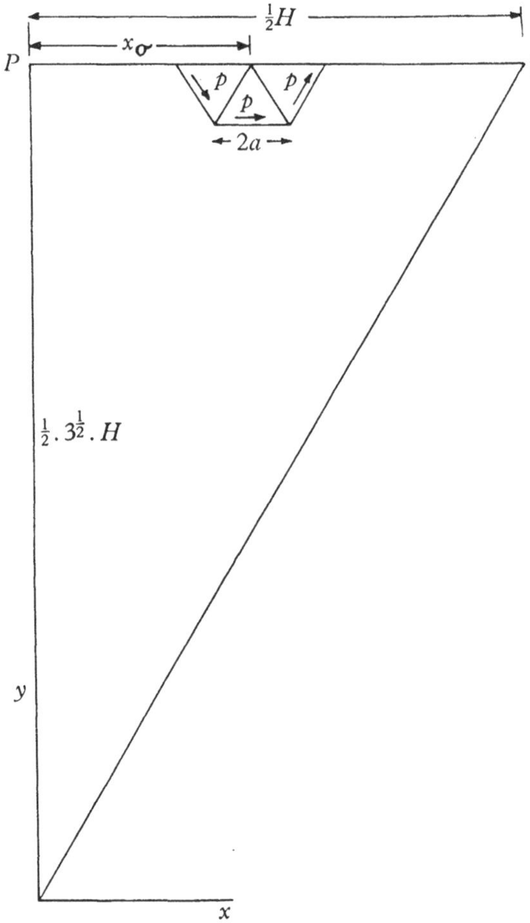

Fig. 4·42. Torsion of regular hexagon. Pyramid F-vector of the second-class $\mathbf{T}_\sigma''$ with centre at x_σ on the half-edge PQ. One-twelfth of complete hexagonal cross-section is shown.

the x-coordinate of the centre, and so (3·618) gives for a weight on PQ

$$3^{\frac{1}{2}}a_\sigma'' = 3^{-\frac{1}{2}}(a_{\sigma 1}'' + a_{\sigma 2}'')_i + \tfrac{1}{2}.3^{-\frac{1}{2}}(a_{\sigma 3}'' + a_{\sigma 4}'')_e + 2ax_\sigma, \qquad (4·409)$$

where the subscript i indicates neighbours centred inside and the subscript e neighbours centred on the edge.

It is convenient to define

$$m_\sigma = x_\sigma/(\tfrac{1}{2}H), \tag{4·410}$$

the fractional distance of the centre of $\mathbf{T}''_\sigma$ from the middle point P of the edge. We define dimensionless *reduced weights* by a formula analogous to (4·402) but different:

$$b''_\sigma = 3^{\frac{1}{2}} a''_\sigma/(aH) = 3^{\frac{1}{2}} . 2na''_\sigma/H^2. \tag{4·411}$$

Then division of (4·409) by aH gives

$$b''_\sigma = \tfrac{1}{3}(b''_{\sigma 1} + b''_{\sigma 2})_i + \tfrac{1}{6}(b''_{\sigma 3} + b''_{\sigma 4})_e + m_\sigma. \tag{4·412}$$

This formula is easy to remember if we regard the first term on the right as due to *four* neighbours, the two actual neighbours inside and two fictitious images of the *same* weights (no reversal of sign) mirrored in the half-edge PQ.

It is most convenient to apply the reduction (4·411) in all cases, and so include (4·407) and (4·412) in a single compendious formula:

$$\boxed{b''_\sigma = \tfrac{1}{6}(b''_{\sigma 1} + b''_{\sigma 2} + \dots + b''_{\sigma 6}) + m_\sigma,} \tag{4·413}$$

where $m_\sigma = 0$ if $\mathbf{T}''_\sigma$ is centred inside the hexagon and is the fractional distance (4·410) when it is centred on the edge PQ; $b''_{\sigma 1}, b''_{\sigma 2}, \dots, b''_{\sigma 6}$ are the reduced weights of the six neighbours when $\mathbf{T}''_\sigma$ is centred inside; when it is centred on the edge the four real neighbours are supplemented by two fictitious ones, obtained by reflexion in the edge.

When the reduced weights have been found from (4·413), the vertex $\mathbf{V}''$ is given by (4·229) as

$$\mathbf{V}'' = \sum_{\sigma=1}^{s} a''_\sigma \mathbf{T}''_\sigma = \tfrac{1}{2} . 3^{-\frac{1}{2}}(H^2/n) \sum_{\sigma=1}^{s} b''_\sigma \mathbf{T}''_\sigma, \tag{4·414}$$

wherein equal and opposite weights are to be attached to symmetrically situated centres. Also, by (4·230), noting that $\mathbf{S}'_0 . \mathbf{T}''_\sigma = 0$ by (4·303) if the centre of $\mathbf{T}''_\sigma$ is inside, we have

$$\mathbf{V}''^2 = \sum_{\sigma=1}^{s} a''_\sigma \mathbf{S}'_0 . \mathbf{T}''_\sigma = 12(2a) \sum_{PQ} a''_\sigma x_\sigma, \tag{4·415}$$

the last summation extending only over the half-edge PQ of Fig. 4·42. (We have used a scalar product from Fig. 4·32.) This reduces to

$$\mathbf{V}''^2 = 3^{\frac{1}{2}}(H^4/n^2) \sum_{PQ} m_\sigma b''_\sigma; \tag{4·416}$$

the summation contains $\frac{1}{2}n-1$ terms if n is even, in view of the zero weights at P and Q resulting from symmetry.

Accordingly (4·231) gives as an upper bound for the torsional rigidity

$$\boxed{\Gamma \leqslant \mathbf{S}_0'^2 - \mathbf{V}''^2 = 3^{\frac{1}{2}}H^4\left(\tfrac{5}{8} - n^{-2}\sum_{PQ} m_\sigma b_\sigma''\right),} \tag{4·417}$$

$$3^{\frac{1}{2}} = 1{\cdot}732051, \quad \tfrac{5}{8} = 0{\cdot}625,$$

since the moment of inertia of the hexagon is

$$\mathbf{S}_0'^2 = I = \tfrac{5}{8}3^{\frac{1}{2}}H^4 = 1{\cdot}082532H^4; \tag{4·418}$$

we recall that m_σ is given by (4·410).

The prominence given to the bounds on Γ might lead the reader to suppose that our aim is merely to bound the torsional rigidity of a regular hexagon. That is not the case. Although the torsional rigidity is of interest in itself, we seek close bounds for it largely because the difference between these bounds $(\mathbf{V}'^2 - \mathbf{V}''^2)$ is equal to the square of the diameter of the hypercircle, as in (3·335). If this difference is small, then $\mathbf{V}'$ or $\mathbf{V}''$ or $\frac{1}{2}(\mathbf{V}'+\mathbf{V}'')$ (the centre of the hypercircle) is a good mean-square approximation to the gradient of the function ψ, the determination of which constitutes the torsion problem; we have also, as in (4·237), good mean-square approximations to the stress distribution. Further, there is the possibility of getting pointwise bounds as in § 3·4, or mean bounds over a mesh triangle as on pp. 185–8, although the computations involved in these methods are rather heavy. The point is that all these approximations depend for their goodness on drawing the bounds of Γ close together, and that is why (apart from intrinsic interest) we pay so much attention to these bounds.

Use of approximate solutions for the weights of the hexagonal pyramid F-vectors

The approximations $n=1$ and $n=2$ are very simple and crude, as will be seen below. For larger values of n, direct solution of (4·403) and (4·413) is no longer feasible, and we have to solve them by arithmetic approximation. So we put in tentative values for the reduced weights (given by an earlier approximation and interpolation) and improve these values by replacing each weight by a weight which satisfies the equation relative to the neighbouring weights. By this process we converge on the solutions, but it is an infinite process, and the best we can do in practice is to find a solution correct to so many significant figures. It is, in fact, impossible to obtain *exact* solutions of (4·403) and (4·413). Yet it has been assumed in the preceding formulae that the weights b', b'' are exact solutions.

Under these circumstances we appeal to (2·797) which gives us reliable bounds on the torsional rigidity no matter what values the weights have. These bounds are

$$-\sum_{\rho=1}^{r} a'_\rho \mathbf{S}'_0 . \mathbf{T}'_\rho - \sum_{\rho=1}^{r} a'_\rho e'_\rho \leqslant \Gamma \leqslant \mathbf{S}'^2_0 - \sum_{\sigma=1}^{s} a''_\sigma \mathbf{S}'_0 . \mathbf{T}''_\sigma + \sum_{\sigma=1}^{s} a''_\sigma e''_\sigma, \quad (4\cdot419)$$

where e'_ρ, e''_σ are the 'errors' as given by (2·792). To use this formula in our present problem, we have to replace the absolute weights a', a'' by the reduced weights b', b'', and that we shall now do, the key being that the right-hand sides of (2·792) are precisely the quantities which we equated to zero in order to get (4·403) and (4·413).

The first of (2·792) may be written, in the notation of (3·615),

$$e'_\rho = 2 . 3^{\frac{1}{2}}[a'_\rho - \tfrac{1}{6}(a'_{\rho 1} + \ldots + a'_{\rho 6})] + \mathbf{S}'_0 . \mathbf{T}'_\rho, \quad (4\cdot420)$$

in which $\mathbf{S}'_0 . \mathbf{T}'_\rho = -4a^2 . 3^{\frac{1}{2}}$ as in (4·302). On the analogy of (4·402), we define the dimensionless *reduced errors*

$$f'_\rho = e'_\rho/(4a^2 . 3^{\frac{1}{2}}), \quad (4\cdot421)$$

and divide (4·420) by $4a^2 . 3^{\frac{1}{2}}$, obtaining

$$f'_\rho = b'_\rho - \tfrac{1}{6}(b'_{\rho 1} + \ldots + b'_{\rho 6}) - 1. \quad (4\cdot422)$$

These quantities can of course be computed for any reduced weights b', and they will be small if the b' are approximate solutions of (4·403). If we now divide the first inequality in (4·419) by $8a^4 . 3^{\frac{1}{2}}$, we get

$$\sum_{\rho=1}^{r} b'_\rho - \sum_{\rho=1}^{r} b'_\rho f'_\rho \leqslant \Gamma/(8a^4 . 3^{\frac{1}{2}}) = 2 . 3^{-\frac{1}{2}}\Gamma(n/H)^4, \quad (4\cdot423)$$

so that we obtain (for any reduced weights b') the following lower bound for Γ:

$$\Gamma/H^4 \geqslant \tfrac{1}{2} . 3^{\frac{1}{2}} . n^{-4}\left(\sum_{\rho=1}^{r} b'_\rho - \sum_{\rho=1}^{r} b'_\rho f'_\rho\right). \quad (4\cdot424)$$

Similarly the second line of (2·792) may be written

$$e''_\sigma = 2 . 3^{\frac{1}{2}}\lambda_\sigma[a''_\sigma - \tfrac{1}{6}(a''_{\sigma 1} + \ldots + a''_{\sigma 6})] - \mathbf{S}'_0 . \mathbf{T}''_\sigma, \quad (4\cdot425)$$

where $\lambda_\sigma = 1$ if the centre is inside and $\lambda_\sigma = \frac{1}{2}$ if it is on the boundary of the cross-section; the ( ) contain the weights of fictitious images if the centre is on the boundary. On the analogy of (4·411) and (4·421), we define a second set of dimensionless reduced errors by

$$f''_\sigma = e''_\sigma/(aH), \tag{4·426}$$

so that division of (4·425) by aH gives

$$f''_\sigma = 2\lambda_\sigma[b''_\sigma - \tfrac{1}{6}(b''_{\sigma 1} + \dots + b''_{\sigma 6})] - \mathbf{S}'_0 . \mathbf{T}''_\sigma/(aH). \tag{4·427}$$

Now $\mathbf{S}'_0 . \mathbf{T}''_\sigma = 0$ if the centre is inside and $\mathbf{S}'_0 . \mathbf{T}''_\sigma = 2ax_\sigma$ (as in Fig. 4·32) if the centre is on the half-edge PQ of Fig. 4·42. Hence the following formula covers the cases where the centre is either inside the cross-section or on the half-edge PQ:

$$f''_\sigma = 2\lambda_\sigma[b''_\sigma - \tfrac{1}{6}(b''_{\sigma 1} + \dots + b''_{\sigma 6})] - m_\sigma, \tag{4·428}$$

where

$\lambda_\sigma = 1, \quad m_\sigma = 0$ if the centre is inside,

$\lambda_\sigma = \tfrac{1}{2}, \quad m_\sigma = x_\sigma/(\tfrac{1}{2}H)$, as in (4·410), if the centre is on PQ,

and in the ( ) fictitious image weights are to be included if the centre is on PQ.

If we now multiply the second inequality of (4·419) by $3^{\frac{1}{2}}/(aH)^2$, we get

$$3^{\frac{1}{2}}\Gamma/(aH)^2 \leqslant \mathbf{S}'^2_0 3^{\frac{1}{2}}/(aH)^2 - 12 \sum_{PQ} b''_\sigma m_\sigma + \sum_{\sigma=1}^{s} b''_\sigma f''_\sigma. \tag{4·429}$$

When we substitute for $\mathbf{S}'^2_0$ from (4·418) and use $2a = H/n$, we get an upper bound for Γ which is incorporated in the following statement: *The torsional rigidity* Γ *of a regular hexagon of side* H *is bounded by*

$$\begin{aligned} \tfrac{1}{2} . 3^{\frac{1}{2}} n^{-4}\left(\sum_{\rho=1}^{r} b'_\rho - \sum_{\rho=1}^{r} b'_\rho f'_\rho\right) &\leqslant \Gamma/H^4 \\ &\leqslant 3^{\frac{1}{2}}\left(\tfrac{5}{8} - n^{-2}\sum_{PQ} b''_\sigma m_\sigma + \tfrac{1}{12} n^{-2} \sum_{\sigma=1}^{s} b''_\sigma f''_\sigma\right); \end{aligned} \tag{4·430}$$

here b'_ρ, b''_σ are any reduced weights consistent with symmetry, n is the number of parts into which the side H is divided in the triangulation, m_σ is as in (4·410), and f'_ρ, f''_σ are the dimensionless reduced errors of (4·422) and (4·428), where we are to remember the interpretation of (4·428) in the lines following it.

If we put $f'_\rho = 0, f''_\sigma = 0$ in (4·430) we get the bounds (4·406), (4·417), as of course we should.

The multiplication process for quick convergence

In using the method of the hypercircle for the torsion of a regular hexagon, the only step presenting any difficulty is the solution of the equations (4·403) for the reduced weights b'_ρ and of the equations (4·413) for the reduced weights b''_σ. This difficulty does not arise when the triangulation mesh is coarse, for then the number of equations is small and we can solve them directly and exactly. But a coarse mesh gives only a crude approximation, and we must consider the case where the mesh is fine and the number of equations so large that a direct solution of them by elementary methods or by matrices is not feasible.

Then we must use a method of successive approximations—a 'relaxation method', in fact. Suppose we have found values of b'_ρ and b''_σ for the approximation $\frac{1}{2}n$. We then bisect the mesh (Fig. 3·64), obtaining roughly four times as many junction-points as before. To start the approximation n, we first insert at the appropriate places the values of b'_ρ and b''_σ from the approximation $\frac{1}{2}n$, *multiplied by* 4 *in the case of* b'_ρ *in view of the factor* n^2 *in* (4·402) *and by* 2 *in the case of* b''_σ *in view of the factor* n *in* (4·411). At the new junction-points generated by the bisection process we insert values obtained by linear interpolation. (The arithmetic is best displayed in a set of boxes, as will be shown later.)

We now have tentative values for b'_ρ and b''_σ in the approximation n. These values do not, of course, satisfy the equations (4·403) and (4·413) respectively. So we set to work to improve the values by means of these equations, replacing each b'_ρ by the arithmetic mean of its six neighbours, increased by unity, and each b''_σ by the arithmetic mean of its six neighbours, with an adjustment at the boundary as explained in connexion with (4·413). This is done systematically, going over the diagram again and again according to some plan, such as from left to right and downwards. This we call the *circling process*. We are in fact solving by *iterations*.

If the values obtained by the circling process settle down and change no more, to the number of significant figures we have decided to use, then we have the required solution to that number of significant figures, and we can complete the arithmetic for the approximation n, investigating the approximate stress and warping and the upper and lower bounds for the torsional rigidity. If we use a small number of significant figures (say two figures), the convergence of the circling process is rapid, but the accumulation of rounding-off errors from many junction points of the mesh destroys accuracy; on the other hand, if we use more significant figures (say four figures), the convergence of the circling process may become very slow. Then we need a device for accelerating convergence, and this will now be described; it is of course of wider applicability in principle than to the particular problem we are considering.

In all the approximations used in the torsion problem for the hexagon, the circling process proved successful in the case of b''_σ; the convergence was reasonably rapid and no special device had to be used to accelerate it. But in the case of b'_ρ the convergence proved so slow as to reduce the computer to despair. This slow convergence was due to the fact that the solution was being approached from one side, all the errors f'_ρ having the same sign. This uniformity of sign was broken by multiplying all the weights b'_ρ by a suitable factor k, determined as follows.

By (4·422) the reduced error f'_ρ is given by

$$6f'_\rho = 6b'_\rho - \Sigma b'_{\rho 1} - 6, \tag{4·431}$$

where the summation is over the six neighbours of the weight b'_ρ. Let us add these equations for all the points inside the boundary B (not merely those in the twelfth part shown in Fig. 4·42). Let N be the number of these points; it is easy to see that in approximation n

$$N = 1 + 6 + 12 + \dots + 6(n-1) = 3n^2 - 3n + 1. \tag{4·432}$$

Now, in adding (4·431) a lot of cancellations occur; we bear in mind that a point has six real neighbours if it is not adjacent to B, and four or three if it is adjacent to B (four if it is not adjacent to a corner of the hexagon and three if it is); hence addition of (4·431) gives

$$6\Sigma f'_\rho = 2\Sigma'_1 b'_\rho + 3\Sigma'_2 b'_\rho - 6N, \tag{4·433}$$

where Σ is a summation over the whole cross-section, Σ'_1 a summation over all points adjacent to B but not to a corner, and Σ'_2 a summation over the six points adjacent to corners. We now make use of symmetry and write (4·433) in the form

$$6\Sigma f'_\rho = 24\Sigma_1 b'_\rho + 18b'(C) - 6N, \tag{4·434}$$

or

$$\Sigma f'_\rho = 4\Sigma_1 b' + 3b'(C) - N, \tag{4·435}$$

where Σ_1 is a summation over the $(\frac{1}{2}n - 1)$ points adjacent to B in the twelfth part of the hexagon shown in Fig. 4·43 and $b'(C)$ the reduced weight at the point C adjacent to the corner Q; the point C is omitted from the summation Σ_1.

Now if we multiply all the weights by a factor k, obtaining new weights $\bar{b}'_\rho = kb'_\rho$, we get new errors $\bar{f}'_\rho$ and their sum over the whole cross-section is, as above,

$$\Sigma \bar{f}'_\rho = 4\Sigma_1 \bar{b}'_\rho + 3\bar{b}(C) - N$$
$$= k[4\Sigma_1 b'_\rho + 3b'(C)] - N. \quad (4\cdot436)$$

The plan is to make this sum zero by choosing

$$k = \frac{N}{4\Sigma_1 b'_\rho + 3b'(C)}. \quad (4\cdot437)$$

Multiplication of all weights by this factor we call the *multiplication process*. Reasonably rapid convergence of the solutions of (4·403) is achieved by alternating the circling process and the multiplication process.

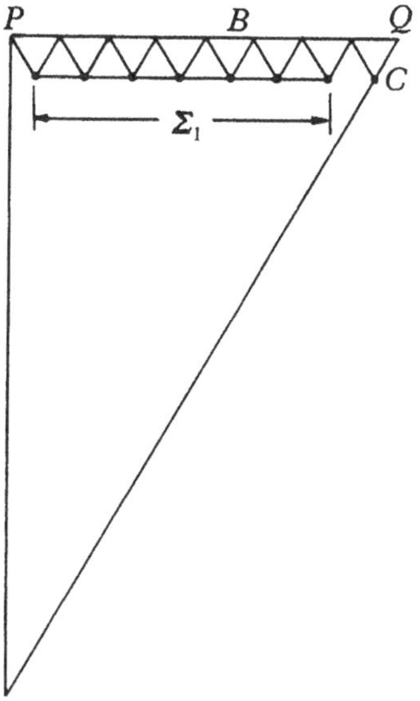

Fig. 4·43. Torsion of a regular hexagon: points involved in the multiplication process. The range for the summation Σ_1 of (4·437) and the point C, shown for the approximation $n = 16$.

It is not hard to see what the multiplication process amounts to. In finding b'_ρ for a fine mesh, we are essentially carrying out a finite difference process for the solution of a boundary value problem

$$\Delta\Psi = \text{const.}, \quad (\Psi)_B = 0.$$

Now, identically,

$$\int (\partial\Psi/\partial n)\, dB = \int \Delta\Psi\, dV,$$

and the right-hand side is known in terms of the constant assigned in the partial differential equation. Thus, out of the totality of functions Ψ we might at once restrict ourselves to those for which $\int (\partial\Psi/\partial n)\, dB$ has the required value, and this condition can always be satisfied if we multiply an arbitrary function by a suitable factor k. Our multiplication process supplies that factor k.

This indicates that the multiplication process is a general one, not specifically connected with the problem of the torsion of a regular hexagon. The idea is to reduce to zero the sum of the errors, suitably defined, and this applies to errors in both types of weights, b'_ρ and b''_σ. In the case of the regular hexagon the latter sum is zero by virtue of the symmetry of the problem, and that is why we have not defined a second factor to make $\Sigma f''_\sigma$ zero. If required, we could without difficulty get a multiplication factor to make $\Sigma f''_\sigma$ zero for summation over the twelfth part shown in Fig. 4·43, but it is more complicated and less necessary than the factor for b'_ρ.

The approximation $n = 1$

This is extremely simple and crude. We have just one pyramid F-vector of the first class ($\mathbf{T}'_1$), based on the whole cross-section (Fig. 4·44). There are none of the second class, since the weights of the seven possible ones vanish by symmetry.

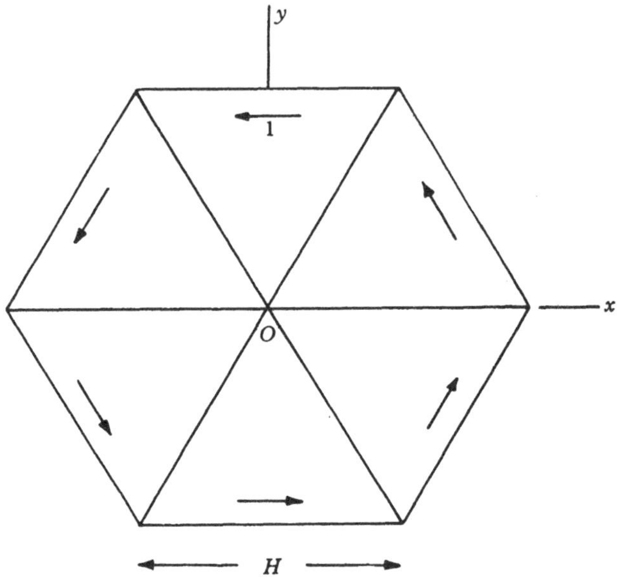

Fig. 4·44. Torsion of a regular hexagon: approximation $n=1$. The arrows represent the approximation to stress on a convenient scale; to get magnitude of stress, multiply by $\mu\alpha H . 3^{-\frac{1}{2}}$, where μ is rigidity and α angle of twist per unit length. Bounds on torsional rigidity: $0{\cdot}8660 < \Gamma/H^4 < 1{\cdot}0825$.

By (4·403) we have $b'_1 = 1$ and so by (4·404) the vertex $\mathbf{V}'$ is

$$\mathbf{V}' = \mathbf{S}'_0 + \tfrac{1}{2}H^2\mathbf{T}'_1. \qquad (4{\cdot}438)$$

Thus $\mathbf{V}' - \mathbf{S}'_0$ corresponds to a P-vector field in the upper triangle of Fig. 4·44 with components $(0, -H . 3^{-\frac{1}{2}})$, and symmetric fields in the other five triangles. Hence by (4·237) we have, as a crude approximation to the stress, a vector field obtained by rotating the above field *clockwise* through a right angle and multiplying it by the factor $\mu\alpha$; this is shown by the arrows in Fig. 4·44.

By (4·406) and (4·417) ($b''_\sigma = 0$ in the latter) we have the following bounds for the torsional rigidity:

$$\tfrac{1}{2} . 3^{\frac{1}{2}} = 0{\cdot}8660 < \Gamma/H^4 < 1{\cdot}0825 = \tfrac{5}{8} . 3^{\frac{1}{2}}. \qquad (4{\cdot}439)$$

The approximation $n=2$

The triangulation is shown in Fig. 4·45. Now we have seven pyramid F-vectors of the first class, with centres at the points marked $a\,1, b\,1, b\,2, \ldots, b\,6$. Again symmetry throws out those of the second class.

By (4·403) we have, in an obvious notation,

$$\left.\begin{aligned} b'(a\,1) &= b'(b\,1)+1, \\ b'(b\,1) &= \tfrac{1}{6}b'(a\,1)+\tfrac{1}{3}b'(b\,1)+1, \end{aligned}\right\} \qquad (4\cdot440)$$

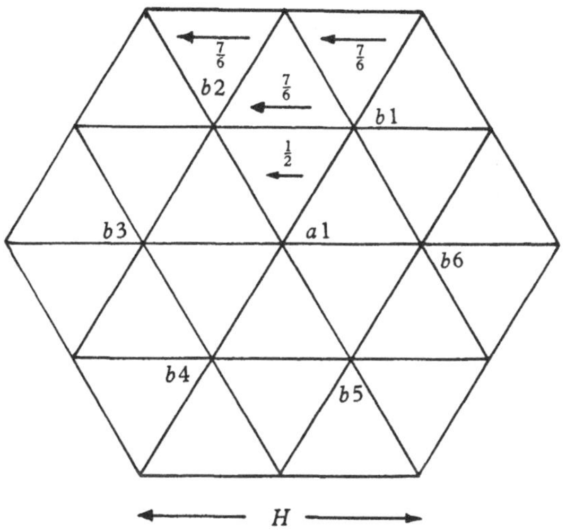

Fig. 4·45. Torsion of a regular hexagon: approximation $n=2$. The arrows represent approximation to stress (multiply numbers by $\mu\alpha H\,.\,3^{-\frac{1}{2}}$). Bounds on torsional rigidity: $0{\cdot}9382 < \Gamma/H^4 < 1{\cdot}0825$.

use being made of symmetry, and so the exact values of the reduced weights are

$$b'(a\,1)=\tfrac{10}{3}, \quad b'(b\,1)=\tfrac{7}{3}, \qquad (4\cdot441)$$

the remaining five weights being also $\frac{7}{3}$. Thus, by (4·404) with $n=2$, the vertex $\mathbf{V}'$ is

$$\mathbf{V}'=\mathbf{S}_0'+(\tfrac{1}{24}H^2)\{10\mathbf{T}'(a\,1)+7[\mathbf{T}'(b\,1)+\ldots+\mathbf{T}'(b\,6)]\}. \qquad (4\cdot442)$$

The F-vector $\mathbf{V}'-\mathbf{S}_0'$ (as above for $n=1$) gives us by (4·237) an approximation to the stress distribution; the result is shown by the arrows in Fig. 4·45, the part not indicated being given by symmetry.

By (4·406) and (4·417) (with $b''_\sigma = 0$) we have the following bounds for the torsional rigidity of the regular hexagon:

$$\tfrac{1}{2}.3^{\frac{1}{2}}.2^{-4}[b'(a\,1)+6b'(b\,1)] = \tfrac{13}{24}.3^{\frac{1}{2}} = 0{\cdot}9382$$
$$< \Gamma/H^4 < 1{\cdot}0825 = \tfrac{5}{8}.3^{\frac{1}{2}}. \quad (4{\cdot}443)$$

The approximation $n=4$

As already remarked, we never need to use more than one-twelfth of the complete hexagonal cross-section, as drawn in Fig. 4·42. Fig. 4·46*a* shows such a twelfth with the centres corresponding to $n=4$ (i.e. the side H is quadrisected).

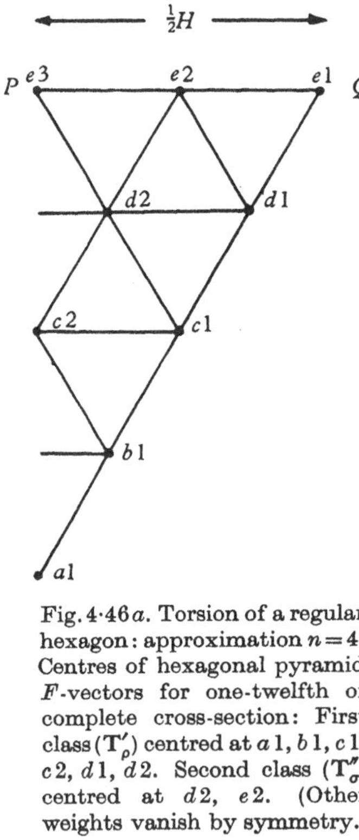

Fig. 4·46*a*. Torsion of a regular hexagon: approximation $n=4$. Centres of hexagonal pyramid F-vectors for one-twelfth of complete cross-section: First class ($\mathbf{T}'_\rho$) centred at $a\,1, b\,1, c\,1, c\,2, d\,1, d\,2$. Second class ($\mathbf{T}''_\sigma$) centred at $d\,2$, $e\,2$. (Other weights vanish by symmetry.)

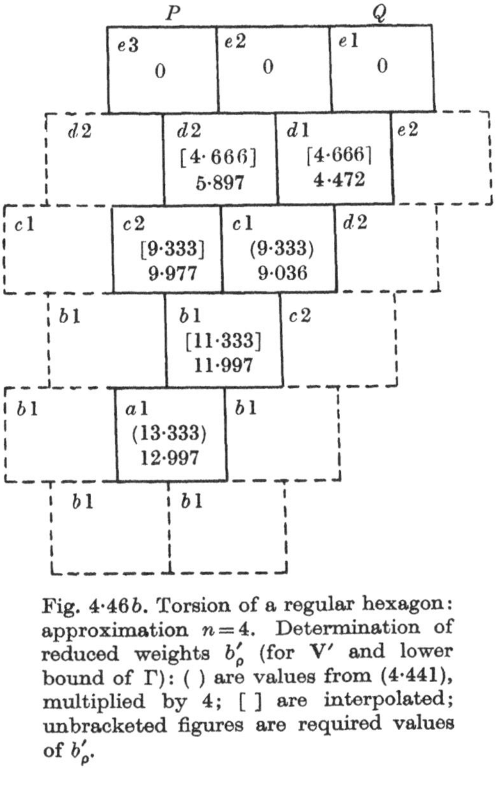

Fig. 4·46*b*. Torsion of a regular hexagon: approximation $n=4$. Determination of reduced weights b'_ρ (for $\mathbf{V}'$ and lower bound of Γ): ( ) are values from (4·441), multiplied by 4; [ ] are interpolated; unbracketed figures are required values of b'_ρ.

For the calculation of $\mathbf{V}'$ (and hence the lower bound of Γ) we need pyramid F-vectors of the first class ($\mathbf{T}'_\rho$) with centres at the points marked

$$a\,1,\quad b\,1,\quad c\,1,\quad c\,2,\quad d\,1,\quad d\,2,$$

and the corresponding symmetric positions, not shown. Six strengths are involved and hence there are six linear equations to solve for them. This is rather many for direct solution and we shall use successive approximations.

On the other hand, for $\mathbf{V}''$ we need only centres of $\mathbf{T}''_\sigma$ at the points marked $d\,2$, $e\,2$, and the corresponding symmetric positions, since all other weights vanish by symmetry; we have only two equations to solve, and this of course we shall do directly.

To carry out the computations for b'_ρ we make a set of boxes as in Fig. 4·46*b*. The boxes $a\,1$, $c\,1$ correspond respectively to $a\,1$, $b\,1$ in Fig. 4·45; in these two boxes we insert in round brackets ( ) the values (4·441), multiplied by 4 on account of the factor n^2 in (4·402), since the weights a'_ρ should not change much under bisection for they represent heights of a polyhedral graph closely connected with the solution of the torsion problem. We put a row of zeros in the top boxes and insert in those which are still empty weights given by linear interpolation; these last are enclosed in square brackets [ ] in Fig. 4·46*b*.

The boxes bounded by broken lines are inserted in Fig. 4·46*b* to remind the reader of the existence of the rest of the hexagon, symmetric correspondents being here indicated by the same reference label. It is unnecessary to write numbers in these boxes, since each at every moment contains the same number as its symmetric correspondent; but it must be remembered that they act as neighbours.

We now improve the initial bracketed values by circling with (4·403), i.e. replacing each weight by the arithmetic mean of its six neighbours, increased by unity, and we alternate this circling process with the multiplication process (4·437). After nineteen operations over the whole set of boxes, the values settle down with a steady up and down oscillation of a unit in the last place. Fig. 4·46*b* shows unbracketed the solutions of our six linear equations obtained in this way.

The vertex $\mathbf{V}'$ is then given by (4·404) with $n=4$. If the values b'_ρ in Fig. 4·46*b* were *accurate* solutions of (4·403), which of course they are not in a strict sense, insertion of them in (4·406) would give a lower bound for Γ. We note that

$$\sum_{\rho=1}^{r} b'_\rho = b'(a\,1)+6[b'(b\,1)+b'(c\,1)+b'(c\,2)+b'(d\,1)] + 12b'(d\,2) = 296{\cdot}653, \qquad (4{\cdot}444)$$

and so, since $\qquad \frac{1}{2}.3^{\frac{1}{2}}.4^{-4} = 3{\cdot}38291\times10^{-3}$,

we get from (4·406) the lower bound

$$\Gamma/H^4 > \mathbf{S}_0'^2 - \mathbf{V}'^2 = \tfrac{1}{2}.3^{\frac{1}{2}}.4^{-4}\sum_{\rho=1}^{r} b'_\rho = 1{\cdot}00355. \qquad (4{\cdot}445)$$

If we are more cautious and refuse to accept the values in Fig. 4·46*b* as accurate solutions of (4·403), we may fall back on (4·430) for a lower bound. For this we prepare Fig. 4·46*c*, which shows b'_ρ (from Fig. 4·46*b*), $6f'_\rho$ from (4·422), and the products $6b'_\rho f'_\rho$. On carrying out the calculations shown in Fig. 4·46*c*, we get

$$\Gamma/H^4 > 1\cdot00344. \qquad (4\cdot446)$$

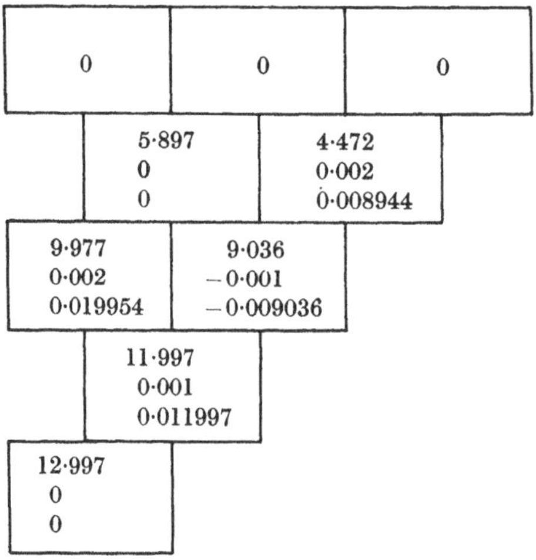

Fig. 4·46*c*. Torsion of a regular hexagon: approximation $n=4$. Values of b'_ρ, $6f'_\rho$, $6b'_\rho f'_\rho$.
Calculations:

$$\sum_{\rho=1}^{r} b'_\rho = 296\cdot653, \quad \sum_{\rho=1}^{r} 6b'_\rho f'_\rho = 0\cdot191154,$$

$$\sum_{\rho=1}^{r} b'_\rho f'_\rho = 0\cdot031859, \quad \sum_{\rho=1}^{r} b'_\rho - \sum_{\rho=1}^{r} b'_\rho f'_\rho = 296\cdot621;$$

$$\Gamma/H^4 > \tfrac{1}{2}.3^{\frac{1}{2}}.4^{-4}\left(\sum_{\rho=1}^{r} b'_\rho - \sum_{\rho=1}^{r} b'_\rho f'_\rho\right)$$

$$= 0\cdot00338291 \times 296\cdot621 = 1\cdot003442.$$

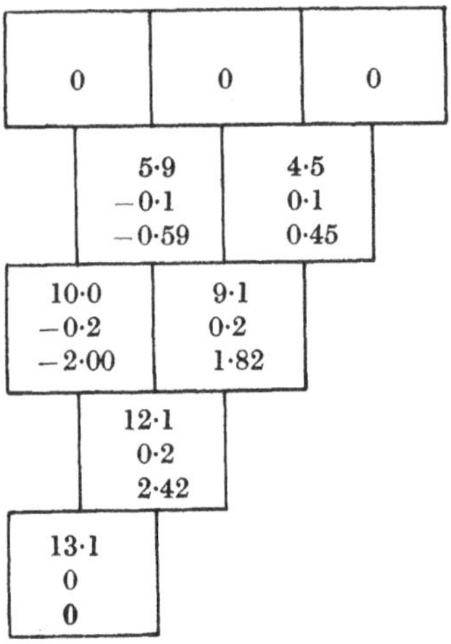

Fig. 4·46*d*. Torsion of a regular hexagon: approximation $n=4$. Values of b'_ρ to one decimal place, with values of $6f'_\rho$ and $6b'_\rho f'_\rho$.
Calculations:

$$\sum_{\rho=1}^{r} b'_\rho = 298\cdot1, \quad \sum_{\rho=1}^{r} 6b'_\rho f'_\rho = 9\cdot06,$$

$$\sum_{\rho=1}^{r} b'_\rho f'_\rho = 1\cdot51, \quad \sum_{\rho=1}^{r} b'_\rho - \sum_{\rho=1}^{r} b'_\rho f'_\rho = 296\cdot59.$$

Unreliable bound

$$\Gamma/H^4 > 0\cdot00338291 \times 298\cdot1 = 1\cdot00844.$$

Reliable bound:

$$\Gamma/H^4 > 0\cdot00338291 \times 296\cdot59 = 1\cdot00334.$$

This is a *reliable* lower bound (cf. p. 123); it does not depend on the assumption that the weights b'_ρ used in calculating it satisfy (4·403). In contrast (4·445) is *unreliable* since it does depend on this assumption.

These two bounds are so close together that it may seem pedantic to distinguish between them, but in other cases the distinction may be important. There is always some uncertainty and vagueness when we solve a set of equations to an assigned number of decimal places, particularly when the number of unknowns is large and errors of rounding-off can accumulate. This vagueness disappears when we use the reliable bounds of (4·430), for we may treat any set of b'_ρ as precise mathematical numbers (i.e. 5·9 means 59/10, and not any number between 5·85 and 5·95) and work with absolute mathematical precision except in the multiplication by the irrational factor ($3^{\frac{1}{2}}$) occurring in (4·430).

Although the multiplication process (4·437) greatly reduces the tedium of the calculations, the solution of (4·403) by successive approximations is bound to be long if we seek b'_ρ to a fairly high degree of accuracy, as we did in Fig. 4·46*c*. It is interesting to see how matters go if we decide to be satisfied with a much cruder solution, say with only one decimal place. The result is shown in Fig. 4·46*d*; stable values of b'_ρ are obtained after only five operations of circling and multiplying. For comparison, the results are collected here:

Torsion of a regular hexagon: approximation $n=4$

No. of decimal places in b'_ρ	No. of operations to secure stable values	Lower bound for Γ/H^4	
		Unreliable	Reliable
3	19	1·00355	1·00344
1	5	1·00844	1·00334

To get an upper bound on Γ, we seek $\mathbf{V}''$. We recall that there are only two non-zero weights in Fig. 4·46*a*, viz. at (d2), (e2). We are to remember the skew-symmetry of the weights of $\mathbf{T}''_\sigma$ under reflexion in lines of symmetry, so that if we passed to the left in Fig. 4·46*a* to the centre (d3) (not shown), we would have $b''(d3) = -b''(d2)$.

The number m_σ of (4·410) is, $m(e2) = \frac{1}{2}$ at (e2); hence (4·413) gives us the two equations

$$\left.\begin{aligned} b''(d2) &= -\tfrac{1}{6}b''(d2) + \tfrac{1}{6}b''(e2), \\ b''(e2) &= \tfrac{1}{3}b''(d2) + \tfrac{1}{2}, \end{aligned}\right\} \tag{4·447}$$

of which the solutions are

$$b''(d2) = \tfrac{3}{40}, \quad b''(e2) = \tfrac{21}{40}. \tag{4·448}$$

All other weights follow by symmetry. Then $\mathbf{V}''$ is given by (4·414) and hence an approximation to the stress distribution by the second line of (4·237).

Using the values (4·448), we obtain an upper bound on Γ by (4·417). The calculations are as follows:

$$\left.\begin{aligned}\sum_{PQ} m_\sigma b''_\sigma = \tfrac{1}{2}b''(e\,2) = \tfrac{21}{80}, \quad n^2 = 16,\\ n^{-2}\sum_{PQ} m_\sigma b''_\sigma = \tfrac{21}{1280},\\ \tfrac{5}{8} - n^{-2}\sum_{PQ} m_\sigma b''_\sigma = \tfrac{779}{1280}\,(=0\cdot608594);\end{aligned}\right\} \qquad (4\cdot449)$$

$$\Gamma/H^4 \leqslant (\mathbf{S}_0'^2 - \mathbf{V}''^2)/H^4 = 3^{\frac{1}{2}}\,.\,\tfrac{779}{1280}. \qquad (4\cdot450)$$

This inequality is true with the right-hand side an exact mathematical number; it is a *reliable* bound since we have used exact solutions of (4·413). Converting to decimals and combining with (4·446), we may write *as reliable bounds on the torsional rigidity of a regular hexagon of side H, obtained by the approximation* $n = 4$,

$$1\cdot0034 < \Gamma/H^4 < 1\cdot0542. \qquad (4\cdot451)$$

(The results for the various approximations are collected on p. 268.)

The approximation $n = 8$

With $n = 8$, or greater, we cannot look for exact solutions of (4·403) or (4·413); successive approximations must be used. The method is the same as that described above for $n = 4$, and it is not necessary to go into details again. Fairly accurate values of b'_ρ, b''_σ are given in Figs. 4·47*b*, *c*, the chart of centres for one-twelfth of the cross-section being shown in Fig. 4·47*a*. We are to remember, in the case of Fig. 4·47*c* for b''_σ, that the strengths are skew-symmetric with respect to reflexion in the vertical through the point P, which is the middle point of the top edge. Note also that when we bisect the mesh (as in passing from $n = 4$ to $n = 8$) and insert initial values from the preceding approximation, we are to *multiply* b' *by* 4 *in view of* (4·402) *but* b'' *by* 2 *in view of* (4·411).

The summations from these values are as follows:

$$\begin{aligned}\sum_{\rho=1}^{r} b'_\rho = b'(a\,1)&\\ &+6[b'(b\,1)+b'(c\,1)+b'(d\,1)+b'(e\,1)\\ &\quad+b'(f\,1)+b'(g\,1)+b'(h\,1)\\ &\quad+b'(c\,2)+b'(e\,3)+b'(g\,4)]\\ &+12[b'(d\,2)+b'(e\,2)+b'(f\,2)+b'(f\,3)\\ &\quad+b'(g\,2)+b'(g\,3)+b'(h\,2)+b'(h\,3)\\ &\quad+b'(h\,4)]\\ &=4856\cdot03 \text{ (exactly)},\end{aligned} \qquad (4\cdot452)$$

$$\sum_{PQ} m_\sigma b''_\sigma = \tfrac{1}{4}[3b''(i\,2) + 2b''(i\,3) + b''(i\,4)]$$
$$= 1{\cdot}52775 \text{ (exactly)}. \qquad (4{\cdot}453)$$

In the sum (4·452), the centre of the complete hexagon occurs once, the lines joining this centre to the corners and the middle points of the sides occur six times, and the remaining points occur twelve times. In (4·453) the summation is along the top edge of Fig. 4·47 *c*, as required in (4·416).

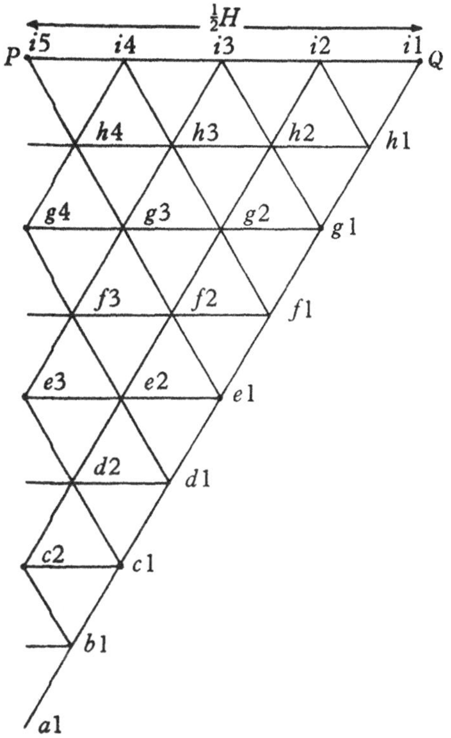

Fig. 4·47 *a*. Torsion of a regular hexagon: approximation $n = 8$. Centres of hexagonal pyramid F-vectors in one twelfth of complete cross section. First class ($\mathbf{T}'_\rho$) have zero weights on i-row. Second class ($\mathbf{T}''_\sigma$) have zero weights at centres vertically above ($a\,1$) and on the line joining it to Q.

If we regard the weights in Figs. 4·47 *b*, *c* as exact solutions, then by (4·406)

$$\Gamma/H^4 \geqslant \tfrac{1}{2}\,.\,3^{\frac{1}{2}}\,.\,8^{-4}\sum_{\rho=1}^{r} b'_\rho = 1{\cdot}02672. \qquad (4{\cdot}454)$$

and by (4·417)

$$\Gamma/H^4 \leqslant 3^{\frac{1}{2}}\left(\tfrac{5}{8} - 8^{-2}\sum_{PQ} m_\sigma b''_\sigma\right) = 1{\cdot}04119. \qquad (4{\cdot}455)$$

These are *unreliable* bounds; to get *reliable* bounds we appeal to (4·430). For this purpose we prepare Figs. 4·47*d*, *e* showing respectively (b'_ρ, $6f'_\rho$, $6b'_\rho f'_\rho$) and (b''_σ, $6f''_\sigma$, $6b''_\sigma f''_\sigma$), the reduced errors being obtained from (4·422) and (4·428) respectively. Here are the results obtained:

Torsion of a regular hexagon: approximation $n=8$.
Bounds on Γ/H^4, taking b'_ρ to 2 decimal places and b''_σ to 3 decimal places.

	Lower bound	Upper bound
Unreliable	1·02672	1·04119
Reliable	1·02641	1·04113

The reliable upper bound is less than the unreliable both in this case and for $n=16$ below.

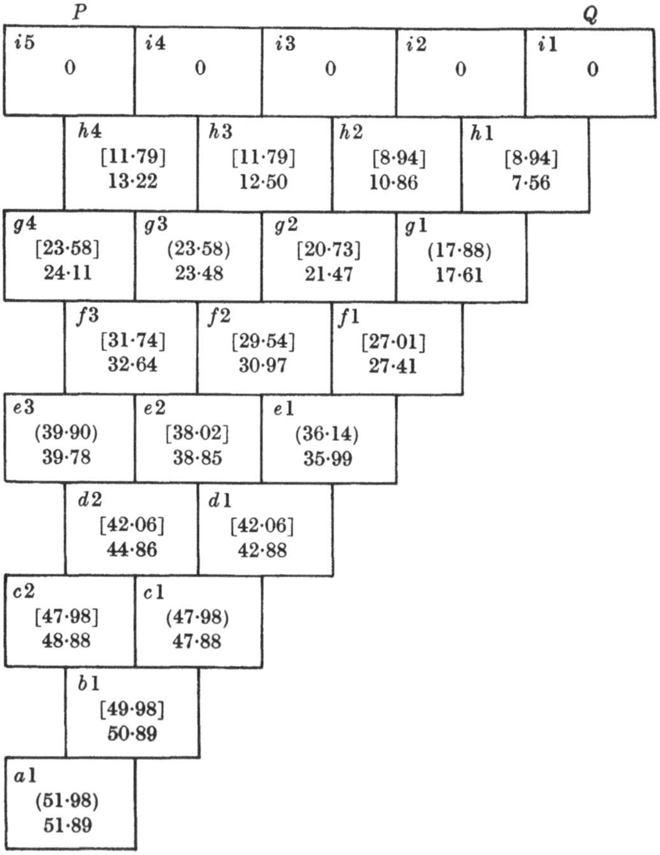

Fig. 4·47*b*. Torsion of a regular hexagon: approximation $n=8$. Block diagram for calculation of b'_ρ (and hence $\mathbf{V}'$). ( ) are values taken from Fig. 4·46*c* ($n=4$), multiplied by 4; [ ] are interpolated; unbracketed figures are values of b'_ρ for $n=8$, obtained by (4·403) and (4·437).

The approximation $n=16$

In view of previous explanations, it is sufficient to show the triangulation in Fig. 4·48*a* (p. 264) and the boxed weights in Figs. 4·48*b*, *c* (pp. 265–6). The results are as follows:

Torsion of a regular hexagon: approximation $n=16$.
Bounds on Γ/H^4, taking b'_ρ to 2 decimal places and b''_σ to 3 decimal places.

	Lower bound	Upper bound
Unreliable	1·03345	1·03703
Reliable	1·03306	1·03701

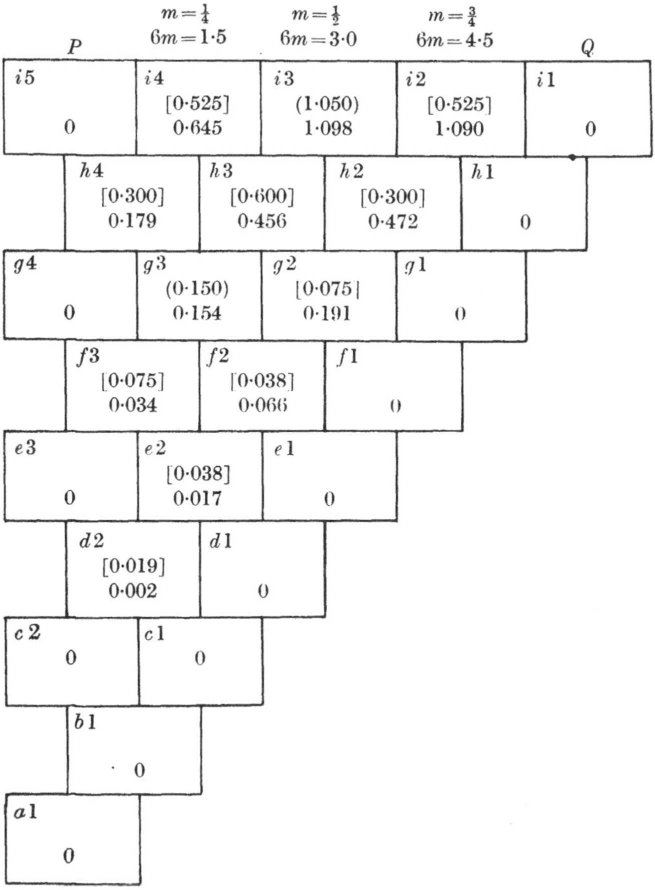

Fig. 4·47*c*. Torsion of a regular hexagon: approximation $n=8$. Block diagram for calculation of b''_σ (and hence $\mathbf{V}''$). ( ) are values taken from (4·448) ($n=4$), multiplied by 2 [cf. (4.411)]; [ ] are interpolated; unbracketed figures are values of b''_σ for $n=8$, obtained by (4·413).

Approximation for stress and warping

We have in (4·237) two approximations for stress and the lines of stress are given approximately by (4·240), or equivalently

$$\sum_{\rho=1}^{r} b'_\rho u'_\rho = \text{const.}, \tag{4·456}$$

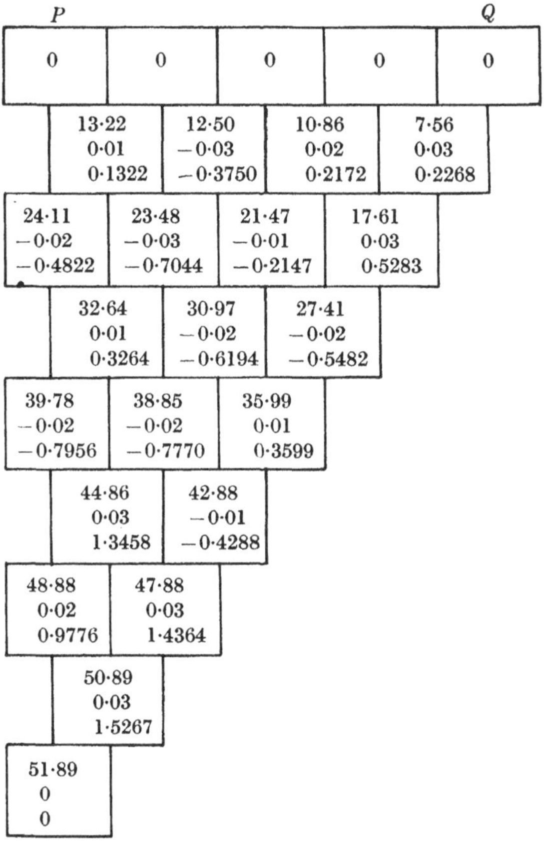

Fig. 4·47*d*. Torsion of a regular hexagon: approximation $n=8$. Calculations for fixing reliable lower bound for Γ. Blocks as in Fig. 4·47*b*; they show in order: b'_ρ from Fig. 4·47*b*; $6f'_\rho$ from (4·422); $6b'_\rho f'_\rho$.

Calculations: $\sum_{\rho=1}^{r} b' = 4856·03$, $\sum_{\rho=1}^{r} 6b'_\rho f'_\rho = 8·7786$,

$$\sum_{\rho=1}^{r} b'_\rho f'_\rho = 1·4631, \quad \sum_{\rho=1}^{r} b'_\rho - \sum_{\rho=1}^{r} b'_\rho f'_\rho = 4854·5669.$$

Reliable lower bound by (4·430):

$$\Gamma/H^4 \geqslant 0·00021143188 \times 4854·5669 = 1·02641.$$

where u'_ρ is the pyramid function of unit height standing on the base of $\mathbf{T}'_\rho$. These lines are in fact, for the approximation $n = 16$, the level lines of the polyhedral function obtained by raising the junction points of the mesh of Fig. 4·48a to the heights b'_ρ of Fig. 4·48b. Fig. 4·48d shows a set of these lines corresponding to equal changes

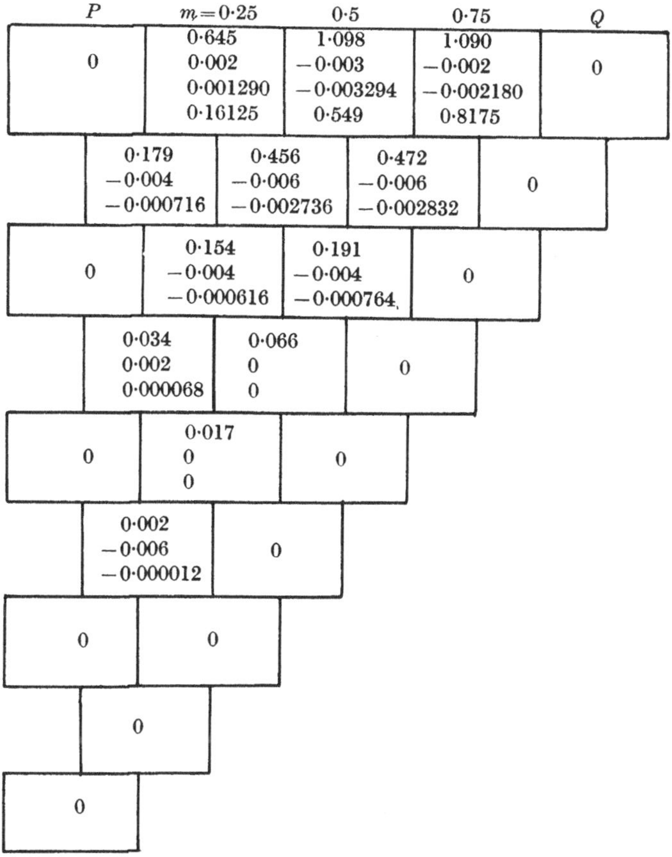
P
m=0·25
0·5
0·75
Q
0
0·645 0·002 0·001290 0·16125
1·098 −0·003 −0·003294 0·549
1·090 −0·002 −0·002180 0·8175
0
0·179 −0·004 −0·000716
0·456 −0·006 −0·002736
0·472 −0·006 −0·002832
0
0
0·154 −0·004 −0·000616
0·191 −0·004 −0·000764
0
0·034 0·002 0·000068
0·066 0 0
0
0
0·017 0 0
0
0·002 −0·006 −0·000012
0
0
0
0
0

Fig. 4·47e. Torsion of a regular hexagon: approximation $n = 8$. Calculations for fixing reliable upper bound for Γ. Blocks as in Fig. 4·47c; they show in order: b''_σ from Fig. 4·47c; $6f''_\sigma$ from (4·428); $6b''_\sigma f''_\sigma$; $m_\sigma b''_\sigma$ in top row.

Calculations: $\sum_{PQ} m_\sigma b''_\sigma = 1{\cdot}52775$, $\quad 6\sum_{\sigma=1}^{8} b''_\sigma f''_\sigma = 0{\cdot}016296 - 0{\cdot}1578$

$$= -0{\cdot}141504 \text{ (exact)};$$

$$\sum_{\sigma=1}^{8} b''_\sigma f''_\sigma = -0{\cdot}023584.$$

Reliable upper bound from (4·430):

$$\Gamma/H^4 \leqslant 3^{\frac{1}{2}}\left(\tfrac{5}{8} - \tfrac{1}{64}(1{\cdot}52775) - \tfrac{1}{768}(0{\cdot}023584)\right)$$
$$= 1{\cdot}732051\ (0{\cdot}625 - 0{\cdot}02387109 - 0{\cdot}00003070) = 1{\cdot}04113.$$

in height; the intensity of the stress is inversely proportional to the distance between the lines.

As in (4·242), the lines of constant warping are given by

$$\sum_{\sigma=1}^{8} b''_\sigma u''_\sigma = \text{const.}, \tag{4·457}$$

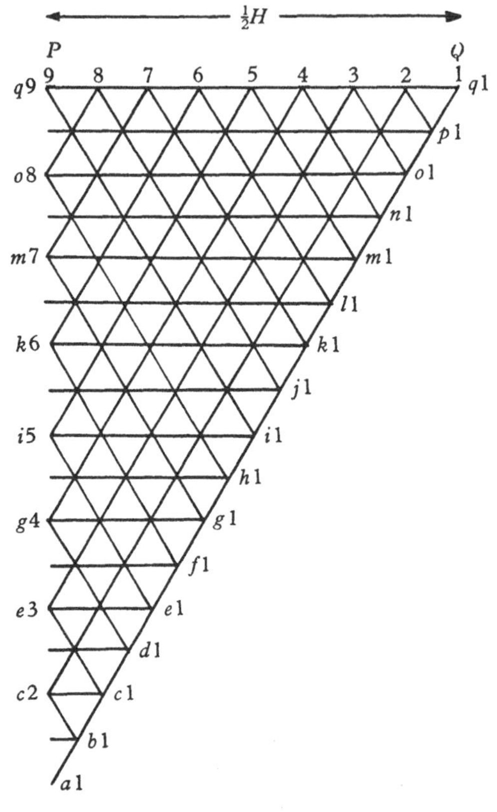

Fig. 4·48*a*. Torsion of a regular hexagon: approximation $n=16$. Centres of hexagonal pyramid F-vectors in one twelfth of complete cross section. First class ($\mathbf{T}'_\rho$) have zero weights on q-row. Second class ($\mathbf{T}''_\sigma$) have zero weights at centres vertically above (a 1) and on the line joining it to Q.

where u''_σ is the pyramid function of unit height standing on the base of $\mathbf{T}''_\sigma$. For the approximation $n=16$, these are the level lines of the polyhedral function obtained by raising the junction points of Fig. 4·48*a* to the heights b''_σ of Fig. 4·48*c*. Fig. 4·48*e* shows a set of such lines, including of course the lines of symmetry of the hexagon, for which the warping is zero; the lines represent equal increments in the warping function ϕ.

Row									
q	0	0	0	0	0	0	0	0	0
p	27·94 0·02	27·58 −0·02	26·85 0·01	25·68 0·01	23·96 −0·03	21·50 0·02	17·85 0·01	11·92 −0·03	
o	53·22 −0·02	52·88 0	51·83 0	50·00 −0·02	47·26 0·01	43·35 0·02	37·82 −0·02	29·85 0·02	
n	75·85 0·01	74·86 0·03	72·81 −0·01	69·59 0	64·97 0·02	58·57 −0·02	49·82 0		
m	96·45 −0·02	95·83 0	93·93 −0·01	90·66 0	85·84 0	79·19 0·01	70·29 0		
l	114·68 0·03	113·21 0	110·22 0·02	105·57 0	99·06 −0·02	90·40 0·01			
k	131·35 0·02	130·53 −0·03	128·07 0·03	123·86 −0·02	117·78 0·03	109·60 0			
j	145·83 −0·01	144·07 0	140·50 −0·03	135·04 0·03	127·52 −0·03				
i	159·07 0·02	158·14 −0·02	155·35 −0·01	150·64 0	143·91 −0·01				
h	170·23 0·01	168·31 0	164·45 0·01	158·59 0·02					
g	180·30 0·02	179·32 0·02	176·37 −0·01	171·43 −0·02					
f	188·34 0·03	186·35 0	182·37 0						
e	195·35 0	194·35 0	191·35 −0·02						
d	200·36 0·03	198·35 −0·03							
c	204·36 0	203·36 0·01							
b	206·36 0								
a	207·36 0								

Fig. 4·48*b*. Torsion of a regular hexagon: approximation $n=16$. Values of reduced weights b'_ρ (for $\mathbf{T}'_\rho$) obtained from (4·403), and corresponding values of $6f'_\rho$, where f'_ρ is the reduced error (4·422).

Calculations:

$$\sum_{\rho=1}^{r} b'_\rho = 78205{\cdot}44, \quad 6\sum_{\rho=1}^{r} b'_\rho f'_\rho = 174{\cdot}5976, \quad \sum_{\rho=1}^{r} b'_\rho f'_\rho = 29{\cdot}0996.$$

Unreliable lower bound for Γ from (4·406), b'_ρ being regarded as accurate solutions of (4·403):

$$\Gamma/H^4 \geqslant \tfrac{1}{2}\,.\,3^{\frac{1}{2}}\,.\,16^{-4}\sum_{\rho=1}^{r} b'_\rho = 1{\cdot}321449\times 10^{-5}\times 78205{\cdot}44$$
$$= 1{\cdot}033445.$$

Reliable lower bound for Γ from (4·430):

$$\Gamma/H^4 \geqslant \tfrac{1}{2}\,.\,3^{\frac{1}{2}}\,.\,16^{-4}\left(\sum_{\rho=1}^{r} b'_\rho - \sum_{\rho=1}^{r} b'_\rho f'_\rho\right) = 1{\cdot}321449\times 10^{-5}\times 78176{\cdot}34 = 1{\cdot}033060.$$

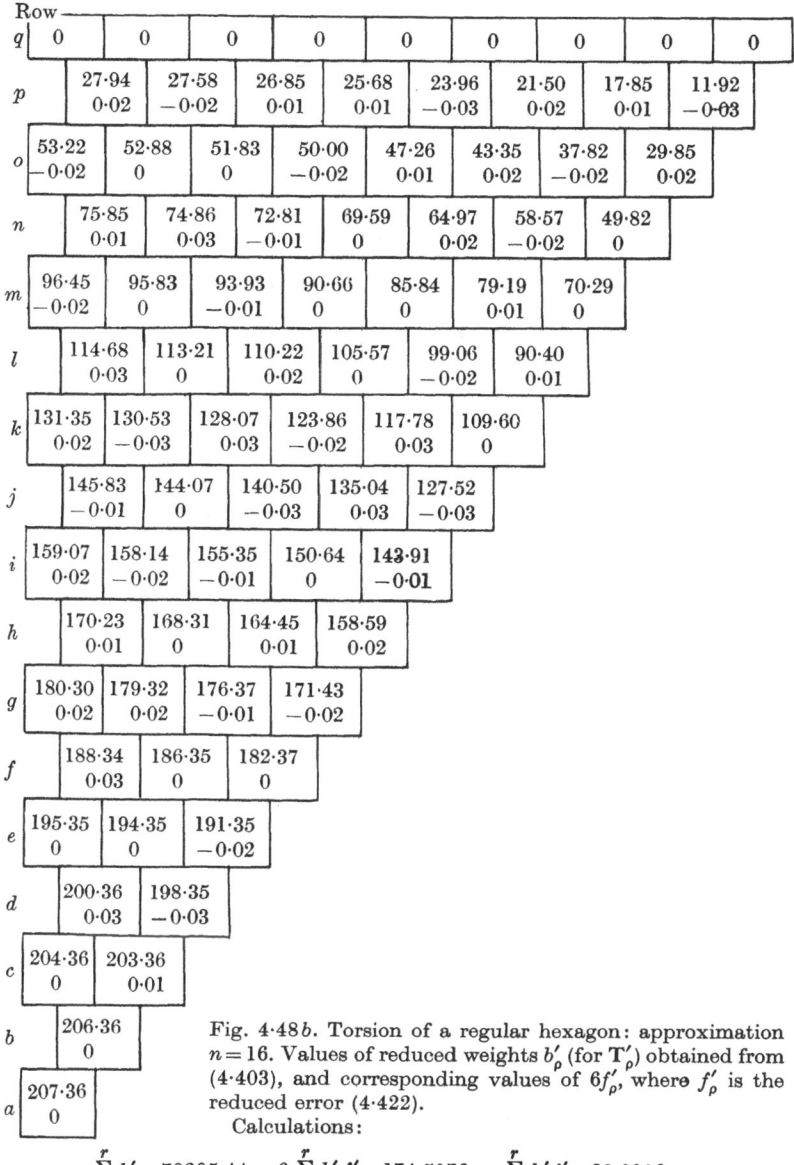

Row	P	$6m_\sigma=0{\cdot}750$	$1{\cdot}500$	$2{\cdot}250$	$3{\cdot}000$	$3{\cdot}750$	$4{\cdot}500$	$5{\cdot}250$	Q
q	0	0·670 0·0005	1·295 −0·001	1·828 0·0005	2·212 0·0005	2·377 0·0015	2·221 −0·0005	1·572 −0·0005	0
p	0·252 −0·002	0·735 −0·001	1·152 +0·001	1·453 −0·001	1·580 −0·002	1·458 +0·001	0·981 +0·001	0	
o	0	0·361 +0·003	0·681 −0·001	0·919 −0·002	1·028 −0·002	0·954 −0·002	0·634 0	0	
n	0·128 −0·002	0·367 −0·002	0·553 −0·002	0·649 0	0·615 −0·001	0·411 0	0		
m	0	0·170 −0·001	0·311 +0·001	0·393 +0·002	0·386 0	0·263 +0·003	0		
l	0·057 +0·003	0·158 0	0·224 −0·002	0·233 +0·002	0·163 −0·001	0			
k	0	0·068 −0·002	0·118 −0·002	0·133 0	0·097 −0·002	0			
j	0·021 +0·001	0·056 +0·001	0·071 0	0·055 0	0				
i	0	0·022 −0·001	0·035 +0·003	0·029 −0·001	0				
h	0·006 0	0·015 +0·002	0·014 −0·001	0					
g	0	0·005 0	0·006 0	0					
f	0·001 −0·001	0·002 −0·001	0						
e	0	0·001 +0·003	0						
d	0	0							
c	0	0							
b	0								
a	0								

Fig. 4·48*c*. Torsion of a regular hexagon: approximation $n=16$. Values of reduced weights b''_σ (for $\mathbf{T}''_\sigma$) obtained from (4·413), and corresponding values of $3f''_\sigma$, where f''_σ is the reduced error (4·428).

Calculations:

$$\sum_{PQ} m_\sigma b''_\sigma = 6{\cdot}725875, \quad 3\sum_{\sigma=1}^{8} b''_\sigma f''_\sigma = -0{\cdot}075840, \quad \sum_{\sigma=1}^{8} b''_\sigma f''_\sigma = -0{\cdot}025280.$$

Unreliable upper bound for Γ from (4·417), b''_σ being regarded as accurate solutions of (4·413):

$$\begin{aligned}\Gamma/H^4 &\leq 3^{\frac{1}{2}}\left(\tfrac{5}{8} - 16^{-2}\sum_{PQ} m_\sigma b''_\sigma\right)\\ &= 1{\cdot}732051\,(0{\cdot}625 - 6{\cdot}725875/256)\\ &= 1{\cdot}037026.\end{aligned}$$

Reliable upper bound for Γ from (4·430):

$$\begin{aligned}\Gamma/H^4 &\leq 3^{\frac{1}{2}}\left(\tfrac{5}{8} - 16^{-2}\sum_{PQ} m_\sigma b''_\sigma + 12^{-1}.\,16^{-2}\sum_{\sigma=1}^{8} b''_\sigma f''_\sigma\right)\\ &= 1{\cdot}732051\,(0{\cdot}625 - 6{\cdot}725875/256 - 0{\cdot}025280/3072)\\ &= 1{\cdot}037012.\end{aligned}$$

The reliable upper bound is actually better than the unreliable one.

Summary of results for the torsion of a regular hexagon

A suitably chosen set of weights b'_ρ gives us, as in (4·404), the vertex $\mathbf{V}'$; and a suitably chosen set of weights b''_σ gives us, as in (4·414), the other vertex $\mathbf{V}''$. Either $\mathbf{V}'$ or $\mathbf{V}''$, or better $\frac{1}{2}(\mathbf{V}'+\mathbf{V}'')$ which is the centre of the hypercircle, may be regarded as an approximation to the solution $\mathbf{S}$ which corresponds to the gradient of the

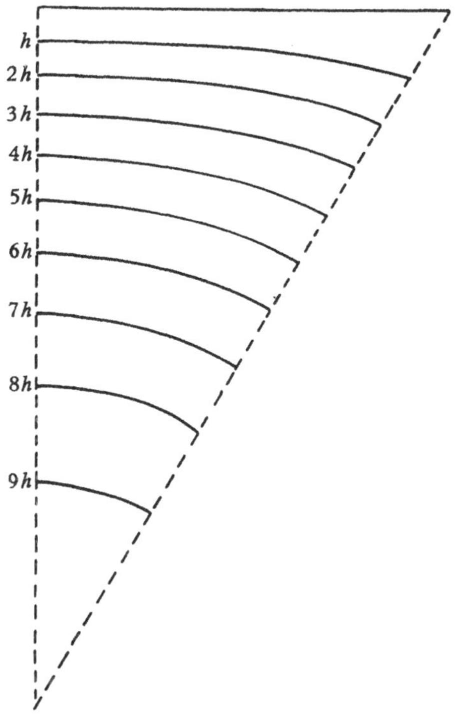

Fig. 4·48*d*. Torsion of a regular hexagon: approximation $n=16$. Approximate lines of stress, being actually the level lines of the polyhedral function based on the triangulation of Fig. 4·48*a* with heights given by the values of b'_ρ displayed in Fig. 4·48*b*. The heights of the lines drawn are 0, h, $2h$, …, $9h$, where $h=20\cdot736$.

function ψ. The best approximation obtained above is that for $n=16$, for which the triangulation is as in Fig. 4·48*a* and for which the weights are shown in Figs. 4·48*b*, *c*. Approximate lines of stress and lines of constant warping are shown in Figs. 4·48*d*, *e* respectively.

To show the convergence of the method, we collect in the table on p. 268 the bounds furnished in the several approximations; these are 'reliable' bounds, based either on exact solutions of (4·403) and (4·413) or on (4·430).

Bounds on Γ/H^4 where Γ is the torsional rigidity of a regular hexagon of side H

Approximation	Lower bound	Upper bound	Reference
$n=1$	0·8660	1·0825	(4·439), Fig. 4·44
$n=2$	0·9382	1·0825	(4·443), Fig. 4·45
$n=4$	1·0034	1·0542	(4·451), Figs. 4·46*a–d*
$n=8$	1·0264	1·0412	Figs. 4·47*a–e*
$n=16$	1·0330	1·0371	Figs. 4·48*a–c*

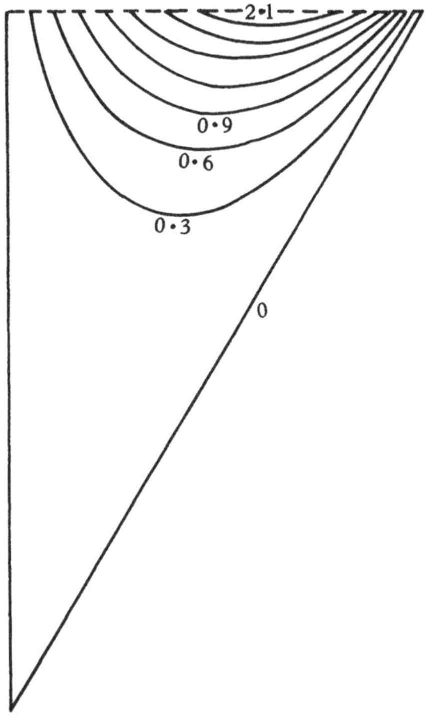

Fig. 4·48*e*. Torsion of a regular hexagon: approximation $n=16$. Approximate level lines of the warping function ϕ, being actually the level lines of the polyhedral function based on the triangulation of Fig. 4·48*a* with heights given by the values of b''_σ displayed in Fig. 4·48*c*. The heights of the lines drawn are:

0, 0·3, 0·6, 0·9, 1·2, 1·5, 1·8, 2·1.

Pólya and Szegö [(1), p. 261] give $2\pi\Gamma A^{-2}=0{\cdot}9643$, where A is the area of the hexagon; equivalently, $\Gamma/H^4=1{\cdot}0360$. The conformal transformation*

$$z=\int_0^w (1-t^6)^{-\frac{1}{3}}\,dt \qquad (4{\cdot}458)$$

* I am indebted to Professor P. M. Quinlan for this formula.

maps the interior of a regular hexagon in the z-plane on the interior of a circle in the w-plane. Hence the solution of the torsion problem for a regular hexagon can be expressed as a contour integral, and the torsional rigidity similarly expressed.*

Exercises

1. Suppose that a set of numbers b_ρ $(\rho=0, 1, \ldots, n)$ are to be found such that $b_0=b_n=0$ and the difference equations

$$b_\rho=\tfrac{1}{2}(b_{\rho-1}+b_{\rho+1})+1 \quad (\rho=1, \ldots, n-1)$$

are satisfied. Observe that if we take $b_0=0$ and any value for b_1, then the difference equations lead to a unique value of b_n, not in general zero and necessarily of the form

$$b_n=ab_1+c,$$

where a and c are numbers independent of b_1. Hence show that the problem may be solved by putting $(b_0=0, b_1=0)$ and finding b_n $(=b_n'$ say), and then putting $(b_0=0, b_1=1)$ and finding b_n $(=b_n''$ say), the values of a and c being then given by

$$c=b_n', \quad a+c=b_n'',$$

and the correct value of b_1 (making $b_n=0$) by

$$b_1=-\frac{c}{a}=\frac{b_n'}{b_n'-b_n''}.$$

Solve the problem numerically for $n=11$, obtaining the following values for b_ρ: 0, 10, 18, 24, 28, 30, 30, 28, 24, 18, 10, 0.

2. Apply the idea of Exercise 1 to the hexagon torsion problem, taking the triangulation for the approximation $n=16$ shown in Fig. 4·48a. Show that if the eight weights

$$A: \quad b'(a1), \quad b'(c2), \quad \ldots, \quad b'(m7), \quad b'(o8)$$

are assigned and $b'(q9)$ put equal to zero, then the difference equation (4·403), combined with symmetry conditions, leads to definite values (in general not zero) for the eight weights

$$B: \quad b'(q1), \quad b'(q2), \quad \ldots, \quad b'(q7), \quad b'(q8),$$

and that the transformation by which the 'vector' B is derived from the 'vector' A is necessarily of the form

$$B=MA+C,$$

where M is an 8×8 matrix and C an 8×1 matrix (or vector), both being independent of the choice of the weights A. Taking for A in turn the values

$$\begin{array}{ccccc}
(0, & 0, & \ldots, & 0, & 0) \\
(1, & 0, & \ldots, & 0, & 0) \\
(0, & 1, & \ldots, & 0, & 0) \\
\ldots & & & & \\
(0, & 0, & \ldots, & 1, & 0) \\
(0, & 0, & \ldots, & 0, & 1)
\end{array}$$

* Cf. Sokolnikoff (1), p. 151.

the vector C and the matrix M can be found arithmetically. The correct value of A (making $B=0$ as required in the torsion problem) is then given by the matrix equation

$$A = -M^{-1}C,$$

or equivalently by solving the eight linear equations represented by

$$MA + C = 0,$$

and when this has been done all the weights in Fig. 4·48 *b* can be calculated.

3. Applying the method of Exercise 2 to the approximation $n=4$ (Fig. 4·46 *a*), taking

$$A:\ b'(a1),\ b'(c2); \qquad B:\ b'(e1),\ b'(e2),$$

show that

$$C = \begin{pmatrix} -748 \\ 125 \end{pmatrix}, \quad M = \begin{pmatrix} 251 & -252 \\ -48 & 50 \end{pmatrix}, \quad M^{-1} = \frac{1}{454}\begin{pmatrix} 50 & 252 \\ 48 & 251 \end{pmatrix},$$

and hence that the correct values for the weights A are

$$b'(a1) = \tfrac{5900}{454} = 12{\cdot}995595, \quad b'(c2) = \tfrac{4529}{454} = 9{\cdot}975771.$$

From these values obtain the weights in Fig. 4·46 *b* to six decimal places.

4. Consider the application of the above method to find b''_σ in Fig. 4·48 *c*, and show that the algebraic problem can be reduced to solving seven simultaneous linear equations.

4·5. THE TORSION OF A HOLLOW SQUARE

In (4·282) bounds were given for the torsional rigidity of a hollow square of outer side $2a$ and inner side $2b$. But the method used there was unsystematic. Now we shall apply the method of pyramid F-vectors to this torsion problem; however, since this method is arithmetical, we must take a definite value for the ratio a/b. We shall choose $a/b=2$, so that the cross-section is as in Fig. 4·51. We shall denote by s the side of the outer square, so that the inner side is $\frac{1}{2}s$, and in the notation of (4·282) we have $a=s/2$, $b=s/4$ so that the bounds already obtained read

$$0{\cdot}1134 < \tfrac{49}{432} \leqslant \Gamma/s^4 \leqslant \tfrac{11}{80} = 0{\cdot}1375. \qquad (4{\cdot}501)$$

It will be interesting to compare these with the bounds obtained below by the method of pyramid F-vectors. In what follows, the symbol a has a different meaning from the above; below it denotes half the side of the base of a square pyramid F-vector as shown in Fig. 3·72, p. 201.

General plan

We shall use square pyramid F-vectors of both classes (homogeneous associated and complementary) and of both types (small base and large base), as described in § 3·7 and shown in Fig. 3·72.

Since the cross-section is doubly connected, we need also a ring F-vector as shown in Fig. 4·31.

Fig. 4·51 shows the triangulation for the lowest approximation we shall take. Here the side s of the outer square is divided into eight equal parts; we call this *the approximation* $n=8$. Higher approximations are given by bisecting the mesh as in Fig. 3·73; this gives successively $n=16, 32, \ldots,$ where n denotes always the number of parts into which the outer side s is divided, so that the small squares are of side s/n.

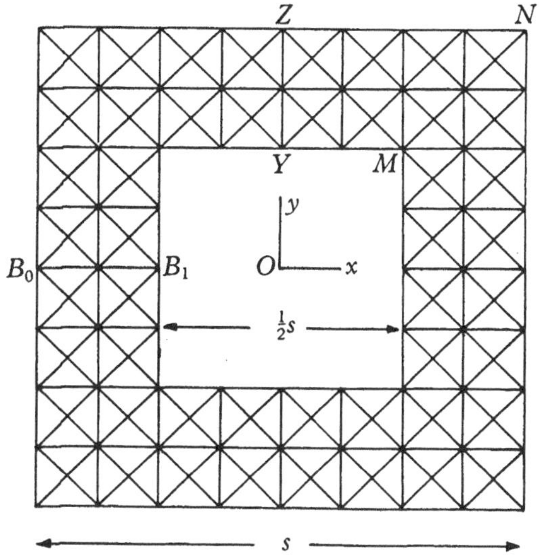

Fig. 4·51. Torsion of a hollow square: outer side s, inner side $\frac{1}{2}s$. Triangulation for approximation $n=8$.

We have to describe here the general procedure, valid for any approximation. To fix the ideas, we shall use the approximation $n=8$ and Fig. 4·51 for reference, but the argument will be general.

In §4·4, in dealing with the torsion of a regular hexagon, symmetry allowed us to work with one-twelfth part of the whole cross-section; similarly for the hollow square we can work with an *octant*. This octant is shown in Fig. 4·52 with the triangulation for $n=8$. The system for labelling the junction-points is obvious.

To get the vertex $\mathbf{V}'$ (and hence a lower bound for Γ), we have to find the weights a'_ρ to satisfy (4·228). These weights are associated with a set of homogeneous associated F-vectors $\mathbf{T}'_\rho$ $(\rho=1, 2, \ldots, r)$; for these we use square pyramid F-vectors of the first class (Fig. 3·72) and a ring F-vector (Fig. 4·31). We recall from Fig. 3·72 the two types of square pyramid F-vector—the small square $\mathbf{P}'$ with side $2a$ and the large square $\mathbf{Q}'$ with side $2a \, . \, 2^{\frac{1}{2}}$; these are normalized by the condition that the pyramid functions which generate them are of unit height.

Thus for the approximation $n=8$ (Figs. 4·51, 4·52) we have for homogeneous associated F-vectors $\mathbf{T}'_\rho$ the following:

(i) Small square pyramid F-vectors of the first class ($\mathbf{P}'$) with centres at $a\,1$, $a\,2$, $a\,3$, $b\,1$, $b\,2$, $b\,3$, $b\,4$ and their symmetric correspondents in the other seven octants.

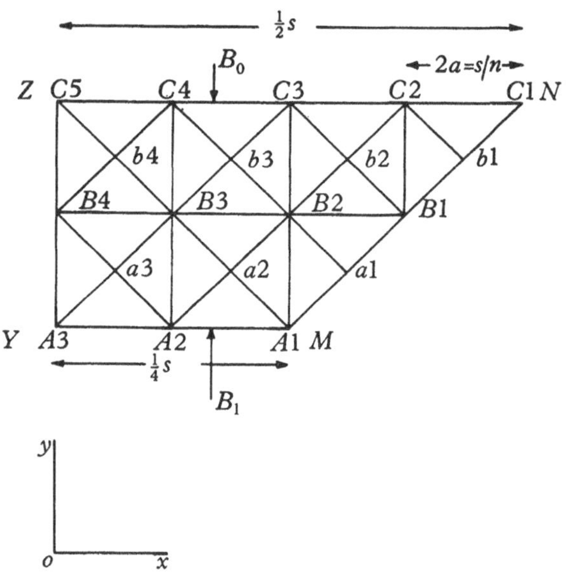

Fig. 4·52. Torsion of a hollow square: approximation $n=8$. Centres of square pyramid F-vectors in an octant: small squares ($\mathbf{P}$) at $a\,1, a\,2, \ldots, b\,3, b\,4$; large squares ($\mathbf{Q}$) at $A\,1, A\,2, \ldots, C\,4, C\,5$. Ring F-vector $\mathbf{R}$ covers the whole cross section.

(ii) Large square pyramid F-vectors of the first class ($\mathbf{Q}'$) with centres at $B\,1$, $B\,2$, $B\,3$, $B\,4$ and their symmetric correspondents in the other seven octants.

(iii) A ring F-vector $\mathbf{R}$ having the whole cross-section for base; the plateau function from which it is generated is of height $\frac{1}{2}$ inside the small square B_1 and zero height on the outer square B_0. This plateau is sketched in Fig. 4·53. This $\mathbf{R}$ is the same for all approximations. [In Fig. 4·31 the base of the ring F-vector covered only a simple ring of triangles; it is more convenient in the present case (and quite legitimate) to take it as described above.]

We shall use $\mathbf{T}'_\rho$ to cover the whole set of $\mathbf{P}'$, $\mathbf{Q}'$, $\mathbf{R}$. Then, referring to the summary of p. 224 and in particular to (4·228), we see that we need the values of the scalar products $\mathbf{T}'_\mu . \mathbf{T}'_\rho$ and $\mathbf{S}'_0 . \mathbf{T}'_\rho$, where $\mathbf{S}'_0$ is as usual the moment of inertia F-vector corresponding to the P-vector field (x, y). (The axes of coordinates are indicated in Figs. 4·51, 4·52.)

By virtue of the normalization used, we have (in approximation n)

$$\text{strength of } P\text{-vector field of } \mathbf{P}' = 2n/s,$$
$$\text{strength of } P\text{-vector field of } \mathbf{Q}' = 2^{\frac{1}{2}}n/s,$$
$$\text{strength of } P\text{-vector field of } \mathbf{R} = 2/s.$$

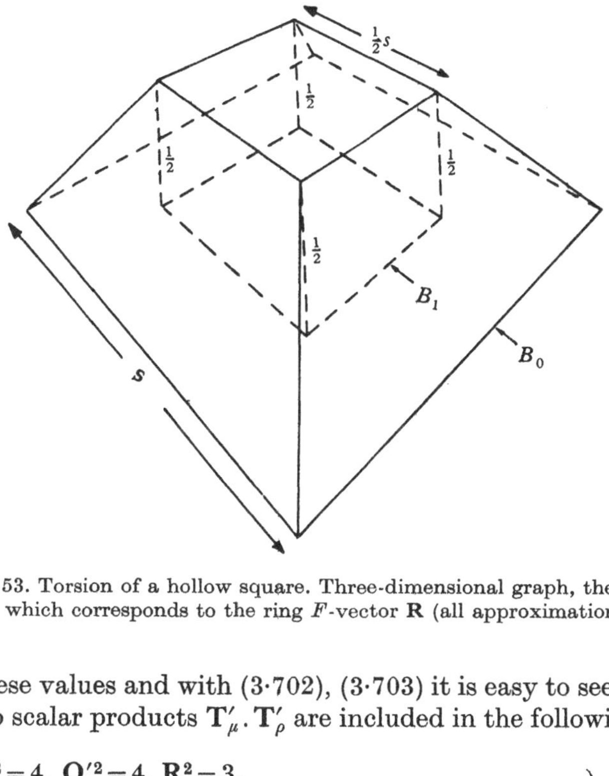

Fig. 4·53. Torsion of a hollow square. Three-dimensional graph, the slope of which corresponds to the ring F-vector $\mathbf{R}$ (all approximations).

With these values and with (3·702), (3·703) it is easy to see that all non-zero scalar products $\mathbf{T}'_\mu . \mathbf{T}'_\rho$ are included in the following list:

$$\left.\begin{aligned} &\mathbf{P}'^2 = 4,\ \mathbf{Q}'^2 = 4,\ \mathbf{R}^2 = 3,\\ &\mathbf{P}'.\mathbf{Q}' = -1 \text{ if bases overlap},\\ &\mathbf{P}'.\mathbf{R} = 2/n \text{ if the centre of } \mathbf{P}' \text{ is on a diagonal of the cross-section [e.g. } a1 \text{ in Fig. 4·52]},\\ &\mathbf{Q}'.\mathbf{R} = 2/n \text{ if the centre of } \mathbf{Q}' \text{ is on a diagonal of the cross section [e.g. } B1 \text{ in Fig. 4·52]}. \end{aligned}\right\} \quad (4·502)$$

Note that **R** is orthogonal to all the square pyramid F-vectors except those with centres on the diagonals.

Making use of (4·301) and (4·304) and remembering that $2a = s/n$, we find

$$\left.\begin{aligned} \mathbf{S}'_0 . \mathbf{P}' = -\tfrac{2}{3}s^2/n^2, \quad \mathbf{S}'_0 . \mathbf{Q}' = -\tfrac{4}{3}s^2/n^2, \\ \mathbf{S}'_0 . \mathbf{R} = -\tfrac{7}{12}s^2. \end{aligned}\right\} \qquad (4{\cdot}503)$$

We might now use (3·705) for the determination of the weights, but it requires modification since **R** is a fifth neighbour for each of the square pyramid F-vectors, and it is simpler to go back to the basic equations in (4·228):

$$\sum_{\mu=1}^{r} a'_\mu \mathbf{T}'_\mu . \mathbf{T}'_\rho + \mathbf{S}'_0 . \mathbf{T}'_\rho = 0. \qquad (4{\cdot}504)$$

By (4·502) and (4·503), these equations may be written, for approximation n, in the form

$$\left.\begin{aligned} &4a'(\mathbf{P}') - \Sigma a'(\mathbf{Q}') + (2/n)\, d(\mathbf{P}')\, a'(\mathbf{R}) - \tfrac{2}{3}(s^2/n^2) = 0, \\ &4a'(\mathbf{Q}') - \Sigma a'(\mathbf{P}') + (2/n)\, d(\mathbf{Q}')\, a'(\mathbf{R}) - \tfrac{4}{3}(s^2/n^2) = 0, \\ &3a'(\mathbf{R}) + (8/n) \sum_{MN} [a'(\mathbf{P}') + a'(\mathbf{Q}')] - \tfrac{7}{12}s^2 = 0. \end{aligned}\right\} \qquad (4{\cdot}505)$$

The notation requires explanation. In the first line, the first term is four times the weight of any $\mathbf{P}'$; $\Sigma a'(\mathbf{Q}')$ is the sum of the weights of its four $\mathbf{Q}'$ neighbours (put weights equal to zero if a neighbour is fictitious, i.e. if its centre is on the boundary B); $a'(\mathbf{R})$ is the weight of **R**; $d(\mathbf{P}') = 1$ if the centre of $\mathbf{P}'$ is on the diagonal MN (Fig. 4·51) and otherwise $d(\mathbf{P}') = 0$. The second line of (4·505) is similarly interpreted. As for the third line, $\sum_{MN}$ means the sum of weights on the diagonal MN, the coefficient being $8/n$ (instead of $2/n$) because there are four diagonals.

To help the computation, we now define dimensionless *reduced weights* by

$$\left.\begin{aligned} b'(\mathbf{P}') = (6n^2/s^2)\, a'(\mathbf{P}'), \quad b'(\mathbf{Q}') = (6n^2/s^2)\, a'(\mathbf{Q}'), \\ b'(\mathbf{R}) = (3n/s^2)\, a'(\mathbf{R}). \end{aligned}\right\} \qquad (4{\cdot}506)$$

Then (4·505) read, on multiplication of the first two by $(\tfrac{3}{2})(n^2/s^2)$ and the last by n/s^2,

$$\left.\begin{aligned} b'(\mathbf{P}') &= \tfrac{1}{4}\Sigma b'(\mathbf{Q}') - d(\mathbf{P}')\, b'(\mathbf{R}) + 1, \\ b'(\mathbf{Q}') &= \tfrac{1}{4}\Sigma b'(\mathbf{P}') - d(\mathbf{Q}')\, b'(\mathbf{R}) + 2, \\ b'(\mathbf{R}) &= -\tfrac{4}{3}n^{-2} \sum_{MN} [b'(\mathbf{P}') + b'(\mathbf{Q}')] + \tfrac{7}{12}n. \end{aligned}\right\} \qquad (4{\cdot}507)$$

These are the formulae for the calculation of the reduced weights b'_ρ by successive approximations.

When these weights have been found, the vertex $\mathbf{V}'$ is given by (4·229) in the form

$$\mathbf{V}' = \mathbf{S}'_0 + \tfrac{1}{6}(s^2/n^2)\,[\Sigma b'(\mathbf{P}')\,\mathbf{P}' + \Sigma b'(\mathbf{Q}')\,\mathbf{Q}'] + \tfrac{1}{3}(s^2/n)\,b'(\mathbf{R})\,\mathbf{R}, \quad (4·508)$$

where the summations are to be taken over the whole cross-section, with symmetrical weights.

As in (4·231), we have as a lower bound for the torsional rigidity, on reference to (4·503) and (4·506),

$$\Gamma \geqslant \mathbf{S}'^2_0 - \mathbf{V}'^2 = -\sum_{\rho=1}^{r} a'_\rho\,\mathbf{S}'_0 \,.\, \mathbf{T}'_\rho$$

$$= \tfrac{1}{9}(s^4/n^4)\,[\Sigma b'(\mathbf{P}') + 2\Sigma b'(\mathbf{Q}')] + \tfrac{7}{36}(s^4/n)\,b'(\mathbf{R}), \quad (4·509)$$

the summations being taken over the whole cross-section.

We now turn our attention to $\mathbf{V}''$ and the upper bound for Γ. We use square pyramid F-vectors of the second class, $\mathbf{P}''$ (small square) and $\mathbf{Q}''$ (large square) as in Fig. 3·72. Zero weights are attached to centres on the lines of symmetry of the cross-section, and the surviving weights are skew-symmetric with respect to these lines.

The weights a''_σ are to satisfy the second line of (4·228), and for that we need the scalar products $\mathbf{T}''_\nu \,.\, \mathbf{T}''_\sigma$ and $\mathbf{S}'_0 \,.\, \mathbf{T}''_\sigma$. We shall give these for the general approximation n, but to fix our ideas we may refer to $n=8$ and Fig. 4·52; there we have to consider square pyramid F-vectors with centres as follows:

small squares ($\mathbf{P}''$) at $a\,2$, $a\,3$, $b\,2$, $b\,3$, $b\,4$;
large squares ($\mathbf{Q}''$) at $A\,2$, $B\,2$, $B\,3$, $C\,2$, $C\,3$, $C\,4$.

For centres not listed here but shown in Fig. 4·52, the weights are zero by symmetry.

With normalization as in (3·701), we have

strength of P-vector field of $\mathbf{P}'' = 2n/s$,
strength of P-vector field of $\mathbf{Q}'' = 2\tfrac{1}{2}n/s$.

For bases contained entirely in the cross-section, we have, as in (3·702),

$$\mathbf{P}''^2 = 4, \quad \mathbf{Q}''^2 = 4, \quad (4·510)$$

and if the centre lies on the boundary (this does not occur for $\mathbf{P}''$)

$$\mathbf{Q}''^2 = 2. \quad (4·511)$$

The only non-zero scalar products $\mathbf{T}''_\nu . \mathbf{T}''_\sigma$ occur when the base of a $\mathbf{P}''$ overlaps the base of a $\mathbf{Q}''$; then in all cases [cf. (3·703)]

$$\mathbf{P}''.\mathbf{Q}'' = -1. \tag{4·512}$$

We have now to consider $\mathbf{S}'_0 . \mathbf{T}''_\sigma$. We already saw in (4·305) that this is zero if the base of $\mathbf{T}''_\sigma$ is contained entirely in the cross-section; we have then only to consider the case where the centre of $\mathbf{T}''_\sigma$ lies on the boundary B, and for this we refer to Fig. 4·33. If x denotes the abscissa of the centre, we may write for the only non-zero scalar products $\mathbf{S}'_0 . \mathbf{T}''_\sigma$

$$\left.\begin{aligned} &\mathbf{S}'_0.\mathbf{Q}'' = 2ax = xs/n, \text{ if } \mathbf{Q}'' \text{ is centred on the outer boundary } B_0, \text{ as } C\,2 \text{ in Fig. 4·52},\\ &\mathbf{S}'_0.\mathbf{Q}'' = -2ax = -xs/n, \text{ if } \mathbf{Q}'' \text{ is centred on the inner boundary } B_1, \text{ as } A\,2 \text{ in Fig. 4·52}.\end{aligned}\right\} \tag{4·513}$$

Accordingly, from (4·228) we have the equations

$$\left.\begin{aligned} &4a''(\mathbf{P}'') - \Sigma a''(\mathbf{Q}'') = 0,\\ &4a''(\mathbf{Q}'') - \Sigma a''(\mathbf{P}'') = 0 \text{ if } \mathbf{Q}'' \text{ is centred inside},\\ &2a''(\mathbf{Q}'') - \Sigma a''(\mathbf{P}'') - xs/n = 0 \text{ if } \mathbf{Q}'' \text{ is centred on the outer boundary } B_0, \text{ as } C\,2 \text{ in Fig. 4·52},\\ &2a''(\mathbf{Q}'') - \Sigma a''(\mathbf{P}'') + xs/n = 0 \text{ if } \mathbf{Q}'' \text{ is centred on the inner boundary } B_1, \text{ as } A\,2 \text{ in Fig. 4·52}.\end{aligned}\right\} \tag{4·514}$$

To explain the notation, in the first equation $a''(\mathbf{P}'')$ is the weight of any $\mathbf{P}''$ and $\Sigma a''(\mathbf{Q}'')$ the sum of the weights of its four $\mathbf{Q}''$ neighbours; in the second equation $a''(\mathbf{Q}'')$ is the weight of any $\mathbf{Q}''$ centred inside and $\Sigma a''(\mathbf{P}'')$ the sum of its four $\mathbf{P}''$ neighbours; in the third equation $a''(\mathbf{Q}'')$ is the weight of any $\mathbf{Q}''$ centred on the outer boundary of the octant shown in Fig. 4·52 and $\Sigma a''(\mathbf{P}'')$ the sum of the weights of its *two* $\mathbf{P}''$ neighbours, while x is the abscissa of the centre of $\mathbf{Q}''$; the explanation of the fourth equation is similar.

We can express these formulae more compactly by introducing dimensionless *reduced weights* b''_σ defined by

$$b''_\sigma = 4na''_\sigma/s^2, \tag{4·515}$$

and numbers $m(\mathbf{Q}'')$ such that

$$\left.\begin{aligned} &m(\mathbf{Q}'') = 0 \text{ if } \mathbf{Q}'' \text{ is centred inside},\\ &m(\mathbf{Q}'') = 2x/s \text{ if } \mathbf{Q}'' \text{ is centred on the outer boundary } B_0 \text{ of the octant of Fig. 4·52},\\ &m(\mathbf{Q}'') = -2x/s \text{ if } \mathbf{Q}'' \text{ is centred on the inner boundary } B_1 \text{ of the octant of Fig. 4·52}.\end{aligned}\right\} \tag{4·516}$$

Then (4·514) are all contained in

$$\left.\begin{aligned} b''(\mathbf{P}'') &= \tfrac{1}{4}\Sigma b''(\mathbf{Q}''), \\ b''(\mathbf{Q}'') &= \tfrac{1}{4}\Sigma b''(\mathbf{P}'') + m(\mathbf{Q}''), \end{aligned}\right\} \qquad (4{\cdot}517)$$

where each summation now includes *four* neighbours, it being understood that if $\mathbf{Q}''$ is centred on the boundary (outer or inner) its two real neighbours are supplemented by two fictitious neighbours, the reflexions of the two real neighbours in the boundary with the same weights (no reversal of sign). *These are the basic formulae for the calculation of the reduced weights* b''_σ *by successive approximations.*

If reduced weights have been found to satisfy (4·517), we get the vertex $\mathbf{V}''$ by (4·229):

$$\mathbf{V}'' = \tfrac{1}{4}(s^2/n)\,\Sigma b''_\sigma \mathbf{T}''_\sigma, \qquad (4{\cdot}518)$$

the summation covering all eight octants, the weights in the seven octants other than that of Fig. 4·52 being supplied by the rule of skew-symmetry. Further, by (4·230) and (4·513) we have

$$\begin{aligned} \mathbf{V}''^2 &= \Sigma a''_\sigma \mathbf{S}'_0 \,.\, \mathbf{T}''_\sigma \\ &= 8\Big(\sum_{ZN} a''_\sigma xs/n - \sum_{YM} a''_\sigma xs/n\Big) \\ &= (s^4/n^2)\Big[\sum_{ZN} m(\mathbf{Q}'')\,b''(\mathbf{Q}'') + \sum_{YM} m(\mathbf{Q}'')\,b''(\mathbf{Q}'')\Big], \end{aligned} \qquad (4{\cdot}519)$$

where these summations are carried out along the half-edges ZN and YM of the outer and inner squares (see Fig. 4·52). Note that $m(\mathbf{Q}'')$ is positive on ZN and negative on YM.

To get an upper bound for the torsional rigidity Γ, we note that the moment of inertia of the cross-section is

$$\mathbf{S}'^2_0 = 5s^4/32, \qquad (4{\cdot}520)$$

and so, by (4·231) and (4·519), we have the upper bound

$$\Gamma \leqslant \mathbf{S}'^2_0 - \mathbf{V}''^2 = s^4\Big[\tfrac{5}{32} - n^{-2}\Big\{\sum_{ZN} m(\mathbf{Q}'')\,b''(\mathbf{Q}'') + \sum_{YM} m(\mathbf{Q}'')\,b''(\mathbf{Q}'')\Big\}\Big]. \qquad (4{\cdot}521)$$

To summarize, the essential formulae for the problem of the torsion of a hollow square (outer side s, inner side $\tfrac{1}{2}s$) in the approximation n (for which the square base of $\mathbf{P}'$ is of side s/n) are as follows:

$$\left.\begin{aligned} b'(\mathbf{P}') &= \tfrac{1}{4}\Sigma b'(\mathbf{Q}') - d(\mathbf{P}')\, b'(\mathbf{R}) + 1, \\ b'(\mathbf{Q}') &= \tfrac{1}{4}\Sigma b'(\mathbf{P}') - d(\mathbf{Q}')\, b'(\mathbf{R}) + 2, \\ b'(\mathbf{R}) &= \tfrac{4}{3}n^{-2}\left[\tfrac{7}{16}n^3 - \sum_{MN}\{b'(\mathbf{P}') + b'(\mathbf{Q}')\}\right] \end{aligned}\right\} \quad (4{\cdot}522)$$

($d(\mathbf{P}') = 1$, $d(\mathbf{Q}') = 1$ on diagonal MN, elsewhere zero);

$$\Gamma/s^4 \geqslant \tfrac{1}{9}n^{-4}[\Sigma b'(\mathbf{P}') + 2\Sigma b'(\mathbf{Q}') + \tfrac{7}{4}n^3 b'(\mathbf{R})] \quad (4{\cdot}523)$$

(summations Σ over all eight octants);

$$\left.\begin{aligned} b''(\mathbf{P}'') &= \tfrac{1}{4}\Sigma b''(\mathbf{Q}''), \\ b''(\mathbf{Q}'') &= \tfrac{1}{4}\Sigma b''(\mathbf{P}'') + m(\mathbf{Q}'') \end{aligned}\right\} \quad (4{\cdot}524)$$

(with explanation following (4·517));

$$\Gamma/s^4 \leqslant n^{-2}\left[\tfrac{5}{32}n^2 - \sum_{ZN} m(\mathbf{Q}'')\, b''(\mathbf{Q}'') - \sum_{YM} m(\mathbf{Q}'')\, b''(\mathbf{Q}'')\right]. \quad (4{\cdot}525)$$

Use of approximate solutions for the weights of the F-vectors

In deriving the above bounds for Γ, we assumed that b'_ρ, b''_σ were *exact* solutions of the linear equations (4·228), or equivalently of (4·522) and (4·524). But in practice they will not be exact solutions, and so the bounds (4·523) and (4·525) must be regarded as *unreliable*; to get *reliable* bounds we must refer to (2·797), as we did on p. 247 in the case of the regular hexagon.

By (2·797) we have for *any* weights

$$-\sum_{\rho=1}^{r} a'_\rho \mathbf{S}'_0 . \mathbf{T}'_\rho - \sum_{\rho=1}^{r} a'_\rho e'_\rho \leqslant \Gamma \leqslant \mathbf{S}'^2_0 - \sum_{\sigma=1}^{s} a''_\sigma \mathbf{S}'_0 . \mathbf{T}''_\sigma + \sum_{\sigma=1}^{s} a''_\sigma e''_\sigma, \quad (4{\cdot}526)$$

the 'errors' being, as in (2·792)

$$\left.\begin{aligned} e'_\rho &= \sum_{\mu=1}^{r} a'_\mu \mathbf{T}'_\mu . \mathbf{T}'_\rho + \mathbf{S}'_0 . \mathbf{T}'_\rho, \\ e''_\sigma &= \sum_{\nu=1}^{s} a''_\nu \mathbf{T}''_\nu . \mathbf{T}''_\sigma - \mathbf{S}'_0 . \mathbf{T}''_\sigma. \end{aligned}\right\} \quad (4{\cdot}527)$$

In the notation of (4·506), the first inequality of (4·526) may be written

$$\begin{aligned} \Gamma \geqslant\; & \tfrac{1}{9}(s^4/n^4)\,[\Sigma b'(\mathbf{P}') + 2\Sigma b'(\mathbf{Q}')] + \tfrac{7}{36}(s^4/n)\, b'(\mathbf{R}) \\ & - \tfrac{1}{6}(s^2/n^2)\,[\Sigma b'(\mathbf{P}')\, e'(\mathbf{P}') + \Sigma b'(\mathbf{Q}')\, e'(\mathbf{Q}')] - \tfrac{1}{3}(s^2/n)\, b'(\mathbf{R})\, e'(\mathbf{R}), \end{aligned} \quad (4{\cdot}528)$$

where Σ runs over all eight octants.

We now introduce dimensionless *reduced errors* f' by

$$\left.\begin{aligned} f'(\mathbf{P}') &= \tfrac{3}{2}(n^2/s^2)\,e'(\mathbf{P}') \\ &= b'(\mathbf{P}') - \tfrac{1}{4}\Sigma b'(\mathbf{Q}') + d(\mathbf{P}')\,b'(\mathbf{R}) - 1, \\ f'(\mathbf{Q}') &= \tfrac{3}{2}(n^2/s^2)\,e'(\mathbf{Q}') \\ &= b'(\mathbf{Q}') - \tfrac{1}{4}\Sigma b'(\mathbf{P}') + d(\mathbf{Q}')\,b'(\mathbf{R}) - 2, \\ f'(\mathbf{R}) &= \tfrac{12}{7}s^{-2}\,e'(\mathbf{R}) \\ &= \tfrac{16}{7}n^{-3}\left[\sum_{MN}\{b'(\mathbf{P}')+b'(\mathbf{Q}')\} + \tfrac{3}{4}n^2 b'(\mathbf{R}) - \tfrac{7}{16}n^3\right]. \end{aligned}\right\} \quad (4{\cdot}529)$$

In the first two the summations are over neighbours and in the last over a diagonal. It is clear that these errors are small if (4·522) are approximately satisfied, but the argument does not depend on such smallness. When these expressions are substituted in (4·528) we get the *reliable* lower bound

$$\Gamma/s^4 \geqslant \tfrac{1}{9}n^{-4}[\Sigma b'(\mathbf{P}') + 2\Sigma b'(\mathbf{Q}') + \tfrac{7}{4}n^3 b'(\mathbf{R}) - \Sigma b'(\mathbf{P}')f'(\mathbf{P}') \\ - \Sigma b'(\mathbf{Q}')f'(\mathbf{Q}') - \tfrac{7}{4}n^3 b'(\mathbf{R})f'(\mathbf{R})], \quad (4{\cdot}530)$$

the summations being over all eight octants. The first part is the expression occurring in (4·523) and the second part a correction which is small if the f' are small.

The second inequality in (4·526) may be written, on substitution from (4·513) and (4·520),

$$\Gamma/s^4 \leqslant \tfrac{5}{32} - n^{-2}\left[\sum_{ZN} m(\mathbf{Q}'')\,b''(\mathbf{Q}'') + \sum_{YM} m(\mathbf{Q}'')\,b''(\mathbf{Q}'')\right] \\ + \tfrac{1}{4}n^{-1}s^{-2}[\Sigma b''(\mathbf{P}'')\,e''(\mathbf{P}'') + \Sigma b''(\mathbf{Q}'')\,e''(\mathbf{Q}'')], \quad (4{\cdot}531)$$

the last two summations being over all eight octants. By (4·527)

$$\left.\begin{aligned} e''(\mathbf{P}'') &= 4a''(\mathbf{P}'') - \Sigma a''(\mathbf{Q}''), \\ e''(\mathbf{Q}'') &= \lambda(\mathbf{Q}'')\,[4a''(\mathbf{Q}'') - \Sigma a''(\mathbf{P}'')] - \tfrac{1}{2}(s^2/n)\,m(\mathbf{Q}''), \end{aligned}\right\} \quad (4{\cdot}532)$$

where the summations are over four neighbours, including two fictitious images if $\mathbf{Q}''$ is centred on the boundary, and

$$\left.\begin{aligned} &\lambda(\mathbf{Q}'') = \tfrac{1}{2} \text{ if } \mathbf{Q}'' \text{ is centred on the boundary,} \\ &\lambda(\mathbf{Q}'') = 1 \text{ if } \mathbf{Q}'' \text{ is centred inside.} \end{aligned}\right\} \quad (4{\cdot}533)$$

We now define dimensionless *reduced errors* f'' by

$$f''_\sigma = n e''_\sigma / s^2, \quad (4{\cdot}534)$$

and obtain from (4·532), on multiplication by n/s^2,

$$\left.\begin{aligned} f''(\mathbf{P}'') &= b''(\mathbf{P}'') - \tfrac{1}{4}\Sigma b''(\mathbf{Q}''), \\ f''(\mathbf{Q}'') &= \lambda(\mathbf{Q}'')\,[b''(\mathbf{Q}'') - \tfrac{1}{4}\Sigma b''(\mathbf{P}'')] - \tfrac{1}{2}m(\mathbf{Q}''). \end{aligned}\right\} \quad (4{\cdot}535)$$

Then (4·531) gives the *reliable* upper bound

$$\Gamma/s^4 \leqslant n^{-2}\Bigg[\tfrac{5}{32}n^2 - \sum_{ZN} m(\mathbf{Q}'')\,b''(\mathbf{Q}'') - \sum_{YM} m(\mathbf{Q}'')\,b''(\mathbf{Q}'')$$
$$+2\Bigg\{\sum_{\text{oct}} b''(\mathbf{P}'')f''(\mathbf{P}'') + \sum_{\text{oct}} b''(\mathbf{Q}'')f''(\mathbf{Q}'')\Bigg\}\Bigg]; \qquad (4\cdot536)$$

the summations ZN and YM are along the edges of the octant of Fig. 4·52 and the other two summations are over the octant (i.e. one-eighth of sums over the whole cross-section). The first part of (4·536) agrees with (4·525) and the last part is a correction which is small when the errors are small.

In (4·530) and (4·536) we have reliable bounds for Γ valid for any weights at all.

The multiplication process for quick convergence

The convergence of the reduced weights b'_ρ by successive approximations based on (4·507) is very slow in the case of a fine mesh, and we need a multiplication process to speed up the convergence, analogous to that connected with (4·437) for the torsion of a regular hexagon.

The equations (4·529) by which the reduced errors are to be calculated may be written

$$\left.\begin{aligned} 4f'(\mathbf{P}') &= 4b'(\mathbf{P}') - \Sigma b'(\mathbf{Q}') + 4d(\mathbf{P}')\,b'(\mathbf{R}) - 4,\\ 4f'(\mathbf{Q}') &= 4b'(\mathbf{Q}') - \Sigma b'(\mathbf{P}') + 4d(\mathbf{Q}')\,b'(\mathbf{R}) - 8,\end{aligned}\right\} \qquad (4\cdot537)$$

these summations being over neighbours. Let us add together these equations for all the junction-points of the mesh in the whole cross-section—not including the boundary of course since the weights are zero there. In the main part of the section (i.e. not adjacent to the boundary) each $\mathbf{P}'$ has four $\mathbf{Q}'$ neighbours and each $\mathbf{Q}'$ has four $\mathbf{P}'$ neighbours; thus much cancellation of terms occurs on addition and we get

$$4[\Sigma f'(\mathbf{P}') + \Sigma f'(\mathbf{Q}')]$$
$$= 2\Sigma'_1 b'(\mathbf{P}') + 3\Sigma'_2 b'(\mathbf{P}') + \Sigma'_3 b'(\mathbf{P}') + 8(n-2)\,b'(\mathbf{R})$$
$$-4N(\mathbf{P}') - 8N(\mathbf{Q}'), \qquad (4\cdot538)$$

where the sums on the left are over the whole cross-section (eight octants), Σ'_1 is a sum over all $\mathbf{P}'$ centres adjacent to the boundary but not lying on a diagonal, Σ'_2 is a sum over the four $\mathbf{P}'$ centres on diagonals and adjacent to outer corners, Σ'_3 is a sum over the four $\mathbf{P}'$ centres on diagonals and adjacent to inner corners; $N(\mathbf{P}')$ and $N(\mathbf{Q}')$ are respectively the numbers of $\mathbf{P}'$ and $\mathbf{Q}'$ centres in the whole cross-section, exclusive of the boundary (inner and outer). With regard to the $b'(\mathbf{R})$ term in (4·538), we have used the fact that in the approximation n there are $\frac{1}{2}n-1$ centres ($\mathbf{P}'$ and $\mathbf{Q}'$) on each of the four diagonals. We note that

$$\left.\begin{aligned} &N(\mathbf{P}')=\tfrac{3}{4}n^2, \quad N(\mathbf{Q}')=\tfrac{3}{4}n(n-4), \\ &4N(\mathbf{P}')+8N(\mathbf{Q}')=9n^2-24n. \end{aligned}\right\} \qquad (4·539)$$

We now use symmetry to replace the summations Σ_1', Σ_2', Σ_3' in (4·538) by summations over an octant. Let Σ_1 denote summation over all $\mathbf{P}'$ centres in the octant adjacent to the boundary but not on a diagonal, as indicated in Fig. 4·54, and let $b'(C_0)$, $b'(C_1)$ denote

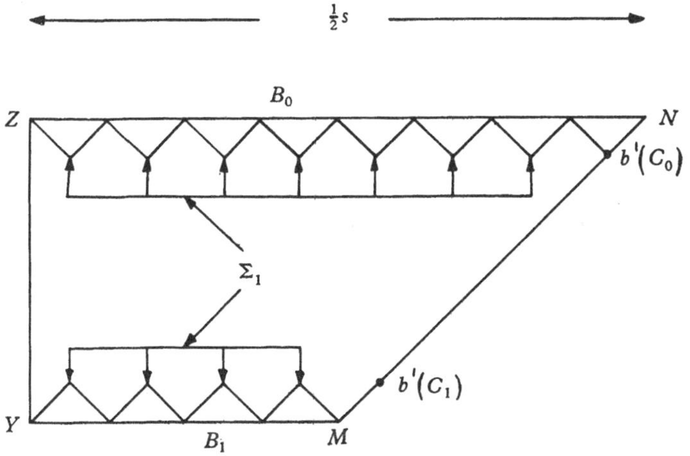

Fig. 4·54. Torsion of a hollow square: the summation Σ_1 and the centres C_0, C_1 for the multiplication process (4·543), drawn for $n=16$.

the weights for the two $\mathbf{P}'$ centres on the diagonal MN adjacent respectively to the outer corner and the inner corner. Then

$$\left.\begin{aligned} &\Sigma_1' b'(\mathbf{P}')=8\Sigma_1 b'(\mathbf{P}'), \\ &\Sigma_2' b'(\mathbf{P}')=4b'(C_0), \quad \Sigma_3' b'(\mathbf{P}')=4b'(C_1), \end{aligned}\right\} \qquad (4·540)$$

and (4·538) may be written

$$\begin{aligned} &\Sigma f'(\mathbf{P}')+\Sigma f'(\mathbf{Q}') \\ &\quad =4\Sigma_1 b'(\mathbf{P}')+3b'(C_0)+b'(C_1)+2(n-2)\,b'(\mathbf{R})-\tfrac{1}{4}(9n^2-24n). \end{aligned} \qquad (4·541)$$

On substituting for $b'(\mathbf{R})$ from (4·507), we get

$$\begin{aligned} &\Sigma f'(\mathbf{P}')+\Sigma f'(\mathbf{Q}') \\ &\quad =4\Sigma_1 b'(\mathbf{P}')+3b'(C_0)+b'(C_1)-\tfrac{8}{3}n^{-2}(n-2)\sum_{MN}[b'(\mathbf{P}')+b'(\mathbf{Q}')] \\ &\qquad -\tfrac{1}{12}(13n^2-44n). \end{aligned} \qquad (4·542)$$

The weights occurring here are to be regarded as arbitrary, although in practice they will be approximate solutions of (4·522).

If we now multiply all the weights $b'(\mathbf{P}')$, $b'(\mathbf{Q}')$ by a common factor k, we get new weights

$$\bar{b}'(\mathbf{P}') = kb'(\mathbf{P}'), \quad \bar{b}'(\mathbf{Q}') = kb'(\mathbf{Q}').$$

Let us at the same time change $b'(\mathbf{R})$ to $\bar{b}'(\mathbf{R})$ as given by the last of (4·522) with $\bar{b}'(\mathbf{P}')$, $\bar{b}'(\mathbf{Q}')$ substituted for $b'(\mathbf{P}')$, $b'(\mathbf{Q}')$. Then for the sum of the new errors $\bar{f}'(\mathbf{P}')$, $\bar{f}'(\mathbf{Q}')$ we get the same formula as (4·542) but with $\bar{b}'$ replacing b'. We choose the factor k to make the sum of the new errors zero:

$$\Sigma\bar{f}'(\mathbf{P}') + \Sigma\bar{f}'(\mathbf{Q}') = 0;$$

the value of k is

$$\left.\begin{aligned} &k = \nu/D, \quad \nu = 13n^2 - 44n, \\ &D = 48\Sigma_1 b'(\mathbf{P}') + 36b'(C_0) + 12b'(C_1) \\ &\qquad - 32n^{-2}(n-2)\sum_{MN}[b'(\mathbf{P}') + b'(\mathbf{Q}')]. \end{aligned}\right\} \quad (4{\cdot}543)$$

Our procedure is to alternate the circling process (4·522) with the multiplication process using the above k.

A similar multiplication process to give quicker convergence for the weights b''_σ is also useful. The reduced errors are as in (4·535), with m as in (4·516) and λ as in (4·533). On account of the skew-symmetry, $\Sigma f''_\sigma$ over the whole cross-section is zero; our aim will be to make this sum vanish for each octant.

If we add together the errors (4·535) for the octant of Fig. 4·52, there is a considerable amount of cancellation. For the triangulation ($n=8$) of Fig. 4·52 we get

$$\begin{aligned} 4\sum_{\text{oct}} f''_\sigma = 2[b''(a3) + b''(b4)] + b''(a2) + b''(B2) \\ + b''(b2) + b''(C2) - 2\Sigma m(\mathbf{Q}''), \end{aligned} \quad (4{\cdot}544)$$

the last summation being along the boundary, $YM + ZN$. It is easy to generalize this to the approximation n. We have

$$\begin{aligned} \Sigma m(\mathbf{Q}'') &= 2n^{-1}[1 + 2 + \ldots + (\tfrac{1}{2}n - 1)] - 2n^{-1}[1 + 2 + \ldots + (\tfrac{1}{4}n - 1)] \\ &= \tfrac{1}{16}(3n - 4), \end{aligned} \quad (4{\cdot}545)$$

and, for the approximation n,

$$4\sum_{\text{oct}} f''_\sigma = 2\Sigma_1 b''_\sigma + \Sigma_2 b''_\sigma - \tfrac{1}{8}(3n-4), \quad (4{\cdot}546)$$

where Σ_1 and Σ_2 are summations over centres adjacent respectively to YZ and MN as shown in Fig. 4·55; note that in Σ_2 the centre on B_0 is included, but not the centre on B_1.

Suppose now that we have any set of reduced weights b''_σ. If we multiply these by a common factor k, we get new weights $\bar{b}''_\sigma = kb''_\sigma$, and a new sum of errors $\bar{f}''_\sigma$ as in (4·546):

$$4\sum_{\text{oct}} \bar{f}''_\sigma = 2\Sigma_1 \bar{b}''_\sigma + \Sigma_2 \bar{b}''_\sigma - \tfrac{1}{8}(3n-4)$$
$$= k(2\Sigma_1 b''_\sigma + \Sigma_2 b''_\sigma) - \tfrac{1}{8}(3n-4). \qquad (4·547)$$

We choose k to make this vanish:

$$k = \nu/D, \quad \nu = \tfrac{1}{4}(3n-4), \quad D = 4\Sigma_1 b''_\sigma + 2\Sigma_2 b''_\sigma. \qquad (4·548)$$

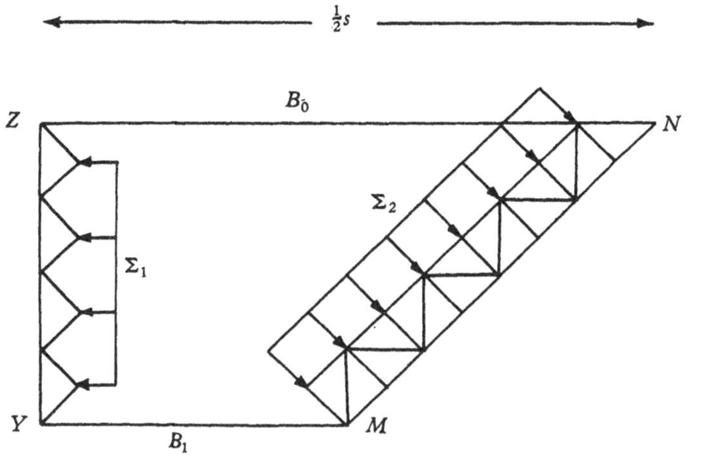

Fig. 4·55. Torsion of a hollow square: the summations Σ_1 and Σ_2 for the multiplication process (4·548), drawn for $n = 16$.

If the convergence of the circling process (4·524) proves slow, we may alternate it with a multiplication process, using the above factor k.

The approximation $n = 8$

The triangulation of an octant for the approximation $n = 8$ is shown in Fig. 4·52, p. 272.

To find the reduced weights b' we have to solve (4·522), viz.

$$\left.\begin{aligned} 4b'(\mathbf{P}') &= \Sigma b'(\mathbf{Q}') - d(\mathbf{P}')\,[4b'(\mathbf{R})] + 4,\\ 4b'(\mathbf{Q}') &= \Sigma b'(\mathbf{P}') - d(\mathbf{Q}')\,[4b'(\mathbf{R})] + 8,\\ 4b'(\mathbf{R}) &= \tfrac{1}{12}\Big(224 - \sum_{MN}\Big). \end{aligned}\right\} \qquad (4·549)$$

To solve these, a set of boxes is drawn as in Fig. 4·56*a*, each box corresponding to a junction point of the triangulation of Fig. 4·52.

We start with zero in every box, and calculate $4b'(\mathbf{R})$ from the last of (4·549); this gives $4b'(\mathbf{R}) = 18{\cdot}66667$. Then, with this value, we use the first two lines of (4·549) to get values for $b'(\mathbf{P}')$ and $b'(\mathbf{Q}')$, working over the boxes from left to right and downwards. We then recalculate $4b'(\mathbf{R})$ and repeat the operation. Convergence is rapid—a matter of ten rounds of the circling process—and it is unnecessary to use the multiplication process except at the very end. By (4·543) the k for the multiplication process is given by

$$k = \nu/D, \quad \nu = 480, \quad D = 12[4\Sigma_1 + 3b'(C_0) + b'(C_1)] - 3\sum_{MN}. \tag{4·550}$$

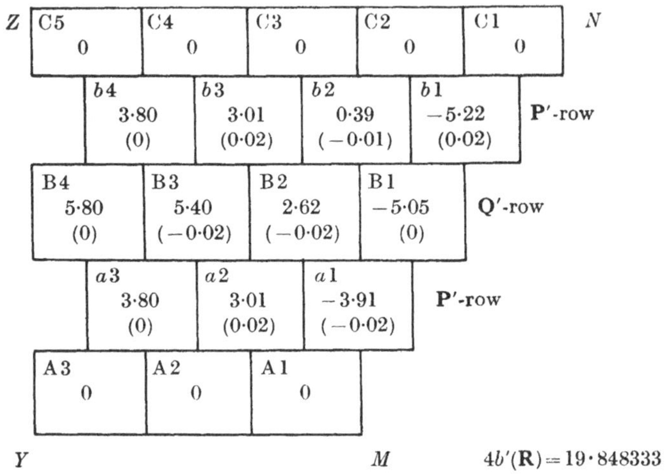

Fig. 4·56*a*. Torsion of a hollow square: approximation $n = 8$. Table of reduced weights $b'(\mathbf{P}')$, $b'(\mathbf{Q}')$, $4b'(\mathbf{R})$ for centres as in Fig. 4·52, satisfying (4·522) approximately. The values of $4f'(\mathbf{P}')$, $4f'(\mathbf{Q}')$, calculated from (4·529), are shown in parentheses.

The values of $b'(\mathbf{P}')$, $b'(\mathbf{Q}')$ and $4b'(\mathbf{R})$, thus obtained, are shown in Fig. 4·56*a*. It should be noted that the four neighbours of a box are the two boxes above it and the two boxes below it; boxes in the same horizontal row are not neighbours. On the edges of the octant we have to take care of neighbours in the adjacent octants, and this is easy to do.

The values so far obtained give only the 'unreliable' bound of (4·523). To get the 'reliable' bound of (4·530), we next calculate the reduced errors f'_ρ from (4·529); we note that $f'(\mathbf{R})$ is zero, since $b'(\mathbf{R})$ is always calculated from the last of (4·549), and so has no error. The values of $4f'_\rho$ are shown in Fig. 4·56*a* in parentheses below the values of b'_ρ.

The calculations are then as follows:

$$\left.\begin{aligned}\Sigma b'(\mathbf{P}') + 2\Sigma b'(\mathbf{Q}') &= 209{\cdot}88 \text{ (summed over all eight octants)},\\ \tfrac{7}{4}n^3 b'(\mathbf{R}) &= 224[4b'(\mathbf{R})] = 4446{\cdot}0266.\end{aligned}\right\} \qquad (4{\cdot}551)$$

By (4·523)

$$\text{unreliable lower bound} = L_u = \tfrac{1}{9}n^{-4}[\Sigma b'(\mathbf{P}') + 2\,\Sigma b'(\mathbf{Q}') + \tfrac{7}{4}n^3 b'(\mathbf{R})]$$
$$= 4655{\cdot}9066/36864 = 0{\cdot}12629954. \qquad (4{\cdot}552)$$

Summing over all eight octants

$$4\Sigma b'_\rho f'_\rho = -0{\cdot}456, \quad \Sigma b'_\rho f'_\rho = -0{\cdot}114. \qquad (4{\cdot}553)$$

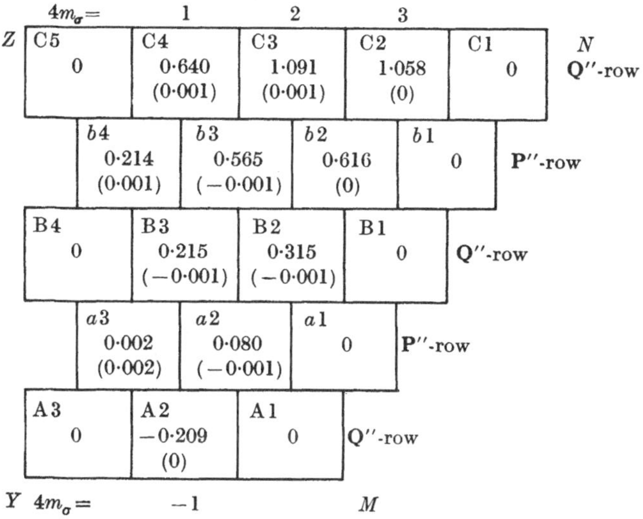

Fig. 4·56*b*. Torsion of a hollow square: approximation $n=8$. Table of reduced weights $b''(\mathbf{P}'')$, $b''(\mathbf{Q}'')$ for centres as in Fig. 4·52, satisfying (4·524) approximately. The values of $4f''(\mathbf{P}'')$, $4f''(\mathbf{Q}'')$, calculated from (4·535), are shown in parentheses. The values of $4m_\sigma$ are also shown.

By (4·530)

$$\text{reliable lower bound} = L_r = L_u - \tfrac{1}{9}n^{-4}\Sigma b'_\rho f'_\rho$$
$$= L_u + 0{\cdot}114/36864 = L_u + 0{\cdot}00003092$$
$$= 0{\cdot}12630263. \qquad (4{\cdot}554)$$

For the reduced weights b''_σ we have to solve (4·524), and for that we use the set of boxes shown in Fig. 4·56*b*. Starting from zero values, stable values are obtained in sixteen rounds; these values are shown in Fig. 4·56*b*, together with $4f''_\sigma$ (in parentheses) obtained from (4·535).

The calculations are then as follows:

$$4 \sum_{ZN+YM} m_\sigma b''_\sigma = 6{\cdot}205, \qquad \sum_{ZN+YM} m_\sigma b''_\sigma = 1{\cdot}55125. \qquad (4{\cdot}555)$$

By (4·525)

$$\begin{aligned} \text{unreliable upper bound} = U_u &= n^{-2}\Big(\tfrac{5}{32}n^2 - \sum_{ZN+YM} m_\sigma b''_\sigma\Big) \\ &= \tfrac{1}{64}(10 - 1{\cdot}55125) \\ &= 0{\cdot}13201171875. \end{aligned} \qquad (4{\cdot}556)$$

Summing over the octant

$$4 \sum_{\text{oct}} b''_\sigma f''_\sigma = 0{\cdot}000774, \qquad \sum_{\text{oct}} b''_\sigma f''_\sigma = 0{\cdot}0001935. \qquad (4{\cdot}557)$$

By (4·536)

$$\begin{aligned} \text{reliable upper bound} = U_r &= U_u + 2n^{-2} \sum_{\text{oct}} b''_\sigma f''_\sigma \\ &= U_u + 0{\cdot}0001935/32 \\ &= U_u + 0{\cdot}0000060046875 \\ &= 0{\cdot}1320177656 25. \end{aligned} \qquad (4{\cdot}558)$$

Thus U_r is very slightly greater than U_u.

The accuracy used is rather more than the crudeness of the triangulation $n = 8$ would seem to warrant, if practical results were our goal. But it is satisfactory to see the close agreement between the 'unreliable' and the 'reliable'. Here are the results, cut down to five decimal places:

Torsion of a hollow square of sides s, $\frac{1}{2}s$: approximation $n = 8$. Bounds on Γ/s^4

	Lower bound	Upper bound
Unreliable	0·12629	0·13202
Reliable	0·12630	0·13202

The approximation $n = 16$

Fig. 4·57*a* shows the triangulation of an octant of the hollow square for the approximation $n = 16$.

After thirty-three rounds of circling and multiplication, steady values of b'_ρ are obtained as shown in Fig. 4·57*b*, which shows also the values of $4f'_\rho$ (reduced errors, multiplied by 4). The calculations are then as follows:

$$\left.\begin{aligned} \Sigma b'(\mathbf{P}') + 2\Sigma b'(\mathbf{Q}') &= 3970{\cdot}40, \\ \tfrac{7}{4}n^3 b'(\mathbf{R}) &= 71674{\cdot}0326. \end{aligned}\right\} \qquad (4{\cdot}559)$$

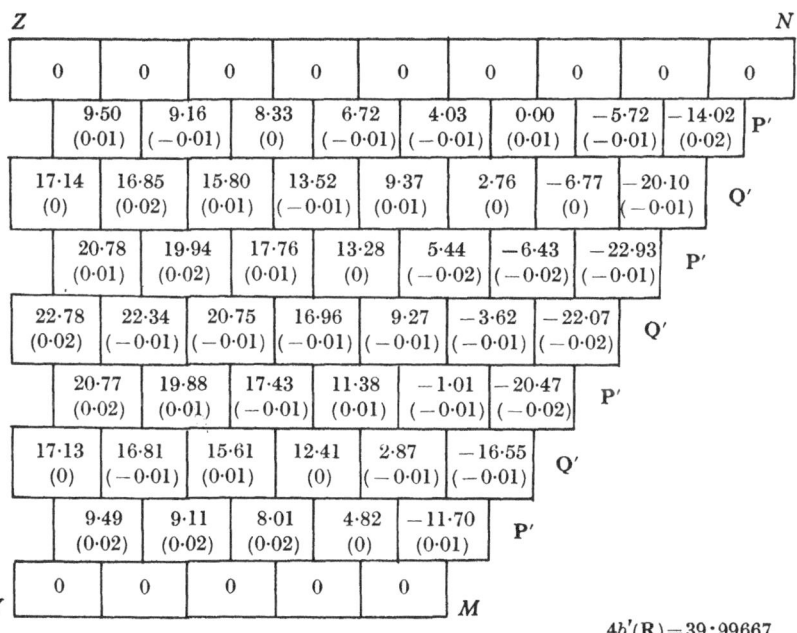

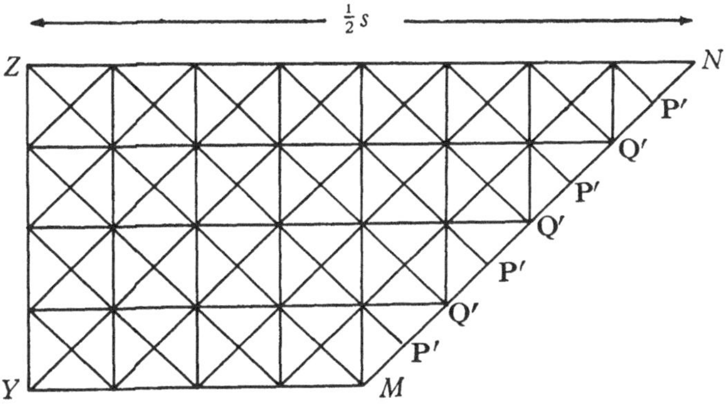

Fig. 4·57*a*. Torsion of a hollow square: approximation $n=16$. Triangulation of an octant.

Z — N

0	0	0	0	0	0	0	0	0	
9·50 (0·01)	9·16 (−0·01)	8·33 (0)	6·72 (−0·01)	4·03 (−0·01)	0·00 (0·01)	−5·72 (−0·01)	−14·02 (0·02)	P′	
17·14 (0)	16·85 (0·02)	15·80 (0·01)	13·52 (−0·01)	9·37 (0·01)	2·76 (0)	−6·77 (0)	−20·10 (−0·01)	Q′	
20·78 (0·01)	19·94 (0·02)	17·76 (0·01)	13·28 (0)	5·44 (−0·02)	−6·43 (−0·02)	−22·93 (−0·01)	P′		
22·78 (0·02)	22·34 (−0·01)	20·75 (−0·01)	16·96 (−0·01)	9·27 (−0·01)	−3·62 (−0·01)	−22·07 (−0·02)	Q′		
20·77 (0·02)	19·88 (0·01)	17·43 (−0·01)	11·38 (0·01)	−1·01 (−0·01)	−20·47 (−0·02)	P′			
17·13 (0)	16·81 (−0·01)	15·61 (0·01)	12·41 (0)	2·87 (−0·01)	−16·55 (−0·01)	Q′			
9·49 (0·02)	9·11 (0·02)	8·01 (0·02)	4·82 (0)	−11·70 (0·01)	P′				
0	0	0	0	0					

Y — M

$4b'(\mathbf{R}) = 39{\cdot}99667$

Fig. 4·57*b*. Torsion of a hollow square: approximation $n=16$. Table of reduced weights $b'(\mathbf{P}')$, $b'(\mathbf{Q}')$, $4b'(\mathbf{R})$ for centres as in Fig. 4·57*a*, satisfying (4·522) approximately. The values of $4f'(\mathbf{P}')$, $4f'(\mathbf{Q}')$, calculated from (4·529), are shown in parentheses.

By (4·523)

$$\text{unreliable lower bound} = L_u = \tfrac{1}{9}n^{-4}[\Sigma b'(\mathbf{P}') + 2\,\Sigma b'(\mathbf{Q}') + \tfrac{7}{4}n^3 b'(\mathbf{R})]$$
$$= 75644{\cdot}4326/589824 = 0{\cdot}12824916. \qquad (4{\cdot}560)$$

Summing over all eight octants,

$$4\Sigma b'_\rho f'_\rho = 18{\cdot}8904, \quad \Sigma b'_\rho f'_\rho = 4{\cdot}7226. \qquad (4{\cdot}561)$$

Z $4m_\sigma =$		0·500	1·000	1·500	2·000	2·500	3·000	3·500	N
Q″	0	0·636 (0)	1·238 (−0·001)	1·761 (−0·001)	2·139 (−0·001)	2·285 (−0·001)	2·092 (0)	1·417 (0·001)	0
P″	0·259 (−0·001)	0·763 (0·001)	1·214 (−0·001)	1·559 (0)	1·720 (−0·001)	1·601 (−0·001)	1·083 (−0·002)	0	
Q″	0	0·401 (−0·002)	0·776 (0)	1·082 (0)	1·254 (0)	1·203 (0)	0·825 (−0·002)	0	
P″	0·148 (0)	0·436 (0)	0·691 (−0·001)	0·864 (−0·001)	0·873 (−0·001)	0·618 (−0·002)	0		
Q″	0	0·191 (0·002)	0·376 (0·002)	0·531 (−0·001)	0·590 (0)	0·446 (0)	0		
P″	0·043 (−0·001)	0·135 (0·002)	0·240 (0·001)	0·330 (0·001)	0·293 (0·001)	0			
Q″	0	−0·018 (0·001)	−0·011 (0)	0·063 (0·001)	0·135 (0·001)	0			
P″	−0·067 (0)	−0·184 (0·002)	−0·235 (0·001)	−0·084 (0)	0				
Q″	0	−0·250 (0·001)	−0·459 (0·001)	−0·534 (0·001)	0				
Y $4m_\sigma =$		−0·500	−1·000	−1·500	M				

Fig. 4·57*c*. Torsion of a hollow square: approximation $n = 16$. Table of reduced weights $b''(\mathbf{P}'')$, $b''(\mathbf{Q}'')$ for centres shown in Fig. 4·57*a*, satisfying (4·524) approximately. The values of $4f''(\mathbf{P}'')$, $4f''(\mathbf{Q}'')$ calculated from (4·535) are shown in parentheses.

By (4·530)

$$\text{reliable lower bound} = L_r = L_u - \tfrac{1}{9}n^{-4}\,\Sigma b'_\rho f'_\rho$$
$$= L_u - 4{\cdot}7226/589824 = L_u - 0{\cdot}00000801$$
$$= 0{\cdot}12824115. \qquad (4{\cdot}562)$$

After forty-eight rounds of circling and multiplying, we obtain values of b''_σ as shown in Fig. 4·57*c*, which shows also in parentheses the quadrupled relative errors, $4f''_\sigma$. The calculations are then as follows:

$$4\sum_{ZN+YM} m_\sigma b''_\sigma = 26{\cdot}8085, \qquad \sum_{ZN+YM} m_\sigma b''_\sigma = 6{\cdot}702125. \qquad (4{\cdot}563)$$

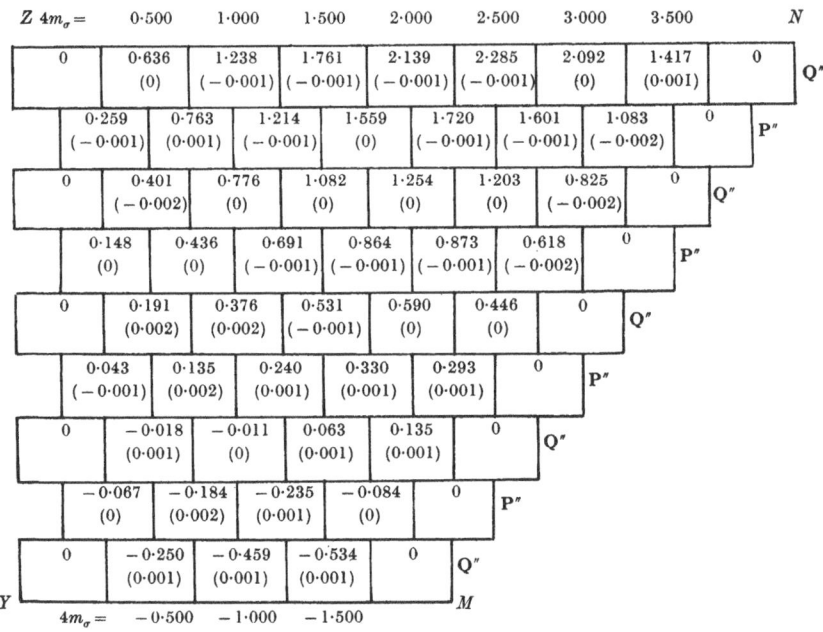

By (4·525)

$$\text{unreliable upper bound} = U_u = n^{-2}\Big(\tfrac{5}{32}n^2 - \sum_{ZN+YM} m_\sigma b''_\sigma\Big)$$
$$= (40 - 6{\cdot}702125)/256$$
$$= 0{\cdot}1300\,6982. \qquad (4{\cdot}564)$$

Summing over the octant,

$$4\sum_{\text{oct}} b''_\sigma f''_\sigma = -0{\cdot}018292, \quad \sum_{\text{oct}} b''_\sigma f''_\sigma = -0{\cdot}004573. \qquad (4{\cdot}565)$$

By (4·536)

$$\text{reliable upper bound} = U_r = U_u + 2n^{-2}\sum_{\text{oct}} b''_\sigma f''_\sigma$$
$$= U_u - 0{\cdot}004573/128$$
$$= 0{\cdot}1300\,6982 - 0{\cdot}0000\,3572$$
$$= 0{\cdot}1300\,3410. \qquad (4{\cdot}566)$$

Approximation for stress and warping

As for the hexagon on p. 262, we now approximate to the stress for the hollow square in torsion, but we must not forget the F-vector **R** which we introduced on account of the connectivity of the section. We go back to the first of (4·237), viz.

$$\tilde{\mathbf{E}}/\mu\alpha \sim \sum_{\rho=1}^{r} a'_\rho \mathbf{T}'_\rho = \Sigma a'(\mathbf{P}')\,\mathbf{P}' + \Sigma a'(\mathbf{Q}')\,\mathbf{Q}' + a'(\mathbf{R})\,\mathbf{R}, \qquad (4{\cdot}567)$$

the sums being over all eight octants. This gives the stress approximately. As in (4·240) the lines of stress are given approximately by

$$\Sigma a'(\mathbf{P}')\,u'(\mathbf{P}') + \Sigma a'(\mathbf{Q}')\,u'(\mathbf{Q}') + a'(\mathbf{R})\,u'(\mathbf{R}) = \text{const.}, \qquad (4{\cdot}568)$$

where $u'(\mathbf{P}')$, $u'(\mathbf{Q}')$ are pyramid functions of unit height and $u'(\mathbf{R})$ is the truncated pyramid function shown in Fig. 4·53. By (4·506) we may write this equation in terms of the reduced weights:

$$\Sigma b'(\mathbf{P}')\,u'(\mathbf{P}') + \Sigma b'(\mathbf{Q}')\,u'(\mathbf{Q}') + 2nb'(\mathbf{R})\,u'(\mathbf{R}) = \text{const.}, \qquad (4{\cdot}569)$$

for the approximation n.

This expression is a linear function of position in each triangle of the network. For $n = 16$ (triangulation as in Fig. 4·57*a*) the values at the junction points are shown in Fig. 4·58*a*. These values are obtained from the values in Fig. 4·57*b* by adding to the bottom row (where $u'(\mathbf{R}) = \tfrac{1}{2}$) the number $16b'(\mathbf{R})$, and to the other rows in proportion. The first row of zeros is unchanged, the next row receives the increment

$$16b'(\mathbf{R})/8 = 2b'(\mathbf{R}) = 39{\cdot}99667/2 = 19{\cdot}99833,$$

and the succeeding rows this number multiplied by 2, 3, ... The approximate lines of stress are the level lines on the resulting polyhedral surface; they are shown in Fig. 4·58*b*.

Z 0	0	0	0	0	0	0	0	0 *N*
29·50	29·16	28·33	26·72	24·03	20·00	14·28	5·98	
57·13	56·84	55·79	53·51	49·36	42·76	33·23	19·90	
80·76	79·92	77·74	73·26	65·43	53·56	37·06		
102·77	102·33	100·74	96·95	89·26	76·37	57·92		
120·76	119·87	117·42	111·37	98·98	79·52			
137·12	136·80	135·60	132·40	122·86	103·44			
149·48	149·10	148·00	144·81	128·29				
Y 159·99	159·99	159·99	159·99	159·99 *M*				

Fig. 4·58*a*. Torsion of a hollow square: approximation $n=16$. Table of values of approximate stress function as in (4·569), obtained from values given in Fig. 4·57*b* by addition of terms arising from $b'(\mathbf{R})$.

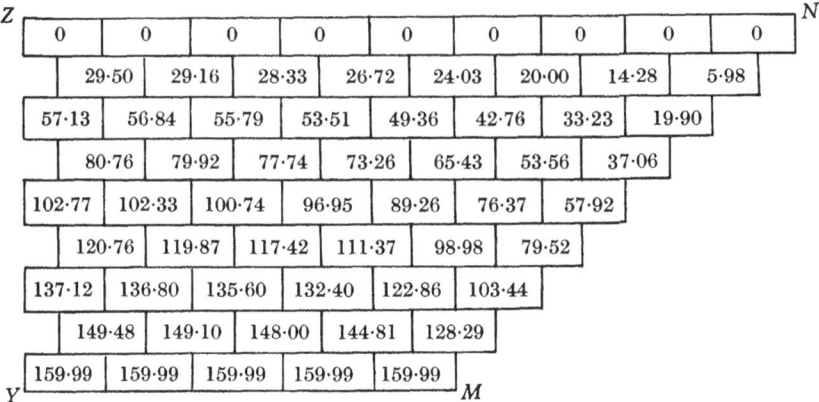

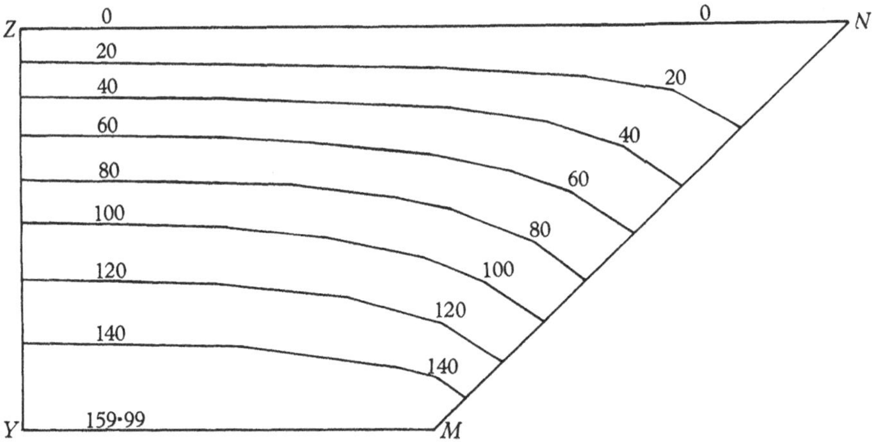

Fig. 4·58*b*. Torsion of a hollow square: approximation $n=16$. Approximate lines of stress, being actually the level lines of the polyhedral function corresponding to Fig. 4·58*a*.

By (4·242) and (4·515) the lines of constant warping are

$$\Sigma b''(\mathbf{P}'')\,u(\mathbf{P}'')+\Sigma b''(\mathbf{Q}'')\,u(\mathbf{Q}'')=\text{const.}, \tag{4·570}$$

where $u(\mathbf{P}'')$, $u(\mathbf{Q}'')$ are again pyramid functions of unit height. This expression represents a polyhedral function with heights as indicated in Fig. 4·57*c*, and its level lines, as shown in Fig. 4·59, give the lines of constant warping approximately.

Summary of results for the torsion of a hollow square

For a hollow square of outer side s and inner side $\frac{1}{2}s$ the following reliable bounds for the torsional rigidity Γ have been obtained:

	Bounds on Γ/s^4	
Approximation	Lower bound	Upper bound
$n=8$	0·12630	0·13202
$n=16$	0·12824	0·13004

For higher approximations calculated on a high-speed computer, see Note A, p. 410.

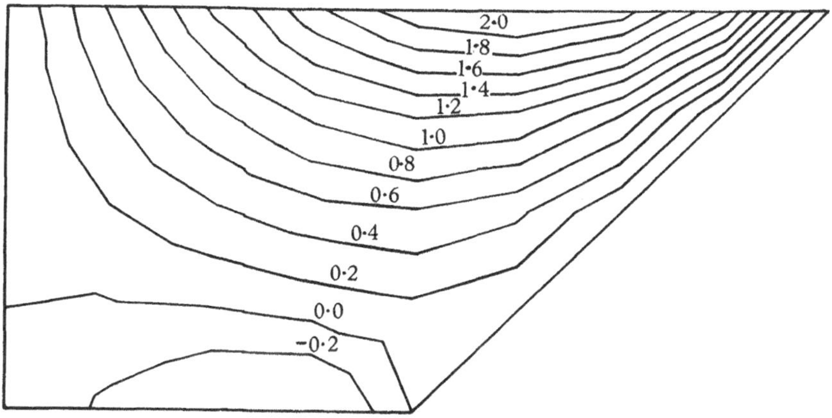

Fig. 4·59. Torsion of a hollow square: approximation $n=16$. Approximate lines of constant warping, being actually the level lines of the polyhedral function corresponding to Fig. 4·57*c*.

Exercises

1. Consider the application of the method of Example 2, p. 269, to the determination of the weights b'_ρ in Fig. 4·52 (approximation $n=8$). Show that if zero weights are assigned to $A\,1$, $A\,2$, $A\,3$, $C\,1$ and any five arbitrary weights to $a\,1$, $a\,2$, $a\,3$, $B\,4$ and to the ring vector $\mathbf{R}$, then application of the difference equations in the first two lines of (4·522) leads directly to the determination of four weights at $C\,2$, $C\,3$, $C\,4$, $C\,5$ and of an error $e'(\mathbf{R})$ given by the last line of (4·522), viz.

$$e'(\mathbf{R}) = b'(\mathbf{R}) - \tfrac{14}{3} + \tfrac{1}{48}\sum_{MN}\{b'(\mathbf{P}') + b'(\mathbf{Q}')\};$$

consequently the determination of the weights satisfying (4·522) can be reduced to solving five linear simultaneous equations.

2. Show that in the approximation n (a multiple of 4) the above method requires the solution of $\frac{1}{2}n+1$ linear simultaneous equations.

CHAPTER 5

VARIOUS BOUNDARY VALUE PROBLEMS

5·1. BOUNDARY VALUE PROBLEMS CONNECTED WITH VARIATIONAL PRINCIPLES

The possibility of applying the hypercircle method to a boundary value problem depends very much on whether the problem corresponds to a variational principle. In fact, for a wide class of variational principles, we can lay down a standard procedure; this is done below, following the method of A. J. McConnell (1). In this procedure we find that indefinite F-metrics occur as well as definite ones. For other more general treatments see P. Cooperman (1) and H. Fujita (1).

A variational principle and corresponding differential equation for a single unknown

Consider, as P-space, a domain V in Euclidean N-space ($N \geqslant 2$). The extension to a Riemannian space is easily made—for simplicity we shall stick to the Euclidean case. But actually the P-metric is important only if transformations of coordinates are used.

We may think of V as a finite domain, with boundary B. But if there is convergence at infinity, V may be infinite with interior boundary B.

Consider the integral

$$I = \int (a^{ij} u_{,i} u_{,j} - cu^2)\, dV, \tag{5·101}$$

where a^{ij} and c are given functions of the coordinates x_i, a^{ij} being symmetric ($a^{ij} = a^{ji}$). Suffixes take the range $1, 2, \ldots, N$ and summation is understood for repeated suffixes. Here u is a function of the coordinates and the comma denotes partial differentiation ($u_{,i} = \partial u / \partial x_i$).

Let us vary the function u. Then

$$\begin{aligned} \delta I &= 2 \int (a^{ij} u_{,j} \delta u_{,i} - cu\,\delta u)\, dV \\ &= 2 \int a^{ij} u_{,j} n_i \delta u\, dB - 2 \int [(a^{ij} u_{,j})_{,i} + cu]\, \delta u\, dV, \end{aligned} \tag{5·102}$$

where n_i is the unit normal to B, drawn outward. The stationary condition (variational principle) $\delta I = 0$ thus leads to the partial differential equation

$$(a^{ij} u_{,j})_{,i} + cu = 0. \tag{5·103}$$

As we shall see, the integral (5·101), from which (5·103) follows by variation, suggests the suitable F-metric to use when we seek to approximate to solutions of (5·103) by the hypercircle method.

Splitting the differential equation

Let us consider, without reference to boundary conditions for the present, the partial differential equation

$$(a^{ij}u_{,j})_{,i}+cu=0. \tag{5·104}$$

We introduce function-space, defining an F-point by

$$\mathbf{S}\leftrightarrow(p_i, u), \tag{5·105}$$

where p_i is a P-vector field and u a P-scalar field in a region V (P-space).

Consider two linear subspaces, defined as follows:

$$\left.\begin{aligned} L_0'\colon\quad \mathbf{S}'\leftrightarrow(p_i', u'),\quad & p_i'=u'_{,i};\\ L_0''\colon\quad \mathbf{S}''\leftrightarrow(p_i'', u''),\quad & (a^{ij}p_j'')_{,i}+cu''=0.\end{aligned}\right\} \tag{5·106}$$

It is clear that any F-point common to L_0' and L_0'' satisfies (5·104), in the sense of the correspondence (5·105) with $p_i=u_{,i}$; there are of course infinitely many solutions since no boundary conditions have so far been imposed.

Note that L_0' and L_0'' both contain the origin of function-space.

The scalar product

To deal with the partial differential equation (5·104) we take, as scalar product of any two F-vectors,

$$\mathbf{S}'\leftrightarrow(p_i', u') \quad\text{and}\quad \mathbf{S}''\leftrightarrow(p_i'', u''),$$

$$\mathbf{S}'.\mathbf{S}''=\int_V (a^{ij}p_i'p_j''-cu'u'')\,dV, \tag{5·107}$$

which gives the metric

$$\mathbf{S}^2=\int_V (a^{ij}p_ip_j-cu^2)\,dV. \tag{5·108}$$

At this point the question of positive-definite character arises. We see that *the F-metric is positive-definite if*

> $a^{ij}p_ip_j$ is a positive-definite quadratic form and
> c is *negative* throughout V.

An indefinite metric can occur in a number of ways, but one is of particular interest because it arises in vibration problems; the F-metric is indefinite if

$a^{ij}p_ip_j$ is a positive-definite quadratic form and
c is positive throughout V.

The case $c=0$ deserves mention, because then the partial differential equation is of the Laplace type, and in fact becomes the Laplace equation if $a^{ij}=\delta_{ij}$, the Kronecker delta. If $c=0$, we revise the formulae (5·104)–(5·108) to read as follows:

Case where $c=0$:

$$\left.\begin{aligned} &(a^{ij}u_{,j})_{,i}=0,\\ &\mathbf{S}\leftrightarrow(p_i),\\ L_0':\quad &\mathbf{S}'\leftrightarrow(p_i'),\quad p_i'=u'_{,i}\quad (u' \text{ some scalar}),\\ L_0'':\quad &\mathbf{S}''\leftrightarrow(p_i''),\quad (a^{ij}p_j'')_{,i}=0,\\ &\mathbf{S}'.\mathbf{S}''=\int a^{ij}p_i'p_j''\,dV,\quad \mathbf{S}^2=\int a^{ij}p_ip_j\,dV. \end{aligned}\right\} \tag{5·109}$$

Now the F-metric is positive-definite if $a^{ij}p_ip_j$ is a positive-definite quadratic form.

Let us compute the scalar product of a vector $\mathbf{T}'$ lying in L_0' and a vector $\mathbf{T}''$ lying in L_0''. We shall treat the general case $c\neq 0$ as in (5·106); the results for $c=0$ will then be obvious. Accordingly, we write

$$\left.\begin{aligned} &\mathbf{T}'\leftrightarrow(p_i',u'),\quad p_i'=u'_{,i},\\ &\mathbf{T}''\leftrightarrow(p_i'',u''),\quad (a^{ij}p_j'')_{,i}+cu''=0. \end{aligned}\right\} \tag{5·110}$$

Using Green's theorem, we find

$$\begin{aligned} \mathbf{T}'.\mathbf{T}''&=\int(a^{ij}p_i'p_j''-cu'u'')\,dV\\ &=\int(a^{ij}u'_{,i}p_j''-cu'u'')\,dV\\ &=\int u'a^{ij}n_ip_j''\,dB-\int[(a^{ij}p_j'')_{,i}+cu'']\,u'\,dV. \end{aligned} \tag{5·111}$$

The last term vanishes by (5·110), but the first does not, because we have not so far imposed boundary conditions. Thus the *splitting* of the problem, which we have partly achieved, is a matter for the most part of the differential equation; the *orthogonality* of the linear subspaces involves the boundary conditions, and these must be considered before we can apply the hypercircle method.

The orthogonal linear subspaces

Consider again the partial differential equation (5·104),

$$(a^{ij}u_{,j})_{,i}+cu=0, \tag{5·112}$$

but now in association with the boundary conditions

$$(u)_{B_1}=f,\quad (a^{ij}u_{,j}n_i)_{B_2}=g. \tag{5·113}$$

This notation means that the boundary B is divided into two portions B_1 and B_2 (symbolically, $B = B_1 + B_2$), the first of the two boundary conditions being satisfied on B_1 and the second on B_2. In particular, either B_1 or B_2 may disappear, leaving us with a boundary value problem of the Neumann or Dirichlet type, respectively.

Using as a guide the splitting of the partial differential equation effected in (5·106), we now *split the problem* in the sense that we reduce the boundary value problem contained in (5·112) and (5·113) to finding the intersection of the following two linear subspaces:

$$\left.\begin{aligned} L'\colon &\quad \mathbf{S}' \leftrightarrow (p'_i, u'), \quad p'_i = u'_{,i}, \quad (u')_{B_1} = f; \\ L''\colon &\quad \mathbf{S}'' \leftrightarrow (p''_i, u''), \quad (a^{ij} p''_j)_{,i} + cu'' = 0, \quad (a^{ij} p''_j n_i)_{B_2} = g. \end{aligned}\right\} \tag{5·114}$$

For F-vectors *lying in* these subspaces we have then

$$\left.\begin{aligned} L'\colon &\quad \mathbf{T}' \leftrightarrow (p'_i, u'), \quad p'_i = u'_{,i}, \quad (u')_{B_1} = 0; \\ L''\colon &\quad \mathbf{T}'' \leftrightarrow (p''_i, u''), \quad (a^{ij} p''_j)_{,i} + cu'' = 0, \quad (a^{ij} p''_j n_i)_{B_2} = 0. \end{aligned}\right\} \tag{5·115}$$

It is essential for the application of the hypercircle method that L' and L'' should be orthogonal. It is easy to see that they are, for if we proceed as in (5·111), the integral over B there obtained vanishes by virtue of (5·115) so that we have

$$\mathbf{T}' . \mathbf{T}'' = 0. \tag{5·116}$$

Thus the problem of solving (5·112) and (5·113) is equivalent to finding the intersection of two *orthogonal* linear subspaces L', L'' as defined in (5·114).

If the F-metric is positive-definite [cf. (5·108)], the method of the hypercircle, as developed in Chapter 2, is immediately available. For detailed discussion of an indefinite metric, see Part III.

The case of several unknowns

We shall now consider the variation of the integral

$$I = \int (a^{ij}_{\alpha\beta} u^\alpha_{,i} u^\beta_{,j} - c_{\alpha\beta} u^\alpha u^\beta)\, dV. \tag{5·117}$$

Here again V is a domain in Euclidean N-space, with boundary B. Latin suffixes take the range $1, 2, \ldots, N$ and Greek suffixes the range $1, 2, \ldots, M$, where M has any value; the summation convention applies for both. The coefficients are functions of the coordinates with the symmetries

$$a^{ij}_{\alpha\beta} = a^{ji}_{\alpha\beta} = a^{ij}_{\beta\alpha}, \quad c_{\alpha\beta} = c_{\beta\alpha}. \tag{5·118}$$

The variational principle $\delta I = 0$ leads at once to the system of partial differential equations

$$(a^{ij}_{\alpha\beta} u^\alpha_{,i})_{,j} + c_{\alpha\beta} u^\alpha = 0 \tag{5·119}$$

for the M variables u^α.

We consider now the problem of solving these partial differential equations with the boundary conditions

$$(u^\alpha)_{B_1} = f^\alpha, \quad (a^{ij}_{\alpha\beta} u^\alpha_{,i} n_j)_{B_2} = g_\beta, \tag{5·120}$$

where B_1 and B_2 are two portions into which the boundary B is divided. This is a generalization to M unknowns of the problem for one unknown set out in (5·112) and (5·113).

We define an F-point or F-vector by

$$\mathbf{S} \leftrightarrow (p^\alpha_i, u^\alpha); \tag{5·121}$$

the number of functions involved here is $M(N+1)$, for we have M P-vectors and M scalars in N-space.

The problem is split as follows:

$$\left.\begin{aligned} L'\colon\quad & \mathbf{S}' \leftrightarrow (p'^\alpha_i, u'^\alpha), \quad p'^\alpha_i = u'^\alpha_{,i}, \quad (u'^\alpha)_{B_1} = f^\alpha; \\ L''\colon\quad & \mathbf{S}'' \leftrightarrow (p''^\alpha_i, u''^\alpha), \quad (a^{ij}_{\alpha\beta} p''^\alpha_i)_{,j} + c_{\alpha\beta} u''^\alpha = 0, \quad (a^{ij}_{\alpha\beta} p''^\alpha_i n_j)_{B_2} = g_\beta. \end{aligned}\right\} \tag{5·122}$$

It is clear that an F-point corresponding to the intersection of these two linear subspaces does in fact correspond to the solution of (5·119) and (5·120) in the sense of (5·121), p^α_i being equal to $u^\alpha_{,i}$.

The form of (5·117) suggests that we take as scalar product

$$\mathbf{S}' . \mathbf{S}'' = \int (a^{ij}_{\alpha\beta} p'^\alpha_i p''^\beta_j - c_{\alpha\beta} u'^\alpha u''^\beta)\, dV, \tag{5·123}$$

so that the F-metric is

$$\mathbf{S}^2 = \int (a^{ij}_{\alpha\beta} p^\alpha_i p^\beta_i - c_{\alpha\beta} u^\alpha u^\beta)\, dV. \tag{5·124}$$

This will be positive-definite if $a^{ij}_{\alpha\beta} p^\alpha_i p^\beta_j$ is positive-definite and $c_{\alpha\beta} u^\alpha u^\beta$ negative-definite.

It is easy to see that L' and L'' are orthogonal, and so the method of the hypercircle applies, provided the F-metric (5·124) is positive-definite; if it is not positive-definite, reference must be made to Part III.

More general case

We have dealt above with two problems: the first, with one unknown, is stated in (5·112) and (5·113), and the second, with M unknowns, is stated in (5·119) and (5·120). In both these cases the partial differential equations involved derivatives of the second order, but no higher. This restriction will now be removed; the highest order of derivatives is raised from 2 to $2Q$, where Q is any integer. This generalization involves some complexity of notation, and the reader who does not care for this sort of thing may very well postpone reading this last part of § 5·1 until such time as he sees the need for it.

We consider a set of functions of the coordinates $x_1, x_2, \ldots, x_N$:

$$\left.\begin{array}{l} c_{\alpha\beta}, \\ c_{\alpha\beta}^{i_1|j_1}, \\ c_{\alpha\beta}^{i_1 i_2|j_1 j_2}, \\ \ldots \\ c_{\alpha\beta}^{i_1 i_2 \ldots i_Q|j_1 j_2 \ldots j_Q}. \end{array}\right\} \tag{5·125}$$

Here α, β take the range $1, 2, \ldots, M$ and the i's and j's the range $1, 2, \ldots, N$ (N is the dimensionality of P-space). Summations over these ranges will be understood for repeated suffixes. The above functions are to have certain symmetries:

(i) symmetry in α, β;

(ii) symmetry in the i's and the j's separately;

(iii) symmetry for i's and j's in corresponding places.

Thus, to illustrate,

$$\begin{aligned} c_{\alpha\beta}^{i_1 i_2 \ldots i_P|j_1 j_2 \ldots j_P} &= c_{\beta\alpha}^{i_1 i_2 \ldots i_P|j_1 j_2 \ldots j_P} \\ &= c_{\alpha\beta}^{i_2 i_1 \ldots i_P|j_1 j_2 \ldots j_P} \\ &= c_{\alpha\beta}^{j_1 i_2 \ldots i_P|i_1 j_2 \ldots j_P}. \end{aligned} \tag{5·126}$$

Now let u^α be a set of functions of the coordinates; we shall denote their partial derivatives by a comma as usual. Write the abbreviations

$$\left.\begin{array}{l} F_0 = c_{\alpha\beta} u^\alpha u^\beta, \\ F_1 = c_{\alpha\beta}^{i_1|j_1} u^\alpha_{,i_1} u^\beta_{,j_1}, \\ F_2 = c_{\alpha\beta}^{i_1 i_2|j_1 j_2} u^\alpha_{,i_1 i_2} u^\beta_{,j_1 j_2}, \\ \ldots \\ F_Q = c_{\alpha\beta}^{i_1 i_2 \ldots i_Q|j_1 j_2 \ldots j_Q} u^\alpha_{,i_1 i_2 \ldots i_Q} u^\beta_{,j_1 j_2 \ldots j_Q}. \end{array}\right\} \tag{5·127}$$

Consider the integral

$$I = \int (F_0 + F_1 + \ldots + F_Q)\, dV. \tag{5·128}$$

Apply a variation to u^α and then integrate by parts again and again so as to throw as much as possible out to the boundary B. Thus

$$\begin{aligned} \delta I = 2\int \{ c_{\alpha\beta} u^\alpha - (c_{\alpha\beta}^{i_1|j_1} u^\alpha_{,i_1})_{,j_1} + (c_{\alpha\beta}^{i_1 i_2|j_1 j_2} u^\alpha_{,i_1 i_2})_{,j_1 j_2} - \ldots \\ + (-1)^Q (c_{\alpha\beta}^{i_1 i_2 \ldots i_Q|j_1 j_2 \ldots j_Q} u^\alpha_{,i_1 i_2 \ldots i_Q})_{,j_1 j_2 \ldots j_Q} \} \delta u^\beta dV + I_B, \end{aligned} \tag{5·129}$$

where I_B is an integral over the boundary B. Thus the variational principle $\delta I = 0$ leads to the following M linear homogeneous partial differential equations of order $2Q$ for the M variables u^α:

$$c_{\alpha\beta} u^\alpha - (c_{\alpha\beta}^{i_1|j_1} u^\alpha_{,i_1})_{,j_1} + (c_{\alpha\beta}^{i_1 i_2|j_1 j_2} u^\alpha_{,i_1 i_2})_{,j_1 j_2} - \dots \\ + (-1)^Q (c_{\alpha\beta}^{i_1 i_2 \dots i_Q|j_1 j_2 \dots j_Q} u^\alpha_{,i_1 i_2 \dots i_Q})_{,j_1 j_2 \dots j_Q} = 0, \quad (5\cdot130)$$

where of course $\beta = 1, 2, \dots, M$.

With these partial differential equations let us associate the following boundary conditions:

$$(u^\alpha)_B = f^\alpha, \quad (u^\alpha_{,i_1})_B = f^\alpha_{i_1}, \quad \dots, \quad (u^\alpha_{,i_1 i_2 \dots i_{Q-1}})_B = f^\alpha_{i_1 i_2 \dots i_{Q-1}}, \quad (5\cdot131)$$

the f's being given functions on B. Then (5·130) and (5·131) contain our problem.

We introduce F-space by the correspondence

$$\mathbf{S} \leftrightarrow (u^\alpha, p^\alpha_{i_1}, p^\alpha_{i_1 i_2}, \dots, p^\alpha_{i_1 i_2 \dots i_Q}), \quad (5\cdot132)$$

the p's being tensors, symmetric in all their subscripts. We split the problem by defining linear subspaces as follows:

$$\left.\begin{array}{l} L': \quad \mathbf{S}' \leftrightarrow (u'^\alpha, p'^\alpha_{i_1}, p'^\alpha_{i_1 i_2}, \dots, p'^\alpha_{i_1 i_2 \dots i_Q}), \\ \quad p'^\alpha_{i_1} = u'^\alpha_{,i_1}, \dots, p'^\alpha_{i_1 i_2 \dots i_Q} = u'^\alpha_{,i_1 i_2 \dots i_Q}, \\ \quad (u'^\alpha)_B = f^\alpha, (u'^\alpha_{,i_1})_B = f^\alpha_{i_1}, \dots, (u'^\alpha_{,i_1 i_2 \dots i_{Q-1}})_B = f^\alpha_{i_1 i_2 \dots i_{Q-1}}; \\ L'': \quad \mathbf{S}'' \leftrightarrow (u''^\alpha, p''^\alpha_{i_1}, p''^\alpha_{i_1 i_2}, \dots, p''^\alpha_{i_1 i_2 \dots i_Q}), \\ \quad c_{\alpha\beta} u''^\beta - (c_{\alpha\beta}^{i_1|j_1} p''^\beta_{i_1})_{,j_1} + (c_{\alpha\beta}^{i_1 i_2|j_1 j_2} p''^\beta_{i_1 i_2})_{,j_1 j_2} - \dots \\ \quad\quad + (-1)^Q (c_{\alpha\beta}^{i_1 i_2 \dots i_Q|j_1 j_2 \dots j_Q} p''^\beta_{i_1 i_2 \dots i_Q})_{,j_1 j_2 \dots j_Q} = 0. \end{array}\right\} \quad (5\cdot133)$$

There are no boundary conditions for $\mathbf{S}''$. It is clear that these subspaces split the problem in the sense that, if they intersect, their point of intersection is the solution.

As scalar product of any two F-vectors $\mathbf{S}'$, $\mathbf{S}''$ we shall take the following expression, suggested by (5·128):

$$\mathbf{S}' . \mathbf{S}'' = \int (c_{\alpha\beta} u'^\alpha u''^\beta + c_{\alpha\beta}^{i_1|j_1} p'^\alpha_{i_1} p''^\beta_{j_1} + \dots \\ + c_{\alpha\beta}^{i_1 i_2 \dots i_Q|j_1 j_2 \dots j_Q} p'^\alpha_{i_1 i_2 \dots i_Q} p''^\beta_{j_1 j_2 \dots j_Q}) \, dV. \quad (5\cdot134)$$

Now let $\mathbf{T}'$, $\mathbf{T}''$ be vectors *lying in* L', L'' respectively. This means that the specification of $\mathbf{T}'$ is the same as that of $\mathbf{S}'$ in (5·133), except that the boundary conditions are now

$$(u'^\alpha)_B = 0, \quad (u'^\alpha_{,i_1})_B = 0, \quad \dots, \quad (u'^\alpha_{,i_1 i_2 \dots i_{Q-1}})_B = 0; \quad (5\cdot135)$$

$\mathbf{T}''$ is the same as $\mathbf{S}''$.

To establish the orthogonality of L' and L'', we calculate $\mathbf{T}'.\mathbf{T}''$ from the formula (5·134). We substitute

$$p'^{\alpha}_{i_1} = u'^{\alpha}_{,i_1}, \quad p'^{\alpha}_{i_1 i_2} = u'^{\alpha}_{,i_1 i_2}, \quad \ldots, \quad p'^{\alpha}_{i_1 i_2 \ldots i_Q} = u'^{\alpha}_{,i_1 i_2 \ldots i_Q},$$

and integrate by parts, throwing terms on to the boundary until all derivatives of u'^{α} have disappeared from the integral over V. The integrand of this integral then contains u'^{α} multiplied by an expression which vanishes on account of the differential equation contained in (5·133) as the specification of $\mathbf{S}''$, or equivalently of $\mathbf{T}''$. As for the integral over B, each term contains as a factor one of the following:

$$u'^{\alpha}, \quad u'^{\alpha}_{,i_1}, \quad \ldots, \quad u'^{\alpha}_{,i_1 i_2 \ldots i_{Q-1}},$$

and these all vanish by (5·135). Hence $\mathbf{T}'.\mathbf{T}'' = 0$, so that L' and L'' are orthogonal.

Since the problem contained in (5·130) and (5·131) is thus reduced to finding the intersection of two orthogonal linear subspaces, the method of the hypercircle is immediately available, provided the F-metric is positive-definite—otherwise we must refer to Part III. This metric is

$$\mathbf{S}^2 = \int (c_{\alpha\beta} u^{\alpha} u^{\beta} + c^{i_1|j_1}_{\alpha\beta} p^{\alpha}_{i_1} p^{\beta}_{j_1} + \ldots \\ + c^{i_1 i_2 \ldots i_Q | j_1 j_2 \ldots j_Q}_{\alpha\beta} p^{\alpha}_{i_1 i_2 \ldots i_Q} p^{\beta}_{j_1 j_2 \ldots j_Q}) \, dV. \quad (5·136)$$

The boundary conditions (5·131) are the natural generalization of the boundary condition of the Dirichlet problem. More general boundary conditions will not be discussed here, but it may be said that such generalization involves switching some of the boundary conditions from L' to L'' in (5·133) in such a way that the problem is still split by (5·133) into the problem of finding the intersection of two orthogonal linear subspaces; the scalar product will remain as in (5·134).

Exercises

1. Consider the boundary value problem in the plane

$$\Delta\Delta u = 0, \quad (u)_B = f, \quad (\partial u/\partial n)_B = g.$$

(This is the *biharmonic* equation.) Show that the following formulae serve for treatment by the method of the hypercircle:

$$\text{(i)} \qquad \mathbf{S} \leftrightarrow p_{ij};$$

$$\text{(ii)} \begin{cases} L': & \mathbf{S}' \leftrightarrow p'_{ij}, \quad p'_{ij} = u'_{,ij}, \\ & (u')_B = f, \quad (\partial u'/\partial n)_B = g; \\ L'': & \mathbf{S}'' \leftrightarrow p''_{ij}, \quad p''_{ij,ij} = 0; \end{cases}$$

$$\text{(iii)} \qquad \mathbf{S}'.\mathbf{S}'' = \int p'_{ij} p''_{ij} \, dV.$$

Show that these formulae serve equally well in a space of N dimensions.

2. Give a similar treatment for

$$\Delta\Delta u - k^2 u = 0, \quad (u)_B = f, \quad (\partial u/\partial n)_B = g.$$

3. As a variant of Ex. 1, consider the boundary value problem

$$\Delta\Delta u = 0, \quad (u)_B = f, \quad (u_{,ij} n_j)_B = g_i.$$

Show that the essential formulae are as follows:

$$\text{(i)} \qquad \mathbf{S} \leftrightarrow p_{ij};$$

$$\text{(ii)} \begin{cases} L': & \mathbf{S}' \leftrightarrow p'_{ij}, \quad p'_{ij} = u'_{,ij}, \quad (u')_B = f; \\ L'': & \mathbf{S}'' \leftrightarrow p''_{ij}, \quad p''_{ij,ij} = 0, \quad (p''_{ij} n_j)_B = g_i; \end{cases}$$

$$\text{(iii)} \qquad \mathbf{S}' . \mathbf{S}'' = \int p'_{ij} p''_{ij} \, dV.$$

5·2. DIRICHLET-NEUMANN PROBLEMS

The general title 'Dirichlet-Neumann problems' may be used to cover the following problems: P-space is a portion of Euclidean N-space bounded by a surface B, and we seek solutions of the Laplace equation

$$\Delta u = 0 \tag{5·201}$$

satisfying one or other of the following boundary conditions:

$$\left.\begin{array}{l} \text{Dirichlet problem:} \\ \quad (u)_B = f, \\ \text{Neumann problem:} \\ \quad (\partial u/\partial n)_B = g, \quad \left(\int g \, dB = 0\right), \\ \text{First mixed problem:} \\ \quad (u)_{B_1} = f, \quad (\partial u/\partial n)_{B_2} = g, \quad B = B_1 + B_2, \\ \text{Second mixed problem:} \\ \quad \left(\alpha u + \beta \dfrac{\partial u}{\partial n}\right)_B = f. \end{array}\right\} \tag{5·202}$$

Here f, g, α, β are given on B. It is clear that the second mixed problem includes the first as a particular case (put $\alpha = 1, \beta = 0$ on B_1, $\alpha = 0$, $\beta = 1$ on B_2), and that the first mixed problem contains the Dirichlet and Neumann problems as particular cases.

In Chapter 3 we dealt at length with the Dirichlet problem in a plane and in Chapter 4 with the extended Dirichlet problem (torsion). The Neumann problem in a plane appeared in §§3·1, 4·1 and 4·2, its intimate connexion with the Dirichlet problem (equivalence for a simply connected domain) being due to the existence of complex variables in a plane. If $N > 2$ there are no complex variables, and this intimate connexion disappears, or at least becomes much less evident.

The systematic approximation to the solution of a plane Dirichlet problem has been based on pyramid F-vectors (p. 168). We shall now indicate how this idea is to be carried into 3-space for the Dirichlet and Neumann problems.

Pyramid F-vectors for three-dimensional P-space

The Dirichlet problem is split by means of the orthogonal linear subspaces

$$\left.\begin{aligned} L'&: \quad \mathbf{S}' \leftrightarrow p_i', \quad p_i' = u'_{,i}, \quad (u')_B = f, \\ L''&: \quad \mathbf{S}'' \leftrightarrow p_i'', \quad p''_{i,i} = 0, \end{aligned}\right\} \tag{5·203}$$

and the Neumann problem by

$$\left.\begin{aligned} L'&: \quad \mathbf{S}' \leftrightarrow p_i', \quad p_i' = u'_{,i}, \\ L''&: \quad \mathbf{S}'' \leftrightarrow p_i'', \quad p''_{i,i} = 0, \quad (p_i'' n_i)_B = g \end{aligned}\right\} \tag{5·204}$$

[cf. (5·114) with $a^{ij} = \delta_{ij}$ and $c = 0$].

To start the hypercircle method, we need some F-point $\mathbf{S}_0'$ on L' in (5·203) and some F-point $\mathbf{S}_0''$ on L'' in (5·204). Let us suppose that such points have been found (it is usually easier for the Dirichlet problem than for the Neumann). Then, in order to get a good approximation to the solution, we need a systematic plan for creating F-vectors $\mathbf{T}'$, $\mathbf{T}''$ lying in L', L'' respectively, such that the gradient of the solution can be approached by a linear combination of $\mathbf{T}'$ or by a linear combination of $\mathbf{T}''$. By (5·203) and (5·204) the basic requirements for $\mathbf{T}'$, $\mathbf{T}''$ are

$$\mathbf{T}' \leftrightarrow p_i', \quad p_i' = u'_{,i}; \quad \mathbf{T}'' \leftrightarrow p_i'', \quad p''_{i,i} = 0; \tag{5·205}$$

these are common to the Dirichlet and Neumann problems, and the boundary conditions are

$$\begin{aligned} \text{Dirichlet:} \quad & (u')_B = 0, \\ \text{Neumann:} \quad & (p_i'' n_i)_B = 0. \end{aligned}$$

There are permissible discontinuities, common to both problems:

$$\left.\begin{aligned} &u' \text{ continuous, } u'_{,i} \text{ piecewise continuous;} \\ &p_i'' n_i \text{ continuous across any surface, } p_i'' \text{ piecewise continuous.} \end{aligned}\right\} \tag{5·206}$$

So far N is arbitrary; we now put $N = 3$ and proceed to develop the three-dimensional analogues of the pyramid F-vectors of

Chapter 3. The theory is due to McMahon (1), who used the method to find a lower bound for the electrostatic capacity of a cube.

Split up all space into tetrahedra. We cannot of course use regular tetrahedra, because they are not space-filling; the simplest plan is to split space into equal cubes and then split each cube into 24 equal tetrahedra, each having the centre of the cube for vertex. But in the present theory we keep the tetrahedra arbitrary.

The tessellation-points (vertices of tetrahedra) are numbered off in any way, 1, 2, ..., so that any tetrahedron has vertices numbered

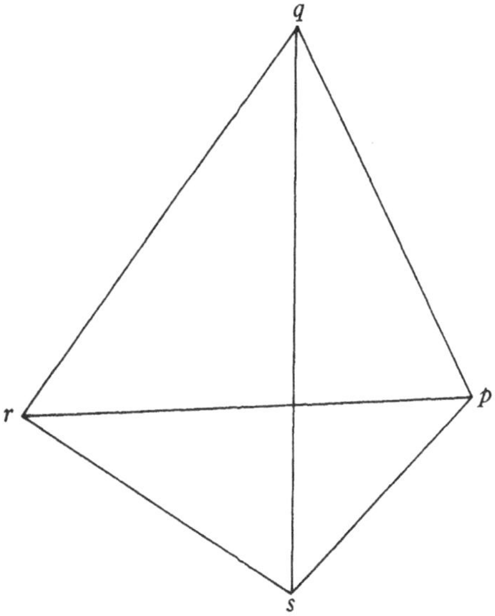

Fig. 5·21. Tetrahedron *rspq* belonging to the cell C_r with centre *r* and to the family F_{rs} with core *rs*.

pqrs. The set of tetrahedra with a common vertex *r* is called the *cell* C_r (*r* is its *centre*). The set of tetrahedra with a common edge *rs* is called the *family* F_{rs} (*rs* is its *core*).

Consider a cell C_r; *rspq* (Fig. 5·21) is one of the tetrahedra belonging to it. We define the *pyramid function* P_r by the following conditions, which obviously define it uniquely as a continuous function in the whole of space:

(i) $P_r = 0$ outside C_r,

(ii) $P_r = 1$ at the point *r*,

(iii) $P_r = 0$ at all the other vertices of C_r,

(iv) P_r is a linear function of the coordinates in each of the tetrahedra belonging to C_r.

Formulae of the type (3·803) are available for P_r; more simply, if A is any point in $rspq$, then

$$P_r(A) = \frac{z_{spq}(A)}{h_{rspq}}, \tag{5·207}$$

where the numerator and denominator are respectively the distances of the points A and r from the plane spq.

It is evident that the F-vector

$$\mathbf{T}'_r \leftrightarrow (P_r)_{,i} \tag{5·208}$$

(this is a partial derivative) satisfies (5·205) and has only permissible discontinuities. If the cell C_r does not cut B, $\mathbf{T}'_r$ satisfies the boundary condition for the Dirichlet problem; there is no such boundary condition for the Neumann problem.

By the method of § 3·8 we can show that, by taking small enough tetrahedra, the solutions of the Dirichlet and Neumann problems can be approached as closely as we like by linear combinations of the following forms:

$$\left.\begin{array}{ll} \text{Dirichlet:} & \mathbf{S} \sim \mathbf{S}'_0 + \Sigma a'_r \mathbf{T}'_r, \\ \text{Neumann:} & \mathbf{S} \sim \Sigma a'_r \mathbf{T}'_r. \end{array}\right\} \tag{5·209}$$

We call $\mathbf{T}'_r$ a *pyramid F-vector of the first class.*

Consider now a family F_{rs} with core rs; $rspq$ is one of the tetrahedra belonging to F_{rs}. In $rspq$ take a constant vector field given by

$$\frac{(pq)_i}{6W_{rspq}}, \tag{5·210}$$

where $(pq)_i$ is the vector joining p to q and W_{rspq} the volume of $rspq$. Take similar constant vector fields in the other tetrahedra belonging to F_{rs} (note that pq is the edge opposite to rs). Take a zero vector field outside F_{rs}. Let $(p''_{rs})_i$ be the vector field defined throughout the whole of space in this way. It is then not hard to show that in the whole of space

$$(p''_{rs})_i = \epsilon_{ijk}(P_r)_{,j}(P_s)_{,k}, \tag{5·211}$$

this being the vector product (cf. p. 406) of the gradients of two pyramid functions, as defined earlier. The components of $(p''_{rs})_i$ are constant in each tetrahedron. We have

$$(p''_{rs})_{i,i} = 0, \tag{5·212}$$

and therefore the F-vector $\mathbf{T}''_{rs} \leftrightarrow (p''_{rs})_i$ (5·213)

satisfies (5·205). No further condition is imposed in the Dirichlet problem; in the Neumann problem the family F_{rs} must not cut B.

It is easy to see that the discontinuities are permissible, the normal component of $(p''_{rs})_i$ being continuous across every plane.

McMahon (1) has shown that the curl of an arbitrary smooth vector field can be approached as closely as we like by a linear combination of $(p''_{rs})_i$. If the problem is such that the solution satisfies $\int (\partial u/\partial n)\, dS = 0$ for every closed surface in the domain V, then $u_{,i}$ is the curl of some vector field, and in this case we can approximate as closely as we like to the solution by linear combinations:

$$\left.\begin{aligned} &\text{Dirichlet:} \quad \mathbf{S} \sim \Sigma a''_{rs}\mathbf{T}''_{rs}, \\ &\text{Neumann:} \quad \mathbf{S} \sim \mathbf{S}''_0 + \Sigma a''_{rs}\mathbf{T}''_{rs}. \end{aligned}\right\} \tag{5·214}$$

If V is such that every closed surface in it can be contracted to a point, the above condition is satisfied and the above approximation is valid. But if every closed surface is not so reducible to a point (e.g. if V is the region between two concentric spheres, or the whole of space outside a sphere), then we may have to supplement the $\mathbf{T}''_{rs}$ with other F-vectors, like the strip F-vectors of p. 184.

We call $\mathbf{T}''_{rs}$ a *pyramid F-vector of the second class*; note its skew-symmetry with respect to r and s.

Naturally pyramid F-vectors in space, as we have defined them above, lead to much more complicated arithmetic than what we met in studying pyramid F-vectors in a plane. But the same general plan may be used, and so they provide a systematic technique for the approximate solution of boundary value problems in three dimensions, differing from finite difference methods in that the error is controlled by the hypercircle. For details see McMahon (1) and, for pyramid vectors in N-space, McMahon (2).

First mixed problem

We now turn to the first mixed problem as set out in (5·201) and (5·202):

$$\Delta u = 0, \quad (u)_{B_1} = f, \quad (\partial u/\partial n)_{B_2} = g, \quad B = B_1 + B_2. \tag{5·215}$$

Like the Neumann problem, this is also a particular case of (5·112) and (5·113), and the essential formulae are

$$\left.\begin{aligned} &\mathbf{S} \leftrightarrow p_i, \quad \mathbf{S}'.\mathbf{S}'' = \int_V p'_i p''_i \, dV, \\ &L': \quad \mathbf{S}' \leftrightarrow p'_i, \quad p'_i = u'_{,i}, \quad (u')_{B_1} = f, \\ &L'': \quad \mathbf{S}'' \leftrightarrow p''_i, \quad p''_{i,i} = 0, \quad (p''_i n_i)_{B_2} = g. \end{aligned}\right\} \tag{5·216}$$

Here again u' is to be continuous but may have only piecewise continuous partial derivatives; p''_i may be only piecewise continuous, but $p''_i n_i$ must be continuous.

Note that neither L' nor L'' contains the origin. For the application of the hypercircle method we have to choose an F-point $\mathbf{S}_0'$ in L' and an F-point $\mathbf{S}_0''$ in L'', and then sets of F-vectors $\mathbf{T}'$ and $\mathbf{T}''$ lying in L' and L'' respectively.

Transfer of boundary conditions from curvilinear to straight boundaries

In order to avoid the awkward calculations involved in computing $\mathbf{S}_0' . \mathbf{T}_\sigma''$ or $\mathbf{S}_0'' . \mathbf{T}_\sigma''$, we may, in some plane problems concerning curved boundaries, be able to find a relation between the Dirichlet integral, $\mathbf{S}_1^2$, for the problem in hand, and the corresponding quantity, $\mathbf{S}_2^2$,

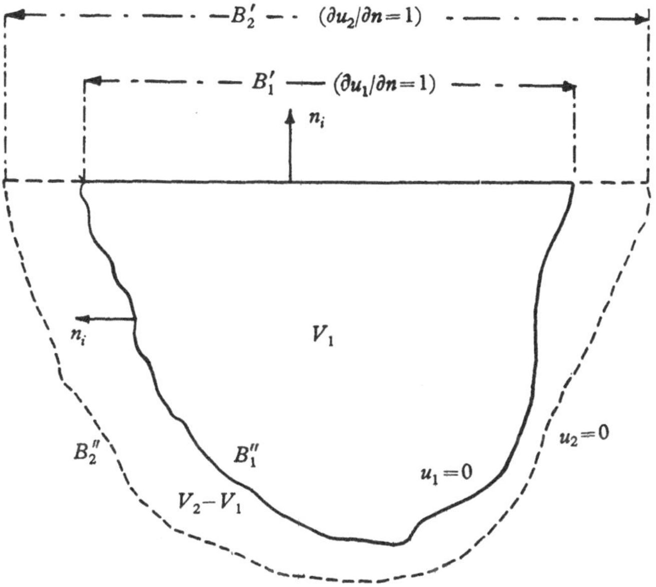

Fig. 5·22. Change of boundary in a mixed boundary value problem.

for a problem involving straight boundaries. The straight boundaries may be chosen so that the region enclosed by them can be completely filled up by hexagonal or square pyramid F-vectors.

A type of boundary value problem to which this idea may be applied runs as follows:

$$\Delta u_1 = 0 \text{ in } V_1; \quad (u_1)_{B_1''} = 0; \quad \left(\frac{\partial u_1}{\partial n}\right)_{B_1'} = 1, \qquad (5·217)$$

where $B_1 = B_1' + B_1''$ is the boundary of V_1, B_1' is a straight line and B_1'' is curved (Fig. 5·22).

To split the problem we take

$$\left.\begin{aligned} L'&: \quad \mathbf{S}' \leftrightarrow p'_i = u'_{,i}, \quad (u')_{B''_1} = 0, \\ L''&: \quad \mathbf{S}'' \leftrightarrow p''_i; \quad p''_{i,i} = 0; \quad (p''_i n_i)_{B'_1} = 1, \end{aligned}\right\} \tag{5·218}$$

and we let

$$\mathbf{S}_1 \leftrightarrow u_{1,i}, \tag{5·219}$$

where u_1 is the solution of (5·217); then $\mathbf{S}_1$ is the intersection of the subspaces L' and L''.

We see that the origin of function-space lies in L', and so

$$\mathbf{V}'^2 \leqslant \mathbf{S}_1^2 \leqslant \mathbf{V}''^2. \tag{5·220}$$

The calculation of the lower bound by means of pyramid F-vectors presents no difficulty, but the upper bound will give rise to awkward calculation on account of the fact that B''_1 is curved.

We note that

$$\begin{aligned} \mathbf{S}_1^2 &= \int_{V_1} u_{1,i} u_{1,i} \, dV_1 = \int_{B'_1 + B''_1} u_1 \frac{\partial u_1}{\partial n} dB \\ &= \int_{B'_1} u_1 \, dB'_1, \end{aligned} \tag{5·221}$$

on account of the boundary conditions.

To avoid the difficulty of calculating $\mathbf{V}''^2$ in (5·220), a second boundary value problem is now considered, namely,

$$\Delta u_2 = 0 \text{ in } V_2, \quad (u_2)_{B''_2} = 0, \quad \left(\frac{\partial u_2}{\partial n}\right)_{B'_2} = 1, \tag{5·222}$$

where B'_2 is a straight line which includes B'_1, B''_2 is any curve which lies outside B''_1, and V_2 is the region bounded by $B_2 = B'_2 + B''_2$. The domain between B_1 and B_2 is called $V_2 - V_1$ (Fig. 5·22).

This problem is now split in the same way as in (5·218), and if we let

$$\mathbf{S}_2 \leftrightarrow u_{2,i}, \tag{5·223}$$

where u_2 is the solution of (5·222), we have

$$\mathbf{S}_2^2 = \int_{V_2} u_{2,i} u_{2,i} \, dV_2. \tag{5·224}$$

It will be noted that the domains of integration in (5·221) and (5·224) are different, one integration being over V_1 and the other over V_2. So far u_1 has been defined throughout V_1 only; we now extend its definition throughout V_2 by putting $u_1 = 0$ in $V_2 - V_1$.

Note that u_1 is a continuous function in V_2. We are thus enabled to write:

$$\mathbf{S}_1 . \mathbf{S}_2 = \int_{V_2} u_{1,i} u_{2,i} dV_2 = \int_{V_1} u_{1,i} u_{2,i} dV_1$$

$$= \int_{V_1} (u_1 u_{2,i})_{,i} dV_1 = \int_{B_1'} u_1 u_{2,i} n_i dB_1', \qquad (5\cdot225)$$

since $\Delta u_2 = 0$ in V_2, and hence in V_1. On applying the boundary conditions on u_1 and u_2 contained in (5·217) and (5·222) we get

$$\mathbf{S}_1 . \mathbf{S}_2 = \int_{B_1'} u_1 dB_1' = \mathbf{S}_1^2 \qquad (5\cdot226)$$

by (5·221). By application of the Schwarz inequality (2·210) we obtain

$$(\mathbf{S}_1^2)^2 = (\mathbf{S}_1 . \mathbf{S}_2)^2 \leqslant \mathbf{S}_1^2 \mathbf{S}_2^2, \qquad (5\cdot227)$$

so that

$$\mathbf{S}_1^2 \leqslant \mathbf{S}_2^2, \qquad (5\cdot228)$$

since $\mathbf{S}_1^2$ is positive. The equality sign holds only when B_2 coincides with B_1.

Now B_2'' is *any* boundary enclosing B_1'' and we may conveniently compose it of straight segments in such a way that V_2 may be completely filled by regular pyramid F-vectors. We then obtain, in the usual way, an upper bound U for $\mathbf{S}_2^2$ in the problem defined by (5·222) and (5·223), the calculation of scalar products being much simplified due to the choice of boundaries. Thus we have

$$\mathbf{S}_1^2 \leqslant \mathbf{S}_2^2 \leqslant U, \qquad (5\cdot229)$$

so that U is also an upper bound for $\mathbf{S}_1^2$. We improve this bound in the successive approximations by making the segments of B_2'' approach closer and closer to B_1''.

This method will be illustrated in the example on viscous flow in §5·3.

Second mixed problem

Consider now the second mixed problem as in (5·202):

$$\Delta u = 0, \quad (\alpha u + \beta \partial u/\partial n)_B = f, \qquad (5\cdot230)$$

where α, β and f are given functions of position on the boundary B. Let P-space be of N dimensions.

As usual we define F-space and the scalar product by

$$\mathbf{S} \leftrightarrow p_i, \quad \mathbf{S}' . \mathbf{S}'' = \int p_i' p_i'' dV. \qquad (5\cdot231)$$

We define two linear subspaces as follows:

$$\left.\begin{aligned} L'&: \quad \mathbf{S}' \leftrightarrow p'_i, \quad p'_i = u'_{,i}; \\ L''&: \quad \mathbf{S}'' \leftrightarrow p''_i, \quad p''_{i,i} = 0. \end{aligned}\right\} \tag{5·232}$$

Note that no mention is made of boundary conditions in the definitions of L' and L''. The solution $\mathbf{S}$ certainly is in both L' and L'', but their intersection is not a unique F-point.

We now think of pairs of points, $\mathbf{S}'$ in L' and $\mathbf{S}''$ in L'', connected by the relation

$$(\alpha u' + \beta p''_i n_i)_B = f, \tag{5·233}$$

n_i being the unit outward normal to B. If we can find such a pair for which the two F-points coincide ($\mathbf{S}' = \mathbf{S}''$), then it is clear that that F-point is a solution of the problem (5·230), i.e. $\mathbf{S} = \mathbf{S}' = \mathbf{S}''$. In that sense, the problem is 'split'.

Consider now a pair $\mathbf{S}'$, $\mathbf{S}''$ linked by (5·233). Let us calculate the scalar product $(\mathbf{S}' - \mathbf{S}) . (\mathbf{S}'' - \mathbf{S})$, where $\mathbf{S}$ is the solution, so that

$$\mathbf{S} \leftrightarrow p_i = u_{,i}, \quad p_{i,i} = 0, \quad (\alpha u + \beta p_i n_i)_B = f. \tag{5·234}$$

We find

$$\begin{aligned} (\mathbf{S}' - \mathbf{S}) . (\mathbf{S}'' - \mathbf{S}) &= \int (p'_i - p_i)(p''_i - p_i)\, dV \\ &= \int (u'_{,i} - u_{,i})(p''_i - p_i)\, dV \\ &= \int (u' - u)(p''_i - p_i)\, n_i dB, \end{aligned} \tag{5·235}$$

since $p''_{i,i} = p_{i,i} = 0$. By (5·230) and (5·233) we have

$$\alpha(u' - u) + \beta(p''_i - p_i)\, n_i = 0 \tag{5·236}$$

on B, and so

$$(\mathbf{S}' - \mathbf{S}) . (\mathbf{S}'' - \mathbf{S}) = -\int (\alpha/\beta)(u' - u)^2\, dB. \tag{5·237}$$

Let us suppose that the quotient α/β is of one sign on B.

Let θ be the angle between the F-vectors $\mathbf{S}' - \mathbf{S}$ and $\mathbf{S}'' - \mathbf{S}$ (Fig. 5·23). Then it is clear from (5·237) that

$$\left.\begin{aligned} &\text{if } \alpha/\beta \geqslant 0, \quad \text{then} \quad \theta \geqslant \tfrac{1}{2}\pi; \\ &\text{if } \alpha/\beta \leqslant 0, \quad \text{then} \quad \theta \leqslant \tfrac{1}{2}\pi. \end{aligned}\right\} \tag{5·238}$$

In the former case we know that the F-point $\mathbf{S}$ is *inside* (or on) the hypersphere with centre and radius given by

$$\mathbf{C} = \tfrac{1}{2}(\mathbf{S}' + \mathbf{S}''), \quad 4R^2 = (\mathbf{S}' - \mathbf{S}'')^2; \tag{5·239}$$

in the latter case $\mathbf{S}$ is *outside* (or on) this hypersphere.

In what follows we shall consider only the former case,* $\alpha/\beta \geqslant 0$. Then the F-point $\mathbf{C}$ may be regarded as an approximation to the

* Using an entirely different method, Pucci (1) has discussed pointwise bounds on the solution for the other case.

solution $\mathbf{S}$, and the smaller R is the better the approximation, since we know that

$$(\mathbf{S}-\mathbf{C})^2 \leqslant R^2. \tag{5·240}$$

Thus, to get a good approximation, we need to draw the F-points $\mathbf{S}'$ and $\mathbf{S}''$ close together, and this we do in the following way.

Choose first a pair of points in L' and L'' respectively,

$$\mathbf{S}_0' \leftrightarrow p'_{(0)i} = u'_{(0),i}, \quad \mathbf{S}_0'' \leftrightarrow p''_{(0)i}, \quad p''_{(0)i,i} = 0, \tag{5·241}$$

linked together, as in (5·233), by

$$(\alpha u'_{(0)} + \beta p''_{(0)i} n_i)_B = f. \tag{5·242}$$

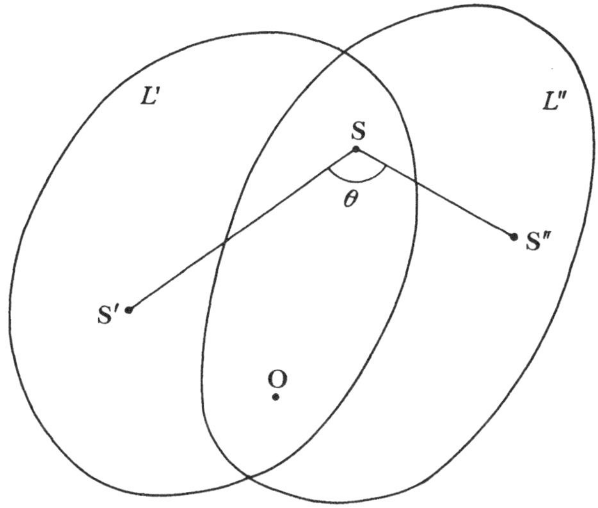

Fig. 5·23. Second mixed problem: $\mathbf{S}'$ and $\mathbf{S}''$ correspond as in (5·233).

Then choose a set of F-vectors $\mathbf{T}'_\rho$ lying in L' and an equal number $\mathbf{T}''_\rho$ lying in L'', say

$$\left.\begin{aligned} &\mathbf{T}'_\rho \leftrightarrow p'_{(\rho)i} = u'_{(\rho),i} \\ &\mathbf{T}''_\rho \leftrightarrow p''_{(\rho)i}, \quad p''_{(\rho)i,i} = 0 \end{aligned}\right\} \quad (\rho = 1, 2, \ldots, r), \tag{5·243}$$

linked together by $\quad (\alpha u'_{(\rho)} + \beta p''_{(\rho)i} n_i)_B = 0. \qquad (5·244)$

(The easiest way to do this is to choose $\mathbf{T}''_\rho$ first, and then make $\mathbf{T}'_\rho$ fit the linkage.)

Then, no matter what values the constants a_ρ may have, the F-points

$$\mathbf{S}' = \mathbf{S}_0' + \sum_{\rho=1}^{r} a_\rho \mathbf{T}'_\rho, \quad \mathbf{S}'' = \mathbf{S}_0'' + \sum_{\rho=1}^{r} a_\rho \mathbf{T}''_\rho \tag{5·245}$$

are in L' and L'' respectively, and they are linked as in (5·233). Then by (5·239) the solution $\mathbf{S}$ is inside (or on) the hypersphere having $\mathbf{S}'$ and $\mathbf{S}''$ for the extremities of a diameter, and we proceed to *minimize* the square of this diameter, viz. $(\mathbf{S}'-\mathbf{S}'')^2$, by proper choice of the a's, i.e. by choosing them to satisfy the r linear equations

$$\sum_{\mu=1}^{r} a_\mu(\mathbf{T}'_\mu-\mathbf{T}''_\mu).(\mathbf{T}'_\rho-\mathbf{T}''_\rho)+(\mathbf{S}'_0-\mathbf{S}''_0).(\mathbf{T}'_\rho-\mathbf{T}''_\rho)=0$$
$$(\rho=1,2,\ldots,r). \quad (5\cdot246)$$

When the a's have been so found and their values inserted, we have for the square of the diameter of the hypersphere the following alternative forms:

$$\begin{aligned}4R^2&=(\mathbf{S}'-\mathbf{S}'')^2\\ &=(\mathbf{S}'_0-\mathbf{S}''_0)^2+\sum_{\rho=1}^{r} a_\rho(\mathbf{S}'_0-\mathbf{S}''_0).(\mathbf{T}'_\rho-\mathbf{T}''_\rho)\\ &=(\mathbf{S}'_0-\mathbf{S}''_0)^2-\sum_{\rho=1}^{r}\sum_{\mu=1}^{r} a_\rho a_\mu(\mathbf{T}'_\rho-\mathbf{T}''_\rho).(\mathbf{T}'_\mu-\mathbf{T}''_\mu),\end{aligned} \quad (5\cdot247)$$

the double summation being positive-definite.

If the hypersphere, with centre $\mathbf{C}$ and radius R as in (5·239), does not contain the origin $\mathbf{O}$ of F-space, then the square $\mathbf{S}^2$ of the solution is bounded below and above, as in (2·515), by

$$\mathbf{C}^2+R^2-2R\,|\,\mathbf{C}\,|\leqslant \mathbf{S}^2\leqslant \mathbf{C}^2+R^2+2R\,|\,\mathbf{C}\,|, \quad (5\cdot248)$$

where R^2 is given by (5·247) and $\mathbf{C}^2$ by

$$\begin{aligned}4\mathbf{C}^2&=(\mathbf{S}'+\mathbf{S}'')^2\\ &=(\mathbf{S}'_0+\mathbf{S}''_0)^2+2\sum_{\rho=1}^{r} a_\rho(\mathbf{S}'_0+\mathbf{S}''_0).(\mathbf{T}'_\rho+\mathbf{T}''_\rho)\\ &\qquad+\sum_{\rho=1}^{r}\sum_{\mu=1}^{r} a_\rho a_\mu(\mathbf{T}'_\rho+\mathbf{T}''_\rho).(\mathbf{T}'_\mu+\mathbf{T}''_\mu).\end{aligned} \quad (5\cdot249)$$

Hence

$$2(\mathbf{C}^2+R^2)=\mathbf{S}_0'^2+\mathbf{S}_0''^2+\sum_{\rho=1}^{r} a_\rho(\mathbf{S}'_0+\mathbf{S}''_0).(\mathbf{T}'_\rho+\mathbf{T}''_\rho)+2\sum_{\rho=1}^{r}\sum_{\mu=1}^{r} a_\rho a_\mu \mathbf{T}'_\rho.\mathbf{T}''_\mu. \quad (5\cdot250)$$

If the hypersphere does contain $\mathbf{O}$, then the upper bound in (5·248) remains valid, but the lower bound does not. For that lower bound as it appeared in (2·515) concerned points *on* the hypersphere, but now all we know is that $\mathbf{S}$ is *inside or on* the hypersphere, and from this information we can deduce no lower bound save zero.

As regards the functions and vector fields involved above, the permissible discontinuities are of the usual form; thus, in (5·232) and similar formulae, u' must be continuous, but with only piecewise continuous partial derivatives, while p_i'' need only be piecewise continuous, provided $p_i''n_i$ is continuous across all surfaces (or curves, if P-space is of two dimensions).

In the next section we discuss examples of the first and second mixed problems.

Exercises

1. The electrostatic capacity of any surface B is

$$C=(4\pi)^{-1}\int_v u_{,i}u_{,i}dV,$$

the integral being taken throughout all space V external to B and u being the solution of the Dirichlet problem

$$\Delta u=0 \text{ in } V, \quad (u)_B=1, \quad (u)_\infty=\text{order } 1/r.$$

Let B be a convex surface with the origin of coordinates inside it; its equation may be written in polar coordinates $r=f(\theta,\phi)$, or equivalently $r=F(x_1,x_2,x_3)$ where F is homogeneous of degree zero, so that $F_{,i}x_i=0$. Then if $u'=F/r$,

$$\mathbf{S}'\leftrightarrow p_i'=u'_{,i}$$

is an F-point on the linear subspace L' of the Dirichlet problem. Deduce the following upper bound for the capacity:

$$C\leqslant(4\pi)^{-1}\int(F+F^{-1}F_{,i}F_{,i})\,d\omega,$$

the integration being taken over a unit sphere (element $d\omega$) with centre at the origin. Show that this bound is attained for a sphere $r=a$ (for which we know that $C=a$).

2. Considering the electrostatic capacity of a surface B as in Example 1, show that if p_i'' is any vector field satisfying $p''_{i,i}=0$ and vanishing like $1/r^2$ at infinity, then the first inequality in (2·765), viz.

$$\mathbf{S}^2\geqslant(\mathbf{S}'.\mathbf{S}'')^2/\mathbf{S}''^2,$$

leads to the following lower bound for the capacity:

$$C\geqslant\frac{1}{4\pi}\frac{\left(\int_B p_i''n_i\,dB\right)^2}{\int_V p_i''p_i''\,dV}.$$

(Here $\mathbf{S}'\leftrightarrow u'_{,i}$ where $u'=1$ on B and u' vanishes like $1/r$ at infinity; we may in particular take $u'=u$, the solution.)

3. B is a convex surface and B' the surface formed by proceeding out from B through a fixed distance R along all its normals, each point P on B thus generating a point P' on B'. Show that corresponding elements of area dB, dB' at P, P' are related by

$$\frac{dB'}{dB}=\frac{(R+R_1)(R+R_2)}{R_1R_2},$$

where R_1 and R_2 are the principal radii of curvature of B at P. Show that if n_i is the unit vector along PP', then the vector field

$$p''_i = n_i \frac{R_1 R_2}{(R+R_1)(R+R_2)},$$

R taking all values from zero to infinity, satisfies $p''_{i,i} = 0$ and vanishes like $1/r^2$ at infinity.

4. Use the results stated in Examples 2 and 3 to establish the following lower bound for the electrostatic capacity of any convex surface B:

$$C \geqslant \frac{1}{4\pi}\frac{B^2}{J},$$

where B is the area of the surface and

$$J = \int_B \frac{\log(1/R_1) - \log(1/R_2)}{(1/R_1)-(1/R_2)}\, dB.$$

Check this result for a sphere $r = a$.

[The reader interested in bounds on capacity should consult Pólya and Szegö (1), Chap. 3.]

5·3. EXAMPLES OF MIXED BOUNDARY VALUE PROBLEMS

First mixed problem: viscous flow in a channel

We consider the steady flow of an incompressible viscous liquid through a horizontal channel, the only body force acting on the fluid being that of gravity.

The origin O of co-ordinates is taken in the free surface PQ of the liquid (Fig. 5·31). The x_3-axis points in the direction of flow along the channel, and the x_1 and x_2 axes lie in the plane of the cross-section with the x_2-axis pointing vertically upwards.

We imagine that the free surface of the fluid is subject to a constant shearing stress S acting in the direction of the x_3-axis. Such a stress may be thought of as due to a wind blowing along the channel, waves produced by the wind being excluded from the mathematical idealization.

The general equations of steady motion are:

$$q_\beta q_{\alpha,\beta} = -\frac{1}{\rho}p_{,\alpha} + R_\alpha + \left(\frac{\mu}{\rho}\right) q_{\alpha,\beta\beta}, \tag{5·301}$$

and

$$q_{\alpha,\alpha} = 0, \tag{5·302}$$

where q_α is the velocity of the fluid, p the pressure, R_α the body force per unit mass, ρ the density and μ the coefficient of viscosity. Greek suffixes have the range 1, 2, 3 and the summation convention applies for repeated suffixes.

We assume
$$q_1 = q_2 = 0, \tag{5·303}$$
and so by (5·302) q_3 is a function of x_1 and x_2 only. Then since $R_1 = R_3 = 0$, $R_2 = -g$, the equations (5·301) reduce to
$$p_{,1} = 0, \tag{5·304}$$
$$p_{,2} = -g\rho, \tag{5·305}$$
$$\Delta q_3 = \frac{1}{\mu} p_{,3}, \tag{5·306}$$
where Δ is the two-dimensional Laplacian.

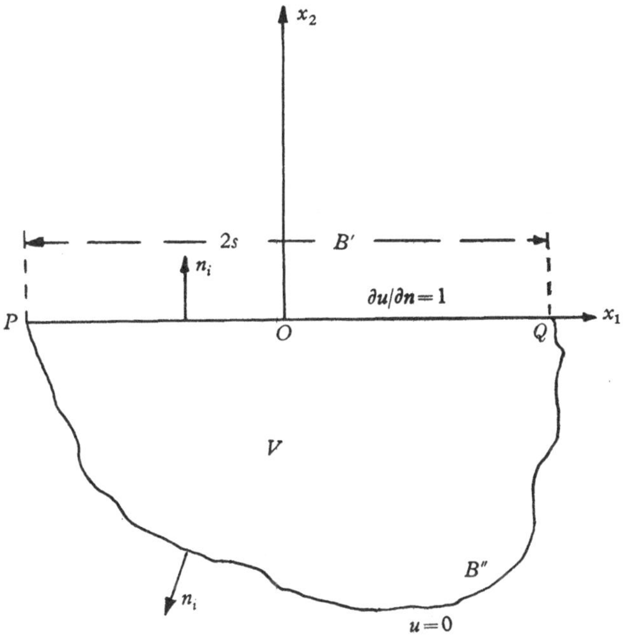

Fig. 5·31. Cross-section of channel in the flow problem.

We note that the left-hand side of (5·306) is a function of x_1 and x_2, and by (5·304) and (5·305) $p_{,3}$ is possibly a function of x_3, but not of x_1 or x_2. Accordingly each side of (5·306) is equal to a constant and we set
$$p_{,3} = -\Pi, \tag{5·307}$$
where Π is a constant. Equations (5·304), (5·305) and (5·307) now give the pressure distribution in the channel as
$$p = p_0 - g\rho x_2 - \Pi x_3, \tag{5·308}$$

where p_0 is a constant. The pressure on the free surface ($x_2 = 0$) must be constant, however, which means that we must set $\Pi = 0$. The equation (5·306) for q_3 then becomes

$$\Delta q_3 = 0. \tag{5·309}$$

The problem is thus reduced to solving Laplace's equation under certain boundary conditions.

The boundary B of the cross-section is composed of two parts, B', the free surface, and B'', the remainder of the perimeter (Fig. 5·31). If we let ∂n represent an element of the outward normal to the free surface, the boundary condition on this surface is

$$\mu \frac{\partial q_3}{\partial n} = S, \tag{5·310}$$

due to the tangential stress S acting on it. The boundary condition on B'' is

$$q_3 = 0, \tag{5·311}$$

since the viscous liquid adheres to the walls of the channel.

The boundary value problem for q_3 is then a mixed problem of the first kind:

$$\Delta q_3 = 0 \text{ in } V, \quad q_3 = 0 \text{ on } B'', \quad \frac{\partial q_3}{\partial n} = \frac{S}{\mu} \text{ on } B'. \tag{5·312}$$

V denotes the domain enclosed by $B = B' + B''$, i.e. the cross-section of the channel.

For convenience we let $$k = S/\mu, \tag{5·313}$$

and $$u = q_3/k, \tag{5·314}$$

so that (5·312) becomes

$$\Delta u = 0 \text{ in } V, \quad u = 0 \text{ on } B'', \quad \frac{\partial u}{\partial n} = 1 \text{ on } B'. \tag{5·315}$$

Flow problem: the Dirichlet integral and flux through the channel

The rate W at which energy is being supplied per unit length of the channel (and dissipated into heat) is given by

$$W = \int_{B'} S q_3 \, dB' = Sk \int_{B'} u \, dB', \tag{5·316}$$

on using the constancy of S and (5·314). By using the boundary conditions (5·315) on u and Green's theorem, we then obtain

$$W = Sk \int_{B'+B''} u \frac{\partial u}{\partial n} dB = Sk \int_V u_{,i} u_{,i} \, dV, \tag{5·317}$$

Latin suffixes having the range 1, 2, with the summation convention. *Thus the Dirichlet integral of the solution to the boundary value problem* (5·315) *represents, to within a constant factor, the rate of dissipation of energy per unit length of the channel.*

It is of interest to consider also the flux of liquid across the section of the channel. The section may have any smooth curve for its boundary B'', but the free surface B' of the liquid is of course plane; we shall make use of this latter fact below. For the axes as in Fig. 5·31, the origin O may be taken anywhere in the free surface B', but for convenience we take it at the centre of B', whose length we denote by $2s$.

The flux F is given by

$$F=\int_V q_3 dV = k\int_V u\,dV, \tag{5·318}$$

where $k=S/\mu$ by (5·313). Noting that $\Delta(\frac{1}{4}r^2)=1$, where $r^2=x_i x_i$, we may write

$$\begin{aligned} F&=k\int_V [u\Delta(\tfrac{1}{4}r^2)-\tfrac{1}{4}r^2\Delta u]\,dV \\ &=k\int_{B=B'+B''}\left[u\frac{\partial}{\partial n}(\tfrac{1}{4}r^2)-\tfrac{1}{4}r^2\frac{\partial u}{\partial n}\right]dB \\ &=-\tfrac{1}{4}k\int_{B'} r^2 dB'-\tfrac{1}{4}k\int_{B''} r^2\frac{\partial u}{\partial n}dB'', \end{aligned} \tag{5·319}$$

since $(u)_{B''}=0$ and, on B',

$$\left(\frac{\partial u}{\partial n}\right)_{B'}=1 \quad \text{and} \quad \frac{\partial r^2}{\partial n}=2x_2=0.$$

But

$$\int_{B'} r^2 dB' = \int_{-s}^{s} x_1^2 dx_1 = \frac{2s^3}{3}, \tag{5·320}$$

and therefore

$$F=k\left[-\frac{s^3}{6}+\frac{1}{4}\int_{B''} r^2\left(-\frac{\partial u}{\partial n}\right)dB''\right]. \tag{5·321}$$

This formula gives the flux in terms of the solution u of the boundary value problem (5·315) *for a general cross-section.* In the case where B'' is a semicircle with centre O and radius s, (5·321) reduces to

$$F=k\left[-\frac{s^3}{6}+\tfrac{1}{4}s^2\int_{B'} dB'\right]=\frac{ks^3}{3}, \tag{5·322}$$

since

$$\int_{B''}\left(-\frac{\partial u}{\partial n}\right)dB''=\int_{B'}\frac{\partial u}{\partial n}dB'=\int_{B'}dB'=2s.$$

The flux through a channel of general cross-section may also be expressed as a scalar product as follows.

We take the F-vector $\mathbf{S}$ corresponding to the solution u of the flow problem (5·315), i.e.

$$\mathbf{S}\leftrightarrow u_{,i}, \tag{5·323}$$

and let **I** be the moment of inertia F-vector as in (4·203) so that

$$\mathbf{I} \leftrightarrow x_i. \tag{5·324}$$

The origin is taken as in Fig. 5·31. Since $x_{i,i} = 2$, we have

$$\begin{aligned} F &= k\int_V u\,dV = k\int_V u(\tfrac{1}{2}x_i)_{,i}\,dV \\ &= k\left[\frac{1}{2}\int_{B=B'+B''} u x_i n_i\,dB - \frac{1}{2}\int_V u_{,i} x_i\,dV\right]; \end{aligned} \tag{5·325}$$

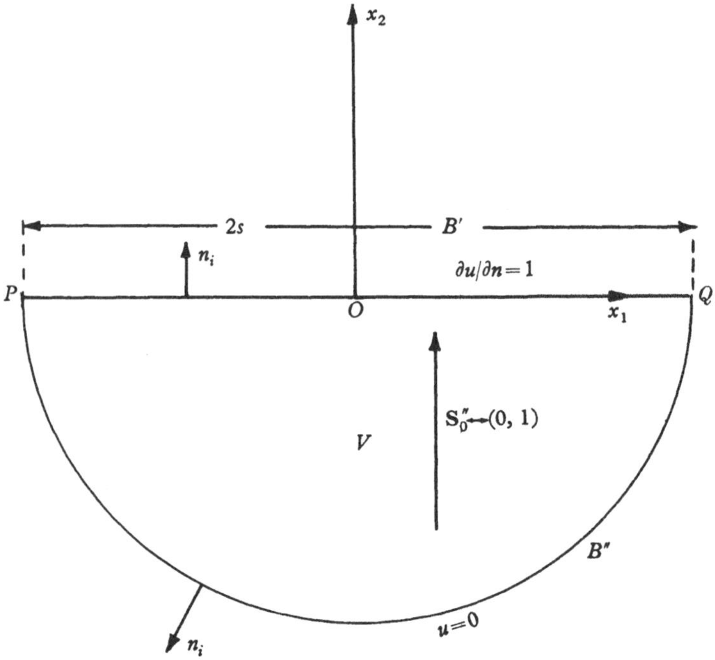

Fig. 5·32. Flow through a semicircular channel.

since $u = 0$ on B'' and $x_i n_i = x_2 = 0$ on B', the integral on B vanishes, and we obtain F as a scalar product:

$$F = -\tfrac{1}{2}k\int_V u_{,i} x_i\,dV = -\tfrac{1}{2}k\mathbf{S}\,.\,\mathbf{I}. \tag{5·326}$$

Flow problem: flow through a channel of semicircular cross-section

So far the pyramid F-vectors of Chapter 3 have not been applied to any numerical problem involving a curved boundary. We shall now apply them to the flow problem (5·315), taking a semicircular cross-section V of radius s (Fig. 5·32). The same technique may be used for any curved section, but the semicircle makes a good illustration and has the advantage that an exact solution is available, providing a check [cf. (5·354)] on the bounds for the flux obtained by the present method. For the use of pyramid F-vectors in the flow problem for a section which can be broken up into squares, see Synge (6).

Referring to (5·216) we note that L' contains the origin of F-space. We choose

$$\mathbf{S}_0' = 0, \quad \mathbf{S}_0'' \leftrightarrow (0, 1), \tag{5·327}$$

this latter being a unit vector-field directed upward, with vanishing divergence and satisfying the required condition on PQ (Fig. 5·32).

We now assemble the appropriate formulae. We need the following F-vectors lying in L', L'' respectively:

$$\left.\begin{array}{llll} \mathbf{T}_\rho' \leftrightarrow p_i', & p_i' = u'_{,i}, & (u')_{B''} = 0 & (\rho = 1, 2, \ldots, r); \\ \mathbf{T}_\sigma'' \leftrightarrow p_i'', & p_{i,i}'' = 0, & (p_i'' n_i)_{B'} = 0 & (\sigma = 1, 2, \ldots, s). \end{array}\right\} \tag{5·328}$$

By (2·775) the vertices are then

$$\mathbf{V}' = \sum_{\rho=1}^{r} a_\rho' \mathbf{T}_\rho', \quad \mathbf{V}'' = \mathbf{S}_0'' + \sum_{\sigma=1}^{s} a_\sigma'' \mathbf{T}_\sigma'', \tag{5·329}$$

where the coefficients are found by minimizing $(\mathbf{V}' - \mathbf{V}'')^2$, i.e. they satisfy

$$\left.\begin{array}{ll} \sum_{\mu=1}^{r} a_\mu' \mathbf{T}_\rho' \cdot \mathbf{T}_\mu' - \mathbf{S}_0'' \cdot \mathbf{T}_\rho' = 0 & (\rho = 1, 2, \ldots, r), \\ \sum_{\nu=1}^{s} a_\nu'' \mathbf{T}_\sigma'' \cdot \mathbf{T}_\nu'' + \mathbf{S}_0'' \cdot \mathbf{T}_\sigma'' = 0 & (\sigma = 1, 2, \ldots, s). \end{array}\right\} \tag{5·330}$$

Also by (2·777)

$$\left.\begin{array}{l} \mathbf{V}'^2 = \sum_{\rho=1}^{r} a_\rho' \mathbf{S}_0'' \cdot \mathbf{T}_\rho', \\ \mathbf{V}''^2 = \mathbf{S}_0''^2 + \sum_{\sigma=1}^{s} a_\sigma'' \mathbf{S}_0'' \cdot \mathbf{T}_\sigma'', \end{array}\right\} \tag{5·331}$$

and the inequalities for $\mathbf{S}^2$, the Dirichlet integral of the solution, are by (2·782)

$$\mathbf{V}'^2 \leqslant \mathbf{S}^2 \leqslant \mathbf{V}''^2. \tag{5·332}$$

(Note that, in passing from the formulae of §2·7, $\mathbf{V}'$ and $\mathbf{V}''$ are interchanged, since the origin in the present problem lies in L'.)

The radius R of the hypercircle is given by

$$4R^2 = \mathbf{V}''^2 - \mathbf{V}'^2. \tag{5·333}$$

Since the boundary is not composed of straight segments, the region V cannot be completely filled by pyramid vectors. This does not complicate the calculation of the lower bound $\mathbf{V}'^2$, but the calculation of the upper bound $\mathbf{V}''^2$ becomes awkward, particularly when the number of pyramid vectors used is large. To avoid this difficulty the method described on pp. 305–7 is employed.

Hexagonal pyramid F-vectors of both classes are used in the problem. In the approximation $n=4$, we divide the radius s of the

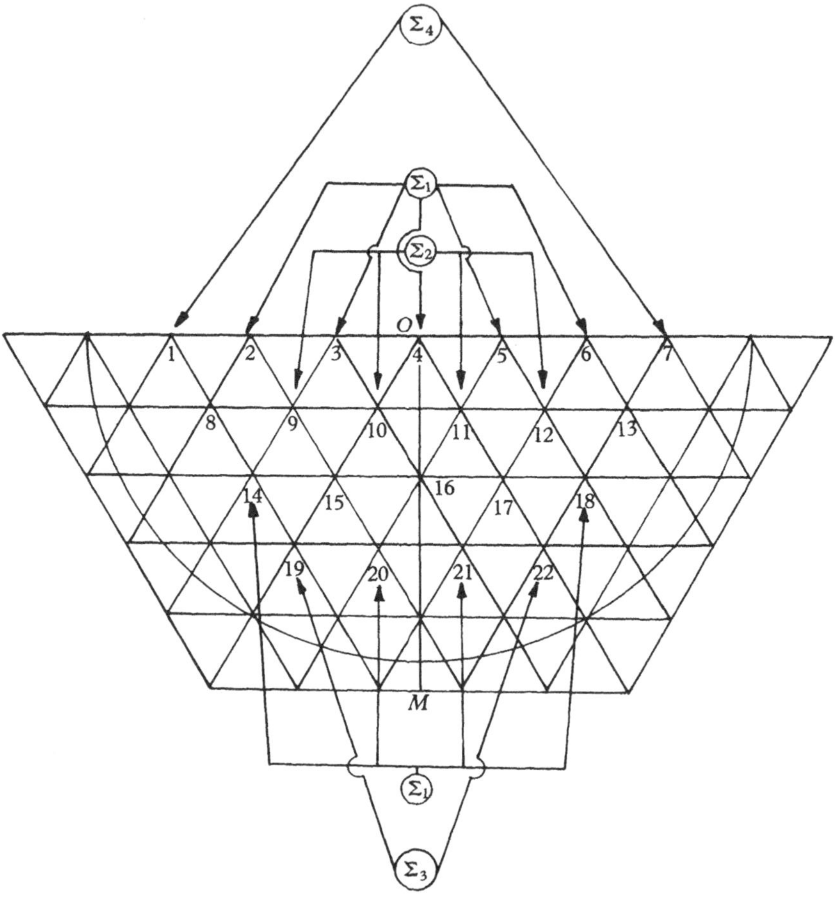

Fig. 5·33. Pyramid F-vectors of the first class (centres numbered 1–22) used in finding a lower bound in the flow problem for a semicircular channel. The diagram also indicates the summations of weights used in calculating the quick convergence factor.

semicircle into four equal parts and then triangulate as in Fig. 5·33. All junction-points shown are used as centres for pyramid vectors, with zero weights attached to some of them due to the boundary condition.

Flow problem: the approximation $n=4$: lower bound for $\mathbf{S}^2$.

Twenty-two pyramid F-vectors $\mathbf{T}'_\rho$ $(\rho=1,2,\ldots,22)$ (Fig. 5·33) are used in the calculation of the lower bound on $\mathbf{S}^2$, but only twelve of the corresponding weights attached to the vectors are independent due to the symmetry of the problem about the line OM. We denote the side of a mesh triangle by $2a$ so that $a=s/8$, and due to the symmetry of the problem we work only with the right-hand half of the semicircle.

Scalar products are as follows:

$$\left.\begin{aligned}
\mathbf{T}'^2_\rho &= \begin{cases} \sqrt{3} \text{ for } \rho=1,2,\ldots,7; \\ 2\sqrt{3} \text{ for } \rho=8,9,\ldots,22; \end{cases} \\
\mathbf{T}'_\rho\,.\,\mathbf{T}'_\mu \text{ (for } \rho\neq\mu) &= \begin{cases} -\dfrac{1}{2\sqrt{3}} \text{ for overlapping bases and } \rho,\mu=1,2,\ldots,7; \\ -\dfrac{1}{\sqrt{3}} \text{ for any other pair of overlapping bases;} \\ 0 \text{ for non-overlapping bases.} \end{cases} \\
\mathbf{S}''_0\,.\,\mathbf{T}'_\rho &= \begin{cases} 2a \text{ for } \rho=1,2,\ldots,7; \\ 0 \text{ for all other values of } \rho. \end{cases}
\end{aligned}\right\} \tag{5·334}$$

We obtain from (5·330)
$$6a'_\rho=\sum_\mu a'_\mu \tag{5·335}$$
when $\mathbf{T}'_\rho$ is in the interior of V, the summation being over all its neighbours, and
$$6a'_\rho=\sum_\mu a'_\mu+4\sqrt{3}\,a \tag{5·336}$$
for all $\mathbf{T}'_\rho$ on the upper boundary, i.e. $\rho=1,2,\ldots,7$, the summation being over neighbours with images included [as described after (4·412)].

We now define reduced weights b'_ρ by
$$b'_\rho=\frac{\sqrt{3}}{2a}a'_\rho; \tag{5·337}$$
then (5·335) and (5·336) can be written as
$$b'_\rho=\tfrac{1}{6}\sum_\mu b'_\mu+F, \tag{5·338}$$
where $F=0$ except for vectors with centres on the upper boundary, in which case $F=1$. The summation is over all neighbours, with

images included in the case of vectors whose centres lie on the upper boundary.

It is easy to find the quick convergence factor [cf. (4·437)] for the case where the radius s is divided into n parts and triangulation is then carried out as in Fig. 5·33; the value is

$$k' = \frac{6(2n-1)}{2(\Sigma_1-\Sigma_2)+3\Sigma_3+4\Sigma_4}, \tag{5·339}$$

and for $n=4$

$$k' = \frac{42}{2(\Sigma_1-\Sigma_2)+3\Sigma_3+4\Sigma_4}; \tag{5·340}$$

Σ_1, Σ_2, Σ_3 and Σ_4 are indicated in the case $n=4$ in Fig. 5·33.

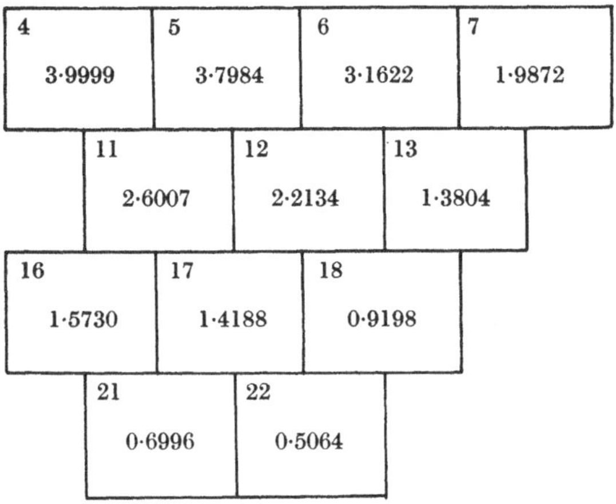

Fig. 5·34. b'_ρ for the flow problem, $n=4$.

Approximate solutions are then obtained for equations (5·338) by applying the circling method and using the quick convergence factor (5·340) as in Chapter 4. These solutions are shown in Fig. 5·34.

The lower bound $\mathbf{V}'^2$ may be written

$$\mathbf{V}'^2 = \frac{2\sqrt{3}\,a}{3}\Sigma b'_\rho \mathbf{S}''_0 . \mathbf{T}'_\rho$$

$$= \frac{\sqrt{3}\,s^2}{48}[b'(4)+2(b'(5)+b'(6)+b'(7))], \tag{5·341}$$

due to (5·331), (5·334) and (5·337). Inserting the numerical values, we obtain $\mathbf{V}'^2 = 0{\cdot}7900s^2$ as the lower bound for the approximation $n=4$.

This bound, and the upper bound which is found below, are 'unreliable' bounds in the sense of p. 123, since the numbers in Fig. 5·34 are treated as exact solutions of the equations.

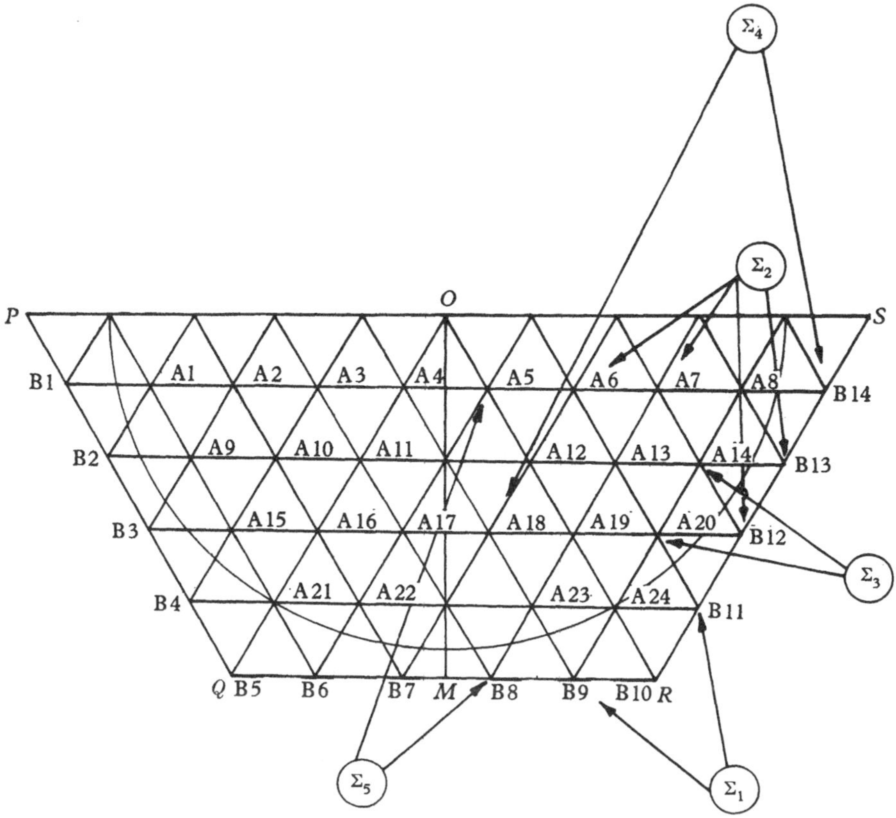

Fig. 5·35. Pyramid F-vectors of the second class (centres numbered A 1–A 24 and B 1–B 14) for the calculation of the upper bound for the flow problem in a semicircular channel. Groups of vectors used in the calculation of the quick convergence factor k'' are also shown.

Flow problem: the approximation $n=4$: upper bound for $\mathbf{S}^2$

We determine the upper bound $\mathbf{V}_2''^2$ for the problem

$$\Delta u = 0 \text{ in } W, \quad u = 0 \text{ on } PQRS, \quad \frac{\partial u}{\partial n} = 1 \text{ on } PS, \tag{5·342}$$

where W is the region enclosed by $PQRS$ in Fig. 5·35. Then by (5·229) we have

$$\mathbf{S}^2 \leqslant \mathbf{V}_2''^2, \tag{5·343}$$

so that $\mathbf{V}_2''^2$ is also an upper bound for the problem relating to the semicircle.

In the calculation of $\mathbf{V}_2''^2$ we use thirty-eight hexagonal pyramid F-vectors of the second class, numbered $A\,1$–$A\,24$, and $B\,1$–$B\,14$ (Fig. 5·35). The corresponding weights a_σ'' are skew-symmetrical about the vertical line OM and they are zero on OM and also on PS due to the requirement $p_i''n_i=0$ on PS for homogeneous F-vectors of the second class. Accordingly we need consider only the right-hand half of the semicircle.

The scalar products are

$$\left.\begin{aligned}
&\mathbf{T}_{Ar}''^2 = 2\sqrt{3} \quad (r=1,2,\ldots,24),\\
&\mathbf{T}_{Br}''^2 = \begin{cases}\sqrt{3} & (r=1,2,3,4,6,7,8,9,11,12,13,14)\\ \tfrac{2}{3}\sqrt{3} & (r=5,10),\end{cases}\\
&\mathbf{T}_\sigma''.\mathbf{T}_\nu'' = \begin{cases}-\dfrac{1}{\sqrt{3}} \text{ for all overlapping bases } (\sigma\neq\nu)\text{, except}\\ \quad\text{when both centres are on the boundary,}\\ -\dfrac{1}{2\sqrt{3}} \text{ when both centres are on the boundary,}\end{cases}\\
&\mathbf{S}_0''.\mathbf{T}_{Ar}'' = 0 \quad (r=1,2,\ldots,24),\\
&\mathbf{S}_0''.\mathbf{T}_{Br}'' = \begin{cases}\sqrt{3}a & (r=1,2,3,4),\\ \dfrac{\sqrt{3}}{2}a & (r=5),\\ 0 & (r=6,7,8,9),\\ -\dfrac{\sqrt{3}}{2}a & (r=10),\\ -\sqrt{3}\,a & (r=11,12,13,14).\end{cases}
\end{aligned}\right\} \qquad (5{\cdot}344)$$

The equations (5·330) then become

$$\left.\begin{aligned}
a''(Ar) &= \tfrac{1}{6}\Sigma a_\sigma'' \quad (r=1,2,\ldots,24),\\
a''(Br) &= \tfrac{1}{6}\Sigma a_\sigma'' \quad (r=8,9),\\
a''(Br) &= \tfrac{1}{6}\Sigma a_\sigma''+a \quad (r=11,12,13,14),\\
a''(B\,10) &= \tfrac{1}{4}(2a''(A\,24)+a''(B\,11)+a''(B\,9))+\tfrac{3}{4}a,
\end{aligned}\right\} \qquad (5{\cdot}345)$$

where the summations are over all neighbours; when the centre is on the edge, they include fictitious images obtained by reflecting interior neighbours in the edge as in (5·336). Defining reduced weights by $b''_\sigma = a''_\sigma/a$, we write (5·345) as

$$\left.\begin{aligned} b''(Ar) &= \tfrac{1}{6}\Sigma b''_\sigma \quad (r=1,2,\ldots,24),\\ b''(Br) &= \tfrac{1}{6}\Sigma b''_\sigma + F\begin{cases}\text{where } F=0 \text{ for } r=8,9,\\ \text{and } F=1 \text{ for } r=11,12,13,14,\end{cases}\\ b''(B\,10) &= \tfrac{1}{4}(2b''(A\,24)+b''(B\,11)+b''(B\,9))+\tfrac{3}{4}. \end{aligned}\right\} \quad (5{\cdot}346)$$

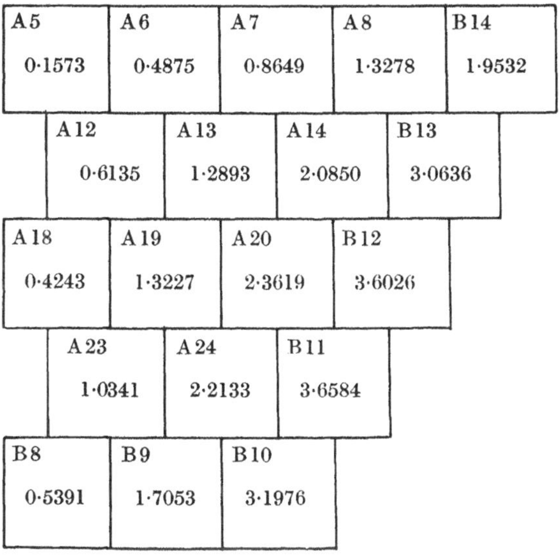

Fig. 5·36. b''_σ for the flow problem, $n=4$.

The quick convergence factor is

$$k'' = 57[2(b''(A\,12)-b''(A\,23))+3\Sigma_1+4(\Sigma_2-\Sigma_3)+6b''(B\,10) + 8(\Sigma_4 - b''(A\,24))+10\Sigma_5]^{-1}, \quad (5{\cdot}347)$$

the summations being indicated in Fig. 5·35. The approximate solutions to the equations (5·346), found by using the circling and multiplying processes alternately, are shown in Fig. 5·36.

The upper bound is found to be

$$\begin{aligned} \mathbf{V}''^2_2 &= \mathbf{S}''^2_0 + a\Sigma b''_\sigma \mathbf{S}''_0 . \mathbf{T}''_\sigma \\ &= \frac{75\sqrt{3}}{64}s^2 - \frac{\sqrt{3}}{64}s^2[b''(B\,10)+2(b''(B\,11)+b''(B\,12)+b''(B\,13) \\ &\quad + b''(B\,14))] \\ &= 1{\cdot}2787s^2. \end{aligned} \quad (5{\cdot}348)$$

Here $\mathbf{S}''^2_0$ is found by integrating over $PQRS$ in Fig. 5·35.

Thus $\mathbf{S}^2$, the Dirichlet integral of the solution of the boundary value problem (5·315) for the case where the curved boundary is a semicircle of radius s, is bounded as follows

$$0{\cdot}7900 \leqslant \mathbf{S}^2/s^2 \leqslant 1{\cdot}2787 \tag{5·349}$$

for the approximation $n=4$ (unreliable bounds).

Flow problem: the approximation $n=8$

On bisecting the mesh of the above approximation ($n=4$), triangulating, and then excluding certain triangles at the edges, we obtain the network shown in Fig. 5·37*a*. We then calculate upper and lower bounds as above. In this approximation we have $a=\frac{1}{16}s$, where $2a$ is the side of a mesh triangle.

One hundred and one hexagonal pyramid F-vectors of the first class, numbered 1–101, are used in the calculation of the lower bound (Fig. 5·37*a*), and $A'B' \ldots Q'R'$ is the outer perimeter for these vectors, all weights on this boundary being zero. Due to symmetry about OM, fifty-three of the weights a'_ρ are independent, and we find approximate solutions by applying the usual method to the equations in the first line of (5·330) which involve these weights. These solutions expressed in terms of the reduced weights b'_ρ, where $b'_\rho = \sqrt{3}\,a'_\rho/2a$, are shown in Fig. 5·37*b*, and the lower bound is then found to be 0·8372.

For the upper bound we consider the boundary value problem

$$\left.\begin{aligned} &\Delta u = 0 \text{ in } W, \quad u=0 \text{ on } A'A''\ldots F''F'G''\ldots L''K'N''\ldots T''R',\\ &\frac{\partial u}{\partial n} = 1 \text{ on } A'R', \end{aligned}\right\} \tag{5·350}$$

where W is the area enclosed by $A'A''\ldots F''F'G''\ldots L''K'N''\ldots T''R'$ (Fig. 5·37*c*), and if $\mathbf{V}_2''^2$ is the upper bound for this problem then by (5·229)

$$\mathbf{S}^2 \leqslant \mathbf{V}_2''^2, \tag{5·351}$$

so that $\mathbf{V}_2''^2$ is also an upper bound for $\mathbf{S}^2$, the Dirichlet integral for the problem (5·315) associated with a semicircular channel. One hundred and thirty-four hexagonal pyramid F-vectors of the second class are used in the calculation of this bound, the vectors being numbered X1–X106 and Z1–Z28 (Fig. 5·37*c*). Sixty-seven of the corresponding weights a''_σ are independent. Reduced weights b''_σ are defined by $b''_\sigma = 8(a''_\sigma/a)$. Making this transformation, we solve the equations in the second line of (5·330) which involve the independent weights by the usual method. In the calculation an oscillation set in in the case of several of the weights after the others had

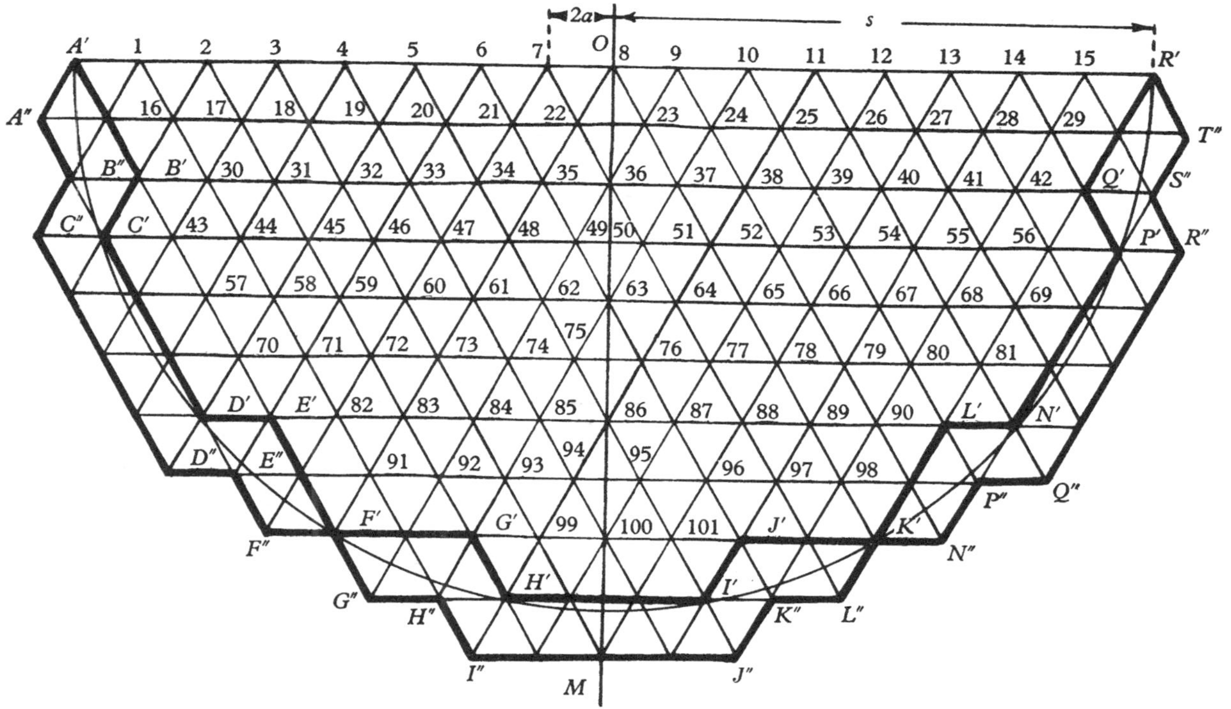

Fig. 5·37*a*. Hexagonal pyramid *F*-vectors of the first class in the approximation $n=8$ for the flow problem in a semicircular channel.

become steady, and two values of the upper bound were found, 1·0317 and 1·0318. The latter value is taken as the upper bound for our problem, and the values of the corresponding weights are shown in Fig. 5·37*d*.

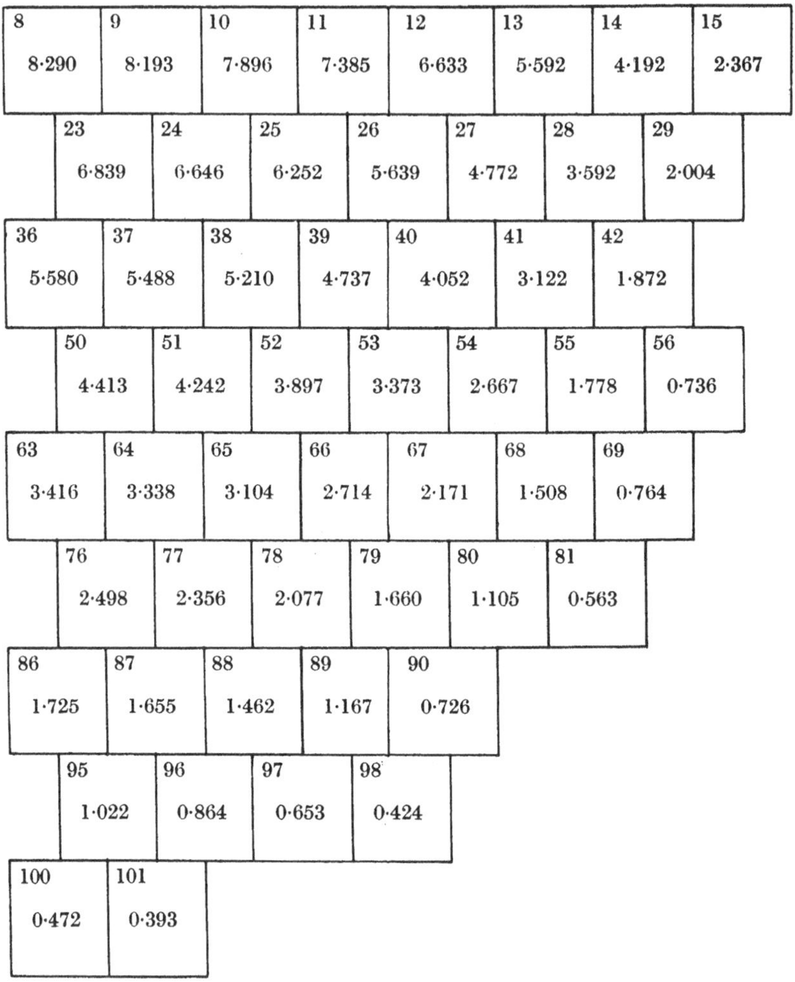

Fig. 5·37*b*. b'_p for the flow problem, $n=8$.

Thus, in the approximation $n=8$, we obtain bounds on $\mathbf{S}^2/s^2$ as follows:

$$0{\cdot}8372 \leqslant \mathbf{S}^2/s^2 \leqslant 1{\cdot}0318, \qquad (5{\cdot}352)$$

the bounds being 'unreliable' in the sense of p. 123.

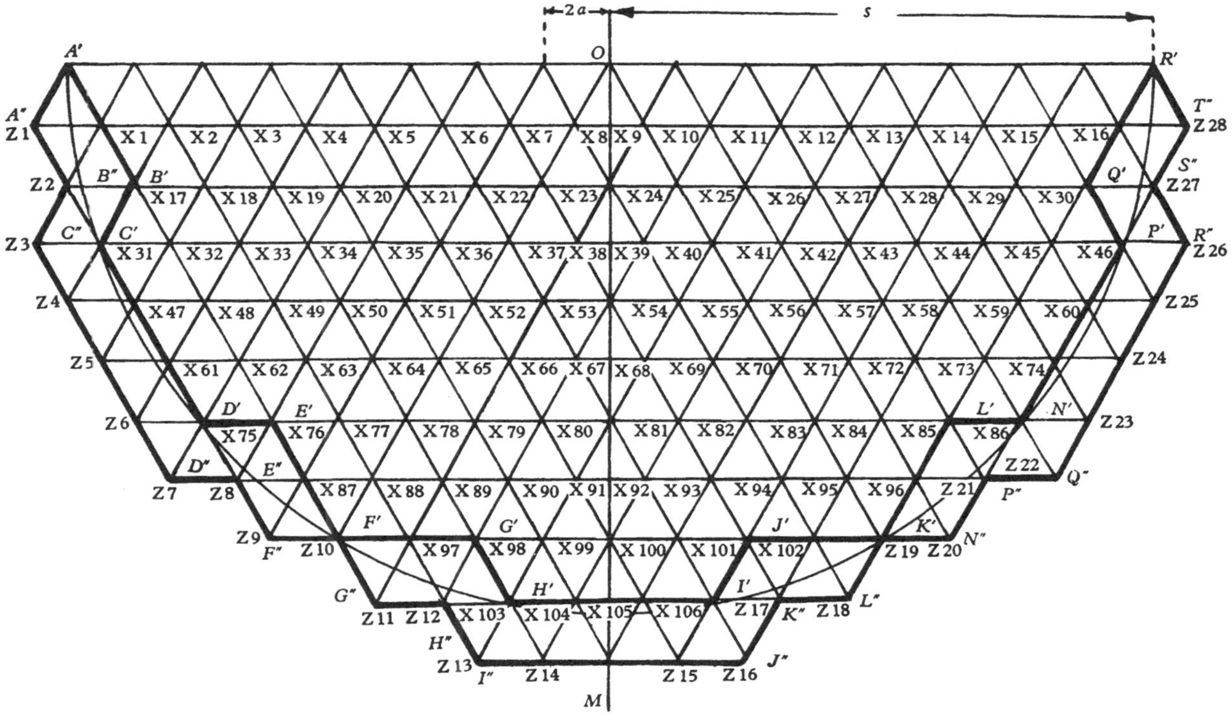

Fig. 5·37*c*. Hexagonal pyramid F-vectors of the second class for approximation $n=8$ in the flow problem for a semicircular channel. All weights on OM are zero.

An exact solution may be found for our problem (5·315), viz.

$$\Delta u = 0 \text{ in } V, \quad (u)_{B''} = 0, \quad \frac{\partial u}{\partial n} = 1 \text{ on } B', \tag{5·353}$$

where B' and B'' are shown in Fig. 5·32. If we denote thė Dirichlet integral by $\mathbf{S}^2$ as usual, this exact solution gives

$$\mathbf{S}^2/s^2 = 0{\cdot}934. \tag{5·354}$$

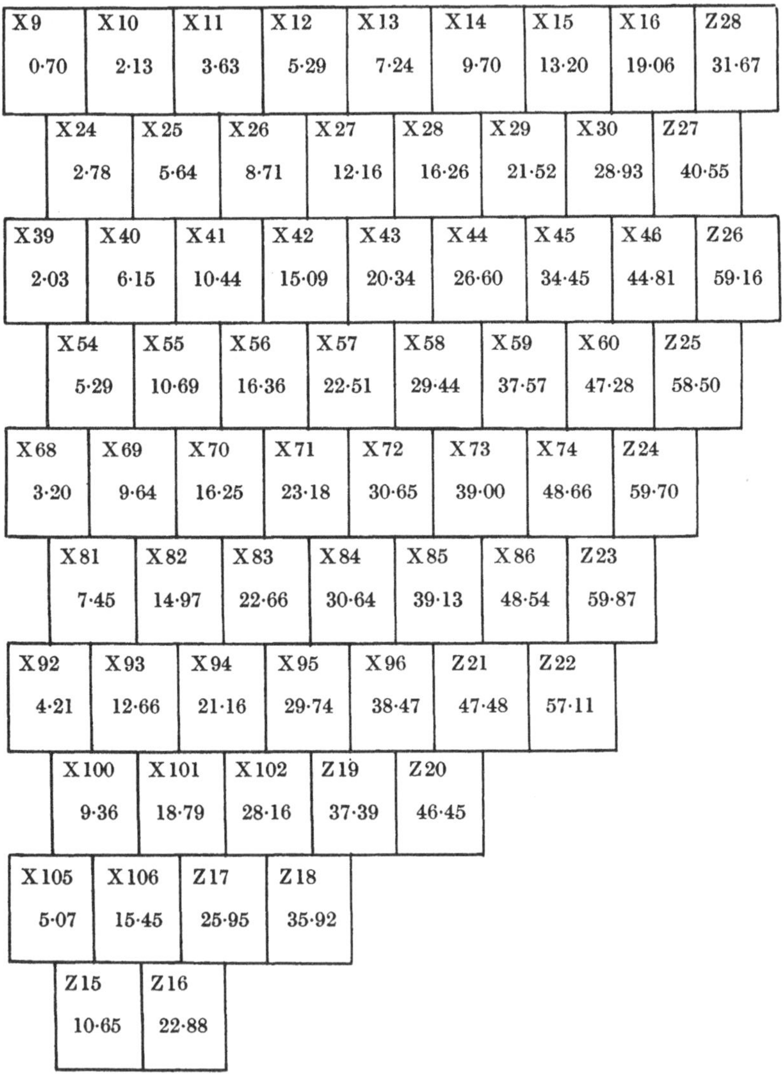
X9 0·70
X10 2·13
X11 3·63
X12 5·29
X13 7·24
X14 9·70
X15 13·20
X16 19·06
Z28 31·67
X24 2·78
X25 5·64
X26 8·71
X27 12·16
X28 16·26
X29 21·52
X30 28·93
Z27 40·55
X39 2·03
X40 6·15
X41 10·44
X42 15·09
X43 20·34
X44 26·60
X45 34·45
X46 44·81
Z26 59·16
X54 5·29
X55 10·69
X56 16·36
X57 22·51
X58 29·44
X59 37·57
X60 47·28
Z25 58·50
X68 3·20
X69 9·64
X70 16·25
X71 23·18
X72 30·65
X73 39·00
X74 48·66
Z24 59·70
X81 7·45
X82 14·97
X83 22·66
X84 30·64
X85 39·13
X86 48·54
Z23 59·87
X92 4·21
X93 12·66
X94 21·16
X95 29·74
X96 38·47
Z21 47·48
Z22 57·11
X100 9·36
X101 18·79
X102 28·16
Z19 37·39
Z20 46·45
X105 5·07
X106 15·45
Z17 25·95
Z18 35·92
Z15 10·65
Z16 22·88

Fig. 5·37*d*. b''_σ for the flow problem, $n=8$.

We may summarize our results in the following table, the bounds being on $\mathbf{S}^2/s^2$:

	Lower bound	Upper bound
$n=4$	0·7900	1·2787
$n=8$	0·8372	1·0318
Exact solution	0·934	

These bounds are much further apart than the bounds obtained in Chapter 4 for meshes of comparable fineness. This coarseness of approximation is due to the curvature of the boundary. By using finer and finer meshes we could of course draw the bounds as close together as we wished, but we would need quite a fine mesh to reduce the error to, say, one per cent, because the polygons of the type used above converge rather slowly to the semicircle.

Second mixed problem: membrane with elastic support

Imagine a membrane stretched to a uniform tension in a plane which we call the neutral plane. A uniform pressure now acts on one side of the membrane which takes up a static deflected position.

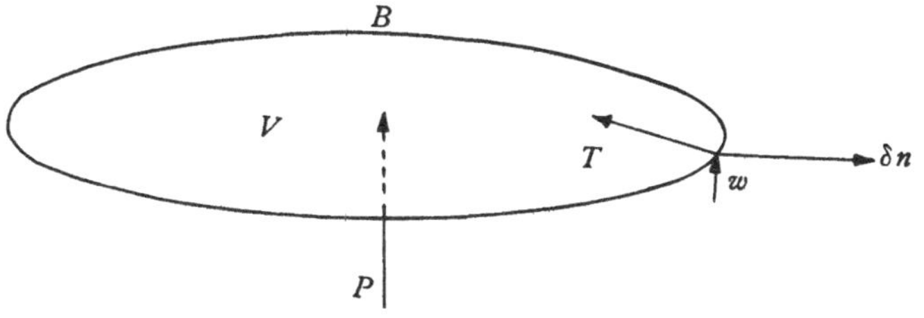

Fig. 5·38*a*. Membrane with edge supported elastically.

We consider only small deflexions so that the tension remains constant over the surface formed. Suppose that the boundary B of the membrane is not fixed but yields elastically in the direction perpendicular to the neutral plane, the yield being proportional to the component in this direction of the force per unit length of B due to the tension in the membrane. Combining the usual partial differential equation for the equilibrium of a membrane with this boundary condition, we arrive at the boundary value problem:

$$\Delta w = -P/T \text{ in } V, \quad (w)_B = -kT(\partial w/\partial n)_B, \tag{5·355}$$

where w is the deflexion, P the pressure, T the tension in the membrane, k an elastic constant and ∂n an element of the outward normal to B in the neutral plane. V is the domain of the membrane (Fig. 5·38*a*).

To get Laplace's equation we now make the transformation

$$u = \tfrac{1}{2}r^2 + 2wT/P, \tag{5·356}$$

where r is the distance from any fixed point in the neutral plane. Then (5·355) becomes

$$\Delta u = 0 \text{ in } V, \quad \left(u + c\frac{\partial u}{\partial n}\right)_B = \tfrac{1}{2}r^2 + c\frac{\partial}{\partial n}(\tfrac{1}{2}r^2), \tag{5·357}$$

where $c = kT$. This is a mixed problem of the second type [cf. (5·230)] with $\alpha/\beta = c^{-1} > 0$. Therefore, with

$$\mathbf{S} \leftrightarrow u_{,i}, \tag{5·358}$$

where u is the solution, the F-point defined by $\mathbf{S}$ may be confined within a hypersphere as in (5·239), and $\mathbf{S}^2$ bounded as in (5·248).

Setting $\mathbf{W} \leftrightarrow w_{,i}$, $\mathbf{I} \leftrightarrow x_i$, we get from (5·356)

$$\mathbf{S} = \mathbf{I} + \frac{2T}{P}\mathbf{W}; \tag{5·359}$$

thus, by confining $\mathbf{S}$ within a hypersphere with centre $\mathbf{C}$ and radius R, we at the same time confine $\mathbf{W}$ within another hypersphere obtained from the first by a translation ($\mathbf{I}$) and a magnification ($P/2T$), the centre and radius being $\mathbf{C}'$ and R' respectively, where

$$\mathbf{C}' = \frac{P}{2T}(\mathbf{C} - \mathbf{I}), \quad R' = \left(\frac{P}{2T}\right)R. \tag{5·360}$$

Provided $\mathbf{C}'^2 > R'^2$, the bounds on $\mathbf{W}^2$ are by (5·248):

$$\mathbf{C}'^2 + R'^2 - 2\,|\,\mathbf{C}'\,|\,R' \leqslant \mathbf{W}^2 \leqslant \mathbf{C}'^2 + R'^2 + 2\,|\,\mathbf{C}'\,|\,R'. \tag{5·361}$$

Since the additional potential energy E in the membrane when deflected out of the neutral plane is expressed in terms of the Dirichlet integral of w by

$$E = \tfrac{1}{2}T\mathbf{W}^2 = \tfrac{1}{2}T\int_V w_{,i}w_{,i}\,dV, \tag{5·362}$$

(5·361) can be used to bound the potential energy in the membrane. Note that this is not the total energy in the system, since energy is also stored in the elastic support.

As a particular example consider a membrane in the form of a right-angled isosceles triangle of hypotenuse $\sqrt{2}\,a$, the origin and axes being taken as in Fig. 5·38b. We seek to confine the F-point $\mathbf{S}$ within a hypersphere of minimum radius.

We select two F-vectors $\mathbf{S}'$ and $\mathbf{S}''$ in L' and L'' respectively satisfying (5·232):

$$\left.\begin{array}{l}\mathbf{S}'\leftrightarrow u'_{,i},\quad u'=\tfrac{1}{2}r^2+\Sigma A_{mn}\left(\dfrac{x_1}{a}\right)^m\left(\dfrac{x_2}{a}\right)^n\quad (A_{mn}=A_{nm}),\\ \mathbf{S}''\leftrightarrow p''_i,\quad p''_1=\partial v''/\partial x_2,\quad p''_2=-\partial v''/\partial x_1,\quad v''=\Sigma B_{mn}\left(\dfrac{x_1}{a}\right)^m\left(\dfrac{x_2}{a}\right)^n\\ \hfill (B_{mn}=-B_{nm}).\end{array}\right\} \tag{5·363}$$

Here Σ means summation for $m, n=0, 1, \ldots$ to infinity, but as we take only a finite number of the coefficients A_{mn}, B_{mn} different from

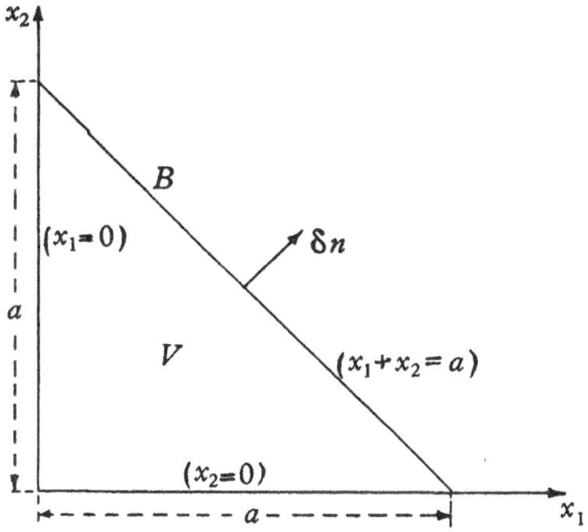

Fig. 5·38b. Elastically supported membrane in the form of an isosceles right-angled triangle.

zero, we are dealing with polynomials and not infinite series. These coefficients must be chosen so that $\mathbf{S}'$ and $\mathbf{S}''$ are connected on the sides and hypotenuse of the triangle by

$$u'+cp''_i n_i=\tfrac{1}{2}r^2+c\frac{\partial}{\partial n}(\tfrac{1}{2}r^2), \tag{5·364}$$

[cf. (5·233)]. These conditions read

$$\left.\begin{array}{l}u'-\tfrac{1}{2}r^2=cp''_1 \text{ on } x_1=0,\\ u'-\tfrac{1}{2}r^2=cp''_2 \text{ on } x_2=0,\\ u'-\tfrac{1}{2}r^2=\dfrac{ca\sqrt{2}}{2}-\dfrac{c\sqrt{2}}{2}(p''_1+p''_2) \text{ on } x_1+x_2=a.\end{array}\right\} \tag{5·365}$$

We substitute the values (5·363) of u' and p''_i into (5·365) and compare coefficients of powers of x_1 and x_2. The first two conditions in (5·365) lead to the relations

$$A_{0n} = -(n+1)\left(\frac{c}{a}\right) B_{n+1,0} \quad (n=0,1,2,\ldots), \qquad (5\cdot366)$$

and the last condition of (5·365) requires A_{mn} and B_{mn} to satisfy more complicated equations.

The vectors $\mathbf{S}'$ and $\mathbf{S}''$ now define a hypersphere, of centre $\mathbf{C}$ and radius R, containing the F-point $\mathbf{S}$, where

$$\left.\begin{aligned} 4\mathbf{C}^2 &= \mathbf{S}'^2 + \mathbf{S}''^2 + 2\mathbf{S}'.\mathbf{S}'', \\ 4R^2 &= \mathbf{S}'^2 + \mathbf{S}''^2 - 2\mathbf{S}'.\mathbf{S}''. \end{aligned}\right\} \qquad (5\cdot367)$$

The number of conditions (5·365) is less than the number of coefficients A_{mn}, B_{mn} and the resulting freedom in the coefficients is used to minimize R^2, the square of the radius of the hypersphere, which is quadratic in these parameters. $\mathbf{S}^2$ is then bounded as in (5·248).

Two approximations are worked out. The problem involves essentially one dimensionless parameter (c/a), and the method of approximation may be used for any value of it; for simplicity we shall take $c/a = 1$.

Membrane problem: first approximation

In the first approximation we take four non-zero coefficients A_{mn} and B_{mn}, defining the functions u' and v'' as

$$\left.\begin{aligned} u' &= \tfrac{1}{2}r^2 + A_{00} + A_{01}(\xi+\eta), \\ v'' &= B_{10}(\xi-\eta) + B_{20}(\xi^2-\eta^2), \end{aligned}\right\} \qquad (5\cdot368)$$

where, for brevity, we have set $x_1/a = \xi$, $x_2/a = \eta$. The conditions (5·365) now give

$$\left.\begin{aligned} A_{00} &= -B_{10}, \quad A_{01} = -2B_{20}, \\ B_{10} + \sqrt{2}\,B_{20} &= -a^2\frac{\sqrt{2}}{2}(\sqrt{2}-1), \end{aligned}\right\} \qquad (5\cdot369)$$

so that there remains one free coefficient.

Using (5·363) and (5·368), $u'_{,i}$ and p''_i are found and $\mathbf{S}'^2$, $\mathbf{S}''^2$ and $\mathbf{S}'.\mathbf{S}''$ calculated with the help of the formula

$$\iint_V \xi^p \eta^q \, dV = \frac{a^2 p!\,q!}{(p+q+2)!} \quad (p,q=0,1,2,\ldots). \qquad (5\cdot370)$$

We find the value of the free parameter which minimizes the expression (5·367) for $4R^2$, and use that value to calculate R^2

and $\mathbf{C}^2$ for the hypersphere enclosing the solution $\mathbf{S}$. The results are

$$\frac{R_1^2}{a^4} = 0{\cdot}013405, \quad \frac{\mathbf{C}_1^2}{a^4} = 0{\cdot}10432, \tag{5·371}$$

the suffix 1 indicating the first approximation.

We may present the result obtained in two ways—first in the form of bounds on $\mathbf{S}^2/a^4$, which by (5·248) are

$$0{\cdot}0429 \leqslant \frac{\mathbf{S}^2}{a^4} \leqslant 0{\cdot}1925, \tag{5·372}$$

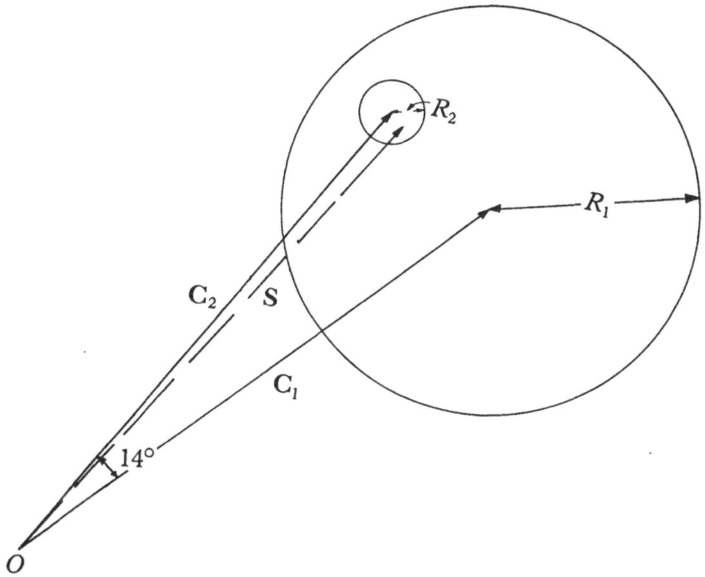

Fig. 5·39. Hyperspheres corresponding to the first and second approximations in the membrane problem.

and secondly as in Fig. 5·39, which shows the ratio $R_1/|\mathbf{C}_1| = 0{\cdot}358$ to scale; this figure is the projection on a plane through the origin and the vector $\mathbf{C}_1$ of the hypersphere containing the F-point $\mathbf{S}$.

Membrane problem: higher approximation

In the second approximation we take eight non-zero coefficients in the definition of u' and v'':

$$\left.\begin{aligned} u' - \tfrac{1}{2}r^2 &= A_{00} + A_{01}(\xi+\eta) + A_{11}\xi\eta \\ &\quad + A_{02}(\xi^2+\eta^2), \\ v'' &= B_{10}(\xi-\eta) + B_{20}(\xi^2-\eta^2) \\ &\quad + B_{21}(\xi^2\eta - \xi\eta^2) + B_{30}(\xi^3-\eta^3). \end{aligned}\right\} \tag{5·373}$$

Then (5·365) imply

$$\left.\begin{aligned} &A_{00} = -B_{10}, \quad A_{01} = -2B_{20}, \quad A_{02} = -3B_{30}, \\ &A_{11} = 3\sqrt{2}\,[B_{21} - (1+\sqrt{2})\,B_{30}], \\ &B_{10} = \frac{\sqrt{2}}{2}\,[(\sqrt{2}-1)\,(B_{21}-a^2) - (3B_{30}+2B_{20})], \end{aligned}\right\} \tag{5·374}$$

thus leaving only three arbitrary parameters.

Following the same process as in the first approximation, we find

$$\begin{aligned} \frac{360R^2}{a^4} = {} & 120(3+2\sqrt{2})\,u^2 + 9(273+177\sqrt{2})\,v^2 \\ & + (229+108\sqrt{2})\,w^2 - 3(416+267\sqrt{2})\,vw \\ & - 24(17+10\sqrt{2})\,wu + 18(96+65\sqrt{2})\,uv - 30u \\ & - 72v - 3(12-5\sqrt{2})\,w + 30(3-2\sqrt{2}), \end{aligned} \tag{5·375}$$

where $u = B_{20}/a^2, \quad v = B_{30}/a^2, \quad w = B_{21}/a^2.$

Minimization of (5·375) with respect to the three parameters yields three simultaneous linear equations for u, v and w; the solutions are

$$\left.\begin{aligned} u &= -0{\cdot}15793237, \\ v &= 0{\cdot}14244165, \\ w &= 0{\cdot}30894374. \end{aligned}\right\} \tag{5·376}$$

With these values of the parameters, R and $|\mathbf{C}|$ for the hypersphere containing the solution are given in this second approximation by

$$\left.\begin{aligned} R_2^2/a^4 &= 0{\cdot}289238 \times 10^{-3}, \\ \mathbf{C}_2^2/a^4 &= 102{\cdot}916 \times 10^{-3}, \end{aligned}\right\} \tag{5·377}$$

and the bounds on $\mathbf{S}^2/a^4$ are by (5·248)

$$0{\cdot}09229 \leqslant \mathbf{S}^2/a^4 \leqslant 0{\cdot}11412. \tag{5·378}$$

Fig. 5·39 is a projection (drawn to scale) on the plane containing the origin of function-space and the vectors $\mathbf{C}_1$, $\mathbf{C}_2$; it shows the two hyperspheres corresponding to the two approximations, and brings out very clearly the improvement obtained in the second one. The vector labelled **S** (drawn tentatively of course, since **S** is not known) is the projection of the solution on this plane, the extremity of the projection certainly lying inside the small circle with centre $\mathbf{C}_2$ and radius R_2.

The angle between $\mathbf{C}_1$ and $\mathbf{C}_2$ is

$$\theta = \cos^{-1}\frac{\mathbf{C}_1 . \mathbf{C}_2}{|\mathbf{C}_1|\,|\mathbf{C}_2|} = 14^\circ\,7', \tag{5·379}$$

calculated from the formulae

$$\mathbf{C}_1=\tfrac{1}{2}(\mathbf{S}_1'+\mathbf{S}_1''),\quad \mathbf{C}_2=\tfrac{1}{2}(\mathbf{S}_2'+\mathbf{S}_2''), \tag{5·380}$$

the suffixes referring to the vectors used in the first and second approximations. From (5·371) and (5·377) we find

$$|\mathbf{C}_1|/|\mathbf{C}_2|=1{\cdot}0067.$$

Higher approximations would confine **S** within hyperspheres of smaller and smaller radii, but perspective must be used if we wish to represent all these approximations on the same diagram.

Exercises

1. The cross-section of a channel is a regular hexagon, the upper side of the hexagon representing the free surface of liquid flowing through the channel under the action of a surface stress. Taking the simple triangulation in which the hexagon is divided into six equilateral triangles, find bounds for the flux from (5·326).

2. Prove that the flux through a channel is increased by enlarging it, i.e. that the flux through the channel bounded by $B_2'+B_2''$ in Fig. 5·22 is greater than the flux through the channel bounded by $B_1'+B_1''$, the surface stress being the same in both cases.

3. Solve the problem (5·357) for a circular membrane and calculate $\mathbf{S}^2$ and $\mathbf{W}^2$.

4. Consider the following approach to the boundary value problem (5·357). Take two linear subspaces L', L'' as follows:

$$L':\quad \mathbf{S}'\leftrightarrow p_i'=u'_{,i},\quad u'+c\,\partial u'/\partial n=\tfrac{1}{2}r^2+c\,\partial(\tfrac{1}{2}r^2)/\partial n \text{ on } B,$$

$$L'':\quad \mathbf{S}''\leftrightarrow p_i''=u''_{,i},\quad \Delta u''=0 \text{ in } V.$$

The solution **S** is on both L' and L'', but L' and L'' are not mutually orthogonal. Regarding any pair $(\mathbf{S}',\mathbf{S}'')$ as a tentative pair of solutions, we may call $(\mathbf{S}'-\mathbf{S}'')^2$ the 'error' of this pair, since $(\mathbf{S}'-\mathbf{S}'')^2=0$ implies $\mathbf{S}'=\mathbf{S}''=\mathbf{S}$. Taking any F-point $\mathbf{S}_0'$ on L', any set of F-vectors $\mathbf{T}_\rho'\ (\rho=1,2,\ldots,r)$ lying in L', and any set of F-vectors $\mathbf{T}_0''\ (\sigma=1,2,\ldots,s)$ lying in L'', then the error of the pair of tentative solutions

$$\mathbf{S}'=\mathbf{S}_0'+\sum_{\rho=1}^{r}a_\rho'\mathbf{T}_\rho',\quad \mathbf{S}''=\sum_{\sigma=1}^{s}a_\sigma''\mathbf{T}_\sigma''$$

is minimized by choosing a_ρ', a_σ'' to satisfy the $(r+s)$ linear equations

$$\sum_{\mu=1}^{r}a_\mu'\mathbf{T}_\mu'.\mathbf{T}_\rho'-\sum_{\nu=1}^{s}a_\nu''\mathbf{T}_\nu''.\mathbf{T}_\rho'+\mathbf{S}_0'.\mathbf{T}_\rho'=0\quad(\rho=1,2,\ldots,r),$$

$$-\sum_{\mu=1}^{r}a_\mu'\mathbf{T}_\mu'.\mathbf{T}_\sigma''+\sum_{\nu=1}^{s}a_\nu''\mathbf{T}_\nu''.\mathbf{T}_\sigma''-\mathbf{S}_0'.\mathbf{T}_\sigma''=0\quad(\sigma=1,2,\ldots,s).$$

5·4. EQUILIBRIUM OF AN ELASTIC BODY

Equations of equilibrium and compatibility

In an elastic body we have to deal with a displacement vector u_i, which we suppose small, so that the components of the strain tensor are

$$e_{ij} = \tfrac{1}{2}(u_{i,j} + u_{j,i}), \tag{5·401}$$

the comma indicating partial differentiation. Here the suffixes have the range 1, 2, 3, the coordinates are rectangular Cartesians, and summation will be understood in the case of repeated suffixes. There is also a symmetric stress tensor $E_{ij}\,(=E_{ji})$, and a body force X_i per unit volume.

The equations of equilibrium are

$$E_{ij,j} + X_i = 0, \tag{5·402}$$

and Hooke's law (in generalized form) reads

$$E_{ij} = c_{ijkl}e_{kl}, \tag{5·403}$$

where c_{ijkl} are elastic coefficients (we shall take them to be constants) satisfying the symmetry conditions

$$c_{ijkl} = c_{jikl} = c_{klij}, \tag{5·404}$$

on account of which the number of coefficients is 21. If we solve (5·403) for e_{ij} we get

$$e_{ij} = C_{ijkl}E_{kl}, \tag{5·405}$$

where

$$C_{ijkl} = C_{jikl} = C_{klij}. \tag{5·406}$$

For an isotropic substance, Hooke's law reads

$$E_{ij} = \lambda\delta_{ij}e_{kk} + 2\mu e_{ij}, \tag{5·407}$$

where λ is Lamé's constant and μ the rigidity, or, equivalently,

$$e_{ij} = \frac{1}{E}\{(1+\sigma)\,E_{ij} - \sigma\delta_{ij}E_{kk}\}, \tag{5·408}$$

where E is Young's modulus, and σ Poisson's ratio.

The stress across a plane with unit normal vector n_i is

$$T_i = E_{ij}n_j. \tag{5·409}$$

In the basic problem of elastic equilibrium for a body of given form, the body force X_i is prescribed throughout the volume V of the body, and conditions are prescribed on the surface B which bounds V. These boundary conditions may consist in the prescription of the displacements u_i on B or of the stress T_i, or of a mixture of these conditions, such as u_i on part of B and T_i on the rest.

There are two ways of attacking such a problem. One way (and the most natural if u_i is prescribed on B) is to change (5·402) into a set of partial differential equations for u_i; this is easily done by means of (5·401) and (5·403). The second way depends on the fact that if e_{ij} is given arbitrarily, then the equations (5·401) do not possess solutions u_i; in fact, these six equations for three unknowns u_i are not compatible. The equations of compatibility (necessary and sufficient) are (see Sokolnikoff (1), p. 25)

$$e_{ij,kl} + e_{kl,ij} - e_{ik,jl} - e_{jl,ik} = 0; \tag{5·410}$$

if e_{ij} satisfies (5·410) then u_i exists satisfying (5·401), and is uniquely determined to within a rigid body displacement. Now if we substitute in (5·410) the values of e_{ij} from (5·405), we get equations of compatibility involving E_{ij}. It is unnecessary to write them out explicitly; we shall denote them by

$$C(E) = 0. \tag{5·411}$$

The second method of attacking a problem in elastic equilibrium is to work with the equations of equilibrium (5·402) and the equations of compatibility (5·411).

The first method is generally easier to use, but the second comes closer to our line of thought. As a matter of fact we do not use either method, because our concern is not with the solving of problems, but rather with approximating to solutions.

Splitting the problem

To solve a problem in elastic equilibrium we must find values of E_{ij} and u_i which satisfy (5·401), (5·402) and (5·403) in V, and are subject to certain boundary conditions on B, which we shall take as follows:

$$(u_i)_{B_1} = f_i, \quad (T_i)_{B_2} = g_i, \tag{5·412}$$

where f_i and g_i are given vector fields on portions of the surface B_1 and B_2, respectively, which together make up the whole surface B.

We write

$$\mathbf{S} \leftrightarrow E_{ij} \tag{5·413}$$

to indicate that a point or vector of function-space corresponds to any distribution of stress in V; it need not satisfy either the equations of equilibrium or those of compatibility, but it must be symmetric ($E_{ij} = E_{ji}$). By e_{ij} we shall always understand a tensor given in terms of E_{ij} by (5·405) so that we may write (5·413) as

$$\mathbf{S} \leftrightarrow e_{ij}, \tag{5·414}$$

if we like.* If E_{ij} satisfies $C(E)=0$, but not otherwise, there corresponds to it, by (5·401) and (5·405), a displacement u_i uniquely determined to within a rigid body displacement.

The scalar product we define by

$$\mathbf{S}.\mathbf{S}'=\int_V E_{ij}e'_{ij}\,dV. \tag{5·415}$$

It is most important that, on account of (5·403) and (5·404), we have the reciprocity relation

$$\begin{aligned} E_{ij}e'_{ij} &= c_{ijkl}e_{kl}e'_{ij} \\ &= c_{klij}e_{kl}e'_{ij} \\ &= c_{klij}e'_{ij}e_{kl} \\ &= E'_{ij}e_{ij}, \end{aligned} \tag{5·416}$$

and hence $$\mathbf{S}.\mathbf{S}'=\mathbf{S}'.\mathbf{S}. \tag{5·417}$$

For the metric we have

$$\mathbf{S}^2=\int_V E_{ij}e_{ij}\,dV=\int_V c_{ijkl}e_{ij}e_{kl}\,dV. \tag{5·418}$$

Now the strain energy per unit volume in the body is

$$w=\tfrac{1}{2}c_{ijkl}e_{ij}e_{kl}=\tfrac{1}{2}E_{ij}e_{ij}, \tag{5·419}$$

and it is assumed that this is always positive unless $E_{ij}=0$, when of course it vanishes. Thus (5·418) defines a positive-definite metric in function-space. For an isotropic body we have, by (5·408),

$$w=\frac{1}{2E}\{(1+\sigma)E_{ij}E_{ij}-\sigma(E_{kk})^2\}. \tag{5·420}$$

We consider now two classes of F-vectors, $\mathbf{S}'$ and $\mathbf{S}''$, defined as follows. Each class forms a linear subspace; we shall indicate them by L' and L''.

$$\left.\begin{aligned} &L'\colon\quad \mathbf{S}'\leftrightarrow E'_{ij},\quad C(E')=0,\quad (u'_i)_{B_1}=f_i, \\ &L''\colon\quad \mathbf{S}''\leftrightarrow E''_{ij},\quad E''_{ij,j}+X_i=0 \text{ in } V,\quad (E''_{ij}n_j)_{B_2}=g_i. \end{aligned}\right\} \tag{5·421}$$

There are permissible discontinuities. We require that E''_{ij} shall be piecewise continuous, with $E''_{ij}n_j$ continuous across every surface drawn in the body, and that u'_i be continuous.

To construct a vector $\mathbf{S}'$, the plan is to select any displacement u'_i satisfying the boundary conditions on B_1 and generate e'_{ij} from it by (5·401). Then $C(E')$ is automatically satisfied. To construct a vector $\mathbf{S}''$, we have to find six quantities E''_{ij} which satisfy the three equilibrium equations and also the boundary condition on B_2.

* Incompressible bodies are here excluded; see p. 345.

It is obvious that the solution **S** of the problem belongs to both L' and L''. In order to use the hypercircle method, all we need now is to establish the orthogonality of L' and L''.

We note that vectors $\mathbf{T}'$, $\mathbf{T}''$ lying in L', L'', respectively, satisfy

$$\left.\begin{aligned} L'&: \quad \mathbf{T}' \leftrightarrow E'_{ij}, \quad C(E')=0, \quad (u'_i)_{B_1}=0,\\ L''&: \quad \mathbf{T}'' \leftrightarrow E''_{ij}, \quad E''_{ij,j}=0, \quad (E''_{ij}n_j)_{B_2}=0. \end{aligned}\right\} \tag{5·422}$$

Then

$$\begin{aligned} \mathbf{T}'.\mathbf{T}'' &= \int_V e'_{ij}E''_{ij}\,dV \\ &= \int_V u'_{i,j}E''_{ij}\,dV \\ &= -\int_V u'_i E''_{ij,j}\,dV + \int_{B_1+B_2} u'_i E''_{ij}n_j\,dB \\ &= 0. \end{aligned} \tag{5·423}$$

Thus L' is orthogonal to L''.

The solution of the problem of elastic equilibrium consists then in finding the intersection of the two orthogonal linear subspaces, L' and L'', defined in (5·421).

The above approach includes the methods of Prager and Synge (1) and Synge (2).

Example: cube deformed by its own weight

In problems in which the displacement u_i is given over the entire surface B, the conditions (5·421) defining the linear subspaces become

$$\left.\begin{aligned} L'&: \quad \mathbf{S}' \leftrightarrow E'_{ij}, \quad C(E')=0, \quad (u'_i)_B=f_i,\\ L''&: \quad \mathbf{S}'' \leftrightarrow E''_{ij}, \quad E''_{ij,j}+X_i=0 \text{ in } V. \end{aligned}\right\} \tag{5·424}$$

As an example let us take the case of a cube of homogeneous isotropic material bounded by the planes

$$x_1 = \pm l, \quad x_2 = \pm l, \quad x_3 = \pm l, \tag{5·425}$$

with gravity acting in the negative sense of the x_3-axis, so that

$$X_1 = X_2 = 0, \quad X_3 = -g\rho, \tag{5·426}$$

where ρ is the density; as boundary condition we require that the faces of the cube be held fixed so that $f_i = 0$ in (5·424).

We shall find bounds for the strain energy in the body. These bounds can be interpreted in another way. Imagine, at first, that no force of gravity acts on the body. The mass centre then coincides with the geometrical centre of the cube. Next, holding the surface B fixed, let gravity act on the body; the interior sags with the result that the mass centre in the deformed state now lies at a distance d below the geometrical centre. (Note that d is not the displacement of the point which lay at the centre in the unstrained state.)

As usual we put
$$\mathbf{S} \leftrightarrow E_{ij}, \tag{5·427}$$
where E_{ij} is the stress set up in the cube by this deformation, so that the strain energy in the body is
$$W = \frac{1}{2}\int_V E_{ij} e_{ij} dV = \tfrac{1}{2}\mathbf{S}^2, \tag{5·428}$$
by (5·418). The work done by gravity as the body takes up its deformed state is stored as strain energy, a fact expressed by
$$8g\rho l^3 d = \tfrac{1}{2}\mathbf{S}^2. \tag{5·429}$$
Thus the bounds on $\mathbf{S}^2$ which will be obtained are also bounds on d, the distance of the mass centre of the deformed body below its geometrical centre.

We shall take $l=1$, $g\rho=1$; general values can be restored by a dimensional argument.

For the displacements, which must vanish on the boundary, we choose three possible sets:
$$\left.\begin{aligned} &u_1 = u_2 = 0, \quad u_3 = \tfrac{1}{2}\phi,\\ &u_1 = u_2 = 0, \quad u_3 = \tfrac{1}{2}x_3^2\phi,\\ &u_1 = x_1 x_3 \phi, \quad u_2 = x_2 x_3 \phi, \quad u_3 = 0,\end{aligned}\right\} \tag{5·430}$$
where $\phi = (x_1^2-1)(x_2^2-1)(x_3^2-1)$. Noting that L' contains the origin of function-space, we use (5·401) and the displacements (5·430) to obtain three F-vectors $\mathbf{T}_1'$, $\mathbf{T}_2'$, $\mathbf{T}_3'$, all lying in L'. Thus
$$\left.\begin{aligned}
&\mathbf{T}_1' \leftrightarrow e_{ij}', \quad \text{where} \quad e_{ij}' = 0,\\
&\text{except} \quad \begin{cases} e_{31}' = \tfrac{1}{2}x_1(x_2^2-1)(x_3^2-1),\\ e_{32}' = \tfrac{1}{2}x_2(x_1^2-1)(x_3^2-1),\\ e_{33}' = x_3(x_1^2-1)(x_2^2-1),\end{cases}\\
&\mathbf{T}_2' \leftrightarrow e_{ij}', \quad \text{where} \quad e_{ij}' = 0,\\
&\text{except} \quad \begin{cases} e_{31}' = \tfrac{1}{2}x_1 x_3^2(x_2^2-1)(x_3^2-1),\\ e_{32}' = \tfrac{1}{2}x_2 x_3^2(x_1^2-1)(x_3^2-1),\\ e_{33}' = x_3(x_1^2-1)(x_2^2-1)(2x_3^2-1),\end{cases}\\
&\mathbf{T}_3' \leftrightarrow e_{ij}',\\
&\text{where} \quad \begin{cases} e_{11}' = x_3(3x_1^2-1)(x_2^2-1)(x_3^2-1),\\ e_{22}' = x_3(x_1^2-1)(3x_2^2-1)(x_3^2-1),\\ e_{33}' = 0,\\ e_{13}' = \tfrac{1}{2}x_1(x_1^2-1)(x_2^2-1)(3x_3^2-1),\\ e_{12}' = x_1 x_2 x_3(x_3^2-1)[(x_1^2-1)+(x_2^2-1)],\\ e_{23}' = \tfrac{1}{2}x_2(x_1^2-1)(x_2^2-1)(3x_3^2-1).\end{cases}
\end{aligned}\right\} \tag{5·431}$$

An F-vector $\mathbf{S}_0''$ is next selected as follows:
$$\mathbf{S}_0'' \leftrightarrow E_{(0)ij}'', \quad \text{where} \quad E_{(0)ij}'' = 0 \text{ except } E_{(0)33}'' = x_3; \tag{5·432}$$

the point $\mathbf{S}_0''$ is in L''. F-vectors $\mathbf{T}''$ lying in L'' are needed and these must be of the type

$$\mathbf{T}'' \leftrightarrow E''_{ij}, \quad E''_{ij,j} = 0. \tag{5·433}$$

We shall take four of them as follows, satisfying (5·433):

$$\left.\begin{aligned}
&\mathbf{T}_1'' \leftrightarrow E''_{ij}, \quad \text{where} \quad E''_{ij} = 0, \\
&\text{except} \quad \begin{cases} E''_{11} = x_1^2 x_3, \\ E''_{22} = x_2^2 x_3, \\ E''_{12} = -2x_1 x_2 x_3, \end{cases} \\
&\mathbf{T}_2'' \leftrightarrow E''_{ij}, \quad \text{where} \quad E''_{ij} = 0, \\
&\text{except} \quad \begin{cases} E''_{11} = x_3, \\ E''_{22} = x_3, \end{cases} \\
&\mathbf{T}_3'' \leftrightarrow E''_{ij}, \quad \text{where} \quad E''_{ij} = 0, \\
&\text{except} \quad \begin{cases} E''_{13} = x_1, \\ E''_{23} = x_2, \\ E''_{33} = -2x_3, \end{cases} \\
&\mathbf{T}_4'' \leftrightarrow E''_{ij}, \quad \text{where} \quad E''_{ij} = 0, \\
&\text{except} \quad \begin{cases} E''_{13} = x_1^3, \\ E''_{23} = x_2^3, \\ E''_{33} = -3x_3(x_1^2 + x_2^2). \end{cases}
\end{aligned}\right\} \tag{5·434}$$

In accordance with the usual hypercircle method the vertices $\mathbf{V}'$ and $\mathbf{V}''$ of L' and L'' respectively are, by (2·775) (interchange L' and L''),

$$\mathbf{V}' = \sum_{\rho=1}^{3} a_\rho' \mathbf{T}_\rho', \quad \mathbf{V}'' = \mathbf{S}_0'' + \sum_{\sigma=1}^{4} a_\sigma'' \mathbf{T}_\sigma'', \tag{5·435}$$

and the weights a_ρ' and a_σ'' are found by minimizing $(\mathbf{V}' - \mathbf{V}'')^2$, i.e. they satisfy

$$\left.\begin{aligned}
&\sum_{\mu=1}^{3} a_\mu' \mathbf{T}_\rho' . \mathbf{T}_\mu' - \mathbf{S}_0'' . \mathbf{T}_\rho' = 0 \quad (\rho = 1, 2, 3), \\
&\sum_{\nu=1}^{4} a_\nu'' \mathbf{T}_\sigma'' . \mathbf{T}_\nu'' + \mathbf{S}_0'' . \mathbf{T}_\sigma'' = 0 \quad (\sigma = 1, 2, 3, 4).
\end{aligned}\right\} \tag{5·436}$$

By (2·777) we have

$$\left.\begin{aligned}
\mathbf{V}'^2 &= \sum_{\rho=1}^{3} a_\rho' \mathbf{S}_0'' . \mathbf{T}_\rho', \\
\mathbf{V}''^2 &= \mathbf{S}_0''^2 + \sum_{\sigma=1}^{4} a_\sigma'' \mathbf{S}_0'' . \mathbf{T}_\sigma'',
\end{aligned}\right\} \tag{5·437}$$

and the radius R of the hypercircle is given by

$$4R^2 = \mathbf{V}''^2 - \mathbf{V}'^2. \tag{5·438}$$

The inequalities for $\mathbf{S}^2$ are $\quad \mathbf{V}'^2 \leqslant \mathbf{S}^2 \leqslant \mathbf{V}''^2. \qquad (5·439)$

In the calculation of $\mathbf{V}'^2$ we use the F-vectors defined in (5·431) and (5·432), and calculate the following scalar products by (5·407) and (5·415):

$$\left.\begin{array}{l}\mathbf{T}_1'^2=\frac{512}{675}(\lambda+4\mu), \quad \mathbf{T}_1'.\mathbf{T}_2'=\frac{512}{23625}(7\lambda+24\mu),\\ \mathbf{T}_2'^2=\frac{512}{70875}(33\lambda+76\mu), \quad \mathbf{T}_1'.\mathbf{T}_3'=-\frac{1024}{3375}(\lambda+\mu),\\ \mathbf{T}_3'^2=\frac{2048}{165375}(49\lambda+143\mu), \quad \mathbf{T}_2'.\mathbf{T}_3'=\frac{1024}{23625}(\lambda+\mu),\\ \mathbf{S}_0''.\mathbf{T}_1'=\frac{32}{27}, \quad \mathbf{S}_0''.\mathbf{T}_2'=\frac{32}{135},\\ \mathbf{S}_0''.\mathbf{T}_3'=0.\end{array}\right\} \tag{5·440}$$

The equations (5·436) for the weights a_ρ' then have the solutions

$$\left.\begin{array}{l}a_1'=175(1701\lambda^2+8795\lambda\mu+10948\mu^2)/D,\\ a_2'=-525(147\lambda^2+49\lambda\mu-568\mu^2)/D,\\ a_3'=50(3087\lambda^3+12593\lambda^2\mu+15925\lambda\mu^2+6419\mu^3)/[(\lambda+\mu)D],\end{array}\right\} \tag{5·441}$$

where $$D=64(2205\lambda^3+24199\lambda^2\mu+77959\lambda\mu^2+77397\mu^3).$$

Thus the lower bound on $\mathbf{S}^2$ is, by (5·437),

$$\mathbf{V}'^2=\frac{70}{27}\left(\frac{2016\lambda^2+10957\lambda\mu+14111\mu^2}{2205\lambda^3+24199\lambda^2\mu+77959\lambda\mu^2+77397\mu^3}\right). \tag{5·442}$$

Now $$\lambda=\frac{\sigma E}{(1+\sigma)(1-2\sigma)}, \quad \mu=\frac{E}{2(1+\sigma)}, \tag{5·443}$$

and so $$\left.\begin{array}{l}\mathbf{V}'^2=\dfrac{1-2\sigma}{E}.f(\sigma),\\ f(\sigma)=\dfrac{140}{27}\,\dfrac{(1+\sigma)(14111-34530\sigma+20680\sigma^2)}{(77397-308464\sigma+401888\sigma^2-171456\sigma^3)}\end{array}\right\}. \tag{5·444}$$

The scalar products of the F-vectors in (5·432) and (5·434) which are used in finding an upper bound to $\mathbf{S}^2$, are, by (5·408) and (5·415),

$$\left.\begin{array}{ll}\mathbf{S}_0''^2=\dfrac{8}{3E}, & \left.\begin{array}{l}\mathbf{T}_2''.\mathbf{T}_3''\\ \mathbf{T}_2''.\mathbf{T}_4''\end{array}\right\}=\dfrac{32}{3}\left(\dfrac{\sigma}{E}\right),\\ \mathbf{T}_1''^2=\dfrac{16}{135}\left(\dfrac{29+15\sigma}{E}\right), & \\ \mathbf{T}_2''^2=\dfrac{16}{3}\left(\dfrac{1-\sigma}{E}\right), & \mathbf{T}_3''.\mathbf{T}_4''=\dfrac{32}{15}\left(\dfrac{8+3\sigma}{E}\right),\\ \mathbf{T}_3''^2=\dfrac{32}{3}\left(\dfrac{2+\sigma}{E}\right), & \mathbf{S}_0''.\mathbf{T}_1''=-\dfrac{16}{9}\left(\dfrac{\sigma}{E}\right),\\ \mathbf{T}_4''^2=\dfrac{32}{105}\left(\dfrac{64+15\sigma}{E}\right), & \mathbf{S}_0''.\mathbf{T}_2''=-\dfrac{16}{3}\left(\dfrac{\sigma}{E}\right),\\ \mathbf{T}_1''.\mathbf{T}_2''=\dfrac{16}{9}\left(\dfrac{1-\sigma}{E}\right), & \left.\begin{array}{l}\mathbf{S}_0''.\mathbf{T}_3''\\ \mathbf{S}_0''.\mathbf{T}_4''\end{array}\right\}=-\dfrac{16}{3E}.\\ \mathbf{T}_1''.\mathbf{T}_3''=\dfrac{32}{9}\left(\dfrac{\sigma}{E}\right), & \\ \mathbf{T}_1''.\mathbf{T}_4''=\dfrac{224}{45}\left(\dfrac{\sigma}{E}\right), & \end{array}\right\} \tag{5·445}$$

The equations (5·436) for the weights a''_σ have the solutions

$$\left.\begin{aligned} a''_1 &= -105\sigma(1-\sigma-2\sigma^2)/D', \\ a''_2 &= \sigma(527+447\sigma-80\sigma^2)/D', \\ a''_3 &= 10(12-23\sigma-13\sigma^2+22\sigma^3)/D', \\ a''_4 &= 35(6-\sigma-17\sigma^2-10\sigma^3)/D', \end{aligned}\right\} \tag{5·446}$$

with $$D' = 2(576-270\sigma-971\sigma^2-125\sigma^3).$$

The upper bound on $\mathbf{S}^2$ is then

$$\left.\begin{aligned} \mathbf{V}''^2 &= \frac{1-2\sigma}{E}g(\sigma), \\ g(\sigma) &= \frac{8}{3}\frac{(246+487\sigma+236\sigma^2-5\sigma^3)}{(576-270\sigma-971\sigma^2-125\sigma^3)}, \end{aligned}\right\} \tag{5·447}$$

by (5·437). Hence by (5·439), (5·444) and (5·447)

$$\frac{1-2\sigma}{E}f(\sigma) \leqslant \mathbf{S}^2 \leqslant \frac{1-2\sigma}{E}g(\sigma). \tag{5·448}$$

Restoring l and $g\rho$ we thus obtain the following bounds on W, the strain energy, and d, the distance of the mass centre below the geometrical centre of the cube:

$$(\tfrac{1}{2}-\sigma)f(\sigma) \leqslant \frac{EW}{g^2\rho^2l^5} = \frac{8Ed}{g\rho l^2} \leqslant (\tfrac{1}{2}-\sigma)\,g(\sigma). \tag{5·449}$$

For hot-rolled copper ($\sigma = 0{\cdot}33$) this gives

$$0{\cdot}4409 \leqslant \frac{EW}{g^2\rho^2l^5} = \frac{8Ed}{g\rho l^2} \leqslant 0{\cdot}5202.$$

Higher accuracy in the bounds can be attained by using a greater number of F-vectors than the seven employed in the above approximation.

For an incompressible body $\sigma = \frac{1}{2}$, and both bounds in (5·449) become zero. For a nearly incompressible body with $\sigma = \frac{1}{2} - \epsilon$ (ϵ small), we get the approximate bounds

$$7{\cdot}111\epsilon \leqslant \frac{EW}{g^2\rho^2l^5} = \frac{8Ed}{g\rho l^2} \leqslant 8\epsilon.$$

Example: beam bent by terminal forces

We next consider the case where the stress is assigned on the entire surface B of a body. The linear subspaces are then

$$\left.\begin{aligned} &L'\colon\quad \mathbf{S}' \leftrightarrow E'_{ij}, \quad C(E') = 0, \\ &L''\colon\quad \mathbf{S}'' \leftrightarrow E''_{ij}, \quad E''_{ij,j} + X_i = 0 \text{ in } V, \quad (E_{ij}n_j)_B = g_i. \end{aligned}\right\} \tag{5·450}$$

Consider a uniform beam of homogeneous isotropic material having a length l and a cross-sectional area A. We take the x_3-axis along the line of centroids of the cross-sections with the origin at one end and Ox_1, Ox_2 as principal axes of the cross-section (Fig. 5·41). The following stress system acts on the surface B:

$$\left.\begin{aligned} &T_i = 0 \text{ on } B_1, \\ &T_1 = T_2 = 0, \quad E_{33} = f(x_1, x_2) - Ex_1/R \text{ on } B_2, \end{aligned}\right\} \tag{5·451}$$

where B_1 and B_2 denote the sides and ends of the beam respectively, $f(x_1, x_2)$ is an arbitrary function and R is a constant (actually the radius of curvature of the beam under the above stress system with $f=0$). E is Young's modulus. We take $X_i=0$. The moment about the x_2-axis on any cross-section is

$$M_2 = EI_2/R - \int_A x_1 f dA, \tag{5·452}$$

where I_2 is the moment of inertia of the cross-section about the x_2-axis.

Noting that L' contains the origin of function-space, we observe from (5·450) that any set of displacements u_i, chosen at random, can be used to generate an F-vector $\mathbf{T}'$ lying in L'. We may therefore define

$$\mathbf{T}' \leftrightarrow e'_{ij}, \tag{5·453}$$

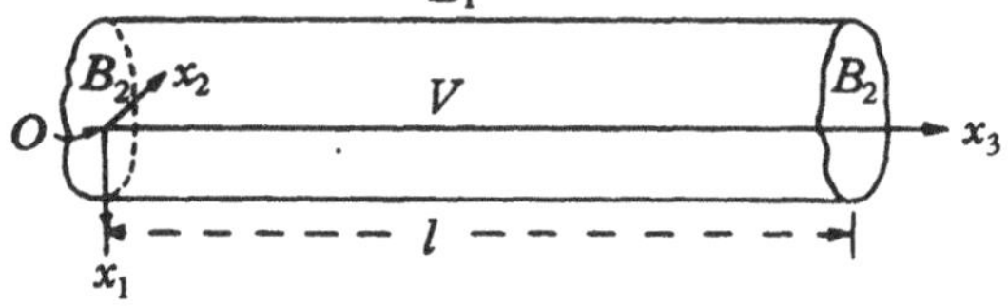

Fig. 5·41. Beam bent by terminal forces.

where e'_{ij} are calculated by (5·401) from the displacements

$$\left.\begin{aligned} u_1 &= \frac{1}{2R}(x_3^2 + \sigma x_1^2 - \sigma x_2^2), \\ u_2 &= \frac{\sigma}{R} x_1 x_2, \\ u_3 &= -\frac{1}{R} x_1 x_3, \end{aligned}\right\} \tag{5·454}$$

which correspond to the case of a beam under the system of stresses (5·451) with $f=0$.

An F-vector $\mathbf{S}''_0$ is defined as follows:

$$\mathbf{S}''_0 \leftrightarrow E''_{(0)ij}, \quad \text{where} \quad E''_{(0)ij}=0, \quad \text{except} \quad E''_{(0)33}=f(x_1, x_2) - Ex_1/R. \tag{5·455}$$

The appropriate equations for the vertices of L' and L'', the single weight a', and the bounds on $\mathbf{S}^2$, are the same as in the last example, i.e. equations (5·435)–(5·439). In the present example the scalar products are, by (5·407), (5·408) and (5·415),

$$\left.\begin{aligned} \mathbf{S}''^2_0 &= \frac{1}{E}\int_V \left(f^2 - \frac{2E}{R}x_1 f\right) dV + \frac{ElI_2}{R^2}, \\ \mathbf{T}'^2 &= \frac{ElI_2}{R^2}, \\ \mathbf{S}''_0 . \mathbf{T}' &= \frac{ElI_2}{R^2} - \frac{1}{R}\int_V x_1 f dV. \end{aligned}\right\} \tag{5·456}$$

Thus the bounds on $\mathbf{S}^2$ (twice the strain energy) are

$$\left[\int_V x_1 f dV - \frac{ElI_2}{R}\right]^2 \leqslant ElI_2 \mathbf{S}^2$$
$$\leqslant lI_2 \int_V \left(f^2 - \frac{2Ex_1 f}{R}\right) dV + \frac{E^2 l^2 I_2^2}{R^2}. \quad (5{\cdot}457)$$

These bounds coincide when $f=0$ and then

$$\mathbf{S}^2 = \frac{ElI_2}{R^2}; \quad (5{\cdot}458)$$

the difference between them will be small if f is small (compared with Ex_1/R).

The lower bound can be improved by using additional $\mathbf{T}'$ vectors generated from arbitrary sets of displacements u_i.

Note that the difference between the bounds in (5·457) is of the second order in f:

$$lI_2 \int_V f^2 dV - \left(\int_V x_1 f dV\right)^2. \quad (5{\cdot}459)$$

If the cross-section is a circle of radius a and $f = bx_1^2$, (5·457) gives

$$1 \leqslant \frac{4R^2\mathbf{S}^2}{\pi a^4 lE} \leqslant 1 + \frac{R^2 a^2 b^2}{2E^2}. \quad (5{\cdot}460)$$

As other examples of the application of the hypercircle method in elasticity, see Greenberg and Truell (1) for the case of an infinite bar of rectangular section compressed between rough rigid parallel supports and Edelman (1) for the case of a short cylinder compressed between rough end-blocks. The method has also been applied by Greenberg and Prager (1) to the determination of bending and twisting moments in thin elastic plates and by Prager (1) to the determination of stresses in frameworks of elastic bars; in this last paper it is shown how the methods of Broglio (1), (2) are related to the hypercircle method.

For a number of applications, see Greenberg (2).

Incompressible elastic bodies

The dilatation of a body is

$$\theta = e_{ii} = u_{i,i}, \quad (5{\cdot}461)$$

and for an isotropic body we have by (5·408)

$$\theta = \frac{1-2\sigma}{E} E_{ii}. \quad (5{\cdot}462)$$

Thus $\sigma = \frac{1}{2}$ is a necessary and sufficient condition for incompressibility ($\theta = 0$).

The elastic constants are connected by

$$\lambda = \frac{E\sigma}{(1+\sigma)(1-2\sigma)}, \quad \mu = \frac{E}{2(1+\sigma)}, \quad (5{\cdot}463)$$

and so λ is infinite for an incompressible body. Thus the first term on the right-hand side of (5·407) becomes indeterminate, and the strain e_{ij} no longer determines the stress E_{ij}. But E_{ij} still determines e_{ij} by (5·408), which now reads (for $\sigma = \frac{1}{2}$)

$$e_{ij} = \frac{3}{2E} F_{ij}, \quad F_{ij} = E_{ij} - \tfrac{1}{3}\delta_{ij} E_{kk}. \tag{5·464}$$

In applying the hypercircle method to elasticity, we have regarded Hooke's law [(5·403) or (5·405)] as a fundamental bridge between strain and stress and, no matter how we have relaxed the elastic problem (by considering stresses which do not satisfy the equations of equilibrium or strains which do not satisfy the equations of compatibility), we have never relaxed Hooke's law. Let us keep to this plan, so that for an incompressible body (5·464) is never to be relaxed.

This implies $e_{ii} = 0$ identically, and so the symmetric tensor e_{ij} has only *five* disposable components, not *six*; E_{ij}, on the other hand, retains six disposable components. There is no longer a one-to-one correspondence between e_{ij} and E_{ij}; (5·464) defines e_{ij} uniquely in terms of E_{ij}, but not conversely. The two function-spaces given by (5·413) ($\mathbf{S} \leftrightarrow E_{ij}$) and (5·414) ($\mathbf{S} \leftrightarrow e_{ij}$) are no longer the same; we must decide to use either the one or the other.

Consider first $\mathbf{S} \leftrightarrow E_{ij}$. Then we may proceed to split the problem as in (5·421). But now to generate $\mathbf{S}'$ starting from u'_i, we do not get E'_{ij} uniquely; by (5·464) we have

$$E'_{ij} - \tfrac{1}{3}\delta_{ij} E'_{kk} = \frac{2E}{3} e'_{ij} = \frac{E}{3}(u'_{i,j} + u'_{j,i}), \tag{5·465}$$

and this leaves considerable freedom in the choice of E'_{ij}. That does not matter. What is important is that (5·465) implies $u'_{i,i} = 0$. Thus for an incompressible body the definitions (5·421) of the linear subspaces must be amended to read

$$\left.\begin{aligned} L'&: \quad \mathbf{S}' \leftrightarrow E'_{ij}, \quad C(E') = 0, \quad u'_{i,i} = 0, \quad (u'_i)_{B_1} = f_i; \\ L''&: \quad \mathbf{S}'' \leftrightarrow E''_{ij}, \quad E''_{ij,j} + X_i = 0, \quad (E''_{ij} n_j)_{B_2} = g_i. \end{aligned}\right\} \tag{5·466}$$

It then follows that the solution $\mathbf{S}$ is the intersection of L' and L'' and that L' and L'' are mutually orthogonal. Actually this applies to an anisotropic incompressible body, although for simplicity we have been considering an isotropic one.

Now consider the plan $\mathbf{S} \leftrightarrow e_{ij}$ for an isotropic incompressible body. Energy is always the key to the F-metric and scalar product, and we note that by (5·464) with $e_{ii} = 0$ the energy is

$$W = \tfrac{1}{2} \int e_{ij} E_{ij} \, dV = \tfrac{1}{2} \int e_{ij} F_{ij} \, dV = \frac{E}{3} \int e_{ij} e_{ij} \, dV. \tag{5·467}$$

This suggests that we should define a general F-vector and the scalar product as follows:

$$\mathbf{S} \leftrightarrow e_{ij}, \quad e_{ij} = e_{ji}, \quad e_{ii} = 0, \quad \mathbf{S}.\mathbf{S}' = \int e_{ij} e'_{ij} dV. \qquad (5\cdot468)$$

This means that an F-point or F-vector corresponds to a symmetric 3×3 matrix with zero trace. We split the problem by defining linear subspaces as follows:

$$\left.\begin{aligned} L'&: \quad \mathbf{S}' \leftrightarrow e'_{ij} = \tfrac{1}{2}(u'_{i,j} + u'_{j,i}), \quad u'_{i,i} = 0, \quad (u'_i)_{B_1} = f_i; \\ L''&: \quad \mathbf{S}'' \leftrightarrow e''_{ij} = p''_{ij} - \tfrac{1}{3}\delta_{ij} p''_{kk}, \quad p''_{ij} = p''_{ji}, \\ &\qquad p''_{ij,j} + \frac{3}{2E} X_i = 0, \quad (p''_{ij} n_j)_{B_2} = \frac{3}{2E} g_i, \end{aligned}\right\} \qquad (5\cdot469)$$

where f_i and g_i are the prescribed values of u_i and $E_{ij} n_j$. In defining the scalar product, we dropped the factor $\frac{1}{3}E$ from (5·467), and so we use the dimensionless p''_{ij} instead of a stress E''_{ij} $(= \frac{2}{3} E p''_{ij})$. It is easy to see that $\mathbf{S}$ is the intersection of L' and L'', and that these subspaces are mutually orthogonal.

The Green's tensor of elasticity

In order to discuss pointwise bounds on the solutions of elastic problems, we need the Green's tensor. Consider an elastic body (in general anisotropic—it may be compressible or incompressible) extending to infinity. Let v be a sphere drawn with centre at some point x'_i. Let a body force X_i per unit volume be applied, with $X_i = A_i$ (a constant vector) inside v and $X_i = 0$ outside v, and let the body be held fixed at infinity ($u_i = 0$). The resulting displacement u_i in equilibrium will depend linearly on A_i and we may write it $u_i = v A_p u^*_{pi}$ where u^*_{pi} is a tensor or 3×3 matrix with components depending on x_i (the coordinates of the point where u_i is measured), on x'_i, on the elastic constants of the body, and on the radius of v. Since the body is in equilibrium, the stress satisfies

$$\int E_{ij} n_j dB = -F_i, \quad \int (\xi_i E_{jk} - \xi_j E_{ik}) n_k dB = 0 \quad (\xi_i = x_i - x'_i), \; (5\cdot470)$$

the integration being over the surface of v or any surface containing v; here $F_i = vA_i$, the total load on the body.

Now, holding x'_i fixed, let the sphere v shrink to zero, A_i increasing to infinity so that $vA_i = F_i$ remains constant; in the limit F_i represents a *concentrated load* at x'_i. The displacement u_i due to this concentrated load is of the form

$$u_i(x) = F_p u_{pi}(x, x'), \qquad (5\cdot471)$$

where u_{pi} is the limit of u^*_{pi} as v tends to zero; this tensor u_{pi}, a function only of x_i and x'_i (and of the elastic constants), is *the Green's tensor of elasticity*.

Note that for the stress due to the concentrated load, (5·470) hold for any surface enclosing x'_i.

Green's tensor has two important symmetries,

$$u_{pi}(x,x') = u_{pi}(x',x), \quad u_{pi}(x,x') = u_{ip}(x,x'); \qquad (5{\cdot}472)$$

these will now be proved.

If we have any two states of equilibrium with displacements u'_i, u''_i under body forces X'_i, X''_i, then by the reciprocity relation (5·416) we have

$$\int e'_{ij}E''_{ij}\,dV = \int e''_{ij}E'_{ij}\,dV. \qquad (5{\cdot}473)$$

Replacing the strains by their expressions in terms of displacement and integrating by parts, we get

$$\int u'_i X''_i\,dV = \int u''_i X'_i\,dV. \qquad (5{\cdot}474)$$

Here the integration is over all space; we have assumed that the displacement vanishes sufficiently rapidly at infinity. Suppose now that the body forces correspond to uniform loadings in spheres v', v'' centred at x'_i, x''_i; then (5·474) gives

$$\int_{v''} u'_i X''_i\,dv'' = \int_{v'} u''_i X'_i\,dv'. \qquad (5{\cdot}475)$$

Letting v' and v'' shrink to points, and denoting by F'_i and F''_i the concentrated loads obtained in the limit, we get from the last equation

$$F''_i F'_p u_{pi}(x'',x') = F'_i F''_p u_{pi}(x',x'').$$

Changing dummy suffixes, we get

$$F'_i F''_p[u_{ip}(x',x'') - u_{pi}(x'',x')] = 0, \qquad (5{\cdot}476)$$

and hence, since the loads are arbitrary,

$$u_{ip}(x',x'') = u_{pi}(x'',x'). \qquad (5{\cdot}477)$$

Note that this symmetry is not one of those shown in (5·472); but we shall now supplement it with another.

Consider an infinite body uniformly loaded throughout a sphere v with centre x'_i; let the displacement be u_i. If we change x'_i to a new position, say $x'_i + a_i$, it is obvious that the effect on the vector field u_i is to leave the pattern of this field unchanged but to shift it through a displacement a_i. The same will hold in the limit $v \to 0$, and we conclude that the tensors u^*_{pi} (for finite sphere of loading) and u_{pi} (for concentrated load) involve the six coordinates x_j, x'_j only in the combinations $x_j - x'_j$; thus we may write

$$u^*_{pi}(x, x') = u^*_{pi}(x - x'), \quad u_{pi}(x, x') = u_{pi}(x - x'). \qquad (5{\cdot}478)$$

We can explore the properties of the functions u^*_{pi} and u_{pi} by taking $x'_i = 0$; then we have the functions $u^*_{pi}(x)$ and $u_{pi}(x)$ to deal with. We are going to show that, for a loaded sphere and also for concentrated load as a limit, the pattern of u_i reflects into itself with respect to the centre, i.e.

$$u^*_{pi}(-x) = u^*_{pi}(x), \quad u_{pi}(-x) = u_{pi}(x). \qquad (5{\cdot}479)$$

To show this, we note that by (5·402) and (5·403) the equations of equilibrium are

$$c_{ijkl} u_{k,lj} + X_i = 0, \qquad (5{\cdot}480)$$

and so, inside and outside a finite loaded sphere, we have

$$F_p c_{ijkl} \frac{\partial^2}{\partial x_l \partial x_j} u^*_{pk}(x) + X_i(x) = 0. \qquad (5{\cdot}481)$$

Let y_i be the reflexion of any point x_i in the origin, so that $y_i = -x_i$. Write $u^*_{pi}(y) = u^*_{pi}(-x) = w^*_{pi}(x)$. Then

$$\frac{\partial u^*_{pi}(y)}{\partial y_l} = -\frac{\partial w^*_{pi}(x)}{\partial x_l}, \quad \frac{\partial^2 u^*_{pi}(y)}{\partial y_l \partial y_j} = \frac{\partial^2 w^*_{pi}(x)}{\partial x_l \partial x_j}.$$

Now (5·481) certainly holds with y substituted for x, for these are current coordinates; further, $X_i(y) = X_i(x)$ on account of the symmetry of the loading. Hence

$$F_p c_{ijkl} \frac{\partial^2}{\partial x_l \partial x_j} w^*_{pk}(x) + X_i(x) = 0.$$

Subtraction of this from (5·481) gives

$$F_p c_{ijkl} \frac{\partial^2}{\partial x_l \partial x_j} [u^*_{pk}(x) - w^*_{pk}(x)] = 0. \qquad (5{\cdot}482)$$

The displacements $u_i = F_p[u^*_{pi}(x) - w^*_{pi}(x)]$ therefore satisfy the equations of equilibrium with no body forces and vanish at infinity. Therefore $u_i = 0$ and we conclude that $u^*_{pi}(x) = w^*_{pi}(x) = u^*_{pi}(-x)$, and hence by (5·478)

$$u^*_{pi}(x, x') = u^*_{pi}(x - x') = u^*_{pi}(x' - x) = u^*_{pi}(x', x). \qquad (5{\cdot}483)$$

Proceeding to the limit of a concentrated load (delete the stars) and combining the result with (5·477), we get the symmetry conditions (5·472).

It is often convenient to write $u_i^{(p)}$ instead of u_{pi}, so that the formula (5·471) for the displacement due to a concentrated load reads

$$u_i = F_p u_i^{(p)}, \tag{5·484}$$

the summation convention being of course understood here as always. We may then think of $u_i^{(1)}$ as the displacement due to a concentrated load with components $(1, 0, 0)$, with similar interpretations for $u_i^{(2)}$ and $u_i^{(3)}$. We may in fact regard $u_i^{(p)}$, not as a tensor, but as three vectors (for $p = 1, 2, 3$), and we may write $e_{ij}^{(p)}$ and $E_{ij}^{(p)}$ for the strain and stress derived from $u_i^{(p)}$ in the usual way. On putting $E_{ij} = F_p E_{ij}^{(p)}$ in (5·470) and remembering that F_p is arbitrary and $E_{ij}^{(p)}$ independent of it, we get

$$\int E_{ij}^{(p)} n_j dB = -\delta_{pi}, \quad \int (\xi_i E_{jk}^{(p)} - \xi_j E_{ik}^{(p)}) n_k dB = 0, \tag{5·485}$$

these integrals being taken over any surface enclosing the point x_i'.

The Green's tensor for an isotropic body was found by Kelvin [cf. Love (1), p. 185; Sokolnikoff (1), p. 337]; it reads

$$\left.\begin{aligned} u_i^{(p)}(x, x') = u_{pi}(x, x') &= K\left[(3-4\sigma)\frac{\delta_{pi}}{r} + \frac{\xi_p \xi_i}{r^3}\right], \\ \xi_j = x_j - x_j', \quad r^2 = \xi_j \xi_j, \quad K &= \frac{1}{8\pi E}\frac{1+\sigma}{1-\sigma}. \end{aligned}\right\} \tag{5·486}$$

We may verify its correctness by showing that, for any fixed p, $u_i^{(p)}$ satisfies the equations of equilibrium

$$u_{k,ki} + (1-2\sigma)\Delta u_i = 0, \tag{5·487}$$

and also (5·485).

The Green's tensor for an anisotropic body is a matter of much greater complexity; it is discussed in Note B, p. 411.

Bounds at a point in elastic equilibrium

Suppose we have succeeded in locating the solution of a problem in elastic equilibrium on a hypercircle. If we take the centre of the hypercircle as an approximation to the solution, the error of this approximation is bounded in what we may call a mean-square sense by the equation $(\mathbf{S} - \mathbf{C})^2 = R^2$. But this places no bound on the error at any given point in the body.

To get bounds at a point we may use a generalization of the method of § 3·4 [cf. Synge (3)], based on the fact that for an isotropic body the dilatation θ is a harmonic function. But we shall follow here a more compact and general method due to Prager (1), based on the Green's tensor we have discussed above. This is applicable to both isotropic and anisotropic bodies.

Consider an elastic body in equilibrium under a body force X_i per unit volume (not a spherical loading, but a general body force), the body occupying a region V with boundary $B = B_1 + B_2$ on which the conditions are

$$(u_i)_{B_1} = f_i, \quad (E_{ij}n_j)_{B_2} = g_i, \tag{5·488}$$

f_i and g_i being given. We seek bounds at a point x'_i in the body.

Let v be a small sphere with centre at x'_i and let $V - v$ be the part of the body outside v. Then by the reciprocity relation (5·416) we have

$$\int_{V-v} e_{ij} E_{ij}^{(p)} dV = \int_{V-v} e_{ij}^{(p)} E_{ij} dV, \tag{5·489}$$

where e_{ij}, E_{ij} refer to the solution of the problem and $e_{ij}^{(p)}, E_{ij}^{(p)}$ are derived from the Green's tensor $u_i^{(p)}$ [i.e. (5·486) if the body is isotropic]. On integration by parts we get

$$\begin{aligned} -\int_b u_i E_{ij}^{(p)} n_j db + \int_B u_i E_{ij}^{(p)} n_j dB - \int_{V-v} u_i E_{ij,j}^{(p)} dV \\ = -\int_b u_i^{(p)} E_{ij} n_j db + \int_B u_i^{(p)} E_{ij} n_j dB - \int_{V-v} u_i^{(p)} E_{ij,j} dV, \end{aligned} \tag{5·490}$$

where b is the surface of the sphere v and n_j the unit outward normal on it and on B. Now $E_{ij,j}^{(p)} = 0$, $E_{ij,j} = -X_i$, and so, proceeding to the limit $v = 0$ and using (5·485), we get

$$\left.\begin{aligned} u_p(x') &= M^{(p)} + N^{(p)}, \\ M^{(p)} &= \int_V u_i^{(p)} X_i dV - \int_{B_1} u_i E_{ij}^{(p)} n_j dB + \int_{B_2} u_i^{(p)} E_{ij} n_j dB, \\ N^{(p)} &= -\int_{B_2} u_i E_{ij}^{(p)} n_j dB + \int_{B_1} u_i^{(p)} E_{ij} n_j dB. \end{aligned}\right\} \tag{5·491}$$

The first integral on the right in (5·490) has disappeared in this limit, because $u_i^{(p)}$ becomes infinite only as $1/r$ when $r \to 0$. This is evident at once from (5·486) in the case of isotropy, and it is true also for anisotropic bodies (Note B, p. 411).

The equation (5·491) expressing the displacement at a point in terms of boundary values (if body force is absent) is due to Somigliana [cf. Love (1), p. 245]. Our reason in writing it in the form shown is that $M^{(p)}$ is known by (5·488).

We have now to transform $N^{(p)}$, and this we do by introducing a continuous displacement $u_i'^{(p)}$ equal to $u_i^{(p)}$ on B_1 and a stress $E_{ij}''^{(p)}$ satisfying

$$E_{ij,j}''^{(p)} = 0 \text{ in } V, \quad E_{ij}''^{(p)} n_j = E_{ij}^{(p)} n_j \text{ on } B_2, \tag{5·492a}$$

with $E''^{(p)}_{ij} n_j$ continuous across any surface; we further restrict $u'^{(p)}_i$ and $E''^{(p)}_{ij}$ by the conditions

$$\int_{B_2} u'^{(p)}_i E_{ij} n_j dB = 0, \quad \int_{B_1} u_i E''^{(p)}_{ij} n_j dB = 0, \qquad (5\cdot 492b)$$

wherein of course $u_i = f_i$, $E_{ij} n_j = g_i$ by (5·488). Then

$$\begin{aligned} N^{(p)} &= -\int_B u_i E''^{(p)}_{ij} n_j dB + \int_B u'^{(p)}_i E_{ij} n_j dB \\ &= -\int_V e_{ij} E''^{(p)}_{ij} dV + \int_V e'^{(p)}_{ij} E_{ij} dV - \int_V u'^{(p)}_i X_i dV. \qquad (5\cdot 493) \end{aligned}$$

Introducing F-vectors $\mathbf{S}$, $\mathbf{S}'^{(p)}$, $\mathbf{S}''^{(p)}$ such that $\mathbf{S}$ corresponds to the solution and

$$\mathbf{S}'^{(p)} \leftrightarrow E'^{(p)}_{ij}, \quad \mathbf{S}''^{(p)} \leftrightarrow E''^{(p)}_{ij}, \qquad (5\cdot 494)$$

we may write (5·491) in the form

$$u_p(x') - M^{(p)} + \int_V u'^{(p)}_i X_i dV = \mathbf{S} \, . \, (\mathbf{S}'^{(p)} - \mathbf{S}''^{(p)}). \qquad (5\cdot 495)$$

If $\mathbf{S}$ has been located on a hypercircle, the scalar product occurring here is bounded as in (2·635), and thus we obtain bounds on $u_p(x')$, the displacement at any point in the body, the other quantities on the left hand side of (5·495) being calculable.

If B_1 is absent, so that $B_2 = B$ and the boundary conditions are for stress only, we cannot find $E''^{(p)}_{ij}$ to satisfy the conditions (5·492a), as we see on integrating the second equation over B and applying the first with (5·485). We are to expect that the method must fail somehow in this case, because the displacement remains undetermined to within a rigid body displacement. We have then to deal with *two* points and bound only the difference between the displacements at them. See Synge (5) for a similar situation in the Neumann problem.

To get pointwise bounds on strain (and hence on stress, by Hooke's law (5·403)) we introduce a symmetric differentiated Green's tensor $u^{(pq)}_i$ by the formula

$$u^{(pq)}_i = -\tfrac{1}{2}(D'_q u^{(p)}_i + D'_p u^{(q)}_i) = \tfrac{1}{2}(D_q u^{(p)}_i + D_p u^{(q)}_i) = u^{(qp)}_i, \quad (5\cdot 496)$$

where $D_p = \partial/\partial x_p$, $D'_p = \partial/\partial x'_p$. It is easy to see that the strain and stress corresponding to the displacement $u^{(pq)}_i$ (with p and q fixed numbers and i the vector index) are

$$e^{(pq)}_{ij} = e^{(qp)}_{ij} = \tfrac{1}{2}(D_q e^{(p)}_{ij} + D_p e^{(q)}_{ij}) = -\tfrac{1}{2}(D'_q e^{(p)}_{ij} + D'_p e^{(q)}_{ij}),$$

and a similar expression with E written for e, $e^{(p)}_{ij}$ and $E^{(p)}_{ij}$ being the strain and stress corresponding to $u^{(p)}_i$.

Operating on (5·485) with D'_q, interchanging p and q, adding the two results, and dividing by 2, we get

$$\int E_{ij}^{(pq)} n_j dB = 0, \quad \int (\xi_i E_{jk}^{(pq)} - \xi_j E_{ik}^{(pq)}) n_k dB = 0; \qquad (5\cdot497)$$

the resultant force and couple vanish.

Applying the same operation to (5·491), we get the following expression for the strain at x'_i:

$$\left.\begin{aligned} e_{pq}(x') &= M^{(pq)} + N^{(pq)}, \\ M^{(pq)} &= -\int_V u_i^{(pq)} X_i dV + \int_{B_1} u_i E_{ij}^{(pq)} n_j dB - \int_{B_2} u_i^{(pq)} E_{ij} n_j dB, \\ N^{(pq)} &= \int_{B_2} u_i E_{ij}^{(pq)} n_j dB - \int_{B_1} u_i^{(pq)} E_{ij} n_j dB. \end{aligned}\right\} \qquad (5\cdot498)$$

Here $M^{(pq)}$ is known and $N^{(pq)}$ may be converted into scalar products and a known integral in the same way as before, introducing $u_i'^{(pq)}$ equal to $u_i^{(pq)}$ on B_1 and $E_{ij}''^{(pq)}$ satisfying

$$E_{ij,j}''^{(pq)} = 0 \text{ in } V, \quad E_{ij}''^{(pq)} n_j = E_{ij}^{(pq)} n_j \text{ on } B_2,$$

with restrictive conditions as in (5·492*b*). Then, with

$$\mathbf{S}'^{(pq)} \leftrightarrow E_{ij}'^{(pq)}, \quad \mathbf{S}''^{(pq)} \leftrightarrow E_{ij}''^{(pq)},$$

we get, instead of (5·495),

$$e_{pq}(x') - M^{(pq)} - \int_V u_i'^{(pq)} X_i dV = -\mathbf{S} \cdot (\mathbf{S}'^{(pq)} - \mathbf{S}''^{(pq)}), \quad (5\cdot499)$$

and so obtain bounds on the strain by (2·635). Bounds on the stress follow from Hooke's law.

This pointwise bounding process holds even if B_1 is absent (conditions on stress only), because the conditions imposed on $E_{ij}''^{(pq)}$ are consistent with (5·497), since the right-hand sides are both zero, unlike (5·485).

Although the combined use of the hypercircle method and the Green's tensor of elasticity is due to Prager, the idea of getting pointwise bounds by means of a Green's function or tensor is due to Greenberg (1) for Laplace's equation and to Diaz and Greenberg (1), (2) for the biharmonic equation. Maple's method for obtaining pointwise bounds in the Dirichlet problem (§3·4) may be regarded as a particular way of using a Green's function, which is $\log(1/r)$ in plane problems and $1/r$ in space problems [see Synge (5)].

For the application of these methods to problems in two and three dimensions (bending of thin plates and torsion of bars), see Washizu (1).

Minimum principles in elastic equilibrium

For the problem of elastic equilibrium under the boundary conditions

$$(u_i)_{B_1} = f_i, \quad (E_{ij}n_j)_{B_2} = g_i, \tag{5·499a}$$

the linear subspaces L', L'' are defined as in (5·421), and we note that

$$\begin{aligned}\mathbf{S}'.\mathbf{S}'' &= \int_V e'_{ij}E''_{ij}\,dV = \int_B u'_i E''_{ij} n_j\,dV - \int_V u'_i E''_{ij,\,j}\,dV \\ &= \int_{B_1} f_i E''_{ij} n_j\,dB + \int_{B_2} u'_i g_i\,dB + \int_V u'_i X_i\,dV. \end{aligned} \tag{5·499b}$$

We refer now to the minimum principles II and III of (2·729), which concern the minimization of $(\mathbf{S}'-\mathbf{S}'')^2$ with either $\mathbf{S}''$ or $\mathbf{S}'$ held fixed. Since

$$(\mathbf{S}'-\mathbf{S}'')^2 = \mathbf{S}'^2 + \mathbf{S}''^2 - 2\mathbf{S}'.\mathbf{S}'' \tag{5·499c}$$

and $\mathbf{S}'.\mathbf{S}''$ is as in (5·499b), we get

$$\tfrac{1}{2}(\mathbf{S}'-\mathbf{S}'')^2 = W' + W'' - \int_{B_1} f_i E''_{ij} n_j\,dB - \int_{B_2} u'_i g_i\,dB - \int_V u'_i X_i\,dV, \tag{5·499d}$$

where W' and W'' are the strain energies for the states $\mathbf{S}'$, $\mathbf{S}''$ respectively.

Hence we have the following minimum principles, commonly called the principle of energy and the principle of complementary energy (W denotes strain energy):

I. ($\mathbf{S}''$ fixed): The equilibrium state minimizes

$$W - \int_{B_2} u_i g_i\,dB - \int_V u_i X_i\,dV \tag{5·499e}$$

for those displacements which satisfy the boundary conditions on displacement.

II. ($\mathbf{S}'$ fixed): The equilibrium state minimizes

$$W - \int_{B_1} f_i E_{ij} n_j\,dB \tag{5·499f}$$

for those states of stress which satisfy the equations of equilibrium and the boundary conditions on stress.

Exercises

1. Consider two equilibrium states of an isotropic cube: in the first it is under tensions applied to a pair of parallel faces, and in the second under tensions of the same magnitude as before applied to a different pair of parallel faces. Show that the cosine of the angle between the corresponding F-vectors is $-\sigma$, where σ is Poisson's ratio.

2. Show that two equilibrium states of an elastic body under no body forces are orthogonal to one another if, and only if, zero work is done by the surface stresses of one state when the points on the surface are given the displacements of the other state.

3. Carry out the verification of the formula (5·486) for the isotropic Green's tensor as indicated in the text.

5·5. THE BIHARMONIC EQUATION

The biharmonic equation ($\Delta\Delta\psi = 0$) may be regarded as a generalization of Laplace's equation ($\Delta\psi = 0$), and the functions which satisfy it (biharmonic functions) as generalizations of harmonic functions. The further generalization is to polyharmonic functions [Nicolesco (1)] which satisfy an equation of the form $\Delta^n\psi = 0$ in a space of any number of dimensions. Here we shall be concerned only with biharmonic functions in a plane, so that our equation is

$$\Delta\Delta\psi = 0 \quad \text{or} \quad \psi_{,1111} + \psi_{,2222} + 2\psi_{,1122} = 0. \tag{5·501}$$

In physics this equation occurs in the slow steady motion of an incompressible viscous fluid and in plane problems of elastic equilibrium. Although there is an economy of thought in setting up a general mathematical theory to deal with biharmonic boundary value problems, no matter what their physical context may be, one easily loses direction in such an abstract theory; the physical problems are suggestive of method and we shall keep closely in touch with them. But first let us establish some fundamental properties of biharmonic functions and their conjugates.

Biharmonic functions and their conjugates

Any analytic function $f(z)$ of the complex variable $z = x + iy$ yields two conjugate harmonic functions, the real and imaginary parts of $f(z)$. For if $f(z) = X + iY$, then its analyticity implies

$$\partial f/\partial x = \partial f/\partial(iy) = f'(z),$$

and hence

$$\left.\begin{aligned} &X_x + iY_x = -i(X_y + iY_y),\\ &X_x = Y_y, \quad X_y = -Y_x,\\ &\Delta X = 0, \quad \Delta Y = 0. \end{aligned}\right\} \tag{5·502}$$

If X is given, then the conjugate Y is determined to within an additive constant in any simply connected domain [cf. p. 131].

We shall now show that if $f(z)$ and $F(z)$ are analytic functions and $\bar{z} = x - iy$, then the real and imaginary parts of $G(z)$, where

$$G(z) = f(z) + \bar{z}F(z), \tag{5·503}$$

are biharmonic functions; and, conversely, that any biharmonic function is the real or imaginary part of an expression of this form. This theorem is due to Goursat; the following compact proof has been supplied by A. E. Schild.

We have first to prove that

$$\Delta\Delta G = 0. \tag{5·504}$$

The following steps are easy to establish:

$$\Delta f = 0, \quad \Delta(\bar{z}F) = 4F', \quad \Delta\Delta(\bar{z}F) = \Delta(4F') = 0. \tag{5·505}$$

Thus (5·504) is proved. (Note that $G(z)$ is of course not analytic unless $F = 0$.)

Now for the converse. Given that $\Delta\Delta\psi = 0$, where ψ is real, we have to establish the existence of analytic functions $f(z)$ and $F(z)$ so that ψ is the real (or imaginary) part of $G(z)$. Since $\Delta\psi$ is harmonic, there exists an analytic function $F(z)$ such that

$$\Delta\psi = \mathrm{Re}\,[4F'(z)], \tag{5·506}$$

where Re means 'real part of'. Then

$$\Delta\{\psi - \mathrm{Re}\,[\bar{z}F(z)]\} = \Delta\psi - \mathrm{Re}\,[4F'(z)] = 0. \tag{5·507}$$

Therefore $\psi - \mathrm{Re}\,[\bar{z}F(z)]$ is harmonic and so there exists an analytic function $f(z)$ such that this harmonic function is $\mathrm{Re} f(z)$. Hence

$$\psi = \mathrm{Re}\,[f(z) + \bar{z}F(z)], \tag{5·508}$$

which proves the required result (we could of course equally well express ψ as the imaginary part of a complex expression with this form).

If real functions ψ, χ satisfy

$$\psi + i\chi = f(z) + \bar{z}F(z), \tag{5·509}$$

with f and F analytic, then both these functions are biharmonic; we say that they are *conjugate*. If ψ is given, there is a certain indeterminacy about χ, greater than the indeterminacy of the conjugate harmonic function, but this is a matter we shall not pursue further here.

There are important relations between the partial derivatives of a pair of conjugate biharmonic functions (ψ, χ). Using the indicial notation $(x_1 = x, x_2 = y)$, differentiation of (5·509) gives

$$\left.\begin{aligned} \psi_{,1} + i\chi_{,1} &= f'(z) + F(z) + \bar{z}F'(z), \\ \psi_{,2} + i\chi_{,2} &= if'(z) - iF(z) + i\bar{z}F'(z), \\ \psi_{,11} + i\chi_{,11} &= f''(z) + 2F'(z) + \bar{z}F''(z), \\ \psi_{,22} + i\chi_{,22} &= -f''(z) + 2F'(z) - \bar{z}F''(z), \\ \psi_{,12} + i\chi_{,12} &= if''(z) + i\bar{z}F''(z); \end{aligned}\right\} \tag{5·510}$$

hence $\psi_{,11} + i\chi_{,11} - \psi_{,22} - i\chi_{,22} = -2i(\psi_{,12} + i\chi_{,12}),$ (5·511)

and therefore

$$\tfrac{1}{2}(\psi_{,11} - \psi_{,22}) = \chi_{,12}, \quad \psi_{,12} = -\tfrac{1}{2}(\chi_{,11} - \chi_{,22}). \tag{5·512}$$

These are the relations between the partial derivatives of two conjugate biharmonic functions. Should we wish to find the indeterminacy of χ when ψ is given, we would do so by integrating these equations, regarded as partial differential equations for χ.

In indicial notation (Greek suffixes = 1, 2), (5·512) read

$$\psi_{,\alpha\rho}\epsilon_{\rho\beta} - \epsilon_{\alpha\rho}\psi_{,\rho\beta} = \chi_{,\alpha\beta} + \epsilon_{\alpha\rho}\chi_{,\rho\sigma}\epsilon_{\sigma\beta}. \tag{5·513}$$

But they are more interesting in matrix notation. As a matrix the permutation symbol ($\epsilon_{11} = \epsilon_{22} = 0, \epsilon_{12} = -\epsilon_{21} = 1$) may be written

$$J = \begin{pmatrix} 0 & 1 \\ -1 & 0 \end{pmatrix}; \tag{5·514}$$

since $J^2 = -E$, where E is the unit matrix, J is in a sense a square root of -1. If we denote the matrices $\psi_{,\alpha\beta}$ and $\chi_{,\alpha\beta}$ by X and Y respectively, then (5·513) reads

$$XJ - JX = Y + JYJ; \tag{5·515}$$

if we multiply by J we get

$$JXJ + X = JY - YJ, \tag{5·516}$$

which is the same as (5·515) with interchange of X and Y and with J changed to $-J$.

When a biharmonic function appears in a physical problem, it is defined in terms of physical quantities; it is interesting then to explore the physical meaning of its conjugate.

The biharmonic equation in hydrodynamics

In a viscous incompressible fluid the stress E_{ij} is given in terms of the pressure p and the velocity u_i by

$$E_{ij} = -p\delta_{ij} + \mu(u_{i,j} + u_{j,i}), \tag{5·517}$$

where μ is the coefficient of viscosity (assumed constant below). These are three-dimensional equations, Latin suffixes taking the values 1, 2, 3; Greek suffixes occurring later take the values 1, 2. Since $u_{i,i} = 0$ by the incompressibility of the fluid, we have $E_{ii} = -3p$. Under a body force X_i per unit volume, the equations of motion are

$$\rho\left(\frac{\partial u_i}{\partial t} + u_j u_{i,j}\right) = E_{ij,j} + X_i, \tag{5·518}$$

where ρ is the density.

We shall now assume the motion steady and slow, so that $\partial u_i/\partial t = 0$ and the quadratic terms are to be discarded. Thus

$$E_{ij,j} + X_i = 0, \tag{5·519}$$

equations formally the same as those of elastic equilibrium (5·402). Substitution from (5·517) gives

$$-p_{,i} + \mu\Delta u_i + X_i = 0. \tag{5·520}$$

If we assume the existence of a potential so that $X_i = -U_{,i}$, then (5·520) implies

$$\Delta(u_{i,j} - u_{j,i}) = 0. \tag{5·521}$$

Passing now to plane motion with $u_3 = 0$ and u_1, u_2 independent of x_3, there exists a stream function ψ such that

$$u_1 = -\psi_{,2}, \quad u_2 = \psi_{,1}, \tag{5·522}$$

and the vorticity is

$$\omega = \tfrac{1}{2}(u_{2,1} - u_{1,2}) = \tfrac{1}{2}\Delta\psi. \tag{5·523}$$

From (5·521) we see that ψ satisfies the biharmonic equation

$$\Delta\Delta\psi = 0. \tag{5·524}$$

From (5·517) we have

$$E_{\alpha\beta} = -p\delta_{\alpha\beta} + \mu(u_{\alpha,\beta} + u_{\beta,\alpha}), \quad p = -\tfrac{1}{2}E_{\gamma\gamma}, \tag{5·525}$$

and if we define $F_{\alpha\beta}$ by

$$F_{\alpha\beta} = E_{\alpha\beta} - \tfrac{1}{2}\delta_{\alpha\beta}E_{\gamma\gamma}, \tag{5·526}$$

we have

$$F_{\alpha\beta} = 2\mu e_{\alpha\beta}, \quad e_{\alpha\beta} = \tfrac{1}{2}(u_{\alpha,\beta} + u_{\beta,\alpha}), \tag{5·527}$$

$e_{\alpha\beta}$ being the rate of strain. Substitution for u_α from (5·522) gives

$$F_{11} = -2\mu\psi_{,12}, \quad F_{12} = \mu(\psi_{,11} - \psi_{,22}), \quad F_{22} = 2\mu\psi_{,12}. \tag{5·528}$$

These equations connect the stress with the stream function.

Problems involving the slow steady motion of an incompressible fluid require the solution of the biharmonic equation (5·524) with boundary conditions on u_α and/or $E_{\alpha\beta}n_\beta$, where n_α is the unit normal on the boundary.

To see how the conjugate biharmonic function enters the hydrodynamical problem, we note that the equations [cf. (5·519)]

$$E_{\alpha\beta,\beta} - U_{,\alpha} = 0 \tag{5·529}$$

imply the existence of a stress function A such that

$$E_{11} - U = A_{,22}, \quad E_{12} = -A_{,12}, \quad E_{22} - U = A_{,11}. \tag{5·530}$$

Then by (5·528) we have

$$\left.\begin{aligned}\tfrac{1}{2}(A_{,11}-A_{,22})=\tfrac{1}{2}(E_{22}-E_{11})=-F_{11}=2\mu\psi_{,12},\\ A_{,12}=-F_{12}=-\mu(\psi_{,11}-\psi_{,22}).\end{aligned}\right\} \qquad (5·531)$$

Comparison with (5·512) shows that the stress function A is biharmonic and proportional to a conjugate χ of ψ:

$$A=-2\mu\chi. \qquad (5·532)$$

Note that ψ is closely connected with the velocity and χ with the stress.

The biharmonic equation in elasticity

We shall consider only isotropic bodies.

By *plane strain* we understand a state with displacement u_i independent of x_3 and with $u_3=0$. This means that $e_{13}=e_{23}=e_{33}=0$ and the surviving components are $e_{\alpha\beta}$, Greek suffixes having the range 1, 2. There is only one compatibility equation in (5·410),

$$e_{11,22}+e_{22,11}-2e_{12,12}=0, \qquad (5·533)$$

the others being identically satisfied.

Since $e_{33}=0$, (5·408) gives

$$E_{33}=\sigma E_{\gamma\gamma}, \quad E_{kk}=(1+\sigma)E_{\gamma\gamma}, \qquad (5·534)$$

and so we have

$$\begin{aligned}e_{\alpha\beta}&=(1/E)\left[(1+\sigma)E_{\alpha\beta}-\sigma(1+\sigma)\delta_{\alpha\beta}E_{\gamma\gamma}\right]\\ &=\frac{1}{2\mu}(E_{\alpha\beta}-\sigma\delta_{\alpha\beta}E_{\gamma\gamma}),\end{aligned} \qquad (5·535)$$

or explicitly

$$\left.\begin{aligned}2\mu e_{11}&=E_{11}-\sigma(E_{11}+E_{22}),\\ 2\mu e_{12}&=E_{12},\\ 2\mu e_{22}&=E_{22}-\sigma(E_{11}+E_{22}).\end{aligned}\right\} \qquad (5·536)$$

Here μ is the rigidity. The compatibility equation (5·533) gives

$$E_{11,22}+E_{22,11}-2E_{12,12}-\sigma\Delta(E_{11}+E_{22})=0. \qquad (5·537)$$

Suppose now that there is equilibrium under a body force derived from a potential, $X_\alpha=-U_{,\alpha}$. Then the equilibrium equations (5·402) give

$$E_{\alpha\beta,\beta}-U_{,\alpha}=0, \qquad (5·538)$$

and it follows that there exists a stress function A (known as the Airy stress function) such that

$$E_{11}-U=A_{,22}, \quad E_{12}=-A_{,12}, \quad E_{22}-U=A_{,11}. \qquad (5·539)$$

Assuming that U is harmonic, the compatibility equation (5·537) leads to the biharmonic equation

$$\Delta\Delta A = 0. \tag{5·540}$$

Thus problems on plane strain involve the solution of the biharmonic equation under boundary conditions on u_α and/or $E_{\alpha\beta}n_\beta$.

In *generalized plane stress* we consider a thin plate of isotropic material, bounded by the planes $x_3 = \pm h$ across which there is no stress, so that

$$E_{13} = E_{23} = E_{33} = 0 \quad \text{for} \quad x_3 = \pm h. \tag{5·541}$$

We suppose the body force X_i to act parallel to the bounding planes, so that $X_3 = 0$. Then, integrating the equilibrium equations (5·402) across the thickness of the plate, dividing by $2h$, and denoting mean values by a bar, we get from the first two equations (Greek letters $= 1, 2$)

$$\bar{E}_{\alpha\beta,\beta} + \bar{X}_\alpha = 0. \tag{5·542}$$

Noting that by (5·402) for $i = 3$ and (5·541) we have $E_{33,3} = 0$ for $x_3 = \pm h$, the mean value of E_{33} is, on integration by parts,

$$\begin{aligned} \bar{E}_{33} = \frac{1}{2h}\int_{-h}^{h} E_{33}\,dx_3 &= -\frac{1}{2h}\int_{-h}^{h} x_3 E_{33,3}\,dx_3 \\ &= \frac{1}{4h}\int_{-h}^{h} x_3^2 E_{33,33}\,dx_3, \end{aligned} \tag{5·543}$$

which is small of order h^2 for a thin plate. We now take mean values across the thickness in (5·408) and discard $\bar{E}_{33}$ as small; thus

$$\bar{e}_{\alpha\beta} = \frac{1}{E}[(1+\sigma)\bar{E}_{\alpha\beta} - \sigma\delta_{\alpha\beta}\bar{E}_{\gamma\gamma}]. \tag{5·544}$$

We have also

$$\bar{e}_{\alpha\beta} = \tfrac{1}{2}(\bar{u}_{\alpha,\beta} + \bar{u}_{\beta,\alpha}), \tag{5·545}$$

and hence a compatibility equation as in (5·533):

$$\bar{e}_{11,22} + \bar{e}_{22,11} - 2\bar{e}_{12,12} = 0. \tag{5·546}$$

Supposing $X_\alpha = -U_{,\alpha}$, where U is harmonic, the argument for plane strain may be applied here, the only differences being a difference between the constants in (5·535) and those in (5·544) and the presence of the bars in the case of generalized plane stress. If we now omit those bars, understanding mean values in the case of generalized plane stress, we can collect all our results together in the following way, including the hydrodynamical theory in the same formulae with a different physical interpretation of symbols:

$$\left.\begin{aligned} &\Delta\Delta A = 0, \\ &E_{11} - U = A_{,22}, \quad E_{12} = -A_{,12}, \quad E_{22} - U = A_{,11}, \\ &e_{\alpha\beta} = \tfrac{1}{2}(u_{\alpha,\beta} + u_{\beta,\alpha}), \\ &e_{\alpha\beta} = k_1 E_{\alpha\beta} - k_2 \delta_{\alpha\beta} E_{\gamma\gamma}, \end{aligned}\right\} \qquad (5\cdot547)$$

where

$$\left.\begin{aligned} \text{for hydrodynamics:}\quad & k_1 = \frac{1}{2\mu}, \quad k_2 = \frac{1}{4\mu}; \\ \text{for plane strain:}\quad & k_1 = \frac{1}{2\mu}, \quad k_2 = \frac{\sigma}{2\mu}; \\ \text{for generalized plane stress:}\quad & k_1 = \frac{1}{2\mu}, \quad k_2 = \frac{\sigma}{1+\sigma}\frac{1}{2\mu}. \end{aligned}\right\} \qquad (5\cdot548)$$

Here μ is viscosity in hydrodynamics and rigidity in elasticity; u_α is velocity in hydrodynamics and displacement in elasticity; $e_{\alpha\beta}$ is rate of strain in hydrodynamics and strain in elasticity. If $\sigma = \frac{1}{2}$, plane strain coincides with hydrodynamics—this is the case of incompressibility. This is the singular case, giving $e_{\gamma\gamma} = 0$; it cannot occur in generalized plane stress, since we cannot have $\sigma = 1$.

Note that, in terms of the permutation symbol, we have

$$E_{\alpha\beta} = U\delta_{\alpha\beta} + \epsilon_{\alpha\gamma}\epsilon_{\beta\delta}A_{,\gamma\delta}. \qquad (5\cdot549)$$

Splitting the biharmonic problem

We can always interpret a biharmonic problem in a plane as a hydrodynamical problem, the biharmonic function being either the stream function ψ or the stress function A. It is essentially the same if we interpret it in terms of the equilibrium of an incompressible elastic body in plane strain. We can also make an interpretation in terms of the plane strain of a compressible elastic body or an interpretation in terms of generalized plane stress; this gives freedom in the choice of the ratio k_2/k_1 and this freedom may be either a convenience or a nuisance.

Adopting the attitude of applied mathematicians, let us rather think that the physical problem has been presented to us for solution. Then, looking at what has been done and comparing it with § 5·4, we recognize merely a reduction in dimensionality from three to two. The former methods are available, but it will make for clarity if we set down some important formulae.

The energy per unit of x_3 (rate of dissipation in hydrodynamics) is

$$W = \tfrac{1}{2}\int e_{\alpha\beta}E_{\alpha\beta}\,dV, \tag{5·550}$$

where dV is an element of area in the plane of the problem. By (5·547) this becomes

$$W = \tfrac{1}{2}\int (k_1 E_{\alpha\beta}E_{\alpha\beta} - k_2 E^2_{\gamma\gamma})\,dV, \tag{5·551}$$

and so, if we define an F-vector by

$$\mathbf{S} \leftrightarrow E_{\alpha\beta}, \tag{5·552}$$

we are led to define the scalar product as

$$\mathbf{S}\,.\,\mathbf{S}' = \int (k_1 E_{\alpha\beta}E'_{\alpha\beta} - k_2 E_{\gamma\gamma}E'_{\delta\delta})\,dV. \tag{5·553}$$

Considerations of energy ensure the positive-definite character of $\mathbf{S}^2$.

If we take

$$\mathbf{S} \leftrightarrow e_{\alpha\beta}, \tag{5·554}$$

we must separate compressibility ($k_2/k_1 \neq \tfrac{1}{2}$) from incompressibility ($k_2/k_1 = \tfrac{1}{2}$). In the former case we solve (5·547) for $E_{\alpha\beta}$, obtaining

$$k_1 E_{\alpha\beta} = e_{\alpha\beta} + \frac{k_2}{k_1 - 2k_2}\,\delta_{\alpha\beta}e_{\gamma\gamma}, \tag{5·555}$$

and (5·550) gives

$$W = \frac{1}{2k_1}\int \left(e_{\alpha\beta}e_{\alpha\beta} + \frac{k_2}{k_1 - 2k_2}\,e^2_{\gamma\gamma}\right)dV; \tag{5·556}$$

this leads us to the scalar product

$$\mathbf{S}\,.\,\mathbf{S}' = \frac{1}{k_1}\int \left(e_{\alpha\beta}e'_{\alpha\beta} + \frac{k_2}{k_1 - 2k_2}\,e_{\gamma\gamma}e'_{\delta\delta}\right)dV. \tag{5·557}$$

In this case of compressibility, (5·553) and (5·557) are merely different ways of looking at the same thing, and it does not matter at all whether we take an F-vector to correspond to stress or to strain. But in the incompressible case ($k_2/k_1 = \tfrac{1}{2}$) (5·557) contains an indeterminate term because $k_1 - 2k_2$ and $e_{\gamma\gamma}$ both vanish. We must either fall back on (5·552) or modify the scalar product as we did in (5·468) in three dimensions. We shall adopt this modification as standard procedure and summarize as follows, showing the definitions of F-vectors, of scalar products, and of the basic linear subspaces which split the problem, the boundary conditions being of the form

$$(u_\alpha)_{B_1} = f_\alpha, \quad (E_{\alpha\beta}n_\beta)_{B_2} = g_\alpha. \tag{5·558}$$

I. Slow steady motion of an incompressible viscous fluid, or plane strain of an incompressible elastic body:

$$\left.\begin{aligned}
&\mathbf{S}\leftrightarrow e_{\alpha\beta},\quad e_{\alpha\beta}=e_{\beta\alpha},\quad e_{\gamma\gamma}=0,\\
&\mathbf{S}.\mathbf{S}'=\int e_{\alpha\beta}e'_{\alpha\beta}dV,\\
&L':\quad \mathbf{S}'\leftrightarrow e'_{\alpha\beta}=\tfrac{1}{2}(u'_{\alpha,\beta}+u'_{\beta,\alpha}),\quad u'_{\alpha,\alpha}=0,\quad (u'_\alpha)_{B_1}=f_\alpha;\\
&L'':\quad \mathbf{S}''\leftrightarrow e''_{\alpha\beta}=\frac{1}{2\mu}(E''_{\alpha\beta}-\tfrac{1}{2}\delta_{\alpha\beta}E''_{\gamma\gamma}),\quad E''_{\alpha\beta}=E''_{\beta\alpha},\\
&\qquad\qquad E''_{\alpha\beta,\beta}-U_{,\alpha}=0,\quad (E''_{\alpha\beta}n_\beta)_{B_2}=g_\alpha.
\end{aligned}\right\}\quad(5{\cdot}559)$$

II. Plane strain of a compressible elastic body, or generalized plane stress:

$$\left.\begin{aligned}
&\mathbf{S}\leftrightarrow E_{\alpha\beta},\quad E_{\alpha\beta}=E_{\beta\alpha},\\
&\mathbf{S}.\mathbf{S}'=\int e_{\alpha\beta}E'_{\alpha\beta}dV=\int e'_{\alpha\beta}E_{\alpha\beta}dV\\
&\qquad=\int(k_1E_{\alpha\beta}E'_{\alpha\beta}-k_2E_{\gamma\gamma}E'_{\delta\delta})\,dV\\
&\qquad=\frac{1}{k_1}\int\left(e_{\alpha\beta}e'_{\alpha\beta}+\frac{k_2}{k_1-2k_2}e_{\gamma\gamma}e'_{\delta\delta}\right)dV,\\
&L':\quad \mathbf{S}'\leftrightarrow E'_{\alpha\beta}=\frac{1}{k_1}\left(e'_{\alpha\beta}+\frac{k_2}{k_1-2k_2}\delta_{\alpha\beta}e'_{\gamma\gamma}\right),\\
&\qquad e'_{\alpha\beta}=\tfrac{1}{2}(u'_{\alpha,\beta}+u'_{\beta,\alpha}),\quad (u'_\alpha)_{B_1}=f_\alpha;\\
&L'':\quad \mathbf{S}''\leftrightarrow E''_{\alpha\beta}=E''_{\beta\alpha},\quad E''_{\alpha\beta,\beta}-U_{,\alpha}=0,\quad (E''_{\alpha\beta}n_\beta)_{B_2}=g_\alpha.
\end{aligned}\right\}\quad(5{\cdot}560)$$

(For k_1, k_2, see (5·548).) Here u'_α is to be continuous and $E''_{\alpha\beta}n_\beta$ continuous across every line. It is easy to verify in each case that the solution is on L' and on L'' and that L', L'' are mutually orthogonal.

In the case of boundary conditions other than (5·558), the general rule is to impose on u'_α the boundary conditions assigned to u_α and to impose on $E''_{\alpha\beta}$ those assigned to $E_{\alpha\beta}$.

The pyramid F-vectors of Chapter 3 do not work for the biharmonic problem, but we can define analogous F-vectors, rather more complicated, which enable us to draw the bounds obtained by the hypercircle method close together in a systematic way. For an account of these biharmonic F-vectors, see Synge (8) and work in progress by V. G. Hart.

Example: viscous flow across a tank

Fig. 5·51 shows the plan of a tank across which an incompressible viscous fluid flows in slow motion, entering through MN and leaving through PQ; these are straight segments parallel to the Ox_1-axis, extending from $x_1=-a$

to $x_1 = a$; their distance apart is b. The boundary B consists of the walls of the tank (B_1) and $B_2 = MN + PQ$. V_2 is the interior of the rectangle $MNPQ$ and V_1 the rest of the tank. There is no body force.

We prescribe the way in which the fluid enters and leaves the tank, writing as boundary conditions

$$\left.\begin{aligned} &u_\alpha = 0 \text{ on } B_1, \\ &u_1 = 0, \quad u_2 = C(a^2 - x_1^2) \text{ on } B_2, \end{aligned}\right\} \tag{5·561}$$

where C is a given constant.

Note that in our designation of B_1 and B_2 we have not used the notation of (5·558); the B_2 of (5·558) is absent in the present problem.

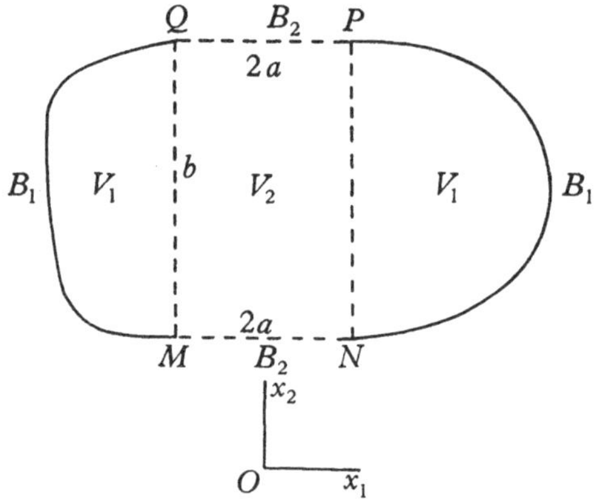

Fig. 5·51. Viscous flow across a tank, or plate under pressure.

We follow the scheme of (5·559), taking

$$\left.\begin{aligned} &\mathbf{S}' \leftrightarrow e'_{\alpha\beta} = \tfrac{1}{2}(u'_{\alpha,\beta} + u'_{\beta,\alpha}), \\ &u'_\alpha = 0 \text{ in } V_1, \\ &u'_1 = 0, \quad u'_2 = C(a^2 - x_1^2) \text{ in } V_2, \end{aligned}\right\} \tag{5·562}$$

so that

$$\left.\begin{aligned} &e'_{\alpha\beta} = 0 \text{ in } V_1, \\ &e'_{11} = e'_{22} = 0, \quad e'_{12} = -Cx_1 \text{ in } V_2. \end{aligned}\right\} \tag{5·563}$$

We take

$$E''_{11} = E''_{22} = \mu C x_2, \quad E''_{12} = -\mu C x_1 \tag{5·564}$$

in $V_1 + V_2$ (these satisfy $E''_{\alpha\beta,\beta} = 0$) and hence

$$\left.\begin{aligned} &\mathbf{S}'' \leftrightarrow e''_{\alpha\beta}, \\ &e''_{11} = -e''_{22} = \frac{1}{4\mu}(E''_{11} - E''_{22}) = 0, \\ &e''_{12} = \frac{1}{2\mu}E''_{12} = -\tfrac{1}{2}Cx_1, \end{aligned}\right\} \tag{5·565}$$

in $V_1 + V_2$.

We have then the scalar products

$$\left.\begin{aligned} \mathbf{S}'^2 &= \int e'_{\alpha\beta} e'_{\alpha\beta} dV = 2\int_{V_2} (Cx_1)^2 dV = K, \\ \mathbf{S}'.\mathbf{S}'' &= \int e'_{\alpha\beta} e''_{\alpha\beta} dV = \int_{V_2} (Cx_1)^2 dV = \tfrac{1}{2}K, \\ \mathbf{S}''^2 &= \int e''_{\alpha\beta} e''_{\alpha\beta} dV = \frac{1}{2}\int_{V_1+V_2} (Cx_1)^2 dV = \tfrac{1}{4}K(1+D), \end{aligned}\right\} \tag{5·566}$$

where

$$K = \tfrac{4}{3}a^3bC^2, \quad D = \frac{3}{2a^3b}\int_{V_1} x_1^2 dV. \tag{5·567}$$

Noting that L'' contains the origin of F-space, we have by (2·772) the inequalities

$$(\mathbf{S}'.\mathbf{S}'')^2/\mathbf{S}''^2 \leqslant \mathbf{S}^2 \leqslant \mathbf{S}'^2, \tag{5·568}$$

and hence

$$\frac{K}{1+D} \leqslant \mathbf{S}^2 \leqslant K. \tag{5·569}$$

This gives bounds on the rate of dissipation of energy,

$$W = \tfrac{1}{2}\int e_{\alpha\beta} E_{\alpha\beta} dV = \mu \int e_{\alpha\beta} e_{\alpha\beta} dV = \mu \mathbf{S}^2. \tag{5·570}$$

These bounds are close together when the walls of the tank lie close to the straight lines MQ, NP, for then D is small. Note that we chose for u'_α, $E''_{\alpha\beta}$ the values for plane Poiseuille flow in $MNPQ$, extending these values suitably into V_1.

Example: plate under pressure

Fig. 5·51 shows a thin plate in generalized plane stress under uniform pressure Π applied to MN and PQ, with B_1 free. Thus the boundary conditions are

$$\left.\begin{aligned} &E_{\alpha\beta} n_\beta = 0 \text{ on } B_1, \\ &E_{12} = 0, \quad E_{22} = -\Pi \text{ on } B_2. \end{aligned}\right\} \tag{5·571}$$

Note that we have changed the meanings of B_1 and B_2 from (5·558), as in the preceding example, but now the B_1 of (5·558) is absent, the boundary conditions being on stress only.

We use the scheme (5·560). We take in $V_1 + V_2$

$$E'_{11} = E'_{12} = 0, \quad E'_{22} = -\Pi, \tag{5·572}$$

but before admitting this as a definition of $\mathbf{S}'$ we have to show the existence of u'_α satisfying

$$\tfrac{1}{2}(u'_{\alpha,\beta} + u'_{\beta,\alpha}) = e'_{\alpha\beta} = k_1 E'_{\alpha\beta} - k_2 \delta_{\alpha\beta} E'_{\gamma\gamma}. \tag{5·573}$$

Explicitly these equations read

$$u'_{1,1} = k_2 \Pi, \quad u'_{1,2} + u'_{2,1} = 0, \quad u'_{2,2} = (k_2 - k_1)\Pi, \tag{5·574}$$

and they have the solution

$$u'_1 = k_2 \Pi x_1, \quad u'_2 = (k_2 - k_1)\Pi x_2. \tag{5·575}$$

Accordingly we take

$$\mathbf{S}' \leftrightarrow E'_{\alpha\beta}, \tag{5·576}$$

where in $V_1 + V_2$ we have

$$\left.\begin{aligned} E'_{11} = E'_{12} = 0, \quad E'_{22} = -\Pi, \\ e'_{11} = k_2\Pi, \quad e'_{12} = 0, \quad e'_{22} = (k_2 - k_1)\,\Pi. \end{aligned}\right\} \tag{5·577}$$

We take

$$\left.\begin{aligned} &\mathbf{S}'' \leftrightarrow E''_{\alpha\beta}, \\ &E''_{\alpha\beta} = 0 \text{ in } V_1, \\ &E''_{11} = E''_{12} = 0, \quad E''_{22} = -\Pi \text{ in } V_2, \end{aligned}\right\} \tag{5·578}$$

thus satisfying $E''_{\alpha\beta,\beta} = 0$ and the continuity condition; the corresponding strain is by (5·547)

$$\left.\begin{aligned} &e''_{\alpha\beta} = 0 \text{ in } V_1, \\ &e''_{11} = k_2\Pi, \quad e''_{12} = 0, \quad e''_{22} = (k_2 - k_1)\,\Pi \text{ in } V_2. \end{aligned}\right\} \tag{5·579}$$

The scalar products are

$$\left.\begin{aligned} \mathbf{S}'^2 &= \int e'_{\alpha\beta} E'_{\alpha\beta}\, dV = (k_1 - k_2)\,(V_1 + V_2)\,\Pi^2, \\ \mathbf{S}'.\mathbf{S}'' &= \int e'_{\alpha\beta} E''_{\alpha\beta}\, dV = (k_1 - k_2)\, V_2 \Pi^2, \\ \mathbf{S}''^2 &= \int e''_{\alpha\beta} E''_{\alpha\beta}\, dV = (k_1 - k_2)\, V_2 \Pi^2. \end{aligned}\right\} \tag{5·580}$$

Interchanging $\mathbf{S}'$ and $\mathbf{S}''$ in (2·772), since now L', not L'', contains the origin, we have

$$(\mathbf{S}'.\mathbf{S}'')^2/\mathbf{S}'^2 \leqslant \mathbf{S}^2 \leqslant \mathbf{S}''^2, \tag{5·581}$$

and so

$$\frac{(k_1 - k_2)\, V_2^2 \Pi^2}{V_1 + V_2} \leqslant \mathbf{S}^2 \leqslant (k_1 - k_2)\, V_2 \Pi^2. \tag{5·582}$$

Here we have bounds for the strain energy W per unit thickness, since

$$W = \tfrac{1}{2}\int e_{\alpha\beta} E_{\alpha\beta}\, dV = \tfrac{1}{2}\mathbf{S}^2. \tag{5·583}$$

By (5·548) and (5·463) we have

$$k_1 - k_2 = \frac{1}{1+\sigma}\, \frac{1}{2\mu} = \frac{1}{E}, \tag{5·584}$$

and so we have the bounds

$$\frac{V_2}{V_1 + V_2} \leqslant \frac{2WE}{\Pi^2 V_2} \leqslant 1; \tag{5·585}$$

here V_1 and V_2 are the areas shown in Fig. 5·51.

Note that the bounds coincide when the plate is reduced to the rectangle *MNPQ*.

Exercises

1. Consider the plane biharmonic problem

$$\Delta\Delta\psi = 0, \quad (\psi_{,\alpha})_B = f_\alpha.$$

Taking an F-vector to correspond to a symmetric tensor ($p_{\alpha\beta} = p_{\beta\alpha}$) and the scalar product to be $\int p_{\alpha\beta} p'_{\alpha\beta}\, dV$, show that the solution $\mathbf{S} \leftrightarrow \psi_{,\alpha\beta}$ is the intersection of the orthogonal linear subspaces

$$\begin{aligned} L'&: \quad \mathbf{S}' \leftrightarrow p'_{\alpha\beta} = \phi_{,\alpha\beta}, \quad (\phi_{,\alpha})_B = f_\alpha; \\ L''&: \quad \mathbf{S}'' \leftrightarrow p''_{\alpha\beta}, \quad p''_{\alpha\beta,\alpha\beta} = 0. \end{aligned}$$

2. Show that the problem

$$\Delta\Delta\psi=0, \quad (\psi_{,\alpha\beta}n_\beta)_B=g_\alpha$$

can be split as in Example 1, with suitable modification.

3. Consider the slow steady motion of an incompressible viscous fluid in an area bounded by a curve $B=B_1+B_2$ with the boundary conditions

$$(u_\alpha n_\alpha)_B=0, \quad (u_\alpha t_\alpha)_{B_1}=0, \quad (E_{\alpha\beta}t_\alpha n_\beta)_{B_2}=f,$$

where n_α and t_α are respectively the unit normal and unit tangent to B. Show that the problem may be split as in (5·559) but with the boundary conditions on u'_α and $E''_{\alpha\beta}$ changed to agree with the above data.

4. Consider the following plane problem:

$$\Delta\Delta\psi=0 \text{ in } V, \quad (\psi)_B=f, \quad (\partial\psi/\partial n)_{B_1}=g,$$

$$[(\psi_{,11}-\psi_{,22})(n_1^2-n_2^2)+4\psi_{,12}n_1n_2]_{B_2}=h;$$

here f, g, h are given boundary values on B $(=B_1+B_2)$, the boundary of V, and n_α is the unit normal to B. If an F-vector is defined by $\mathbf{S}\leftrightarrow(p_1, p_2)$, with the scalar product

$$\mathbf{S}'.\mathbf{S}''=\int(p'_1p''_1+p'_2p''_2)\,dV,$$

show that the solution ψ is such that the F-point $\mathbf{S}\leftrightarrow(\psi_{,11}-\psi_{,22}, 2\psi_{,12})$ corresponding to it is the intersection of the orthogonal linear subspaces L', L'' defined as follows:

$$L':\mathbf{S}'\leftrightarrow(p'_1,p'_2); \quad p'_1=u'_{,11}-u'_{,22}, \quad p'_2=2u'_{,12}, \quad (u')_B=f \quad (\partial u'/\partial n)_{B_1}=g;$$

$$L'':\mathbf{S}''\leftrightarrow(p''_1,p''_2); \quad p''_{1,11}-p''_{1,22}+2p''_{2,12}=0 \quad [p''_1(n_1^2-n_2^2)+2p''_2n_1n_2]_{B_2}=h.$$

[Cf. review by H. F. Weinberger (*Math. Rev.* **17** (1956), 373) of Synge (8).]

PART III

INDEFINITE METRIC

CHAPTER 6

GEOMETRY OF FUNCTION-SPACE WITH INDEFINITE METRIC

6·1. NULL VECTORS, NULL CONES, ORTHOGONALITY, MINKOWSKIAN F-SPACES

Résumé of available facts

In dealing with function-space in Part I, no metric was assumed; everything stated there holds when a metric is imposed. Let us recall some basic formulae.

A point in F-space corresponds to a set of functions in P-space, which is of any number of dimensions:

$$\mathbf{S} \leftrightarrow (s_1, s_2, \ldots, s_M). \tag{6·101}$$

We know the meanings of $\mathbf{S}+\mathbf{T}$ and $a\mathbf{S}$. A straight line is defined by

$$\mathbf{X} = a\mathbf{A} + b\mathbf{B}, \quad a+b=1, \tag{6·102}$$

and a linear subspace by the condition that the straight line joining any two points on it is contained completely in it. A linear n-space has the equation

$$\mathbf{X} = \mathbf{X}_0 + a_1\mathbf{A}_1 + \ldots + a_n\mathbf{A}_n, \tag{6·103}$$

the a's being arbitrary parameters.

In Chapter 2 we have already introduced the idea of a scalar product $\mathbf{S}\,.\,\mathbf{S}'$ obeying the basic rules (2·101) and giving an *indefinite metric* $\mathbf{S}^2$, by which we mean that $\mathbf{S}^2$ may be positive, zero or negative according to the choice of $\mathbf{S}$, and $\mathbf{S}^2=0$ does not imply $\mathbf{S}=\mathbf{O}$. To fix our ideas we may consider, as an example, a case included in (5·107):

$$\left.\begin{aligned} &\mathbf{S} \leftrightarrow (p_i, u), \\ &\mathbf{S}\,.\,\mathbf{S}' = \int (p_i p_i' - k^2 u u')\, dV, \end{aligned}\right\} \tag{6·104}$$

where P-space may have any number of dimensions, but we may think of two or three. Here an F-point corresponds to a P-vector field plus a P-scalar field. We have

$$\mathbf{S}^2 = \int (p_i p_i - k^2 u^2)\, dV, \tag{6·105}$$

and so $\mathbf{S}^2$ is positive if $u=0$ and negative if $p_i=0$, these conditions for positive and negative character being of course sufficient but not necessary. This example will serve for illustrative purposes, but the following theory applies to any indefinite metric.

Null F-vectors

A *null* F-vector satisfies

$$\mathbf{S}^2 = 0. \tag{6·106}$$

Thus for the metric (6·105), $\mathbf{S} \leftrightarrow (p_i, u)$ is null if

$$\int p_i p_i dV = \int k^2 u^2 dV. \tag{6·107}$$

We do not meet null vectors in Euclidean geometry, but they occur in the special theory of relativity for which the expression

$$x^2 + y^2 + z^2 - c^2t^2 \tag{6·108}$$

is analogous to our $\mathbf{S}^2$. Borrowing the terminology of relativity, we shall call $\mathbf{S}$ *spacelike* if $\mathbf{S}^2 > 0$ and *timelike* if $\mathbf{S}^2 < 0$.

It is convenient to introduce an *indicator* ϵ chosen equal to 1 or -1 to make $\epsilon\mathbf{S}^2$ positive, and to define the *magnitude* $|\mathbf{S}|$ of $\mathbf{S}$ by

$$|\mathbf{S}|^2 = \epsilon\mathbf{S}^2, \quad |\mathbf{S}| > 0. \tag{6·109}$$

This is for a spacelike or timelike $\mathbf{S}$; if it is null, $|\mathbf{S}| = 0$. The positive (or zero) number $|\mathbf{S}|$ may also be called the *length* or *norm* of $\mathbf{S}$ and $|\mathbf{A} - \mathbf{B}|$ the *distance* between the points $\mathbf{A}$ and $\mathbf{B}$. Two points may be at zero distance from one another without coinciding. Note that $\epsilon = 1$ for a spacelike vector and $\epsilon = -1$ for a timelike one.

Null cones

The set of F-points $\mathbf{X}$ for which

$$\mathbf{X}^2 = 0 \tag{6·110}$$

form a *null cone*. Here the *vertex* is at the origin. More generally the null cone with vertex $\mathbf{C}$ has the equation

$$(\mathbf{X} - \mathbf{C})^2 = 0. \tag{6·111}$$

This cone divides F-space into two regions which we shall call *exterior* and *interior*:

$$\left.\begin{aligned} \text{exterior:} &\quad (\mathbf{X} - \mathbf{C})^2 > 0, \\ \text{interior:} &\quad (\mathbf{X} - \mathbf{C})^2 < 0. \end{aligned}\right\} \tag{6·112}$$

We cannot pass from the interior to the exterior without passing through the null cone, because we cannot change a number continuously from negative to positive without passing through zero.

In the relativistic theory connected with the expression (6·108), we get a null cone by equating that expression to zero. Physically it represents the history of a spherical light wave, which contracts to a point and then expands. In this case the interior of the null cone is itself divided into two parts, the past and the future, and we cannot go continuously from the one to the other without going through the null cone. This is mentioned here to contrast it with what happens in F-space; there any two points of the interior are accessible to one another without crossing the null cone, and the same is true for two points of the exterior.

Let us investigate the intersection of a straight line (6·102) with a null cone, taking (6·110) for simplicity. The existence of an intersection requires that a and b satisfy

$$a^2\mathbf{A}^2 + 2ab\mathbf{A}.\mathbf{B} + b^2\mathbf{B}^2 = 0, \quad a + b = 1. \tag{6·113}$$

There are four possibilities:

(i) the straight line is contained in the null cone; this happens if $\mathbf{A}$ and $\mathbf{B}$ are null vectors with a common direction;

(ii) the straight line cuts the null cone in two points; this happens if

$$(\mathbf{A}.\mathbf{B})^2 > \mathbf{A}^2\mathbf{B}^2; \tag{6·114}$$

(iii) the straight line touches the null cone; this happens if

$$(\mathbf{A}.\mathbf{B})^2 = \mathbf{A}^2\mathbf{B}^2 \neq 0; \tag{6·115}$$

(iv) the straight line does not meet the null cone; this happens if

$$(\mathbf{A}.\mathbf{B})^2 < \mathbf{A}^2\mathbf{B}^2. \tag{6·116}$$

According to the Schwarz inequality (2·210) the inequality (6·114) is impossible; but the Schwarz inequality does not apply in the case of an indefinite metric.

Note that if $\mathbf{A}$ and $\mathbf{B}$ have opposite indicators, one vector being spacelike and the other timelike, then $\mathbf{A}^2\mathbf{B}^2$ is negative and (6·114) holds; thus *the linear 2-space containing two vectors drawn from the origin* $\mathbf{O}$, *one spacelike and the other timelike, cuts the null cone with vertex* $\mathbf{O}$ *in two null lines*, these null lines being the generators of the null cone drawn from $\mathbf{O}$ to the two points of intersection we have been discussing.

Orthogonality and orthogonal projection

In the case of a positive-definite metric we made extensive use of the idea of angle. But when the metric is indefinite, this idea fails through the lack of the Schwarz inequality, and the only angles we shall think of are the particular angles 0, $\frac{1}{2}\pi$, π. The first and last are trivial, corresponding respectively to vectors having the same direction and opposite directions ($\mathbf{A} = k\mathbf{B}$ where k is positive or negative, respectively), but the angle $\frac{1}{2}\pi$ is important. It occurs when

$$\mathbf{A}.\mathbf{B}=0; \tag{6·117}$$

then the vectors $\mathbf{A}$ and $\mathbf{B}$ are *orthogonal*. Every null vector is orthogonal to itself.

Let us now consider the *orthogonal projection* of an F-point $\mathbf{P}$ on a linear subspace L, starting with the case where L is a straight line L_1, with equation (equivalent to (6·102))

$$\mathbf{X}=\mathbf{X}_0+c\mathbf{A}, \tag{6·118}$$

where $\mathbf{X}_0$ and $\mathbf{A}$ are fixed and c a parameter ranging from $-\infty$ to $+\infty$. To say that the point $\mathbf{N}$ (Fig. 6·11) is the orthogonal projection

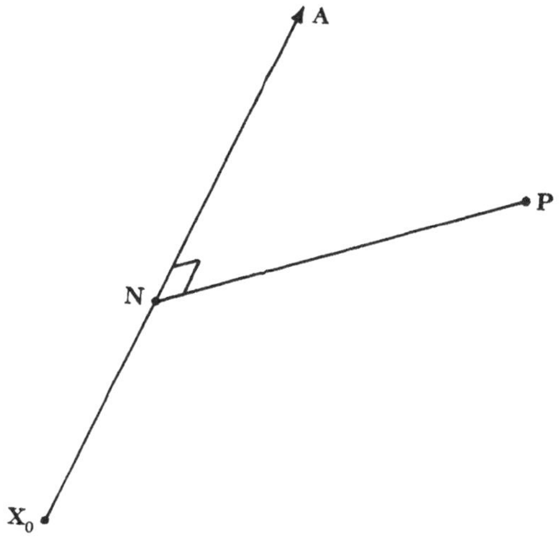

Fig. 6·11. Orthogonal projection of $\mathbf{P}$ on L_1.

of $\mathbf{P}$ on L_1 means that the joining vector $\mathbf{N}-\mathbf{P}$ is orthogonal to $\mathbf{A}$, which lies in L_1, i.e.

$$(\mathbf{N}-\mathbf{P}).\mathbf{A}=0. \tag{6·119}$$

But $\mathbf{N}$ satisfies (6·118) and so

$$c\mathbf{A}^2+(\mathbf{X}_0-\mathbf{P}).\mathbf{A}=0. \tag{6·120}$$

Consider first the case where the line is null ($\mathbf{A}^2=0$). Then either (i) (6·120) is satisfied for all c; this happens if $(\mathbf{X}_0-\mathbf{P}).\mathbf{A}=0$; or (ii) (6·120) is not satisfied by any c; this happens if $(\mathbf{X}_0-\mathbf{P}).\mathbf{A}\neq 0$. In the former case the orthogonal projection of $\mathbf{P}$ on L_1 is not determined; all points on L_1 are orthogonal projections. In the second case the orthogonal projection does not exist.

In the general case ($\mathbf{A}^2 \neq 0$), (6·120) is satisfied by a unique value of c; the orthogonal projection of $\mathbf{P}$ exists and is unique.

When the metric is positive-definite, orthogonal projection (equivalently, dropping the normal) gives us the shortest distance from $\mathbf{P}$ to L_1. For an indefinite metric we have, as $\mathbf{X}$ ranges on L_1,

$$\begin{aligned}(\mathbf{X}-\mathbf{P})^2 &= (\mathbf{X}_0 + c\mathbf{A} - \mathbf{P})^2 \\ &= (\mathbf{X}_0 - \mathbf{P})^2 + 2c\mathbf{A}.(\mathbf{X}_0 - \mathbf{P}) + c^2\mathbf{A}^2. \qquad (6{\cdot}121)\end{aligned}$$

Suppose $\mathbf{A}^2 \neq 0$. Then the distance $|\mathbf{X}-\mathbf{P}| \to \infty$ as $c \to \pm\infty$, as in the case of a positive-definite metric, and if

$$[\mathbf{A}.(\mathbf{X}_0 - \mathbf{P})]^2 < \mathbf{A}^2(\mathbf{X}_0 - \mathbf{P})^2, \qquad (6{\cdot}122)$$

the distance behaves in an ordinary way, sinking to a minimum at the point $\mathbf{N}$. But if the inequality (6·122) is reversed, the behaviour is quite different, for the distance vanishes twice and takes a relative maximum at $\mathbf{N}$. The graph of $(\mathbf{X}-\mathbf{P})^2$ is a parabola as in Fig. 6·12; this is drawn for $\mathbf{A}^2 > 0$ (for $\mathbf{A}^2 < 0$, turn the graph upsidedown).

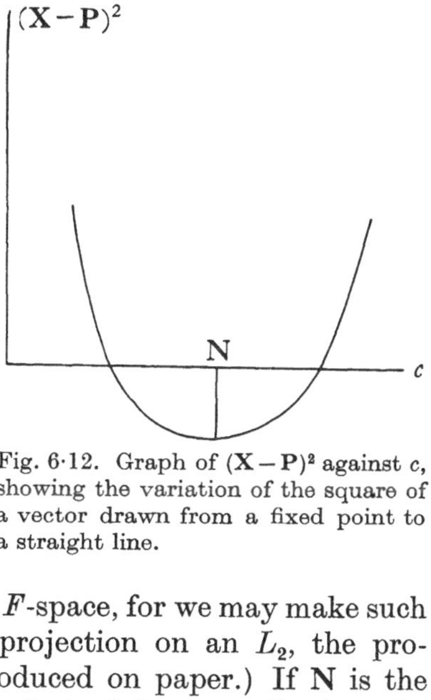

Fig. 6·12. Graph of $(\mathbf{X}-\mathbf{P})^2$ against c, showing the variation of the square of a vector drawn from a fixed point to a straight line.

Consider now the orthogonal projection of a point $\mathbf{P}$ on a linear n-space L_n with equation

$$\mathbf{X} = \mathbf{X}_0 + a_1\mathbf{T}_1 + \ldots + a_n\mathbf{T}_n. \qquad (6{\cdot}123)$$

Here $\mathbf{X}_0$ is a point on L_n and $\mathbf{T}_1, \ldots, \mathbf{T}_n$ vectors lying in it. (For $n=2$ this question of projection is of interest in connexion with the preparation of accurate sketches of F-space, for we may make such accurate sketches by orthogonal projection on an L_2, the projections of points being then reproduced on paper.) If $\mathbf{N}$ is the orthogonal projection of $\mathbf{P}$ (i.e. the foot of the normal), then the equations

$$(\mathbf{N}-\mathbf{P}).\mathbf{T}_\rho = 0 \quad (\rho = 1, 2, \ldots, n) \qquad (6{\cdot}124)$$

are satisfied, these equations expressing the orthogonality of $\mathbf{N}-\mathbf{P}$ to all vectors lying in L_n. But $\mathbf{N}$ is of the form (6·123), and so we get

$$\sum_{\sigma=1}^{n} a_\sigma \mathbf{T}_\sigma . \mathbf{T}_\rho + (\mathbf{X}_0 - \mathbf{P}).\mathbf{T}_\rho = 0 \quad (\rho = 1, 2, \ldots, n) \qquad (6{\cdot}125)$$

as equations to be satisfied by the values of the parameters a_ρ for the orthogonal projection **N**.

There are various possibilities depending on the nature of the matrix $\mathbf{T}_\rho . \mathbf{T}_\sigma$, but it would be tedious to go into them. In the general case for which

$$\det(\mathbf{T}_\rho . \mathbf{T}_\sigma) \neq 0, \tag{6·126}$$

the equations (6·125) have unique solutions, giving a unique orthogonal projection. But, as in the case of the line, the normal joining **P** and **N** need not be a line of shortest distance joining **P** to L_n. It has, however, a stationary character, for the square

$$(\mathbf{X}-\mathbf{P})^2 = (\mathbf{X}_0 + a_1 \mathbf{T}_1 + \ldots + a_n \mathbf{T}_n - \mathbf{P})^2 \tag{6·127}$$

has a stationary value, with respect to variation of the a's, for those values of the a's which satisfy (6·125). It is typical of the change from a positive-definite metric to an indefinite one that what was formerly a *minimum* value becomes merely a *stationary* value.

In the case of a linear subspace L of infinite dimensionality the orthogonal projection of **P** on L is defined in the same way as above, i.e. if **N** is the orthogonal projection, then $\mathbf{N}-\mathbf{P}$ is to be orthogonal to all vectors lying in L. We cannot prove the existence of **N**, but we can show that if it exists it is unique, provided L contains no null vector. For if there are two orthogonal projections, **M** and **N**, then

$$(\mathbf{M}-\mathbf{P}) . \mathbf{T} = 0, \quad (\mathbf{N}-\mathbf{P}) . \mathbf{T} = 0, \tag{6·128}$$

for every vector **T** lying in L. Subtraction gives

$$(\mathbf{M}-\mathbf{N}) . \mathbf{T} = 0. \tag{6·129}$$

But we may choose $\mathbf{T} = \mathbf{M}-\mathbf{N}$, since this vector lies in L, and hence

$$(\mathbf{M}-\mathbf{N})^2 = 0, \tag{6·130}$$

which contradicts the hypothesis that no null vector lies in L.

The orthogonal projection on L gives a stationary value to the distance $|\mathbf{X}-\mathbf{P}|$ for **X** ranging on L. For, varying **X** in L, we have

$$\delta(\mathbf{X}-\mathbf{P})^2 = 2(\mathbf{X}-\mathbf{P}) . \delta\mathbf{X} = 0, \tag{6·131}$$

since $\delta\mathbf{X}$ lies in L.

Orthonormalization

For a positive-definite metric, the Gram–Schmidt process of orthonormalization was described on pp. 53–5, starting with equation (2·311). It can never fail in the case of a positive-definite metric because the extended Schwarz inequality (2·333) guarantees that we never have to divide by zero in (2·325), and that is the only place at which the process could possibly break down. With an indefinite metric this inequality does not hold, and the process may fail.

To illustrate this failure, take

$$\mathbf{S}_1 = \mathbf{N} + \mathbf{T}, \quad \mathbf{S}_2 = \mathbf{N} - \mathbf{T}, \tag{6·132}$$

where **N** and **T** satisfy

$$\mathbf{N}^2 = 0, \quad \mathbf{N}\,.\,\mathbf{T} = 0, \quad \mathbf{T}^2 = 1. \tag{6·133}$$

Then

$$\mathbf{S}_1^2 = 1, \quad \mathbf{S}_1\,.\,\mathbf{S}_2 = -1, \quad \mathbf{S}_2^2 = 1, \quad |\,\mathbf{S}_1\,| = 1, \quad |\,\mathbf{S}_2\,| = 1. \tag{6·134}$$

The normalization (2·311) gives $\mathbf{I}_1 = \mathbf{S}_1$, and we proceed to form $\mathbf{I}_2$ as in (2·312), writing

$$\mathbf{I}_2 = a_{21}\mathbf{I}_1 + a_{22}\mathbf{S}_2, \tag{6·135}$$

and imposing the orthonormality conditions

$$\mathbf{I}_1\,.\,\mathbf{I}_2 = 0, \quad |\,\mathbf{I}_2\,| = 1. \tag{6·136}$$

The first gives

$$a_{21} = -a_{22}\mathbf{I}_1\,.\,\mathbf{S}_2 = a_{22}, \tag{6·137}$$

and the second gives

$$\begin{aligned} \pm 1 = \mathbf{I}_2^2 &= a_{21}^2\mathbf{I}_1^2 + 2a_{21}a_{22}\mathbf{I}_1\,.\,\mathbf{S}_2 + a_{22}^2\mathbf{S}_2^2 \\ &= a_{21}^2 - 2a_{21}a_{22} + a_{22}^2 \\ &= 0, \end{aligned} \tag{6·138}$$

by (6·137). Thus the process breaks down. But it breaks down in the normalization, not in the orthogonalization, and we can easily obtain a pair of vectors $\mathbf{S}_1'$, $\mathbf{S}_2'$ orthogonal to one another and lying in the linear 2-space of $\mathbf{S}_1$ and $\mathbf{S}_2$. Choose $\mathbf{S}_1' = \mathbf{S}_1$ and put

$$\mathbf{S}_2' = \mathbf{S}_1 + a\mathbf{S}_2, \tag{6·139}$$

choosing a to satisfy $\mathbf{S}_1\,.\,\mathbf{S}_2' = 0$, i.e.

$$1 - a = 0, \quad a = 1. \tag{6·140}$$

In fact the vectors

$$\mathbf{S}_1' = \mathbf{S}_1, \quad \mathbf{S}_2' = \mathbf{S}_1 + \mathbf{S}_2 \tag{6·141}$$

are orthogonal, as is immediately verified.

The existence of a set of mutually orthogonal vectors $\mathbf{S}_\rho'$ $(\rho = 1, 2, \ldots, n)$ in the linear n-space of a given set $\mathbf{S}_\rho$ $(\rho = 1, 2, \ldots, n)$ (linearly independent but not mutually orthogonal) is shown as follows. Since $\mathbf{S}_\rho\,.\,\mathbf{S}_\sigma = \mathbf{S}_\sigma\,.\,\mathbf{S}_\rho$, a well-known theorem in algebra tells us that the roots λ of the determinantal equation

$$\det(\mathbf{S}_\rho\,.\,\mathbf{S}_\sigma - \lambda\delta_{\rho\sigma}) = 0 \tag{6·142}$$

are real. Let us assume these roots $\lambda_1, \lambda_2, \ldots, \lambda_n$ distinct. Then the equations

$$\sum_{\sigma=1}^{n} (\mathbf{S}_\rho\,.\,\mathbf{S}_\sigma - \lambda_\nu\delta_{\rho\sigma})X_\nu^\sigma = 0 \quad (\rho, \nu = 1, 2, \ldots, n) \tag{6·143}$$

possess solutions X_ν^σ. Multiplying by X_μ^ρ, summing with respect to ρ, then interchanging μ and ν and subtracting, we get

$$(\lambda_\mu - \lambda_\nu) \sum_{\rho=1}^{n} X_\mu^\rho X_\nu^\rho = 0 \quad (\mu, \nu = 1, 2, \ldots, n). \tag{6·144}$$

Hence
$$\sum_{\rho=1}^{n} X_\mu^\rho X_\nu^\rho = 0 \quad (\mu \neq \nu). \tag{6·145}$$

Define vectors $\mathbf{S}'_\mu$ by

$$\mathbf{S}'_\mu = \sum_{\rho=1}^{n} X_\mu^\rho \mathbf{S}_\rho \quad (\mu = 1, 2, \ldots, n). \tag{6·146}$$

Then
$$\mathbf{S}'_\mu \,.\, \mathbf{S}'_\nu = \sum_{\rho=1}^{n} \sum_{\sigma=1}^{n} X_\mu^\rho X_\nu^\sigma \mathbf{S}_\rho \,.\, \mathbf{S}_\sigma$$

$$= \lambda_\nu \sum_{\rho=1}^{n} X_\mu^\rho X_\nu^\rho \tag{6·147}$$

by (6·143). But this vanishes by (6·145) if $\mu \neq \nu$, and so we get the orthogonality equations

$$\mathbf{S}'_\mu \,.\, \mathbf{S}'_\nu = 0 \quad (\mu \neq \nu). \tag{6·148}$$

To illustrate this procedure, take the example (6·132). The equation (6·142) reads

$$\begin{vmatrix} 1-\lambda & -1 \\ -1 & 1-\lambda \end{vmatrix} = 0, \tag{6·149}$$

and the roots are $\lambda_1 = 0$ and $\lambda_2 = 2$. For $\nu = 1$, (6·143) gives

$$\left.\begin{aligned} (1-0)\,X_1^1 - X_1^2 &= 0, \\ -X_1^1 + (1-0)\,X_1^2 &= 0, \end{aligned}\right\} \tag{6·150}$$

and so
$$X_1^1 = X_1^2 \quad (= a \text{ say}), \tag{6·151}$$

and for $\nu = 2$, (6·143) gives

$$\left.\begin{aligned} (1-2)\,X_2^1 - X_2^2 &= 0, \\ -X_2^1 + (1-2)\,X_2^2 &= 0, \end{aligned}\right\} \tag{6·152}$$

whence
$$X_2^1 = -X_2^2 \quad (= b \text{ say}). \tag{6·153}$$

Accordingly (6·146) gives

$$\left.\begin{aligned} \mathbf{S}'_1 &= X_1^1 \mathbf{S}_1 + X_1^2 \mathbf{S}_2 = a\mathbf{S}_1 + a\mathbf{S}_2 = 2a\mathbf{N}, \\ \mathbf{S}'_2 &= X_2^1 \mathbf{S}_1 + X_2^2 \mathbf{S}_2 = b\mathbf{S}_1 - b\mathbf{S}_2 = 2b\mathbf{T}. \end{aligned}\right\} \tag{6·154}$$

Clearly these are orthogonal since $\mathbf{N}\,.\,\mathbf{T} = 0$. We can normalize $\mathbf{S}'_2$ by taking $b = \frac{1}{2}$, but we cannot normalize $\mathbf{S}'_1$ because it is a null vector.

We have introduced this example as a warning, but the failure of the Gram–Schmidt process, if it fails, is due to the vanishing of certain numbers and that is to be regarded as an exceptional case. In general it works, even with an indefinite metric. And even if it fails in normalization, it works in orthogonalization.

Minkowskian F-spaces

The line element $$dx^2 + dy^2 + dz^2 - c^2dt^2, \tag{6·155}$$

assigned to space-time by Minkowski, is an indefinite algebraic form. But it is the difference of two positive-definite forms. A similar decomposition occurs in the F-spaces suited to certain physical problems, as we shall see later, and we shall use the word *Minkowskian* for a certain type of F-space described below.

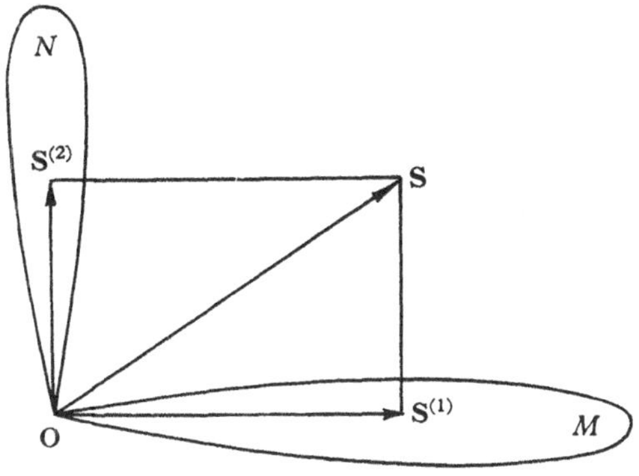

Fig. 6·13. Minkowskian F-space.

A Minkowskian F-space contains two linear subspaces, M and N (Fig. 6·13), which are orthogonal to one another, all vectors lying in M being spacelike ($\mathbf{T}^2 > 0$) and all vectors lying in N being timelike ($\mathbf{T}^2 < 0$); further, any vector $\mathbf{S}$ is the sum

$$\mathbf{S} = \mathbf{S}^{(1)} + \mathbf{S}^{(2)} \tag{6·156}$$

of a vector $\mathbf{S}^{(1)}$ lying in M and a vector $\mathbf{S}^{(2)}$ lying in N.

We note that (6·104) is a Minkowskian F-space, M consisting of vectors

$$\mathbf{S}^{(1)} \leftrightarrow (p_i, 0) \tag{6·157}$$

and N of vectors $$\mathbf{S}^{(2)} \leftrightarrow (0, u). \tag{6·158}$$

By (6·156) we have in general, since $\mathbf{S}^{(1)} . \mathbf{S}^{(2)} = 0$,

$$\mathbf{S}^2 = \mathbf{S}^{(1)2} + \mathbf{S}^{(2)2} = |\,\mathbf{S}^{(1)}\,|^2 - |\,\mathbf{S}^{(2)}\,|^2, \tag{6·159}$$

which reminds us of the Minkowskian form (6·155).

Note that the points $\mathbf{S}^{(1)}$ and $\mathbf{S}^{(2)}$ in Fig. 6·13 are the orthogonal projections of the point $\mathbf{S}$ on M and N respectively.

For an indefinite metric it is very difficult to bound solutions to physical problems. Suppose that $\mathbf{A}$ and $\mathbf{B}$ are two F-points, for which we know that $(\mathbf{A}-\mathbf{B})^2$ is small. In the case of a positive-definite metric this assures us that $\mathbf{A}$ and $\mathbf{B}$ are close together in an easily interpretable sense, but this is not true for an indefinite metric. We might have $(\mathbf{A}-\mathbf{B})^2=0$, i.e. $\mathbf{A}-\mathbf{B}$ a null vector, and yet the points $\mathbf{A}$ and $\mathbf{B}$ might not be close together at all.

The situation is improved in the case of a Minkowskian F-space. Let $\mathbf{A}$, $\mathbf{B}$ be two points, $\mathbf{A}^{(1)}$, $\mathbf{B}^{(1)}$ their orthogonal projections on M and $\mathbf{A}^{(2)}$, $\mathbf{B}^{(2)}$ their orthogonal projections on N. Since all vectors lying in M are spacelike, it has a positive-definite metric; similarly N has a negative-definite metric. Therefore

$$(\mathbf{A}^{(1)}-\mathbf{B}^{(1)})^2=0 \tag{6·160}$$

implies $\mathbf{A}^{(1)}=\mathbf{B}^{(1)}$, and

$$(\mathbf{A}^{(2)}-\mathbf{B}^{(2)})^2=0 \tag{6·161}$$

implies $\mathbf{A}^{(2)}=\mathbf{B}^{(2)}$. If both (6·160) and (6·161) hold, then $\mathbf{A}=\mathbf{B}$. Even if (6·160) and (6·161) are not satisfied, yet if these squares are small we can assert that $\mathbf{A}$ is near to $\mathbf{B}$. In fact the smallness of

$$(\mathbf{A}^{(1)}-\mathbf{B}^{(1)})^2-(\mathbf{A}^{(2)}-\mathbf{B}^{(2)})^2 \tag{6·162}$$

(note the minus sign) would assure us of the nearness of $\mathbf{A}$ and $\mathbf{B}$.

Exercises

1. If $\mathbf{S}\leftrightarrow(f,g)$ and $\mathbf{S}'\leftrightarrow(f',g')$ (the primes denoting other functions, not derivatives), where f, g, f', g' are functions of x for $0\leqslant x\leqslant 1$, and if $\mathbf{S}.\mathbf{S}'=\int_0^1(ff'-gg')\,dx$, find a so that $\mathbf{S}\leftrightarrow(x,ax^2)$ is a null vector.

2. For the F-space of Exercise 1, show that the F-points $\mathbf{S}_m\leftrightarrow(x^{m+1},x^m)$ for all m greater than $\frac{1}{2}$, lie in the interior of the null cone with vertex $\mathbf{O}$, and that the same is true of $\mathbf{T}_n\leftrightarrow(-x^{n+1},-x^n)$ $(n>\frac{1}{2})$. Does the linear 2-space containing the vectors $\mathbf{S}_m,\mathbf{T}_n$ cut this null cone?

3. Referring to Fig. 6·13, show that the orthogonal projection of the point $\mathbf{S}$ on the straight line joining the points $\mathbf{S}^{(1)}$, $\mathbf{S}^{(2)}$ is the point

$$\mathbf{X}=[\mathbf{S}^{(1)}(\mathbf{S}^{(1)2})+\mathbf{S}^{(2)}(\mathbf{S}^{(2)2})]/\mathbf{S}^2.$$

6·2. HYPERPLANES, PSEUDOHYPERSPHERES, PSEUDOHYPERCIRCLES

Hyperplanes

The definitions of hyperplanes given on p. 69 hold good for an indefinite metric and we have [cf. (2·404)]

$$\mathbf{X}.\mathbf{S}_1=b_1 \tag{6·201}$$

as the equation of a hyperplane of class 1 and [cf. (2·406)]

$$\mathbf{X}.\mathbf{S}_\rho = b_\rho \quad (\rho = 1, 2, \ldots, n) \tag{6·202}$$

as the equations of a hyperplane of class n. We shall not trouble to go over the statements of §2·4 to see which hold for an indefinite metric, but merely note that the minimum properties discussed there now become stationary properties.

If $\mathbf{S}_1$ is a null vector ($\mathbf{S}_1^2 = 0$), the hyperplane H_1 with equation

$$\mathbf{X}.\mathbf{S}_1 = 0 \tag{6·203}$$

has an interesting relationship to the null cone C with equation $\mathbf{X}^2 = 0$. We note first that the straight line $\mathbf{X} = k\mathbf{S}_1$ (k arbitrary) is contained in both H_1 and C. Secondly, we note that a point $\mathbf{X} = \mathbf{S}_1 + \mathbf{A}$ is on H_1 if $\mathbf{A}$ satisfies

$$\mathbf{A}.\mathbf{S}_1 = 0, \tag{6·204}$$

and on C if $\mathbf{A}$ satisfies

$$2\mathbf{A}.\mathbf{S}_1 + \mathbf{A}^2 = 0. \tag{6·205}$$

If $\mathbf{A}$ is infinitesimal, the principal parts of (6·204) and (6·205) give the same equation, and therefore we say that H_1 with equation (6·203) is *the tangent hyperplane to the null cone C along the generator* $\mathbf{X} = k\mathbf{S}_1$.

Pseudohyperspheres

Consider all F-points with position vectors $\mathbf{X}$ satisfying

$$\mathbf{X}^2 = R^2, \tag{6·206}$$

where R is a real number, not zero. If the metric were positive-definite, this would be a hypersphere; for an indefinite metric we call it a *pseudohypersphere*, or PHS for short. The equation

$$\mathbf{X}^2 = -R^2 \tag{6·207}$$

defines a PHS conjugate to (6·206). These PHS have their centres at the origin; for centre $\mathbf{C}$ the equations of conjugate PHS are

$$(\mathbf{X} - \mathbf{C})^2 = R^2, \quad (\mathbf{X} - \mathbf{C})^2 = -R^2. \tag{6·208}$$

Since the first equation makes $\mathbf{X} - \mathbf{C}$ spacelike and the second makes it timelike, we see that two conjugate PHS cannot intersect. Note that if we put $R = 0$, both of these PHS degenerate into the same null cone.

Supposing F-space Minkowskian, we may project as in Fig. 6·13 and write

$$\mathbf{X} = \mathbf{X}^{(1)} + \mathbf{X}^{(2)}, \quad \mathbf{C} = \mathbf{C}^{(1)} + \mathbf{C}^{(2)}. \tag{6·209}$$

Here the vectors labelled (1) are orthogonal to those labelled (2), and the equations (6·208) become

$$(\mathbf{X}^{(1)}-\mathbf{C}^{(1)})^2+(\mathbf{X}^{(2)}-\mathbf{C}^{(2)})^2=R^2, \tag{6·210}$$

$$(\mathbf{X}^{(1)}-\mathbf{C}^{(1)})^2+(\mathbf{X}^{(2)}-\mathbf{C}^{(2)})^2=-R^2. \tag{6·211}$$

Let us examine these subspaces by projecting them orthogonally on the linear subspaces M and N of Fig. 6·13.

The projection of (6·210) on M gives

$$(\mathbf{X}^{(1)}-\mathbf{C}^{(1)})^2=R^2-(\mathbf{X}^{(2)}-\mathbf{C}^{(2)})^2, \tag{6·212}$$

and so, since the metric of M is positive-definite and that of N is negative-definite, we have

$$|\mathbf{X}^{(1)}-\mathbf{C}^{(1)}|\geqslant R. \tag{6·213}$$

Thus the projection of the PHS (6·210) *on M covers that part of M outside a hypersphere with centre* $\mathbf{C}^{(1)}$ *and radius R.* On the other hand if we project (6·210) on N, we get

$$(\mathbf{X}^{(2)}-\mathbf{C}^{(2)})^2=R^2-(\mathbf{X}^{(1)}-\mathbf{C}^{(1)})^2, \tag{6·214}$$

and so
$$|\mathbf{X}^{(2)}-\mathbf{C}^{(2)}|^2=-R^2+|\mathbf{X}^{(1)}-\mathbf{C}^{(1)}|^2. \tag{6·215}$$

Since the right-hand side takes all positive values, *the projection of* (6·210) *on N covers N.*

The situation here is like what we see when we project a one-sheeted hyperboloid of revolution on its equatorial plane and on a plane orthogonal to that plane; in the first case the projection covers all the plane outside the equator, and in the second case it covers all the plane.

In the projections of (6·211) the roles of M and N are reversed. The projection on M is

$$(\mathbf{X}^{(1)}-\mathbf{C}^{(1)})^2=-R^2-(\mathbf{X}^{(2)}-\mathbf{C}^{(2)})^2, \tag{6·216}$$

so that
$$|\mathbf{X}^{(1)}-\mathbf{C}^{(1)}|^2=-R^2+|\mathbf{X}^{(2)}-\mathbf{C}^{(2)}|^2, \tag{6·217}$$

and the projection covers M. The projection on N is

$$(\mathbf{X}^{(2)}-\mathbf{C}^{(2)})^2=-R^2-(\mathbf{X}^{(1)}-\mathbf{C}^{(1)})^2, \tag{6·218}$$

so that
$$|\mathbf{X}^{(2)}-\mathbf{C}^{(2)}|^2=R^2+|\mathbf{X}^{(1)}-\mathbf{C}^{(1)}|^2, \tag{6·219}$$

and the projection covers that part of N outside a hypersphere with centre $\mathbf{C}^{(2)}$ and radius R.

With regard to the geometry of N, which has a negative-definite metric, it should be noted that this is essentially the same as geometry for a positive-definite metric; for a positive-definite metric we put $|\mathbf{A}|^2 = \mathbf{A}^2$ and for a negative-definite metric $|\mathbf{A}|^2 = -\mathbf{A}^2$, $|\mathbf{A}|$ being always (even for an indefinite metric) a real non-negative quantity.

In an F-space with positive-definite metric, the hypersphere is the analogue of the ordinary sphere, and our familiar intuitions are extremely valuable. The PHS is more elusive. Sometimes it is useful to think of a hyperboloid, regarding a pair of conjugate PHS as a pair of hyperboloids with a common asymptotic cone, one of one sheet and the other of two sheets. Or we may think of a pair of conjugate PHS as an analogue of the space-time loci

$$x^2+y^2+z^2-c^2t^2 = R^2, \quad x^2+y^2+z^2-c^2t^2 = -R^2. \tag{6·220}$$

Note that the equations (6·208) for conjugate PHS may also be written in parametric form

$$\left.\begin{aligned} \mathbf{X} &= \mathbf{C}+R\mathbf{J}, \quad \mathbf{J}^2 = 1; \\ \mathbf{X} &= \mathbf{C}+R\mathbf{J}, \quad \mathbf{J}^2 = -1. \end{aligned}\right\} \tag{6·221}$$

Pseudohypercircles

A pseudohypercircle (PHC for short) of class n is the intersection of a PHS with a hyperplane of class n. Thus a PHC has equations of the form

$$(\mathbf{X}-\mathbf{C}_0)^2 = \epsilon_0 R_0^2, \quad \mathbf{X}.\mathbf{T}_\rho = a_\rho \quad (\rho = 1, 2, \ldots, n), \tag{6·222}$$

where $\mathbf{C}_0$ is the centre of the PHS, R_0 its radius, $\epsilon_0 = \pm 1$, $\mathbf{T}_\rho$ fixed vectors and a_ρ fixed numbers. Formally this agrees with the equations (2·615) for a hypercircle, except that here the vectors $\mathbf{T}_\rho$ are not orthonormalized, there being, as we have seen, possible difficulties in orthonormalization when the metric is indefinite.

As we changed (2·615) to parametric form, so we can change (6·222). Define $\mathbf{C}$ by

$$\mathbf{C} = \mathbf{C}_0 + \sum_{\rho=1}^{n} b_\rho \mathbf{T}_\rho, \tag{6·223}$$

with the b's chosen to satisfy

$$\sum_{\sigma=1}^{n} b_\sigma \mathbf{T}_\sigma.\mathbf{T}_\rho = a_\rho - \mathbf{C}_0.\mathbf{T}_\rho \quad (\rho = 1, 2, \ldots, n). \tag{6·224}$$

Then by (6·222)

$$(\mathbf{X}-\mathbf{C}).\mathbf{T}_\rho = 0, \tag{6·225}$$

and hence

$$(\mathbf{X}-\mathbf{C}).(\mathbf{C}-\mathbf{C}_0) = 0. \tag{6·226}$$

We can now replace (6·222) by

$$(\mathbf{X}-\mathbf{C})^2 = \epsilon R^2, \quad (\mathbf{X}-\mathbf{C}).\mathbf{T}_\rho = 0 \quad (\rho = 1, 2, \ldots, n), \tag{6·227}$$

where
$$\epsilon R^2 = \epsilon_0 R_0^2 - (\mathbf{C} - \mathbf{C}_0)^2$$
$$= \epsilon_0 R_0^2 - \sum_{\rho=1}^{n} \sum_{\sigma=1}^{n} b_\rho b_\sigma \mathbf{T}_\rho \, . \, \mathbf{T}_\sigma$$
$$= \epsilon_0 R_0^2 + \sum_{\rho=1}^{n} b_\rho (\mathbf{C}_0 \, . \, \mathbf{T}_\rho - a_\rho), \tag{6·228}$$

ϵ being chosen equal to 1 or -1 to make R^2 positive. Thus, in analogy with (2·625), we can write the equations of a PHC in parametric form

$$\left.\begin{aligned} &\mathbf{X} = \mathbf{C} + R\mathbf{J}, \quad \mathbf{J}^2 = \epsilon = \pm 1, \\ &\mathbf{J} \, . \, \mathbf{T}_\rho = 0 \quad (\rho = 1, 2, \ldots, n). \end{aligned}\right\} \tag{6·229}$$

Here the parametric vector $\mathbf{J}$ is to be regarded as free except for the stated conditions; note that all $\mathbf{J}$'s are spacelike or all timelike for any given PHC.

We can also generate a PHC by cutting a null cone by a hyperplane. In that case we put $R_0 = 0$ in (6·222); the argument is not affected.

Bounds connected with PHS

In the case of a positive-definite metric, the equation

$$(\mathbf{X} - \mathbf{C})^2 = R^2 \tag{6·230}$$

sets bounds on $\mathbf{X}^2$ and $\mathbf{X} \, . \, \mathbf{G}$, where $\mathbf{G}$ is arbitrary [cf. (2·515) and (2·517)]. But it is not so when the metric is indefinite. In a Minkowskian F-space some rather meagre bounds are obtainable. For the PHS with equation (6·230), we have, as in (6·213),

$$|\mathbf{X}^{(1)} - \mathbf{C}^{(1)}| \geqslant R, \tag{6·231}$$

and for the PHS conjugate to (6·230), we have as in (6·219)

$$|\mathbf{X}^{(2)} - \mathbf{C}^{(2)}| \geqslant R. \tag{6·232}$$

These inequalities place the projections of $\mathbf{X}$ on M and N *outside* hyperspheres.

Consider the PHS (6·230). The hypersphere in M with equation

$$(\mathbf{X}^{(1)} - \mathbf{C}^{(1)})^2 = R^2 \tag{6·233}$$

contains the origin $\mathbf{O}$ if $|\mathbf{C}^{(1)}| < R$. In that case $|\mathbf{X}^{(1)}|$ is bounded below by

$$|\mathbf{X}^{(1)}| \geqslant R - |\mathbf{C}^{(1)}|. \tag{6·234}$$

Under like circumstances a similar bound holds for $|\mathbf{X}^{(2)}|$ in the case of the PHS conjugate to (6·230).

No bounds are available for scalar products of the type $\mathbf{X}.\mathbf{G}$ or $\mathbf{X}^{(1)}.\mathbf{G}^{(1)}$ or $\mathbf{X}^{(2)}.\mathbf{G}^{(2)}$.

This account of bounds has been given to destroy any false hopes which might exist as to the possibility of carrying over to an indefinite metric the methods found so useful for a positive-definite metric. As we shall see in Chapter 7, F-space with indefinite metric throws light on certain problems of mathematical physics; but it does not provide bounds.

Exercises

1. Taking the F-space of Exercise 1, p. 380, show that the points (x, x^2), (x^2, x) are on opposite sides of the hyperplane $\mathbf{X}.\mathbf{S}_1 = 0$ where $\mathbf{S}_1 \leftrightarrow (1, 1)$.
2. For the same F-space find a so that the point (x, ax) is on the PHS $\mathbf{X}^2 = R^2$.
3. Find the conditions that the straight line $\mathbf{X} = \mathbf{X}_0 + c\mathbf{A}$ should meet the PHS $\mathbf{X}^2 = R^2$ (i) nowhere, (ii) in one point, (iii) in two points, (iv) in an infinite number of points.

6·3. THE METHOD OF THE PSEUDOHYPERCIRCLE

As we shall see in Chapter 7, certain physical problems may be stated in this form: find the point of intersection of two orthogonal linear subspaces in an F-space with indefinite metric. As in the earlier part of the book, we are interested not so much in infinite processes which converge to the solution as in finite processes approximating to it.

The vertices $\mathbf{V}'$, $\mathbf{V}''$ *of two non-intersecting linear subspaces* L', L''

In § 2·7 we discussed the vertices of two linear subspaces, the metric being positive-definite. These vertices $(\mathbf{V}', \mathbf{V}'')$ were defined as the points of closest approach of the subspaces, but we might equally well have defined them by the condition that $\mathbf{V}' - \mathbf{V}''$ should be orthogonal to both the subspaces. The two definitions are equivalent for a positive-definite metric, but not for an indefinite one, on account of the existence of null lines. Instead we have the following equivalent definitions:

(i) $(\mathbf{V}' - \mathbf{V}'')^2$ has a stationary value;

(ii) $\mathbf{V}' - \mathbf{V}''$ is orthogonal to both the subspaces.

Let us examine these definitions. We shall not assume the linear subspaces L', L'' to be mutually orthogonal.

Let $\mathbf{S}'$ and $\mathbf{S}''$ be general points on L' and L'' respectively. The stationary definition (i) means that $\mathbf{V}' = \mathbf{S}'$, $\mathbf{V}'' = \mathbf{S}''$ are to satisfy the equations

$$(\mathbf{S}' - \mathbf{S}'').\delta\mathbf{S}' = 0, \quad (\mathbf{S}' - \mathbf{S}'').\delta\mathbf{S}'' = 0, \tag{6·301}$$

where $\delta\mathbf{S}'$, $\delta\mathbf{S}''$ are infinitesimal variations in L', L'' respectively. On account of the linear character of L' and L'', we can replace these infinitesimals by finite vectors $\mathbf{T}'$, $\mathbf{T}''$ lying in the two subspaces, and the equations to determine $\mathbf{V}'$, $\mathbf{V}''$ become

$$(\mathbf{V}'-\mathbf{V}'').\mathbf{T}'=0, \quad (\mathbf{V}'-\mathbf{V}'').\mathbf{T}''=0, \tag{6·302}$$

$\mathbf{T}'$, $\mathbf{T}''$ being arbitrary vectors lying in L', L'' respectively. These equations are the conditions of orthogonality of $\mathbf{V}'-\mathbf{V}''$ to the two subspaces, and so we reconcile the definitions (i) and (ii) of vertices.

The vertices $\mathbf{V}'$, $\mathbf{V}''$ *of two non-intersecting linear subspaces of finite dimensionality* L'_r, L''_s

What has been said above applies in particular to linear subspaces of finite dimensionality with the following equations:

$$\left.\begin{aligned} L'_r:\quad \mathbf{X}&=\mathbf{S}'_0+\sum_{\rho=1}^{r} a'_\rho \mathbf{T}'_\rho,\\ L''_s:\quad \mathbf{X}&=\mathbf{S}''_0+\sum_{\sigma=1}^{s} a''_\sigma \mathbf{T}''_\sigma.\end{aligned}\right\} \tag{6·303}$$

By (6·302) the values of a'_ρ, a''_σ corresponding to the vertices $\mathbf{V}'$, $\mathbf{V}''$ are to satisfy the linear equations

$$\left.\begin{aligned} (\mathbf{S}'_0-\mathbf{S}''_0).\mathbf{T}'_\mu+\sum_{\rho=1}^{r} a'_\rho \mathbf{T}'_\rho.\mathbf{T}'_\mu-\sum_{\sigma=1}^{s} a''_\sigma \mathbf{T}''_\sigma.\mathbf{T}'_\mu=0 \quad (\mu=1,\ldots,r),\\ (\mathbf{S}'_0-\mathbf{S}''_0).\mathbf{T}''_\nu+\sum_{\rho=1}^{r} a'_\rho \mathbf{T}'_\rho.\mathbf{T}''_\nu-\sum_{\sigma=1}^{s} a''_\sigma \mathbf{T}''_\sigma.\mathbf{T}''_\nu=0 \quad (\nu=1,\ldots,s).\end{aligned}\right\} \tag{6·304}$$

If L'_r is orthogonal to L''_s these equations separate into two sets, one set for a'_ρ and the other for a''_σ, but in general they do not separate.

The pseudohypercircle

Suppose now that there are two *orthogonal* linear subspaces L', L'' which intersect at some unknown point $\mathbf{S}$ which we wish to locate. Take a point $\mathbf{S}'$ on L' and a point $\mathbf{S}''$ on L''. Then by the orthogonality of L' and L'',

$$(\mathbf{S}-\mathbf{S}').(\mathbf{S}-\mathbf{S}'')=0, \tag{6·305}$$

or equivalently

$$\left.\begin{aligned} &(\mathbf{S}-\mathbf{C})^2=\epsilon R^2,\\ &\mathbf{C}=\tfrac{1}{2}(\mathbf{S}'+\mathbf{S}''), \quad \epsilon R^2=\tfrac{1}{4}(\mathbf{S}'-\mathbf{S}'')^2, \quad \epsilon=\pm 1.\end{aligned}\right\} \tag{6·306}$$

Thus $\mathbf{S}$ is located on a PHS with centre $\mathbf{C}$ and radius R.

If L'' contains $\mathbf{O}$, we may put $\mathbf{S}''=\mathbf{O}$ in (6·305) and get

$$(\mathbf{S}-\mathbf{S}').\mathbf{S}=0, \tag{6·307}$$

and hence by (6·305) for the previously chosen $\mathbf{S}''$

$$(\mathbf{S}-\mathbf{S}').\mathbf{S}''=0. \tag{6·308}$$

This locates $\mathbf{S}$ on a hyperplane of class 1, or, combining this with (6·306), on a PHC of class 1.

Except for the factor ϵ, the algebra is exactly as it was for a positive-definite metric.

Take now a point $\mathbf{S}'_0$ on L' and vectors $\mathbf{T}'_\rho$ $(\rho=1,2,\ldots,r)$ lying in L'; also a point $\mathbf{S}''_0$ on L'' and vectors $\mathbf{T}''_\sigma$ $(\sigma=1,2,\ldots,s)$ lying in L''. Thus we get linear subspaces L'_r, L''_s contained in L', L'' and we can find their vertices in the form

$$\mathbf{V}'=\mathbf{S}'_0+\sum_{\rho=1}^{r} a'_\rho \mathbf{T}'_\rho, \quad \mathbf{V}''=\mathbf{S}''_0+\sum_{\sigma=1}^{s} a''_\sigma \mathbf{T}''_\sigma, \tag{6·309}$$

where the coefficients are chosen to satisfy

$$\left.\begin{aligned} &\sum_{\rho=1}^{r} a'_\rho \mathbf{T}'_\rho.\mathbf{T}'_\mu = -(\mathbf{S}'_0-\mathbf{S}''_0).\mathbf{T}'_\mu \quad (\mu=1,2,\ldots,r),\\ &\sum_{\sigma=1}^{s} a''_\sigma \mathbf{T}''_\sigma.\mathbf{T}''_\nu = \quad (\mathbf{S}'_0-\mathbf{S}''_0).\mathbf{T}''_\nu \quad (\nu=1,2,\ldots,s).\end{aligned}\right\} \tag{6·310}$$

Then by the same algebraic processes as those which led to (2·743), we locate $\mathbf{S}$ on a PHC of class $r+s$ with equations

$$\left.\begin{aligned} &\mathbf{X}=\mathbf{C}+R\mathbf{J}, \quad \mathbf{J}^2=\epsilon \quad (=\pm 1),\\ &\mathbf{J}.\mathbf{T}'_\rho=0 \quad (\rho=1,2,\ldots,r),\\ &\mathbf{J}.\mathbf{T}''_\sigma=0 \quad (\sigma=1,2,\ldots,s),\\ &\mathbf{C}=\tfrac{1}{2}(\mathbf{V}'+\mathbf{V}''), \quad \epsilon R^2=\tfrac{1}{4}(\mathbf{V}'-\mathbf{V}'')^2.\end{aligned}\right\} \tag{6·311}$$

This is a general argument; it may happen that (6·310) have no solutions. Further, if $(\mathbf{V}'-\mathbf{V}'')^2=0$ we must not conclude that $\mathbf{X}=\mathbf{C}$; in that case $(\mathbf{X}-\mathbf{C})^2=0$, so that $\mathbf{X}-\mathbf{C}$ is a null vector.

Leaving these singular cases aside, we have now a plan for approximating to the solution $\mathbf{S}$, the point of intersection of L' and L''. It is the same plan, formally, as we used successfully in the case of a positive-definite metric. We take points $\mathbf{S}'_0$, $\mathbf{S}''_0$ on L', L'' and vectors $\mathbf{T}'_\rho$, $\mathbf{T}''_\sigma$ lying in L', L'', solve the linear equations (6·310) and obtain $\mathbf{V}'$, $\mathbf{V}''$ by (6·309). We know then that $\mathbf{S}$ is on the HPC (6·311). We take its centre $\mathbf{C}$ as an approximation to $\mathbf{S}$.

But we are now on much less sure ground than we were in the case of a positive-definite metric. Then, when we had located **S** on a hypercircle with centre **C** and radius R, we had the solution bounded in a mean-square sense when we applied the abstract geometry to a physical problem. But now, instead of a hypercircle, we have a PHC and this is of hyperbolic type, extending to infinity, and a small value of R does not ensure that **S** is close to **C**. Indeed, the concept of *closeness* remains vague for an indefinite metric in general; in §6·1 we were able to clarify this concept for a Minkowskian F-space, and in §6·4 we shall use that idea as a basis of approximation.

Stationary principles

The equation (6·305) yields certain stationary principles. Write

$$I = (\mathbf{S}' - \mathbf{S}'')^2, \tag{6·312}$$

where $\mathbf{S}'$ and $\mathbf{S}''$ are respectively on L' and L'', the two orthogonal linear subspaces which intersect at **S**. Holding $\mathbf{S}''$ fixed and varying $\mathbf{S}'$, we get

$$\delta I = 2(\mathbf{S}' - \mathbf{S}'') . \delta \mathbf{S}'. \tag{6·313}$$

If $\mathbf{S}'$ happens to be **S**, then

$$\delta I = 2(\mathbf{S} - \mathbf{S}'') . \delta \mathbf{S}'. \tag{6·314}$$

But by varying $\mathbf{S}'$ in (6·305) we get

$$(\mathbf{S} - \mathbf{S}'') . \delta \mathbf{S}' = 0, \tag{6·315}$$

and so $\delta I = 0$. Thus *the stationary principle* $\delta I = 0$ *is satisfied by* $\mathbf{S}' = \mathbf{S}$ *for* $\mathbf{S}'$ *free in* L' *and* $\mathbf{S}''$ *fixed in* L''. Similarly, $\delta I = 0$ is satisfied by $\mathbf{S}'' = \mathbf{S}$ for $\mathbf{S}''$ free in L'' and $\mathbf{S}'$ fixed in L'.

Note that, with an indefinite metric, it is a *stationary* principle, not a *minimum* one.

Exercises

1. Consider the two straight lines

$$L_1': \quad \mathbf{S}' = \mathbf{S}_0' + a'\mathbf{T}',$$

$$L_1'': \quad \mathbf{S}'' = \mathbf{S}_0'' + a''\mathbf{T}'',$$

$\mathbf{S}'$ and $\mathbf{S}''$ being current points on the lines. The square $(\mathbf{S}' - \mathbf{S}'')^2$ is then a quadratic function of the parameters a', a''. Show that if

$$\mathbf{T}'^2 > 0, \quad (\mathbf{T}' . \mathbf{T}'')^2 < \mathbf{T}'^2 \mathbf{T}''^2$$

there exists a pair of values (a', a'') for which $(\mathbf{S}' - \mathbf{S}'')^2$ is a minimum. Under what condition is this minimum positive?

2. For the F-space of Exercise 1, §6·1 (p. 380), consider the straight line L_1' through the points $(0, 0)$, $(0, x)$ and the straight line L_1'' through the points $(1, 1)$, $(x, 0)$. Show that the vertices are

$$\mathbf{V}' \leftrightarrow (0, -3x), \quad \mathbf{V}'' \leftrightarrow (3x-2, -2).$$

Check these by (6·302).

6·4. APPROXIMATION BY ORTHOGONAL PROJECTION IN MINKOWSKIAN F-SPACE

A Minkowskian F-space (see p. 379) has a positive-definite metric in M and a negative-definite metric in N (Fig. 6·13). We shall use these definite metrics to help us in approximating to the intersection of two orthogonal linear subspaces.

The separation of two straight lines

Consider two straight lines with equations

$$\left.\begin{aligned} L_1'\colon \quad \mathbf{X} &= \mathbf{S}_0' + a'\mathbf{T}', \\ L_1''\colon \quad \mathbf{X} &= \mathbf{S}_0'' + a''\mathbf{T}'', \end{aligned}\right\} \tag{6·401}$$

where a', a'' are parameters. The F-space being Minkowskian, we project as in (6·156), writing for L_1'

$$\mathbf{X}^{(1)} + \mathbf{X}^{(2)} = \mathbf{S}_0^{(1)\prime} + \mathbf{S}_0^{(2)\prime} + a'\mathbf{T}^{(1)\prime} + a'\mathbf{T}^{(2)\prime}. \tag{6·402}$$

Since the vectors labelled (1) and (2) are orthogonal, this gives the two equations

$$\mathbf{X}^{(1)} = \mathbf{S}_0^{(1)\prime} + a'\mathbf{T}^{(1)\prime}, \tag{6·403}$$

$$\mathbf{X}^{(2)} = \mathbf{S}_0^{(2)\prime} + a'\mathbf{T}^{(2)\prime}. \tag{6·404}$$

These represent straight lines in M and N respectively.* Thus the straight line L_1' projects on M into a straight line $L_1^{(1)\prime}$ with equation (6·403), and it projects on N into a straight line $L_1^{(2)\prime}$ with equation (6·404). Similarly for L_1''.

The upshot of this is that, instead of thinking of a pair of straight lines L_1', L_1'' in the full F-space, we think of two pairs, $(L_1^{(1)\prime}, L_1^{(1)\prime\prime})$ in M and $(L_1^{(2)\prime}, L_1^{(2)\prime\prime})$ in N. If L_1' and L_1'' intersect at some point $\mathbf{P}$, then the first projected pair intersect at $\mathbf{P}^{(1)}$ and the second pair at $\mathbf{P}^{(2)}$, where these points are the projections of $\mathbf{P}$, so that

$$\mathbf{P} = \mathbf{P}^{(1)} + \mathbf{P}^{(2)}. \tag{6·405}$$

Since the metric of F-space is indefinite, it is not easy to say what we mean by two straight lines passing *close* to one another without

* One of the projections might be a single point; but this particular occurrence does not spoil the argument.

intersecting. On the other hand, since M has a definite metric we do know what we mean if we say that $L_1^{(1)\prime}$ and $L_1^{(1)\prime\prime}$ pass close to one another: we mean that the shortest distance between them is small. And of course the same is true for $L_1^{(2)\prime}$ and $L_1^{(2)\prime\prime}$. We can throw these facts back on L_1' and L_1'', saying (as a definition of closeness) that L_1' *passes close to* L_1'' if $L_1^{(1)\prime}$ passes close to $L_1^{(1)\prime\prime}$ and $L_1^{(2)\prime}$ passes close to $L_1^{(2)\prime\prime}$.

In fact, if $\mathbf{V}^{(1)\prime}$, $\mathbf{V}^{(1)\prime\prime}$ are the vertices of $L_1^{(1)\prime}$, $L_1^{(1)\prime\prime}$ and $\mathbf{V}^{(2)\prime}$, $\mathbf{V}^{(2)\prime\prime}$ those of $L_1^{(2)\prime}$, $L_1^{(2)\prime\prime}$, we may define the *separation* of L_1' and L_1'' as the positive number D with square

$$\begin{aligned} D^2 &= |\mathbf{V}^{(1)\prime}-\mathbf{V}^{(1)\prime\prime}|^2+|\mathbf{V}^{(2)\prime}-\mathbf{V}^{(2)\prime\prime}|^2 \\ &= (\mathbf{V}^{(1)\prime}-\mathbf{V}^{(1)\prime\prime})^2-(\mathbf{V}^{(2)\prime}-\mathbf{V}^{(2)\prime\prime})^2, \end{aligned} \tag{6·406}$$

this square being of course positive. The separation D vanishes if, and only if, L_1' intersects L_1'', for the intersection of these two lines implies (and is implied by) the pair of intersections, $L_1^{(1)\prime}$ with $L_1^{(1)\prime\prime}$ and $L_1^{(2)\prime}$ with $L_1^{(2)\prime\prime}$.

We have said nothing about the orthogonality of the lines involved. Orthogonality is not preserved under orthogonal projection (not even in Euclidean space), and so, even if the original lines were orthogonal, the pair of projections on M would not be mutually orthogonal, nor would those on N (although of course the projections on M are orthogonal to the projections on N). This means that in finding the vertices $\mathbf{V}^{(1)\prime}$, $\mathbf{V}^{(1)\prime\prime}$, $\mathbf{V}^{(2)\prime}$, $\mathbf{V}^{(2)\prime\prime}$ we must use equations of the form (6·304).

Thus for the lines in M,

$$\mathbf{X}^{(1)}=\mathbf{S}_0^{(1)\prime}+a'\mathbf{T}^{(1)\prime}, \quad \mathbf{X}^{(1)}=\mathbf{S}_0^{(1)\prime\prime}+a''\mathbf{T}^{(1)\prime\prime}, \tag{6·407}$$

the vertices are

$$\mathbf{V}^{(1)\prime}=\mathbf{S}_0^{(1)\prime}+a'\mathbf{T}^{(1)\prime}, \quad \mathbf{V}^{(1)\prime\prime}=\mathbf{S}_0^{(1)\prime\prime}+a''\mathbf{T}^{(1)\prime\prime}, \tag{6·408}$$

where a' and a'' are chosen to satisfy

$$\left.\begin{aligned} (\mathbf{S}_0^{(1)\prime}-\mathbf{S}_0^{(1)\prime\prime}).\mathbf{T}^{(1)\prime}+a'\mathbf{T}^{(1)\prime}.\mathbf{T}^{(1)\prime}-a''\mathbf{T}^{(1)\prime\prime}.\mathbf{T}^{(1)\prime}=0, \\ (\mathbf{S}_0^{(1)\prime}-\mathbf{S}_0^{(1)\prime\prime}).\mathbf{T}^{(1)\prime\prime}+a'\mathbf{T}^{(1)\prime}.\mathbf{T}^{(1)\prime\prime}-a''\mathbf{T}^{(1)\prime\prime}.\mathbf{T}^{(1)\prime\prime}=0. \end{aligned}\right\} \tag{6·409}$$

If L_1' and L_1'' are orthogonal, then $\mathbf{T}'.\mathbf{T}''=0$, or

$$(\mathbf{T}^{(1)\prime}+\mathbf{T}^{(2)\prime}).(\mathbf{T}^{(1)\prime\prime}+\mathbf{T}^{(2)\prime\prime})=0; \tag{6·410}$$

hence

$$\mathbf{T}^{(1)\prime}.\mathbf{T}^{(1)\prime\prime}+\mathbf{T}^{(2)\prime}.\mathbf{T}^{(2)\prime\prime}=0, \tag{6·411}$$

but in general

$$\mathbf{T}^{(1)\prime}.\mathbf{T}^{(1)\prime\prime}\neq 0, \quad \mathbf{T}^{(2)\prime}.\mathbf{T}^{(2)\prime\prime}\neq 0. \tag{6·412}$$

The separation of two linear subspaces of finite dimensionality L'_r, L''_s

Let us now replace the straight lines L'_1, L''_1 of the above argument by linear subspaces of finite dimensionality, L'_r, L''_s. It is easy to see that on M they project into linear subspaces $L^{(1)'}_{r_1}$, $L^{(1)''}_{s_1}$ and on N into linear subspaces $L^{(2)'}_{r_2}$, $L^{(2)''}_{s_2}$; here r_1, r_2 are less than or equal to r, and s_1, s_2 less than or equal to s, because dimensions may be lost in projection (cf. the edge-on projection of a plane in Euclidean 3-space).

Having so projected L'_r, L''_s, we can find vertices $\mathbf{V}^{(1)'}$, $\mathbf{V}^{(1)''}$, $\mathbf{V}^{(2)'}$, $\mathbf{V}^{(2)''}$ on their projections, using (6·304) with suitable suffixes (1), (2), and hence define the separation D of L'_r and L''_s as in (6·406). An intersection of L'_r with L''_s implies, and is implied by, a pair of intersections, $L^{(1)'}_{r_1}$ with $L^{(1)''}_{s_1}$ and $L^{(2)'}_{r_2}$ with $L^{(2)''}_{s_2}$.

Plan for approximation

We can now set down a plan for approximation to the intersection $\mathbf{S}$ of two orthogonal linear subspaces L', L'' in a Minkowskian F-space, the existence and uniqueness of $\mathbf{S}$ being assumed.

We choose two linear subspaces of finite dimensionalities (r and s, the bigger the better) contained in L', L'' respectively:

$$\left.\begin{aligned} L'_r:\quad \mathbf{X} &= \mathbf{S}'_0 + \sum_{\rho=1}^{r} a'_\rho \mathbf{T}'_\rho, \\ L''_s:\quad \mathbf{X} &= \mathbf{S}''_0 + \sum_{\sigma=1}^{s} a''_\sigma \mathbf{T}''_\sigma. \end{aligned}\right\} \tag{6·413}$$

Here $\mathbf{S}'_0$, $\mathbf{S}''_0$ are points on L', L'' and $\mathbf{T}'_\rho$, $\mathbf{T}''_\sigma$ vectors lying in L', L''.

We now project L'_r, L''_s orthogonally on M and N (Fig. 6·13), obtaining $L^{(1)'}_{r_1}$, $L^{(1)''}_{s_1}$ in M and $L^{(2)'}_{r_2}$, $L^{(2)''}_{s_2}$ in N. We find the vertices $\mathbf{V}^{(1)'}$, $\mathbf{V}^{(1)''}$, $\mathbf{V}^{(2)'}$, $\mathbf{V}^{(2)''}$ as described above, and calculate D by (6·406) as a measure of the separation of L'_r and L''_s. Fig. 6·41 shows the situation; it is drawn for $r=s=1$ for simplicity.

Finally we define $\mathbf{\Gamma}^{(1)}$, $\mathbf{\Gamma}^{(2)}$, $\mathbf{\Gamma}$ by

$$\left.\begin{aligned} \mathbf{\Gamma}^{(1)} &= \tfrac{1}{2}(\mathbf{V}^{(1)'} + \mathbf{V}^{(1)''}), \\ \mathbf{\Gamma}^{(2)} &= \tfrac{1}{2}(\mathbf{V}^{(2)'} + \mathbf{V}^{(2)''}), \\ \mathbf{\Gamma} &= \mathbf{\Gamma}^{(1)} + \mathbf{\Gamma}^{(2)}, \end{aligned}\right\} \tag{6·414}$$

and accept $\mathbf{\Gamma}$ as an approximation to the intersection $\mathbf{S}$ of L' and L''.

This is a much more empirical procedure than the method of the hypercircle in a space with positive-definite metric. All we are sure of is that if L'_r and L''_s do intersect, then $\mathbf{\Gamma}$ is their intersection. But in practice this is most unlikely to happen, and we base our confidence in $\mathbf{\Gamma}$ as a good approximation to $\mathbf{S}$ on the smallness of the separation D. It is easy to see that if we increase the dimensionalities of L'_r and L''_s by adding more vectors $\mathbf{T}'$, $\mathbf{T}''$, linearly independent of those already used, we draw the projections $L^{(1)\prime}_{r_1}$, $L^{(1)\prime\prime}_{s_1}$ closer together, and also $L^{(2)\prime}_{r_2}$, $L^{(2)\prime\prime}_{s_2}$. Thus D decreases monotonically under this increase in dimensionality, and by proper systematic choice of

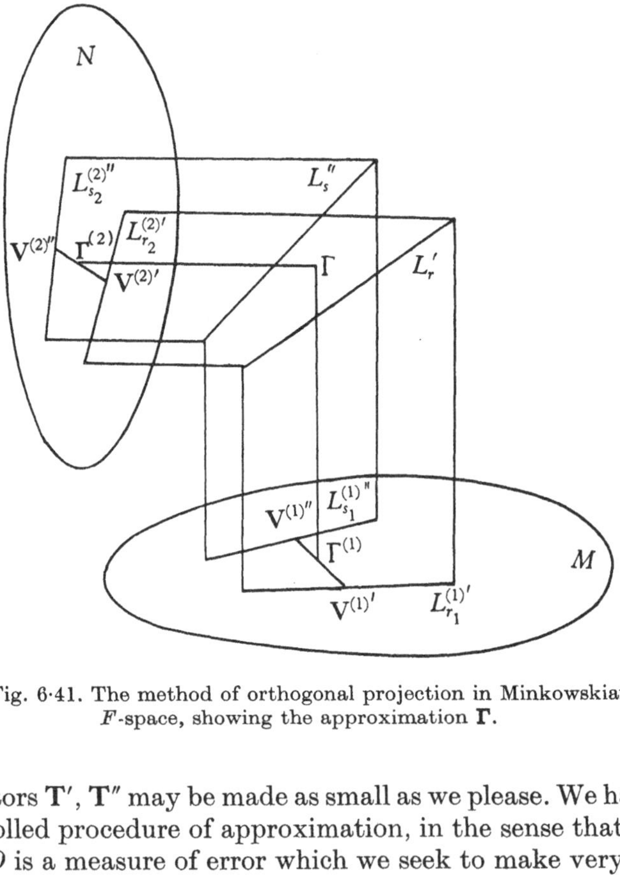

Fig. 6·41. The method of orthogonal projection in Minkowskian F-space, showing the approximation $\mathbf{\Gamma}$.

the vectors $\mathbf{T}'$, $\mathbf{T}''$ may be made as small as we please. We have here a controlled procedure of approximation, in the sense that, at any stage, D is a measure of error which we seek to make very small.

Exercises

1. For an F-space with negative-definite metric, show that the Schwarz inequality

$$(\mathbf{A}.\mathbf{B})^2 \leqslant \mathbf{A}^2\mathbf{B}^2$$

holds good.

2. In the F-space of (6·104) consider the two straight lines, L'_1 passing through the points (p'_i, u'), (q'_i, v') and L''_1 passing through the points (p''_i, u''),

(q''_i, v''). Here p'_i, u', etc. are functions of the coordinates in a domain V of Euclidean 3-space, the suffix i taking the values 1, 2, 3. Show that the condition (necessary and sufficient) for the intersection of L'_1 and L''_1 is that constants c', c'' exist so that

$$p'_i + c'(q'_i - p'_i) = p''_i + c''(q''_i - p''_i), \quad u' + c'(v' - u') = u'' + c''(v'' - u'')$$

throughout V.

3. Supposing that the condition stated in Exercise 2 is not satisfied, find, in terms of integrals, the separation D (6·406) between L'_1 and L''_1.

CHAPTER 7

VIBRATION PROBLEMS

7·1. SCALAR VIBRATIONS

The scalar wave equation

For a vibrating string, the displacement v satisfies the partial differential equation

$$T\frac{\partial^2 v}{\partial x^2}-\rho\frac{\partial^2 v}{\partial t^2}=0, \tag{7·101}$$

where T is the tension and ρ the linear density. For a vibrating membrane the equation is

$$T\Delta v-\rho\frac{\partial^2 v}{\partial t^2}=0, \tag{7·102}$$

where Δ is the two-dimensional Laplacian, T the tension and ρ the surface density. We meet the same type of equation in the study of sound waves in a gas, and it is convenient to consider the equation

$$\Delta v-\frac{1}{c^2}\frac{\partial^2 v}{\partial t^2}=0 \tag{7·103}$$

as a *scalar wave equation* in one, two or three dimensions, Δ being replaced by $\partial^2/\partial x^2$ in one dimension and being the Laplacian in two or three dimensions; c is some function of position (perhaps a constant), called the *speed of propagation.*

If distributed forces act, we replace the zero on the right-hand side of (7·103) by some assigned function of position, and perhaps of time.

As far as the general approach is concerned, it does not matter whether we think in one, two or three dimensions; for definiteness we shall keep a vibrating membrane in mind, so that P-space is a domain V of the Euclidean plane. The boundary conditions usually occurring in physical problems assign the displacement v on the boundary B, or else the component of the tension perpendicular to the plane of the membrane, and this is proportional to the normal derivative $\partial v/\partial n$ in the plane of the membrane.

We shall confine our attention to simple harmonic vibrations, writing

$$v=u\cos\omega t, \tag{7·104}$$

where u is a function of position only and ω a constant. Then, with $k=\omega/c$, the equation (7·103) becomes

$$\Delta u+k^2u=0; \tag{7·105}$$

with this we associate the boundary conditions

$$(u)_{B_1}=f, \quad (\partial u/\partial n)_{B_2}=g, \tag{7·106}$$

where $B=B_1+B_2$ and the functions f and g are assigned; k is a function of the coordinates, perhaps constant.

Splitting the problem

The problem stated in (7·105) and (7·106) is a particular case of (5·112) and (5·113), and we follow the plan indicated there. A point or vector in F-space is defined by

$$\mathbf{S}\leftrightarrow(p_i, u), \tag{7·107}$$

i.e. a vector field p_i and a scalar field u, and the scalar product is defined as

$$\mathbf{S}.\mathbf{S}'=\int_V (p_i p_i' - k^2 uu')\,dV, \tag{7·108}$$

which gives the indefinite metric

$$\mathbf{S}^2=\int_V (p_i p_i - k^2 u^2)\,dV. \tag{7·109}$$

Note that we have a Minkowskian F-space.

A positive-definite metric is so much easier to use than an indefinite one that there is a temptation to force a positive-definite metric on the problem by changing $-k^2$ to k^2 in (7·108). But this does not work, because the splitting of the problem demands that we use (7·108) as it stands. The reason why we get a positive-definite metric in statical problems and an indefinite metric in vibration problems is that equilibrium corresponds to stationary potential energy, which is positive-definite for small displacements from (stable) equilibrium, whereas dynamics is controlled by Hamilton's principle

$$\delta\int (T-V)\,dt=0, \tag{7·110}$$

in which the integrand is not positive-definite.

The splitting of the problem is effected by means of the two linear subspaces:

$$\left.\begin{aligned} L'&: \quad \mathbf{S}'\leftrightarrow(p_i', u'), \quad p_i'=u'_{,i}, \quad (u')_{B_1}=f,\\ L''&: \quad \mathbf{S}''\leftrightarrow(p_i'', u''), \quad p''_{i,i}+k^2u''=0, \quad (p_i''n_i)_{B_2}=g, \end{aligned}\right\} \tag{7·111}$$

with the following conditions of continuity:

u' continuous, but $u'_{,i}$ need only be piecewise continuous;
$p_i''n_i$ continuous across every line, but p_i'' and u'' need only be piecewise continuous. (7·112)

For the solution we set
$$\mathbf{S} \leftrightarrow (u_{,i}, u); \tag{7·113}$$
then $\mathbf{S}$ is on L' and on L'', and the orthogonality of L' and L'' is established as in (5·111). Thus the problem is split in the sense that the solution is the point of intersection of two orthogonal linear subspaces.

Eigenvalue problems

The problem set out in (7·105) and (7·106) is a problem in forced vibrations, the motion being due to the application on B of periodic disturbances of given frequency. We would also get forced vibrations by keeping the boundary fixed and applying a periodic distributed force; the corresponding equations read
$$\Delta u + k^2 u = X, \quad (u)_B = 0, \tag{7·114}$$
where X is the distributed force with the time factor taken out. The essential feature of forced vibrations is the *inhomogeneity* of the problem (differential equation plus boundary conditions).

On the other hand, the problem
$$\Delta u + k^2 u = 0, \quad (u)_B = 0 \tag{7·115}$$
is homogeneous. It corresponds to the vibration of a membrane with fixed boundary, and in general possesses no solution, except the trivial solution $u = 0$. The problem of finding those constants λ for which
$$\Delta u + \lambda u = 0, \quad (u)_B = 0 \tag{7·116}$$
possesses solutions is the *eigenvalue problem* for the membrane, the values of λ being the eigenvalues (or characteristic values or proper values).*

We shall not be concerned with the determination of eigenvalues. For such problems the geometry of a suitably chosen function-space is important, the determination of eigenvalues being essentially the problem of finding the principal axes of a quadric surface [cf. Weinstein (1)], but our present approach through the splitting of the problem supposes that k is given; in fact we are dealing with forced vibrations of given frequency and not with free vibrations for which the problem is that of determining the frequency.

Nevertheless, it is interesting to look briefly at the geometry of eigenfunctions in the F-space with metric (7·109) in which k is a given function of the coordinates. We shall take the eigenvalue problem with homogeneous boundary conditions similar to (7·106), viz.
$$\Delta u + \lambda u = 0, \quad (u)_{B_1} = 0, \quad (\partial u/\partial n)_{B_2} = 0. \tag{7·117}$$

* λ is an eigenvalue of the operator $-\Delta$ for the assigned boundary condition.

For an eigen F-vector we have

$$\mathbf{S}_\lambda \leftrightarrow (p_i, u), \quad p_i = u_{,i}, \quad p_{i,i} + \lambda u = 0, \quad (u)_{B_1} = 0, \quad (p_i n_i)_{B_2} = 0. \tag{7·118}$$

Note that for any value of the constant a, $a\mathbf{S}_\lambda$ is an eigen F-vector, and so we recognize *eigenlines* through the origin $\mathbf{O}$, the position vector of any point on such a line being an eigenvector.

For the square we have

$$\begin{aligned} \mathbf{S}_\lambda^2 &= \int (p_i p_i - k^2 u^2)\, dV \\ &= \int (u_{,i} p_i - k^2 u^2)\, dV \\ &= -\int u\,(p_{i,i} + k^2 u)\, dV \\ &= \int (\lambda - k^2)\, u^2 dV, \end{aligned} \tag{7·119}$$

the integral over B vanishing by the boundary conditions in (7·118). Here k^2 is a given positive function of position. If $\lambda > k^2_{\text{max.}}$, then $\mathbf{S}_\lambda$ is spacelike; if $\lambda < k^2_{\text{min.}}$, then $\mathbf{S}_\lambda$ is timelike. In the case where k is a constant, we get the following classification of eigenvectors:

$$\left.\begin{array}{ll} \lambda > k^2, & \mathbf{S}_\lambda \text{ spacelike,} \\ \lambda < k^2, & \mathbf{S}_\lambda \text{ timelike,} \\ \lambda = k^2, & \mathbf{S}_\lambda \text{ null.} \end{array}\right\} \tag{7·120}$$

In other words, $\mathbf{S}_\lambda$ lies inside the null cone $\mathbf{X}^2 = 0$ if $\lambda < k^2$, outside it if $\lambda > k^2$, and on it if $\lambda = k^2$.

For two eigen F-vectors,

$$\mathbf{S}_\lambda \leftrightarrow (p_i, u), \quad \mathbf{S}_\mu \leftrightarrow (p'_i, u'), \tag{7·121}$$

we have

$$\mathbf{S}_\lambda . \mathbf{S}_\mu = \int (p_i p'_i - k^2 u u')\, dV. \tag{7·122}$$

Replacing p'_i by $u'_{,i}$ and integrating by parts, and then doing the same with p_i replaced by $u_{,i}$, we find

$$\mathbf{S}_\lambda . \mathbf{S}_\mu = \int (\lambda - k^2)\, u u'\, dV = \int (\mu - k^2)\, u u'\, dV. \tag{7·123}$$

If $\lambda \neq \mu$, it follows that

$$\int u u'\, dV = 0, \tag{7·124}$$

and also

$$\mathbf{S}_\lambda . \mathbf{S}_\mu = -\int k^2 u u'\, dV. \tag{7·125}$$

Note that (7·124) is the usual Hilbert condition of orthogonality for eigenfunctions. The eigen F-vectors are not in general orthogonal for the F-metric (7·109); but if k is a constant, then (7·124) implies

$$\mathbf{S}_\lambda . \mathbf{S}_\mu = 0 \quad (\lambda \neq \mu). \tag{7·126}$$

Thus in the case where k is constant, we get a picture of the eigen F-vectors as an infinite set of mutually orthogonal vectors (or lines), a few of them lying inside the null cone and the rest outside it. By increasing k^2 we trap more eigenvectors inside the null cone, and by decreasing k^2 we drive them out, until, when k^2 is reduced below the smallest eigenvalue, there are no eigenvectors left inside the null cone.

In cases of degeneracy (coincidence of eigenvalues), the eigenvectors are partly indeterminate, but the general picture remains the same.

Triangulation in membrane problems

If the homogeneous (eigenvalue) problem has been solved, then the solution of the inhomogeneous (forced vibration) problem can be obtained exactly. But solutions of the eigenvalue problem are available only for certain simple domains, and it is of interest to set up a technique for direct approximation to solutions for forced vibrations, which may be used for any domain.

Let us consider the application of the method of triangulation (the pyramid F-vectors of § 3·5) to the approximate solution of the membrane problem

$$\Delta u + k^2 u = 0, \quad (u)_{B_1} = f, \quad (\partial u/\partial n)_{B_2} = g. \tag{7·127}$$

We start by covering the plane with a network of triangles as in Fig. 3·54. In practice we would use one of the simple networks, with equilateral triangles as in Fig. 3·61 or isosceles right-angled triangles as in Fig. 3·71. Each junction point is the 'centre' of a polygonal base composed of triangles meeting at the centre.

We pick out those centres (say $P_1, P_2, \ldots, P_r$) for which the corresponding bases do not cut B_1, and in these bases we define pyramid functions $u'_{(\rho)}$ $(\rho = 1, 2, \ldots, r)$ by the linear interpolation formula (3·801), the value of each pyramid function being unity at the centre and zero on the edge of the base.

Next we pick out those centres (say $Q_1, Q_2, \ldots, Q_s$) for which the corresponding bases do not cut B_2. Taking a unit vector $U_{(\sigma)i}$ at the centre Q_σ and zero vectors on the edge of the corresponding base, linear interpolation as in (3·801), applied in turn to the two components of a vector, gives a *pyramid vector field* over the base; this field is parallel to $U_{(\sigma)i}$. Taking $U_{(\sigma)i}$ in two different directions, preferably orthogonal, we get *two* pyramid vector fields, say $p''^{\mathrm{I}}_{(\sigma)i}$ and $p''^{\mathrm{II}}_{(\sigma)i}$.

Note that the scalar fields $u'_{(\rho)}$ and the vector fields just described (each field vanishes outside its base) are determined by the geometry of the network and the choice of directions of $U_{(\sigma)i}$; they are independent of the form of the boundary B and of the value of k in (7·127).

To split the problem (7·127) we choose $\mathbf{S}_0'$, $\mathbf{S}_0''$ as follows:

$$\left.\begin{aligned} \mathbf{S}_0' &\leftrightarrow (p'_{(0)i}, u'_{(0)}), \quad p'_{(0)i} = u'_{(0),i}, \quad (u'_{(0)})_{B_1} = f, \\ \mathbf{S}_0'' &\leftrightarrow (p''_{(0)i}, u''_{(0)}), \quad p''_{(0)i,i} + k^2 u''_{(0)} = 0, \quad (p''_{(0)i} n_i)_{B_2} = g. \end{aligned}\right\} \quad (7·128)$$

This means that we are to choose any continuous function $u'_{(0)}$ satisfying the boundary condition on B_1, and construct $\mathbf{S}_0'$ from it. Similarly $\mathbf{S}_0''$ is constructed from any vector field $p''_{(0)i}$ which satisfies the boundary condition on B_2, for we simply *define* $u''_{(0)}$ by

$$u''_{(0)} = -k^{-2} p''_{(0)i,i}. \quad (7·129)$$

The continuity conditions (7·112) are to be observed.

We have then two points, $\mathbf{S}_0'$ and $\mathbf{S}_0''$, on L' and L'' respectively [cf. (7·111)]. The next thing is to get F-vectors lying in L' and L'', and for these we choose

$$\left.\begin{aligned} \mathbf{T}_\rho' &\leftrightarrow (p'_{(\rho)i}, u'_{(\rho)}), \quad p'_{(\rho)i} = u'_{(\rho),i} \quad (\rho = 1, 2, \ldots, r), \\ \mathbf{T}_\sigma'' &\leftrightarrow (p''_{(\sigma)i}, u''_{(\sigma)}) \quad (\sigma = 1, 2, \ldots, 2s), \end{aligned}\right\} \quad (7·130)$$

where $u'_{(\rho)}$ are the pyramid functions and $p''_{(\sigma)i}$ the whole set of pyramid vector fields, including those in both directions previously denoted by $p''^{\mathrm{I}}_{(\sigma)i}$ and $p''^{\mathrm{II}}_{(\sigma)i}$; $u''_{(\sigma)}$ is defined by

$$u''_{(\sigma)} = -k^{-2} p''_{(\sigma)i,i}. \quad (7·131)$$

Note that the number of F-vectors $\mathbf{T}_\sigma''$ is twice the number of junction-points Q.

The vectors $\mathbf{T}_\rho'$, $\mathbf{T}_\sigma''$ lie in L' and L'' respectively and satisfy the continuity requirements.

We have now the apparatus available to locate the solution on a PHC of class $r + 2s$. But we want to show that the solution can be approached as closely as we like in L' and in L'' by using a sufficiently fine triangulation. To this end we note that by the argument of §3·8 (which does not involve the F-metric at all) we can make the following approximations as good as we like by choosing a_ρ' suitably and taking a fine enough triangulation, the solution being $\mathbf{S} \leftrightarrow (p_i, u)$ where $p_i = u_{,i}$:

$$\left.\begin{aligned} u &\sim u'_{(0)} + \sum_{\rho=1}^{r} a_\rho' u'_{(\rho)}, \\ u_{,i} &\sim u'_{(0),i} + \sum_{\rho=1}^{r} a_\rho' u'_{(\rho),i} = u'_{(0),i} + \sum_{\rho=1}^{r} a_\rho' p'_{(\rho)i}. \end{aligned}\right\} \quad (7·132)$$

The important point is that the polyhedral function approximates well to a given function *and its first derivatives*. We can then write

$$\mathbf{S} \sim \mathbf{S}_0' + \sum_{\rho=1}^{r} a_\rho' \mathbf{T}_\rho'. \quad (7·133)$$

Also we can get good approximations

$$\left.\begin{aligned} p_i &\sim p''_{(0)i} + \sum_{\sigma=1}^{2s} a''_\sigma p''_{(\sigma)i}, \\ p_{i,j} &\sim p''_{(0)i,j} + \sum_{\sigma=1}^{2s} a''_\sigma p''_{(\sigma)i,j}; \end{aligned}\right\} \tag{7·134}$$

hence

$$p_{i,i} \sim p''_{(0)i,i} + \sum_{\sigma=1}^{2s} a''_\sigma p''_{(\sigma)i,i}, \tag{7·135}$$

or by (7·127), (7·129) and (7·131)

$$u \sim u''_{(0)} + \sum_{\sigma=1}^{2s} a''_\sigma u''_{(\sigma)}. \tag{7·136}$$

Combining this with (7·134), we may write

$$\mathbf{S} \sim \mathbf{S}''_0 + \sum_{\sigma=1}^{2s} a''_\sigma \mathbf{T}''_\sigma. \tag{7·137}$$

By (7·133) and (7·137) we are assured that by taking a fine enough triangulation we can get as near as we like to the solution $\mathbf{S}$ in a pointwise sense, approximating either in L' or in L'', provided of course that $\mathbf{S}$ exists. It will not exist if k^2 is a constant equal to one of the eigenvalues of the homogeneous problem

$$\Delta u + \lambda u = 0, \quad (u)_{B_1} = 0, \quad (\partial u/\partial n)_{B_2} = 0. \tag{7·138}$$

Having made a triangulation and set up our pyramid functions and vector fields, we have a choice of two procedures, the pseudo-hypercircle (p. 385) and orthogonal projection (p. 389).

In the PHC method we obtain the vertices $\mathbf{V}'$, $\mathbf{V}''$ in the form

$$\left.\begin{aligned} \mathbf{V}' &= \mathbf{S}'_0 + \sum_{\rho=1}^{r} a'_\rho \mathbf{T}'_\rho, \\ \mathbf{V}'' &= \mathbf{S}''_0 + \sum_{\sigma=1}^{2s} a''_\sigma \mathbf{T}''_\sigma, \end{aligned}\right\} \tag{7·139}$$

the coefficients being chosen to satisfy

$$\left.\begin{aligned} (\mathbf{S}'_0 - \mathbf{S}''_0) \,.\, \mathbf{T}'_\mu + \sum_{\rho=1}^{r} a'_\rho \mathbf{T}'_\rho \,.\, \mathbf{T}'_\mu = 0 \quad (\mu = 1, 2, \ldots, r), \\ (\mathbf{S}'_0 - \mathbf{S}''_0) \,.\, \mathbf{T}''_\nu - \sum_{\sigma=1}^{2s} a''_\sigma \mathbf{T}''_\sigma \,.\, \mathbf{T}''_\nu = 0 \quad (\nu = 1, 2, \ldots, 2s). \end{aligned}\right\} \tag{7·140}$$

We take the centre of the PHC

$$\mathbf{C} = \tfrac{1}{2}(\mathbf{V}' + \mathbf{V}'') \tag{7·141}$$

as an approximation to the solution $\mathbf{S}$, but we have no measure of the goodness of approximation.

Since two pyramid F-vectors are orthogonal if their bases do not overlap, the equations (7·140) are of finite-difference type, each equation involving no more than seven coefficients for the equilateral triangulation, or five for the triangulation by squares and their diagonals. Thus the calculations are of the same general nature as those for the torsion problem (Chapter 4) but not so simple for two reasons. In the first place we have the parameter k, in general a function of position; even if it is a constant, we have either to carry it algebraically or give it a numerical value, and in the latter case it will spoil the simplicity of the arithmetic. Secondly, $\mathbf{T}'_\rho$ and $\mathbf{T}''_\sigma$ no longer correspond to vector fields which are constant in each triangle. We have $\mathbf{T}'_\rho \leftrightarrow (p'_{(\rho)i}, u'_{(\rho)})$, where the vector field is constant over each triangle as before but the scalar field is a linear function of the coordinates in each triangle; $\mathbf{T}''_\sigma \leftrightarrow (p''_{(\sigma)i}, u''_{(\sigma)})$, where the scalar field is constant in each triangle but the vector field is linear. These complications make the calculations harder [Synge (7)].

To see how k is involved in (7·140), in the case where k is constant, let us write these equations out in terms of integrals, multiplying the second line across by k^2 for convenience. We have then

$$\left.\begin{aligned} A'_\mu - k^2 B'_\mu + \sum_{\rho=1}^{r} a'_\rho (A'_{\rho\mu} - k^2 B'_{\rho\mu}) = 0 \quad (\mu = 1, 2, \ldots, r), \\ A''_\nu - k^2 B''_\nu + \sum_{\sigma=1}^{2s} a''_\sigma (A''_{\sigma\nu} - k^2 B''_{\sigma\nu}) = 0 \quad (\nu = 1, 2, \ldots, 2s), \end{aligned}\right\} \tag{7·142}$$

where, by (7·129) and (7·131),

$$\left.\begin{aligned} A'_\mu &= \int [(p'_{(0)i} - p''_{(0)i})\, p'_{(\mu)i} - p''_{(0)i,i}\, u'_{(\mu)}]\, dV, \\ B'_\mu &= \int u'_{(0)} u'_{(\mu)}\, dV, \\ A'_{\rho\mu} &= \int p'_{(\rho)i} p'_{(\mu)i}\, dV, \\ B'_{\rho\mu} &= \int u'_{(\rho)} u'_{(\mu)}\, dV, \\ A''_\nu &= \int p''_{(0)i,i}\, p''_{(\nu)j,j}\, dV, \\ B''_\nu &= -\int [(p'_{(0)i} - p''_{(0)i})\, p''_{(\nu)i} + u'_{(0)} p''_{(\nu)i,i}]\, dV, \\ A''_{\sigma\nu} &= \int p''_{(\sigma)i,i}\, p''_{(\nu)j,j}\, dV, \\ B''_{\sigma\nu} &= \int p''_{(\sigma)i} p''_{(\nu)i}\, dV. \end{aligned}\right\} \tag{7·143}$$

These quantities depend only on the domain V, the boundary conditions, and the triangulation; they are independent of k. We note that

$$\left.\begin{aligned} A'_\rho - k^2 B'_\rho &= (\mathbf{S}'_0 - \mathbf{S}''_0) \,.\, \mathbf{T}'_\rho, \\ A''_\sigma - k^2 B''_\sigma &= k^2 (\mathbf{S}'_0 - \mathbf{S}''_0) \,.\, \mathbf{T}''_\sigma, \\ A'_{\rho\mu} - k^2 B'_{\rho\mu} &= \mathbf{T}'_\rho \,.\, \mathbf{T}'_\mu, \\ A''_{\sigma\nu} - k^2 B''_{\sigma\nu} &= -k^2 \mathbf{T}''_\sigma \,.\, \mathbf{T}''_\nu. \end{aligned}\right\} \tag{7·144}$$

Approximations to the eigenvalues λ of the homogeneous problem (7·138) are given by solving either of the determinantal equations

$$\det(\mathbf{T}'_{\rho}\,.\,\mathbf{T}'_{\mu})=0, \quad \det(\mathbf{T}''_{\sigma}\,.\,\mathbf{T}''_{\nu})=0, \tag{7·145}$$

and putting $\lambda=k^2$, or equivalently by solving either of

$$\det(A'_{\rho\mu}-\lambda B'_{\rho\mu})=0, \quad \det(A''_{\sigma\nu}-\lambda B''_{\sigma\nu})=0. \tag{7·146}$$

In the method of orthogonal projection, we project all F-vectors on M and N as in Fig. 6·13, p. 379, and proceed as in Fig. 6·41, p. 392, finding the vertices $\mathbf{V}^{(1)'}, \mathbf{V}^{(1)''}, \mathbf{V}^{(2)'}, \mathbf{V}^{(2)''}$, getting the approximation $\mathbf{\Gamma}$ by (6·414), and testing the goodness of the approximation by the smallness of D as in (6·406). We shall not trouble to write out formulae explicitly.

Exercises

1. Referring to (6·312), show that for the membrane problem (7·127), the integral
$$I=\int[(u'_{,i}-p''_i)(u'_{,i}-p''_i)-k^2(u'+k^{-2}p''_{i,i})^2]\,dV,$$
for any chosen vector field p''_i satisfying $(p''_i n_i)_{B_2}=g$, has a stationary value when $u'=u$ (the solution) for all variations of u' subject to $(u')_{B_1}=f$.

2. Show that the integral I of Exercise 1 for any chosen function u' satisfying $(u')_{B_1}=f$ has a stationary value when $p''_i=u_{,i}$ (the gradient of the solution) for all variations of p''_i subject to $(p''_i n_i)_{B_2}=g$.

3. Show also for the membrane problem (7·127) that the solution u gives a stationary value to
$$\int_V (u_{,i}u_{,i}-k^2u^2)\,dV-2\int_{B_2} ug\,dB$$
calculated for all functions u which satisfy $(u)_{B_1}=f$, and that the gradient of the solution $(p_i=u_{,i})$ gives a stationary value to
$$\int_V (p_ip_i-k^{-2}p_{i,i}p_{j,j})\,dV-2\int_{B_1} fp_in_i\,dB$$
calculated for all vector fields p_i which satisfy $(p_in_i)_{B_2}=g$.

4. For the forced vibrations of a square membrane with the boundary condition $(u)_B=1$ (B_2 absent), take the simple triangulation in which the square is divided into four congruent triangles, and apply the methods of the pseudohypercircle and orthogonal projection to obtain the approximations $\mathbf{C}=\tfrac{1}{2}(\mathbf{V}'+\mathbf{V}'')$ and $\mathbf{\Gamma}$.

7·2. ELASTIC AND ELECTROMAGNETIC VIBRATIONS

The elastic wave equations

Consider an elastic body in vibration. Let u^*_i and E^*_{ij} be displacement and stress and X^*_i body force per unit volume. By Hooke's law (5·403) we have

$$E^*_{ij}=c_{ijkl}e^*_{kl}, \quad e^*_{ij}=\tfrac{1}{2}(u^*_{i,j}+u^*_{j,i}). \tag{7·201}$$

The equations of motion are

$$\rho \frac{\partial^2 u_i^*}{\partial t^2} = E_{ij,j}^* + X_i^*, \tag{7·202}$$

where ρ is the density; in terms of displacement we have

$$\rho \frac{\partial^2 u_i^*}{\partial t^2} = c_{ijkl} u_{k,lj}^* + X_i^*. \tag{7·203}$$

We shall now restrict ourselves to simple harmonic vibrations of given frequency $\omega/2\pi$, writing

$$u_i^* = u_i \cos \omega t, \quad E_{ij}^* = E_{ij} \cos \omega t, \quad X_i^* = X_i \cos \omega t, \tag{7·204}$$

where u_i, E_{ij} and X_j are functions of position. Then (7·202) becomes

$$E_{ij,j} + X_i + k^2 u_i = 0, \tag{7·205}$$

where $k^2 = \rho\omega^2$; (7·201) gives

$$E_{ij} = c_{ijkl} e_{kl}, \quad e_{ij} = \tfrac{1}{2}(u_{i,j} + u_{j,i}), \tag{7·206}$$

and by (7·203)

$$c_{ijkl} u_{k,lj} + X_i + k^2 u_i = 0. \tag{7·207}$$

To solve a problem in forced vibrations we can use either (7·205) with (7·206) or (7·207) alone; they are equivalent. We recognize in (7·207) a particular case of (5·119) (except for the inclusion of X_i), but we shall here carry out the splitting of the problem a little more physically. For curvilinear coordinates, see McConnell (1).

Splitting elastic vibrations

Let us set down again the equations

$$E_{ij,j} + X_i + k^2 u_i = 0, \tag{7·208}$$

$$E_{ij} = c_{ijkl} e_{kl}, \tag{7·209}$$

$$e_{ij} = \tfrac{1}{2}(u_{i,j} + u_{j,i}), \tag{7·210}$$

and take as boundary conditions

$$(u_i)_{B_1} = f_i, \quad (E_{ij} n_j)_{B_2} = g_i, \tag{7·211}$$

where $B_1 + B_2 = B$, the bounding surface of the elastic body.

We define a point or vector in F-space by

$$\mathbf{S} \leftrightarrow (E_{ij}, u_i), \quad E_{ij} = E_{ji}, \tag{7·212}$$

i.e. a state of stress and a displacement, in general quite unconnected. For the scalar product we take

$$\mathbf{S}.\mathbf{S}' = \int_V (e_{ij} E'_{ij} - k^2 u_i u'_i)\, dV$$

$$= \int_V (e'_{ij} E_{ij} - k^2 u_i u'_i)\, dV, \tag{7·213}$$

where V is the region occupied by the body; stress and strain are related by Hooke's law (7·209), which is never relaxed in this work. This gives us the metric

$$\mathbf{S}^2 = \int_V (e_{ij}E_{ij} - k^2 u_i u_i)\, dV, \tag{7·214}$$

which is indefinite but Minkowskian, on account of the positive-definite character of strain energy. It is easy to see that

$$\mathbf{S}^2 = 4(W-K), \tag{7·215}$$

where W is the time average of strain energy (calculated from E^*_{ij}) and K the time average of kinetic energy (calculated from u^*_i).

For simplicity we shall suppose the body compressible, since otherwise the strain would have to satisfy $e_{ii}=0$ [cf. p. 345].

We define linear subspaces L', L'' as follows:

$$\left.\begin{array}{lllll} L': & \mathbf{S}' \leftrightarrow (E'_{ij}, u'_i), & e'_{ij} = \frac{1}{2}(u'_{i,j} + u'_{j,i}), & (u'_i)_{B_1} = f_i; \\ L'': & \mathbf{S}'' \leftrightarrow (E''_{ij}, u''_i), & E''_{ij,j} + X_i + k^2 u''_i = 0, & (E''_{ij}n_j)_{B_2} = g_i, \end{array}\right\} \tag{7·216}$$

where u'_i is to be continuous and $E''_{ij}n_j$ continuous across every surface. This means that we are to choose any displacement u'_i satisfying the boundary condition on displacement and take for E'_{ij} the corresponding stress given by Hooke's law; we are to choose any stress E''_{ij} satisfying the boundary condition on stress and obtain u''_i from the equations of motion.

It is clear that the solution

$$\mathbf{S} \leftrightarrow (E_{ij}, u_i) \tag{7·217}$$

is on L' and on L''. To show that L' and L'' are orthogonal, we take F-vectors $\mathbf{T}' \leftrightarrow (E'_{ij}, u'_i)$ and $\mathbf{T}'' \leftrightarrow (E''_{ij}, u''_i)$ lying in them, so that

$$(u'_i)_{B_1} = 0, \quad (E''_{ij}n_j)_{B_2} = 0, \quad E''_{ij,j} + k^2 u''_i = 0. \tag{7·218}$$

Then

$$\begin{aligned} \mathbf{T}'.\mathbf{T}'' &= \int_V (e'_{ij}E''_{ij} - k^2 u'_i u''_i)\, dV \\ &= \int_V (u'_{i,j}E''_{ij} - k^2 u'_i u''_i)\, dV \\ &= \int_B u'_i E''_{ij} n_j \, dB - \int_V u'_i (E''_{ij,j} + k^2 u''_i)\, dV \\ &= 0. \end{aligned} \tag{7·219}$$

Having thus shown that the solution is the intersection of two orthogonal linear subspaces, we have available the two methods of approximation, the pseudohypercircle (§6·3) and orthogonal projection (§6·4).

Stationary principles for elastic vibrations

We get stationary principles for elastic vibrations by the argument of p. 388; this tells us that

$$(\mathbf{S}'-\mathbf{S}'')^2=\int_V [(e'_{ij}-e''_{ij})(E'_{ij}-E''_{ij})-k^2(u'_i-u''_i)(u'_i-u''_i)]\,dV \quad (7\cdot220)$$

has a stationary value for $\mathbf{S}'=\mathbf{S}$ when $\mathbf{S}'$ is free in L' and $\mathbf{S}''$ fixed in L'', and also a stationary value for $\mathbf{S}''=\mathbf{S}$ when $\mathbf{S}''$ is free in L'' and $\mathbf{S}'$ fixed in L'.

These stationary principles can be expressed more simply. By a reduction as in (7·219) we find

$$\mathbf{S}'.\mathbf{S}''=\int_{B_1} f_i E''_{ij} n_j\,dB+\int_{B_2} u'_i g_i\,dB+\int_V u'_i X_i\,dV, \quad (7\cdot221)$$

and if we write $\quad (\mathbf{S}'-\mathbf{S}'')^2=\mathbf{S}'^2+\mathbf{S}''^2-2\mathbf{S}'.\mathbf{S}'' \quad (7\cdot222)$

and use (7·215), we get the following two stationary principles:

I. ($\mathbf{S}''$ fixed): The forced vibrations of an elastic body give a stationary value to

$$2(W-K)-\int_{B_2} u_i g_i\,dB-\int_V u_i X_i\,dV \quad (7\cdot223)$$

for the class of displacements satisfying the boundary conditions on displacement, W and K being the time averages of strain energy and kinetic energy calculated from the displacement by use of Hooke's law.

II. ($\mathbf{S}'$ fixed): The forced vibrations of an elastic body give a stationary value to

$$2(W-K)-\int_{B_1} f_i E_{ij} n_j\,dB \quad (7\cdot224)$$

for the class of stresses satisfying the boundary conditions on stress, W and K being the time averages of strain energy and kinetic energy, with W calculated directly from stress and K calculated for the displacement

$$u^*_i=-k^{-2}(E^*_{ij,j}+X^*_i). \quad (7\cdot225)$$

The first of these stationary principles is essentially Hamilton's principle (7·110) applied to a continuum in simple harmonic motion; the second principle is due to Reissner (1).

The factor 2 which multiplies $W-K$ in these expressions comes from taking time averages; thus, the time average of strain energy is

$$W=\frac{\omega}{2\pi}\int_0^{2\pi/\omega}\cos^2\omega t\,dt\int_V \tfrac{1}{2}e_{ij}E_{ij}\,dV$$
$$=\tfrac{1}{4}\int_V e_{ij}E_{ij}\,dV, \quad (7\cdot226)$$

whereas the strain energy in statical problems is given by this formula with $\frac{1}{2}$ instead of $\frac{1}{4}$. The integrals in (7·223) and (7·224) correspond to rates of working of surface and body forces, and if we were to state the principles in terms of time averages of these rates of working, the factor 2 would divide out. Remember that the strains and stresses are e^*_{ij} and E^*_{ij}, not e_{ij} and E_{ij}, which are merely the coefficients of the time factor in (7·204).

The dynamical stationary principles should be compared with the statical minimum principles (5·499e) and (5·499f) of p. 354.

Maxwell's equations

If E^*_i and H^*_i are the electric and magnetic vectors, Maxwell's equations in vacuo read

$$\left.\begin{aligned} \frac{1}{c}\frac{\partial E^*_i}{\partial t} &= \epsilon_{ijk} H^*_{k,j}, \quad E^*_{i,i} = 0, \\ -\frac{1}{c}\frac{\partial H^*_i}{\partial t} &= \epsilon_{ijk} E^*_{k,j}, \quad H^*_{i,i} = 0, \end{aligned}\right\} \tag{7·227}$$

where c is the speed of light and ϵ_{ijk} the permutation symbol, defined by the rules that it vanishes unless the suffixes are different from one another, that $\epsilon_{123} = 1$, and that every permutation of the suffixes reverses the sign. Thus $\epsilon_{ijk} H^*_{k,j}$ is in fact the usual curl H^*; the indicial notation is a little easier to use.

For simple harmonic electromagnetic vibrations we put

$$E^*_i = E_i \sin \omega t, \quad H^*_i = H_i \cos \omega t, \tag{7·228}$$

where E_i and H_i are functions of position only. Then (7·227) give

$$\left.\begin{aligned} kE_i &= \epsilon_{ijk} H_{k,j}, \quad E_{i,i} = 0, \\ kH_i &= \epsilon_{ijk} E_{k,j}, \quad H_{i,i} = 0, \end{aligned}\right\} \tag{7·229}$$

where $k = \omega/c$.

We think of electromagnetic vibrations inside a closed cavity V with bounding surface B, and take as boundary conditions that the tangential component of E_i is assigned on part of B and the tangential component of H_i on the rest of B. These conditions may be written

$$(\epsilon_{ijk} E_j n_k)_{B_1} = f_i, \quad (\epsilon_{ijk} H_j n_k)_{B_2} = g_i, \tag{7·230}$$

where $B_1 + B_2 = B$, n_i is the unit normal to B, and f_i, g_i are assigned vector fields on B_1, B_2 respectively, tangential to the surface.

Splitting electromagnetic vibrations

For a point or vector in F-space we take

$$\mathbf{S} \leftrightarrow (E_i, H_i), \tag{7·231}$$

a pair of vector fields, in general independent of one another. For the scalar product we take

$$\mathbf{S}.\mathbf{S}' = \int_V (E_i E'_i - H_i H'_i)\, dV; \tag{7·232}$$

this gives the metric
$$\mathbf{S}^2 = \int_V (E^2 - H^2)\, dV, \tag{7·233}$$

which is indefinite but Minkowskian.

Linear subspaces are defined as follows:

$$\left.\begin{aligned} L'&: \quad \mathbf{S}' \leftrightarrow (E'_i, H'_i), \quad kH'_i = \epsilon_{ijk} E'_{k,j}, \quad (\epsilon_{ijk} E'_j n_k)_{B_1} = f_i; \\ L''&: \quad \mathbf{S}'' \leftrightarrow (E''_i, H''_i), \quad kE''_i = \epsilon_{ijk} H''_{k,j}, \quad (\epsilon_{ijk} H''_j n_k)_{B_2} = g_i. \end{aligned}\right\} \tag{7·234}$$

Note that to get an F-vector $\mathbf{S}'$ we have merely to choose any vector field E'_i satisfying the boundary condition on B_1 and obtain H'_i from it by differentiation; we have then $H'_{i,i} = 0$, but not necessarily $E'_{i,i} = 0$. We get $\mathbf{S}''$ similarly. We impose as continuity conditions (needed in (7·237) below) the continuity across any surface of the tangential components of E'_i and H''_i.

Clearly the solution $\mathbf{S} \leftrightarrow (E_i, H_i)$ is on L' and on L''. To discuss the orthogonality of the subspaces, we take vectors $\mathbf{T}'$, $\mathbf{T}''$ lying in them, so that

$$(\epsilon_{ijk} E'_j n_k)_{B_1} = 0, \quad (\epsilon_{ijk} H''_j n_k)_{B_2} = 0. \tag{7·235}$$

Their scalar product is

$$\mathbf{T}'.\mathbf{T}'' = \int_V (E'_i E''_i - H'_i H''_i)\, dV. \tag{7·236}$$

The first part of this integral is

$$\begin{aligned} \int_V E'_i E''_i\, dV &= k^{-1} \int_V E'_i \epsilon_{ijk} H''_{k,j}\, dV \\ &= k^{-1} \int_B E'_i \epsilon_{ijk} H''_k n_j\, dB - k^{-1} \int_V E'_{i,j} \epsilon_{ijk} H''_k\, dV. \end{aligned} \tag{7·237}$$

Now
$$E'_i \epsilon_{ijk} n_j = \epsilon_{kij} E'_i n_j, \tag{7·238}$$

which vanishes on B_1 by (7·235), and by (7·234)

$$E'_{i,j} \epsilon_{ijk} = -\epsilon_{kij} E'_{j,i} = -kH'_k; \tag{7·239}$$

thus (7·237) may be written

$$\mathbf{T}'.\mathbf{T}'' = k^{-1} \int_{B_2} H''_k \epsilon_{kij} E'_i n_j\, dB. \tag{7·240}$$

But this vanishes, since by the second of (7·235) we have

$$H''_k \epsilon_{kij} n_j = 0 \tag{7·241}$$

on B_2. The orthogonality of L' and L'' is established.

Having thus located the solution at the intersection of two orthogonal linear subspaces in a Minkowskian F-space, the method of the pseudohypercircle (§6·3) or that of orthogonal projection (§6·4) is available for approximate solution.

Stationary principles for electromagnetic vibrations

Turning to (6·312), we recognize that the integral

$$(\mathbf{S}'-\mathbf{S}'')^2=\int_V [(E'_i-E''_i)(E'_i-E''_i)-(H'_i-H''_i)(H'_i-H''_i)]\,dV \quad (7{\cdot}242)$$

takes a stationary value for $\mathbf{S}'=\mathbf{S}$ with $\mathbf{S}'$ free in L' and $\mathbf{S}''$ fixed in L'', and a stationary value for $\mathbf{S}''=\mathbf{S}$ with $\mathbf{S}''$ free in L'' and $\mathbf{S}'$ fixed in L'. But as in the case of elastic vibrations, we can get more interesting results by calculating $\mathbf{S}'.\mathbf{S}''$. We have

$$\mathbf{S}'.\mathbf{S}''=\int_V (E'_iE''_i-H'_iH''_i)\,dV, \quad (7{\cdot}243)$$

and we proceed as in (7·237), paying attention to the fact that the boundary conditions are now as in (7·234) and not as in (7·235). We get

$$\mathbf{S}'.\mathbf{S}''=k^{-1}\int_B E'_i\epsilon_{ijk}H''_k n_j\,dB$$

$$=k^{-1}\int_{B_1} f_iH''_i\,dB-k^{-1}\int_{B_2} g_iE'_i\,dB. \quad (7{\cdot}244)$$

Hence, writing $(\mathbf{S}'-\mathbf{S}'')^2=\mathbf{S}'^2+\mathbf{S}''^2-2\mathbf{S}'.\mathbf{S}''$, (7·245)

we obtain the two following stationary principles:

I. ($\mathbf{S}''$ fixed): For forced electromagnetic vibrations under the boundary conditions (7·230), the actual vibration gives a stationary value to the expression

$$\int_V (E^2-H^2)\,dV+2k^{-1}\int_{B_2} g_iE_i\,dB, \quad (7{\cdot}246)$$

when considered in the class of fields (E_i, H_i) such that E_i satisfies the boundary condition on E_i and H_i is calculated by $kH_i=\epsilon_{ijk}E_{k,j}$.

II. ($\mathbf{S}'$ fixed): For forced electromagnetic vibrations under the boundary conditions (7·230), the actual vibration gives a stationary value to the expression

$$\int_V (E^2-H^2)\,dV-2k^{-1}\int_{B_1} f_iH_i\,dB, \quad (7{\cdot}247)$$

when considered in the class of fields (E_i, H_i) such that H_i satisfies the boundary condition on H_i and E_i is calculated by $kE_i=\epsilon_{ijk}H_{k,j}$.

But indeed it is unnecessary to write out II on account of the symmetry of the problem in regard to E_i and H_i.

The above theory for elastic and electromagnetic vibrations was presented in a lecture at the Eighth Symposium on Applied Mathematics of the American Mathematical Society, Chicago, 12 April 1956. The Symposium was devoted to the Calculus of Variations and its Applications, and the Proceedings (to be published by the McGraw-Hill Book Company) will contain this lecture together with a discussion by H. F. Weinberger, in which he suggests a different approach, using a positive-definite metric and abandoning the orthogonality of the linear subspaces L' and L''.

Exercises

1. Verify that the F-vector $\mathbf{T}'' \leftrightarrow (E''_{ij}, u''_i)$ lies in the linear subspace L'' of (7·216) if

$$E''_{ij} = \phi_{,ij}, \quad u''_i = \phi_{,i},$$

where ϕ is any function satisfying

$$\Delta\phi + k^2\phi = 0, \quad (\phi_{,ij} n_j)_{B_2} = 0.$$

2. For an incompressible isotropic elastic body modify the treatment given above by using a scalar product analogous to (5·468).

3. For the electromagnetic subspaces of (7·234), show that $\mathbf{T}' \leftrightarrow (E'_i, H'_i)$ lies in L' if

$$E'_i = \phi_{,i}, \quad H'_i = 0,$$

where ϕ is any function which is constant over B_1.

4. Set up a function-space treatment for electromagnetic vibrations in a crystalline medium for which

$$D_i = a_{ij} E_j, \quad B_i = b_{ij} H_j$$

(a_{ij} and b_{ij} are constant symmetric tensors) and the equations (7·229) are replaced by

$$kD_i = \epsilon_{ijk} H_{k,j}, \quad D_{i,i} = 0,$$

$$kB_i = \epsilon_{ijk} E_{k,j}, \quad B_{i,i} = 0.$$

NOTE A

THE TORSION OF A HOLLOW SQUARE

All the calculations completed when this book went to press were done on desk calculators (Facit and Madas). Since then Mr W. F. Cahill, of the National Bureau of Standards, Washington, D.C., has used the electronic computer (SEAC) of the Bureau to improve the bounds given on p. 291 for the torsional rigidity of a hollow square (outer side s, inner side $\frac{1}{2}s$).*

The method used was direct iteration or circling, based on the equations (4·522) and (4·524), without use of the quick convergence factor (p. 280). The following bounds were obtained for Γ/s^4; they are unreliable bounds, in the technical sense of p. 278:

Approximation n	L_u	U_u	$U_u - L_u$	No. of iterations or circlings: L_u	No. of iterations or circlings: U_u
8	0·1263 0626	0·1320 1773	0·0057 1147	50	60
16	0·1282 4044	0·1300 3396	0·0017 9352	180	270
32	0·1288 4044	0·1294 2496	0·0005 8452	680	1070
48	0·1289 7704	0·1292 8697	0·0003 0993	1400	2200

In the approximation n, the number of simultaneous equations to be solved is $\frac{3}{16}n^2$ in the case of the lower bound, and one less in the case of the upper bound. Thus, for $n = 48$, there were 432 and 431 equations respectively.

* This work was done while I was Consultant at the National Bureau of Standards, under contract with the American University. The work was supported by the United States Air Force, through the Office of Scientific Research of the Research and Development Command. A more complete account will appear in *Quarterly of Applied Mathematics*.

NOTE B

THE GREEN'S TENSOR OR FUNDAMENTAL SOLUTION FOR THE EQUILIBRIUM OF AN ANISOTROPIC ELASTIC BODY

For an anisotropic elastic body in equilibrium under a body force X_p per unit volume, the displacement u_p satisfies the partial differential equations (cf. p. 336)

$$c_{pqrs} u_{r,sq} + X_p = 0. \tag{1}$$

We seek the solution corresponding to a concentrated load A_p at the origin; if this solution is written in the form

$$u_r = A_t u_{tr}, \tag{2}$$

then u_{tr} is the Green's tensor (cf. p. 350), or equivalently u_{tr} for fixed t is the *fundamental solution* of

$$c_{pqrs} u_{tr,sq} = 0. \tag{3}$$

To find this Green's tensor or fundamental solution, we shall use Fourier transforms in space, with the notation

$$\left.\begin{aligned} Ff(x) &= \frac{1}{(2\pi)^{\frac{3}{2}}} \int f(y) \exp(ix_a y_a)\, dy, \\ F^{-1}f(x) &= \frac{1}{(2\pi)^{\frac{3}{2}}} \int f(y) \exp(-ix_a y_a)\, dy, \end{aligned}\right\} \tag{4}$$

where $dy = dy_1 dy_2 dy_3$; the integrals run over all space. We have then

$$FF^{-1}f(x) = F^{-1}Ff(x) = f(x). \tag{5}$$

Applying the operator F to (1), integrating by parts and throwing away the integrals over the infinite sphere, we get

$$K_{pr}(x)\, Fu_r(x) = FX_p(x), \tag{6}$$

where

$$K_{pr}(x) = c_{pqrs} x_q x_s. \tag{7}$$

Let $K^*_{pr}(x)$ be the reciprocal matrix, defined by

$$K^*_{pt}(x)\, K_{pr}(x) = \delta_{tr}, \tag{8}$$

the positive-definite character of strain energy ensuring its existence, except at the origin. Multiplication of (6) by $K^*_{pq}(x)$ gives

$$Fu_q(x) = K^*_{pq}(x)\, FX_p(x), \tag{9}$$

and hence, on application of the operator F^{-1}, we have the following expression for the displacement at the point x due to any loading:

$$u_q(x) = F^{-1}[K^*_{pq}(x)\, FX_p(x)]. \tag{10}$$

By (4) this means

$$(2\pi)^3\, u_q(x) = \int X_p(z)\, dz \int \exp\,[iy_a(z_a - x_a)]\, K^*_{pq}(y)\, dy. \tag{11}$$

To deal with a concentrated load at the origin, we suppose $X_p(x)$ to vanish except in the neighbourhood of the origin, and to tend to infinity there so as to give the finite limit

$$\lim \int X_p(x)\, dx = A_p, \tag{12}$$

this being the concentrated load. In this limit, (11) gives

$$u_q(x) = A_p\, u_{pq}(x), \tag{13}$$

where

$$(2\pi)^3\, u_{pq}(x) = \int \exp\,(-ix_a y_a)\, K^*_{pq}(y)\, dy; \tag{14}$$

this is the Green's tensor we have been seeking, and it remains only to express it in a simpler form.

To do this, we transform to polar coordinates by writing $y_p = \rho\eta_p$, where η_p is a unit vector; then the volume element is $dy = \rho^2\, d\rho\, d\omega$ where $d\omega$ is an elementary solid angle. Noting that $K_{pr}(x)$ is homogeneous of degree 2, and consequently $K^*_{pr}(x)$ homogeneous of degree -2, we may write (14) as follows:

$$\begin{aligned}(2\pi)^3\, u_{pq}(x) &= \lim_{R\to\infty} \int K^*_{pq}(\eta)\, d\omega \int_{\rho=0}^{R} \cos\,(\rho\eta_a x_a)\, d\rho \\ &= \lim_{R\to\infty} \int K^*_{pq}(\eta)\, \frac{\sin R\eta_a x_a}{\eta_b x_b}\, d\omega. \end{aligned} \tag{15}$$

To carry out the integration over the unit sphere, we use spherical polar angles (θ, ϕ) with θ measured from the direction of the vector x_p. Put $\cos\theta = \mu$, $x_a x_a = r^2$, $K^*_{pq}(\eta) = k_{pq}(\mu, \phi)$; since $d\omega = -d\mu\, d\phi$, (15) gives

$$\begin{aligned}(2\pi)^3\, r u_{pq}(x) &= \lim_{R\to\infty} \int_0^{2\pi} d\phi \int_{-1}^{1} k_{pq}(\mu, \phi)\, \frac{\sin R r\mu}{\mu}\, d\mu \\ &= \pi \int_0^{2\pi} k_{pq}\,(0, \phi)\, d\phi, \end{aligned} \tag{16}$$

by the famous theorem of Dirichlet. Thus the Green's tensor corresponding to a concentrated load at the origin is

$$u_{pq}(x) = \frac{1}{8\pi^2 r} \oint K^*_{pq}(\eta)\, ds, \tag{17}$$

the integral being taken round the unit circle (ds = element of arc) which has its centre at the origin and lies in the plane perpendicular to the vector x_a; it is clear that the integral depends only on the ratios $x_1:x_2:x_3$, so that $u_{pq}(x)$ is homogeneous of degree -1.

In the case of isotropy, Kelvin's formula (p. 350) can be obtained from (17) by a simple calculation.

The study of fundamental solutions of partial differential equations has created a formidable body of mathematical theory, from which, on account of its generality, it is not easy to pick out a result so special as that required above. The equations (3) form a homogenous elliptic system with constant coefficients; for the appropriate general theory, see John (1), p. 72. Starting from a formula of Fredholm, Kröner (1) has used a different approach to the fundamental solution, establishing some explicit results in the case of transverse isotropy (equivalently, hexagonal symmetry).

BIBLIOGRAPHY

Barta, J.
(1) On the estimation of torsional rigidity. *K. Akad. Wet. Amst.* B **58**, (1955), 80–9.

Basu, N. M.
(1) On the application of the new methods of the calculus of variations to some problems in the theory of elasticity. *Phil. Mag.* (7), **10** (1930), 886–96.
(2) On the torsion problem of the theory of elasticity. *Phil. Mag.* (7), **10** (1930), 896–904.

Broglio, L.
(1) Alcuni teoremi sintetici di elasticità e di fisica matematica. *Mongr. sci. Aero., Roma*, no. 8 (1948).
(2) Some synthetic theorems of elasticity and of mathematical physics. *Proc. 7th International Congress of Applied Mechanics, London*, **1**, (1948), 84–97.

Cooperman, P.
(1) An extension of the method of Trefftz for finding local bounds on the solutions of boundary value problems, and on their derivatives. *Quart. Appl. Math.* **10** (1952), 359–73.

Courant, R.
(1) Variational methods for the solution of problems of equilibrium and vibrations. *Bull. Amer. Math. Soc.* **49** (1943), 1–23.

Diaz, J. B.
(1) Upper and lower bounds for quadratic functionals. *Proc. of the Symposium on Spectral Theory and Differential Problems.* The Mathematics Department, Oklahoma Agricultural and Mechanical College, Stillwater, Oklahoma (1951), pp. 279–89.
(2) Upper and lower bounds for quadratic functionals. *Coll. Math., Barcelona,* **4** (1951), 1–50.
(3) On the estimation of torsional rigidity and other physical quantities. *Proc. of the 1st National Congress of Applied Mechanics, Amer. Soc. Mechanical Engineers* (1952), pp. 259–63.

Diaz, J. B. and Greenberg, H. J.
(1) Upper and lower bounds for the solution of the first biharmonic boundary value problem. *J. Math. Phys.* **27** (1948), 193–201.
(2) Upper and lower bounds for the solution of the first boundary value problem of elasticity. *Quart. Appl. Math.* **6** (1948), 326–31.

Diaz, J. B. and Roberts, R. C.
(1) On the numerical solution of the Dirichlet problem for Laplace's difference equation. *Quart. Appl. Math.* **9** (1951), 355–60.
(2) Upper and lower bounds for the numerical solution of the Dirichlet difference boundary value problem. *J. Math. Phys.* **31** (1952), 184–91.

DIAZ, J. B. and WEINSTEIN, A.

(1) Schwarz' inequality and the methods of Rayleigh–Ritz and Trefftz. *J. Math. Phys.* **26** (1947), 133–6.

(2) The torsional rigidity and variational methods. *Amer. J. Math.* **70** (1948), 107–16.

DIRICHLET, P. G. L.

(1) *Vorlesungen über die im umgekehrten Verhältniss des Quadrats der Entfernung wirkenden Kräfte.* Leipzig, Teubner (1876).

EDELMAN, F.

(1) On the compression of a short cylinder between rough end-blocks. *Quart. Appl. Math.* **7** (1949), 334–7.

FRIEDRICHS, K.

(1) Ein Verfahren der Variationsrechnung das Minimum eines Integrals als das Maximum eines anderen Ausdruckes darzustellen. *Ges. Wiss. Göttingen, Nachr.* (1929), pp. 13–20.

FUJITA, H.

(1) Contribution to the theory of upper and lower bounds in boundary value problems. *J. Phys. Soc. Japan,* **10** (1955), 1–8.

GRAM, J. P.

(1) Ueber die Entwickelung reeler Functionen in Reihen mittelst der Methode der kleinsten Quadrate. *J. Math.* (Crelle), **94** (1883), 41–73.

GREENBERG, H. J.

(1) The determination of upper and lower bounds for the solution of the Dirichlet problem. *J. Math. Phys.* **27** (1948), 161–82.

(2) *The method of inequalities and related studies.* Stencilled Summary Report, Brown University (1948).

GREENBERG, H. J. and PRAGER, W.

(1) Direct determination of bending and twisting moments in thin elastic plates. *Amer. J. Math.* **70** (1948), 749–63.

GREENBERG, H. J. and TRUELL, R.

(1) On a problem in plane strain. *Quart. Appl. Math.* **6** (1948), 53–62.

HADAMARD, J.

(1) *The Psychology of Invention in the Mathematical Field.* Princeton University Press (1949).

HALMOS, P. R.

(1) *Introduction to Hilbert Space and the Theory of Spectral Multiplicity.* Chelsea Publishing Company, New York (1951).

(2) *Finite Dimensional Vector Spaces.* Princeton University Press (1955).

HOBSON, E. W.

(1) *Plane Trigonometry.* Cambridge University Press (1897).

JOHN, F.

(1) *Plane Waves and Spherical Means applied to Partial Differential Equations.* New York, Interscience Publishers Inc. (1955).

KELLOGG, O. D.

(1) *Foundations of Potential Theory.* Berlin, Springer (1929).

KRÖNER, E.

(1) Das Fundamentalintegral der anisotropen elastischen Differentialgleichungen. *Z. Phys.* **136** (1953/4), 402–10.

LAMB, Sir H.
(1) *Hydrodynamics*. Cambridge University Press (1932).

LOVE, A. E. H.
(1) *Mathematical Theory of Elasticity*. Cambridge University Press (1934).

McCONNELL, A. J.
(1) The hypercircle method of approximation for a system of partial differential equations of the second order. *Proc. R. Irish Acad.* **54** A (1951), 263–90.

McMAHON, J.
(1) Lower bounds for the electrostatic capacity of a cube. *Proc. R. Irish Acad.* **55** A (1953), 133–67.
(2) Lower bounds for the Dirichlet integral in Euclidean n-space. *Proc. R. Irish Acad.* **58** A (1956), 1–12.

MAPLE, C. G.
(1) The Dirichlet problem: bounds at a point for the solution and its derivatives. *Quart. Appl. Math.* **8** (1950), 213–28.

MICHAL, A. D.
(1) Function space-time manifolds. *Proc. Nat. Acad. Sci., Wash.*, **17** (1931), 217–25.
(2) The vibrations of elastic strings as studies in geodesics. *Actas Acad. Cienc. Lima*, **9** (1946), 3–27.

NICOLESCO, M.
(1) *Les fonctions polyharmoniques*. Paris, Hermann (1936).

PAYNE, L. E. and WEINBERGER, H. F.
(1) Upper and lower bounds for harmonic functions, Dirichlet integrals, and biharmonic functions. *Tech. Note, Univ. of Maryland, Inst. for Fluid Dynamics and Applied Mathematics*, BN–21 (1954).
(2) New bounds in harmonic and biharmonic problems. *J. Math. Phys.* **33** (1955), 291–307.

PEACH, M. O.
(1) Simplified technique for constructing orthonormal functions. *Bull. Amer. Math. Soc.* **50** (1944), 556–64.

PÓLYA, G.
(1) Estimates for eigenvalues. *Studies in Mathematics and Mechanics presented to Richard von Mises*. New York, Academic Press Inc. (1954), pp. 200–7.

PÓLYA, G. and SZEGÖ, G.
(1) Isoperimetric inequalities in mathematical physics. *Annals of Mathematics Studies*, no. 27. Princeton University Press (1951).

PÓLYA, G. and WEINSTEIN, A.
(1) On the torsional rigidity of multiply connected cross-sections. *Ann. Math.* **52** (1950), 154–63.

PRAGER, W.
(1) The extremum principles of the mathematical theory of elasticity and their use in stress analysis. *Bull. Univ. Wash., Engng Exp. Sta.* no. 119 (1950).

PRAGER, W. and SYNGE, J. L.
(1) Approximations in elasticity based on the concept of function space. *Quart. Appl. Math.* **5** (1947), 241–69.

PUCCI, C.
(1) Bounds for solutions of Laplace's equation satisfying mixed conditions. *J. Rat. Mech. Anal.* **2** (1953), 299–302.

REISSNER, E.
(1) Note on the method of complementary energy. *J. Math. Phys.* **27** (1948), 159–60.

SCHMIDT, E.
(1) Zur Theorie der linearen und nichtlinearen Integralgleichungen. I. Teil: Entwicklung willkürlicher Funktionen nach Systemen vorschriebener. *Math. Ann.* **63** (1907), 443–72.

SOKOLNIKOFF, I. S.
(1) *Mathematical Theory of Elasticity.* New York, McGraw-Hill (1956).

STERNBERG, W. J. and SMITH, T. L.
(1) *The Theory of Potential and Spherical Harmonics.* University of Toronto Press (1944).

STONE, M. H.
(1) Linear transformations in Hilbert space. *Amer. Math. Soc. Colloq. Publ.* **15** (1932).

SYNGE, J. L.
(1) The method of the hypercircle in function-space for boundary-value problems. *Proc. Roy. Soc.* A, **191** (1947), 447–67.
(2) The method of the hypercircle in elasticity when body forces are present. *Quart. Appl. Math.* **6** (1948), 15–19.
(3) Upper and lower bounds for the solutions of problems in elasticity. *Proc. R. Irish Acad.* **53** A (1950), 41–64.
(4) Approximations in boundary-value problems by the method of the hypercircle in function-space. *R.C. Mat. Univ. Roma*, (5), **10** (1951), 24–44.
(5) Pointwise bounds for the solutions of certain boundary-value problems. *Proc. Roy. Soc.* A, **208** (1951), 170–5.
(6) Flow of viscous liquid through pipes and channels. *Proceedings of the Fourth Symposium in Applied Mathematics* (1951). McGraw-Hill, New York (1953), pp. 141–65.
(7) Triangulation in the hypercircle method for plane problems. *Proc. R. Irish Acad.* **54** A (1952), 341–67.
(8) A technique for the solution of the biharmonic equation. *Atti del Convegno internazionale sulle Equazioni alle derivate parziali*, Trieste, agosto 1954, 39–53. Cremonese, Roma (1955).

TREFFTZ, E.
(1) Ein Gegenstuck zum Ritzschen Verfahren. *Proceedings of the 2nd International Congress for Applied Mechanics, Zürich* (1926), pp. 131–7.

WASHIZU, K.
(1) Bounds for solutions of boundary value problems in elasticity. *J. Math. Phys.* **32** (1953), 117–28.

WEINBERGER, H. F.

(1) Upper and lower bounds for torsional rigidity. *J. Math. Phys.* **32** (1953), 54–62.

WEINSTEIN, A.

(1) Les vibrations et le calcul des variations. *Portug. Math.* **2** (1941), 36–55.

(2) New methods for the estimation of torsional rigidity. *Proceedings of the Third Symposium in Applied Mathematics* (1949). McGraw-Hill, New York (1950), pp. 141–61.

WEYL, H.

(1) The method of orthogonal projection in potential theory. *Duke Math. J.* **7** (1940), 411–44.

INDEX